Methods in Molecular Biology

Series Editor
John M. Walker
School of Life Sciences
University of Hertfordshire
Hatfield, Hertfordshire, AL10 9AB, UK

For further volumes:
http://www.springer.com/series/7651

Shotgun Proteomics

Methods and Protocols

Edited by

Daniel Martins-de-Souza

Laboratory of Neuroproteomics, Department of Biochemistry, Institute of Biology, State University of Campinas (UNICAMP), Campinas, Brazil; Max Planck Institute of Psychiatry, Munich, Germany; Research Group of Proteomics, Department of Psychiatry and Psychotherapy, Ludwig-Maximilians-University of Munich, Munich, Germany; Laboratory of Neurosciences (LIM-27), Institute of Psychiatry, University of Sao Paulo, Sao Paulo, Brazil

Humana Press

Editor
Daniel Martins-de-Souza
Laboratory of Neuroproteomics
Department of Biochemistry
Institute of Biology, State University of Campinas (UNICAMP)
Campinas, Brazil

Max Planck Institute of Psychiatry
Munich, Germany

Research Group of Proteomics
Department of Psychiatry and Psychotherapy
Ludwig-Maximilians-University of Munich,
Munich, Germany

Laboratory of Neurosciences (LIM-27)
Institute of Psychiatry
University of Sao Paulo
Sao Paulo, Brazil

Additional material to this book can be downloaded from http://extras.springer.com

ISSN 1064-3745 ISSN 1940-6029 (electronic)
ISBN 978-1-4939-0684-0 ISBN 978-1-4939-0685-7 (eBook)
DOI 10.1007/978-1-4939-0685-7
Springer New York Heidelberg Dordrecht London

Library of Congress Control Number: 2014936840

Printed on acid-free paper

Humana Press is a brand of Springer
Springer is part of Springer Science+Business Media (www.springer.com)

Preface

Less than 20 years ago, the term "proteome" (1, 2) launched unintentionally one of the most fast-growing sciences of the twenty-first century: proteomics. Thanks to the quick development of mass spectrometry techniques, proteomics has become an influential topic in molecular biology.

The greatest hallmark of proteomics was in 1999, when the concept of "shotgun proteomics" was introduced (3, 4). From then on, thousands of proteins could be simultaneously identified, quantified, and characterized in a high-throughput manner.

This MiMB edition serves as the ultimate source of shotgun proteomics protocols, including most aspects of proteomics developed in reference labs of this field. Enjoy!

Campinas, Brazil
Munich, Germany
Sao Paulo, Brazil

Daniel Martins-de-Souza

References

1. Wasinger VC, Cordwell SJ, Cerpa-Poljak A, Yan JX, Gooley AA, Wilkins MR, Duncan MW, Harris R, Williams KL, Humphery-Smith I (1995) Progress with gene-product mapping of the Mollicutes: Mycoplasma genitalium. Electrophoresis 16.
2. Wilkins MR, Pasquali C, Appel RD, Ou K, Golaz O, Sanchez JC, Yan JX, Gooley AA, Hughes G, Humphery-Smith I, Williams KL, Hochstrasser DF (1996) From proteins to proteomes: large scale protein identification by two-dimensional electrophoresis and amino acid analysis. Biotechnology (NY) 14.
3. Link AJ, Eng J, Schieltz DM, Carmack E, Mize GJ, Morris DR, Garvik BM, Yates JR 3rd (1999) Direct analysis of protein complexes using mass spectrometry. Nat Biotechnol 17.
4. Wolters DA, Washburn MP, Yates JR 3rd (2001) An automated multidimensional protein identification technology for shotgun proteomics. Anal Chem 73.

Contents

Contributors

DANIELE ALBUQUERQUE • *Department of Biochemistry, Ribeirão Preto Medical School, University of São Paulo, Ribeirão Preto, SP, Brazil*

ALFONSO MARTINEZ ARIAS • *Department of Genetics, University of Cambridge, Cambridge, UK*

LIISA ARIKE • *Competence Center of Food and Fermentation Technologies, Tallinn, Estonia; Proteomics Core Laboratory, Institute of Technology, University of Tartu, Tartu, Estonia*

JUSTINE V. ARRINGTON • *Department of Chemistry, Purdue University, West Lafayette, IN, USA*

MARCUS BANTSCHEFF • *Cellzome GmbH, Heidelberg, Germany*

HARALD BARSNES • *Proteomics Unit, Department of Biomedicine, University of Bergen, Bergen, Norway*

ISABELLE BECHER • *Cellzome GmbH, Heidelberg, Germany*

FRODE S. BERVEN • *Proteomics Unit, Department of Biomedicine, University of Bergen, Bergen, Norway*

FRANCISCO J. BLANCO • *Rheumatology Division, ProteoRed/ISCIII Proteomics Group, INIBIC – Hospital Universitario de A Coruña, A Coruña, Spain; RIER-RED de Inflamación y Enfermedades Reumáticas, INIBIC-CHUAC, A Coruña, Spain*

MARTIN E. BOEHM • *Molecular Structure Analysis, German Cancer Research Center (DKFZ), Heidelberg, Germany*

CHRISTOPH H. BORCHERS • *University of Victoria - Genome BC Proteomics Centre, University of Victoria, Victoria, BC, Canada; Department of Biochemistry and Microbiology, University of Victoria, Victoria, BC, Canada*

GABRIELA N.S. CANCHAYA • *Department of Biochemistry, Ribeirão Preto Medical School, University of São Paulo, Ribeirão Preto, SP, Brazil*

ANGEL MAURICIO CASTRO-GAMERO • *Department of Pediatrics, Ribeirão Preto Medical School, University of São Paulo, Ribeirão Preto, Brazil*

ANDY CHRISTOFOROU • *Department of Biochemistry, Cambridge Centre for Proteomics, University of Cambridge, Cambridge, UK*

IVAN K. CHU • *Department of Chemistry, The University of Hong Kong, Hong Kong, China*

LENNARD J.M. DEKKER • *Department of Neurology, Erasmus University Medical Center, Rotterdam, The Netherlands*

CLARISSA DICKHUT • *Leibniz-Institut für Analytische Wissenschaften – ISAS – e.V., Dortmund, Germany*

ANTJE DITTMANN • *Cellzome GmbH, Heidelberg, Germany*

GILBERTO B. DOMONT • *Proteomics Unit, Institute of Chemistry, Federal University of Rio de Janeiro (UFRJ), Rio de Janeiro, RJ, Brazil*

MARTIJN M. VAN DUIJN • *Department of Neurology, Erasmus University Medical Center, Rotterdam, The Netherlands*

MONIKA DZIECIATKOWSKA • *Biochemistry and Molecular Genetics, University of Colorado Denver, Aurora, CO, USA*

WILLIAM C. EDELMAN • *Department of Genome Sciences, University of Washington, Seattle, WA, USA*
VERA L. EPIFÂNIO • *Department of Biochemistry, Ribeirão Preto Medical School, University of São Paulo, Ribeirão Preto, SP, Brazil*
MURAT ERAVCI • *Institute of Chemistry and Biochemistry, Freie Universität Berlin, Berlin, Germany*
VITOR M. FAÇA • *Department of Biochemistry, Ribeirão Preto Medical School, University of São Paulo, Ribeirão Preto, SP, Brazil*
PATRICIA FERNÁNDEZ-PUENTE • *Rheumatology Division, ProteoRed/ISCIII Proteomics Group, INIBIC – Hospital Universitario de A Coruña, A Coruña, Spain*
SINA GHAEMMAGHAMI • *Department of Biology, University of Rochester, Rochester, NY, USA*
SONJA GHIDELLI-DISSE • *Cellzome GmbH, Heidelberg, Germany*
LYRIS M.F. DE GODOY • *Fundação Oswaldo Cruz, Instituto Carlos Chagas, Curitiba, PR, Brazil*
LIVIA GOTO-SILVA • *Proteomics Unit, Institute of Chemistry, Federal University of Rio de Janeiro (UFRJ), Rio de Janeiro, Brazil*
MARIANA L. GRASSI • *Department of Biochemistry, Ribeirão Preto Medical School, University of São Paulo, Ribeirão Preto, SP, Brazil*
KELSEY M. HAAS • *Department of Genome Sciences, University of Washington, Seattle, WA, USA*
BETTINA HAHN • *Molecular Structure Analysis, German Cancer Research Center (DKFZ), Heidelberg, Germany*
KIRK C. HANSEN • *Biochemistry and Molecular Genetics, University of Colorado Denver, Aurora, CO, USA*
RYAN HILL • *Biochemistry and Molecular Genetics, University of Colorado Denver, Aurora, CO, USA*
CARSTEN HOPF • *Cellzome GmbH, Heidelberg, Germany*
JOANNE I. HSU • *Department of Genome Sciences, University of Washington, Seattle, WA, USA*
CLARICE IZUMI • *Department of Molecular and Cellular Biology, Protein Chemistry Center, Hemotherapy Center of Ribeirão Preto, Ribeirão Preto Medical School, University of São Paulo, Ribeirão Preto, Brazil*
MAGNO JUNQUEIRA • *Proteomics Unit, Institute of Chemistry, Federal University of Rio de Janeiro (UFRJ), Rio de Janeiro, Brazil*
ANNA KONOPKA • *Molecular Structure Analysis, German Cancer Research Center (DKFZ), Heidelberg, Germany*
MAGGIE P.Y. LAM • *Department of Chemistry, The University of Hong Kong, Hong Kong, China*
CHUN HIN LAW • *Department of Chemistry, The University of Hong Kong, Hong Kong, China*
ROBERT T. LAWRENCE • *Department of Genome Sciences, University of Washington, Seattle, WA, USA*
MARIA LEBAR • *Max Planck Institute of Psychiatry, Munich, Germany*
WOLF D. LEHMANN • *Molecular Structure Analysis, German Cancer Research Center (DKFZ), Heidelberg, Germany*
YISHAI LEVIN • *Israel National Center for Personalized Medicine, Weizmann Institute of Science, Rehovot, Israel*
KATHRYN S. LILLEY • *Department of Biochemistry, Cambridge Centre for Proteomics, University of Cambridge, Cambridge, UK*

NING QING LIU • *Department of Medical Oncology, Laboratory of Breast Cancer Genomics and Proteomics, Erasmus University Medical Center, Erasmus MC Cancer Institute, Rotterdam, The Netherlands; Netherlands Proteomics Centre, Utrecht, The Netherlands*

GIUSEPPINA MACCARRONE • *Max Planck Institute of Psychiatry, Munich, Germany*

KARL A.T. MAKEPEACE • *University of Victoria - Genome BC Proteomics Centre, University of Victoria, Victoria, BC, Canada*

ZOLTAN MALIGA • *Laboratory of Systems Pharmacology, Department of Systems Biology, Harvard Medical School, Boston, MA, USA*

LENNART MARTENS • *Department of Biochemistry, Ghent University, Ghent, Belgium; Department of Medical Protein Research, Ghent University, Ghent, Belgium*

DANIEL MARTINS-DE-SOUZA • *Laboratory of Neuroproteomics, Department of Biochemistry, Institute of Biology, State University of Campinas (UNICAMP), Campinas, Brazil; Max Planck Institute of Psychiatry, Munich, Germany; Research Group of Proteomics, Department of Psychiatry and Psychotherapy, Ludwig-Maximilians-University of Munich, Munich, Germany; Laboratory of Neurosciences (LIM-27), Institute of Psychiatry, University of Sao Paulo, Sao Paulo, Brazil*

JESÚS MATEOS • *Rheumatology Division, ProteoRed/ISCIII Proteomics Group, INIBIC – Hospital Universitario de A Coruña, A Coruña, Spain*

DANIEL B. MCCLATCHY • *Department of Chemical Physiology, The Scripps Research Institute, La Jolla, CA, USA*

JIMMY RODRIGUEZ MURILLO • *Proteomics Unit, Institute of Chemistry, Federal University of Rio de Janeiro (UFRJ), Rio de Janeiro, Brazil*

FABIO CESAR SOUSA NOGUEIRA • *Proteomics Unit, Institute of Chemistry, Federal University of Rio de Janeiro (UFRJ), Rio de Janeiro, RJ, Brazil*

DARRAGH P. O'BRIEN • *Cancer Proteomics Laboratory, EGA Institute for Women's Health, University College London, London, UK*

EDUARDO B. DE OLIVEIRA • *Department of Biochemistry, Ribeirão Preto Medical School, University of São Paulo, Ribeirão Preto, SP, Brazil*

BRUNO MENEZES DE OLIVEIRA • *Muscle Biology Unit, Department of Experimental Medical Science, University of Lund, Lund, Sweden; Catarinense Federal Institute, Videira Campus, Bairro Campo Experimental, Videira - Santa Catarina, Brazil*

FRANCISCO M. BASTOS DE OLIVEIRA • *Institute of Biophysics, Federal University of Rio de Janeiro, Av Carlos Chagas Filho RJ, Brazil*

CAMILA S. PALMA • *Department of Biochemistry, Ribeirão Preto Medical School, University of São Paulo, Ribeirão Preto, SP, Brazil*

SHENG PAN • *Department of Medicine, University of Washington, Seattle, WA, USA*

LAURI PEIL • *Proteomics Core Laboratory, Institute of Technology, University of Tartu, Tartu, Estonia; Wellcome Trust Centre for Cell Biology, University of Edinburgh, Edinburgh, UK*

EVGENIY V. PETROTCHENKO • *University of Victoria - Genome BC Proteomics Centre, University of Victoria, Victoria, BC, Canada*

JOHN C. PRICE • *Department of Chemistry and Biochemistry, Brigham Young University, Provo, UT, USA*

QUAN QUAN • *Department of Chemistry, The University of Hong Kong, Hong Kong, China*

SONJA RADAU • *Leibniz-Institut für Analytische Wissenschaften – ISAS – e.V., Dortmund, Germany*

STEFAN RECKOW • *Proteomics and Biomarkers, Max Planck Institute of Psychiatry, Munich, Germany*

JOSÉ CÉSAR ROSA • *Department of Molecular and Cellular Biology, Protein Chemistry Center, Hemotherapy Center of Ribeirão Preto, Ribeirão Preto Medical School, University of São Paulo, Ribeirão Preto, Brazil*

CRISTINA RUIZ-ROMERO • *Rheumatology Division, ProteoRed/ISCIII Proteomics Group, INIBIC – Hospital Universitario de A Coruña, A Coruña, Spain; CIBER-BBN Instituto de Salud Carlos III, INIBIC-CHUAC, A Coruña, Spain*

ALON SAVIDOR • *Israel National Center for Personalized Medicine, Weizmann Institute of Science, Rehovot, Israel*

MARIA FÄLTH SAVITSKI • *Cellzome GmbH, Heidelberg, Germany*

KAI SCHEFFLER • *Thermo Fisher Scientific, Dreieich, Germany*

MATTHIAS SELBACH • *Max Delbrück Center for Molecular Medicine, Berlin, Germany*

JASON J. SERPA • *University of Victoria - Genome BC Proteomics Centre, University of Victoria, Victoria, BC, Canada*

MIKOLAJ SLABICKI • *Department of Translational Oncology, National Center for Tumor Diseases, Heidelberg, Germany*

MARCUS B. SMOLKA • *Department of Molecular Biology and Genetics, Weill Institute for Cell and Molecular Biology, Cornell University, Ithaca, NY, USA*

CHRISTIAN SOMMER • *Max Delbrück Center for Molecular Medicine, Berlin, Germany*

W. ANDY TAO • *Department of Chemistry, Purdue University, West Lafayette, IN, USA; Department of Biochemistry, Purdue University, West Lafayette, IN, USA; Department of Medicinal Chemistry and Molecular Pharmacology, Purdue University, West Lafayette, IN, USA; Purdue University Center for Cancer Research, Purdue University, West Lafayette, IN, USA*

JOHN F. TIMMS • *Cancer Proteomics Laboratory, EGA Institute for Women's Health, University College London, London, UK*

ARZU UMAR • *Department of Medical Oncology, Laboratory of Breast Cancer Genomics and Proteomics, Erasmus University Medical Center, Erasmus MC Cancer Institute, Rotterdam, The Netherlands; Netherlands Proteomics Centre, Utrecht, The Netherlands, Center for Translational Molecular Medicine, Rotterdam, The Netherlands*

MARC VAUDEL • *Proteomics Unit, Department of Biomedicine, University of Bergen, Bergen, Norway*

JUDIT VILLÉN • *Department of Genome Sciences, University of Washington, Seattle, WA, USA*

CHRISTIAN WEBHOFER • *Proteomics and Biomarkers, Max Planck Institute of Psychiatry, Munich, Germany*

CHRISTINA WILD • *Molecular Structure Analysis, German Cancer Research Center (DKFZ), Heidelberg, Germany*

LIANG XUE • *Department of Biochemistry, Purdue University, West Lafayette, IN, USA*

JOHN R. YATES III • *Department of Chemical Physiology, The Scripps Research Institute, La Jolla, CA, USA*

RENÉ P. ZAHEDI • *Leibniz-Institut für Analytische Wissenschaften – ISAS – e.V., Dortmund, Germany*

YUN ZHAO • *Department of Chemistry, The University of Hong Kong, Hong Kong, China*

NICO ZINN • *Molecular Structure Analysis, German Cancer Research Center (DKFZ), Heidelberg, Germany*

Part I

Introduction

Chapter 1

Survey of Shotgun Proteomics

Fabio Cesar Sousa Nogueira and Gilberto B. Domont

Abstract

Proteins provide the verbs to biology, and proteomics provides the nouns for their analytical and discovery-driven studies. The term proteomics was coined in the 1990s and deals with the protein complement of the genome—the proteome. Following the classical proteomics era, the development of new mass spectrometric methods for peptide analysis permitted the identification of proteins in peptide mixtures obtained by proteolytic digestion of complex samples, e.g., shotgun proteomics. Since its introduction, shotgun proteomics became the standard technique for the analysis of protein hydrolyzates in a high-throughput way. In this chapter, we provide a survey in shotgun proteomics highlighting instruments and techniques used in modern second and third proteomics generation.

Key words Shotgun proteomics, Liquid chromatography, Quantitative proteomics, Mass analyzers, Tandem mass spectrometry

1 Introduction

Proteins provide the verbs to biology [1], and proteomics provides the nouns for their analytical and discovery-driven studies. Proteomics deals with proteins and proteomes, a term first used in 1995 as *the protein complement of the genome* [2]. A post-genomic discipline, proteomics conceptually encompasses identification and quantification of all protein components of a proteome including primary and covalent structures, conformation (structural proteomics), posttranslational modifications, expression (isoforms, alternative splicing, polymorphisms), localization, and turnover as a function of stimuli, space, time, and cell type.

Proteomics can be made top-down, middle-down, or bottom-up. The first, best performed using electron collision (ECD) or transfer (ETD) dissociations still under development, has been successfully applied to characterize intact proteins by several groups, e.g., Neil Kelleher's (http://www.kelleher.northwestern.edu/research/top-down- proteomics), Joshua Coon's (http://www.chem.wisc.edu/~ coon/), and Donald Hunt's (http://chem.virginia.edu/faculty-research/faculty/donald-f-hunt/). It has advantages over the

Daniel Martins-de-Souza (ed.), *Shotgun Proteomics: Methods and Protocols*, Methods in Molecular Biology, vol. 1156, DOI 10.1007/978-1-4939-0685-7_1,

shotgun approach for the determination and localization of posttranslational modifications (PTMs) and characterization of isoforms; disadvantages are difficulties to fractionate, ionize, and fragment proteins in the gas phase. Middle-down proteomics is an intermediate approach that uses large peptides obtained by partial or limited proteolytic digests for protein identification. Finally, shotgun—a term coined at Yates lab by analogy with shotgun genomics [3]—is conceptually accepted as the use of enzymatic digests to characterize proteins in a mixture [3] by tandem mass spectrometry, the bottom-up approach.

Historically, the new biochemical strategy introduced with the use of two-dimensional electrophoresis (2DE) [4] and mass spectrometry for direct identification of proteins resulted in different insights to understand the complexity of biological systems. At the dawn of proteomics, ionization techniques could not be applied to peptides and proteins because they could not be converted into intact ions. This was solved when in 1988 Karas and Hillenkamp collected 10 kDa protein spectra [5] and Tanaka et al. [6] improved this matrix-assisted laser desorption ionization (MALDI) technique to detect 100 kDa specimens, and Fenn (1989) showed that proteins could be ionized at picomole levels by an electrospray system [7]. It was the development of soft ionization of thermally unstable peptides and proteins (i.e., MALDI and electrospray ionization, ESI) that brought classical proteomics into existence obscuring traditional protein techniques for structural determination of peptides and proteins, like Edman's sequence method as described by Matsuidara [8].

At the beginning, identification of proteins fractionated by SDS-PAGE or 2DE and analyzed by mass spectrometry was performed by peptide mass fingerprinting (PMF) [9–12] and uninterpreted mass spectra [13]. Processing of mass spectral data for protein identification was carried out by algorithms that matched the data to the theoretical peptide and/or fragment ion masses available in public sequence databases. These expanded procedures used all around the world for identification of proteins have, for two decades, created increasing demand for new techniques for the identification of proteins in mixtures. This demand was satisfied connecting enzymatic digests, peptide fractionation by liquid chromatography, tandem mass spectrometry, and bioinformatics tool algorithms. However, this second-generation proteomics [14] also demanded the development of mass spectrometers with higher sensitivity, resolution, and accuracy forcing the introduction and maturation of new techniques such as multidimensional protein identification technology (MudPIT) [15, 16], PTM characterization [17–23], and labeled and label-free quantification [24–31]. It was the ripe time for proteomics to assume a "systems biology" approach, considering that every biological system is finite and can be described in terms of bio-molecules, interactions, and pathways [14].

Shotgun proteomics has been intensively used for analysis of peptides and proteins with goals of "proteome profiling, protein quantification, protein modification, and protein–protein interaction [32]." In this review, we try to provide an update survey of this area, centered in concepts, mass spectrometers, and techniques that laid the ground for its foundation, development, and application to help nonspecialized scientists to better understand shotgun proteomics. The reader is referred to reviews on mass spectrometry history [33, 34], shotgun proteomics [3, 31, 35], and quantification [36].

2 Tandem Mass Spectrometry

Generically, mass spectrometry is an analytical technique to selectively detect and determine the amount, composition, and structure of analytes. By the end of the 1980s two ionization techniques able to routinely ionize biomolecules set the pace for peptide and protein new mass spectrometry: ESI [7]—now nanoESI [31]—and MALDI [5, 6]. Once in the gas phase ions are transferred to an analyzer where they will be separated by their mass-to-charge ratio. The need to improve dynamic range, resolution, sensitivity, and accuracy for peptide analysis pushed the development of new mass spectrometers. This imposed the creation of multiple architectures for new hybrid instruments combining different ion optics, fragmentation sources, and mass analyzers.

Mass spectrometry-based proteomics depends on the fragmentation of peptides and/or proteins to gain information on their structures using mass spectrometers in tandem able to perform MS and MS/MS modes. In other words, after acquiring ion precursor mass spectrum the fragment ions obtained by dissociation are analyzed generating MS/MS spectra.

3 Mass Analyzer and Mass Spectrometers

For peptide analysis the most successful mass analyzers are the scanning or the ion beam type, time of flight (ToF), and quadrupole or trapping, ion trap (IT), Orbitrap, and Fourier transform-ion cyclotron resonance (FT-ICR) [31, 37]. These analyzers can be compared according to such characteristics as:

- Mass resolution: Measurement of the separation between two peaks, normally measured at half-height.
- Mass accuracy: Comparison between measured and theoretical masses typically expressed as ppm.
- Mass range: Difference between measured major and minor masses.

- Detection limits: The least amount that can be measured with a signal/noise ratio of 1/3.
- Dynamic range: Molar difference between the less and the most abundant analyte that can be detected in an experiment.
- Scan speed: Acquisition time of a mass spectrum by an analyzer.
- Tandem mass spectrometry (MS^2): Ability of a mass spectrometer to isolate and fragment a selected ion for structural investigation.

3.1 Scanning/Ion Beam Mass Spectrometers: ToF, Q-ToF, and QqQ

The scanning/ion beam-ToF separates ions based on their velocities; it is the simplest mass analyzer. The ions formed at the same time and place in the ionization source are accelerated in the flying free tube under a fixed potential and low pressure. Because ions acquire the same kinetic energy after being accelerated, low *m/z* ions travel faster than higher *m/z* and arrive at different times at the detector; flight time is proportional to the peptide *m/z*. The flight path of an ion can be increased inserting a *reflectron* device (an ion mirror that reflects the ions toward the detector) at the end of the flight tube to correct for minor kinetic energy differences among ions with the same *m/z*, improving resolution. ToF analyzers can be coupled to MALDI or ESI ion sources. Structural information from ToF MS is obtained from fragmentation events. In MALDI, from metastable decomposition, a post-source decay [38, 39] occurs at the field-free region of the flight tube and by fast metastable decay in the ion source if a long delay occurs before acceleration of the ions [40]. Another peptide ion fragmentation process can be made by inserting a collision cell (collision-induced-dissociation, CID) to induce peptide fragmentation [41].

Quadrupole analyzers take advantage of the stability of ion paths in oscillating electric fields to separate them according to their *m/z*. Mass analysis is a consequence of a stabilized electric field generated by the simultaneous application of a direct current (DC) and oscillating voltages (alternate current, AC, and radio frequencies) on the four parallel quadrupole rods, thus becoming a function of the radiofrequency (RF) and DC voltages applied to the poles [32]. In a Q-ToF the continuous mass filtering and fragmentation of peptides make possible the determination of their intact and fragment ion masses. This type of instrument uses quadrupoles to focus and guide ions in the first Q, accelerate them and let peptide ions pass freely through Q_1 for precursor mass measurement, and finally detect in a ToF or else to be selected in the first Q, to be fragmented in the collision cell (Q_1), and for fragment mass measurements and detection in the ToF. Triple quadrupole (QqQ) type, a linear set of three quadrupoles, uses the first and third quadrupoles as mass scanning and the second as collision cell. Peptide fragments obtained in Q_2 are separated in Q_3 and detected.

When peptide ions pass freely through Q_2, that is, when Q_2 is not used as collision cell, the precursor peptide m/z is measured [42].

QqToF mass spectrometers have high accuracy and resolution in both MS and MS/MS modes. In MS mode the quadrupoles guide ions to the ToF analyzer to measure m/z, whereas in the MS/MS mode precursor ions are selected in the first Q (Q) and fragmented by CID/CAD in the second (q) which are then analyzed in the ToF. SYNAPT® series instruments introduce a different step for separation and analysis according to analyte size, shape, and charge combining a high sensitivity of ion transfer with ion mobility. Differently from the ToF analyzer in the ion mobility flight tube all ions move under the influence of an electric field through a collision gas that flows in the reverse direction. Ion separation is dependent on the size and cross area of the ions [43].

3.2 Trapping Mass Spectrometers: IT, Orbitrap, and FT-ICR

The most common trapping analyzer used for shotgun proteomics is the IT. Whereas in triple quadrupoles the electric field is bidimensional, ITs have three-dimensional fields for trapping peptide ions. By properly oscillating (RF) the electric field, this detector confines the ions in a small volume between one circular and two hyperbolic electrodes. Varying the potentials applied to the hyperbolic electrodes makes possible to trap ions with a specific m/z range as well as above or below defined m/z values. Trapping, stabilization, and excitation of the precursor ions and product ejection depend on the variation of different potentials applied to both hyperbolic and circular electrodes. Collision with helium weakens the kinetic energy of the ions compressing their trajectories to the center of the ion trap and stabilizing them. Peptide ions thus trapped in the IT are scanned by varying the radiofrequency voltage at the trap. After determining the m/z of the precursor ions the trap is filled, undesired ions are ejected, and the population of isolated ions is excited prior to collisions that converts translational to vibrational energy and leads to fragmentation. Fragments are scanned out the IT, and the process is repeated with a new population of entrapped ions [44, 45].

Fourier transform ion cyclotron resonance mass spectrometry (FTICR-MS) based on the circular movement of charged particles in a strong magnetic field provides the highest available resolution. The periodic movement of ion packs is recorded and converted to a frequency spectrum by Fourier transform, which is converted to a mass spectrum after calibration; cyclotron frequency depends directly on the mass-to-charge ratio of the ions. The ultrahigh attainable resolution mass spectrometer achieves 650,000 FWHM at $m/z=400$ and 1-s scan time, with 2 ppm mass accuracy in routine work opening new frontiers in proteomics [46, 47].

The LTQ-Orbitrap® has two analyzers: a linear ion trap-quadrupole (LTQ) and an Orbitrap®. The LTQ is made of four hyperbolic rods, each one divided into three successive axial sections that are able to capture and trap ions along the Z-axis in the central section when submitted to discreet DC voltages. Two other RF voltages applied in pairs to the X and Y rods are necessary for ion imprisonment as well as an AC voltage applied to the X rod for isolation, excitation, and ejection of the ions. The captured ions are ejected through two-side apertures in the central section of the rods in the X-position and detected by two multipliers. Advantages of the LTQ over the 3D ITs are reduced charge space effects because of higher volumes for ion storage and enhanced sensitivity for externally injected ions due to a better trapping efficiency [48–50]. Announced in 2000 [51] the Orbitrap analyzer became commercially available in 2005 [52]. It consists of one external electrode divided in the middle by a ceramic ring and one internal central electrode. Ions tangentially injected in the electric field between the two electrodes are trapped and stay orbiting the central internal electrode because of the balance between the electrostatic attraction of the internal electrode and the centrifugal force created by the tangential initial velocity of the ion when injected. Ions also oscillate along axial Z-axis; their oscillations are detected as current images that are resolved as individual oscillating frequencies using Fourier transform [52, 53]. A hybrid mass spectrometer combining an Orbitrap and a LTQ in the same instrument profits from the high resolution and accuracy of the Orbitrap and the sensitivity of the LTQ: LTQ, selects and fragments ions; a C-trap stores and ejects ions to the Orbitrap; and the Orbitrap detects ions in high resolution [49, 52].

Recently, the characterization of the yeast proteome was completed in 1 h (4,000 proteins identified) employing improvements in chromatographic separation and mass analysis in a new high-field Orbitrap hybrid mass spectrometer (Q-OT-Qit, model fusion) equipped with mass filter, collision cell, and a dual-cell LIT analyzer [54].

4 Modes for Acquiring Mass Spectra

There are several ways to acquire mass spectra. The main modes available for mass spectrometry are *product ion scanning*, *precursor ion scanning*, *neutral loss*, and *selected/multiple reaction monitoring*.

- *Production scanning* is commonly used in proteomics experiments. In this mode, the first analyzer is set to select one specific precursor ion. This ion is fragmented (CID), and the resulting fragments are analyzed in the second analyzer in a repeating process for different precursor ions.
- *Precursor ion scanning* uses the second analyzer set to transmit only one specific fragment ion to the detector. MS is used to detect all precursor ions that generate this fragment. It is typically

used to detect peptides that possess one specific functional group, as phosphate or carbohydrate.

- *Neutral loss scanning* describes the mode in which both analyzers operate in concerted action such that mass differences between the ions that fly through the analyzers remain constant. This difference of mass is the neutral loss originated from precursor fragment. This scanning mode is used to detect peptides that have labile functional groups attached to specific amino acids such as serine or threonine that lose phosphoric acid upon fragmentation.
- *Selected/multiple reaction monitoring*: In this mode the precursor ion and the specific fragment that it characterizes are selected by the first and second analyzers, respectively. The mass spectrometer looks for a series of transitions (precursor–fragment pair) to acquire their signal as a function of column retention times. It is used for the detection of specific analytes whose fragmentation properties in complex mixtures are well known.

5 Peptide Fragmentation

Peptide fragmentation is performed to generate complementary ion pieces to cover protein sequence and enable quantification. Fragmentation techniques implemented on mass spectrometers have general and particular aspects. They can be collision induced in an ion trap, a beam-type collision cell in a quadrupole, as well as ECD and up-to-date ETD; they vary according to the instrument. The most common and robust fragmentation technique for peptide analysis is collision-induced/activated dissociation (CID/CAD). Selected precursor ions are accelerated to increase their kinetic energy before collision with atoms or molecules of a neutral gas (N_2, Ar, or He). After this collision the kinetic energy of the ions is converted to vibrational internal energy resulting in bond cleavage. In a linear trap almost single-fragmentation events occur and the predominant ions formed are of the b and y type formed by cleavages at the peptide bond leaving the positive charge on the N- or the C-terminal fragment, respectively [55, 56]. Fragmentation may be explained by the mobile proton model [57]. Collision energy mobilizes the proton from any peptide basic site where it seats with variable efficiencies—e.g., N-terminal, ε-NH_2, and guanidine—to a heteroatom of the polypeptide chain stimulating charge-directed fragmentation. This model explains the propensity of fragmentation in specific amino acids like proline that has a secondary amine group with strong basic character in the gas phase.

When peptides have PTMs the preferred fragmentation pathways may change, for example, neutral loss of a phosphoric acid from phosphorylated serine and threonine residues. Ions with only one residue are denominated immonium with masses varying from 30 to 159 m/z and general structure HN = CH-R. They may be used to indicate the presence of individual amino acids. Satellite peaks connected to individual ion fragments may be observed in MS^2 spectra due to ammonia loss (−17 Da) mostly from Asn, Gln, Lys, and Arg side chains or water (−18 Da) from the C-terminal carboxyl group or from the side chain of Asp, Glu, Ser, and Thr residues [58].

A majority of y-type ions is generated in triple quadrupoles when ions passing through the quadrupoles are fragmented as a result of beam-type collisional activation. Recently beam-type fragmentation, higher energy CID (HCD) [59], has also been demonstrated on hybrid instruments with dedicated collision cells. HCD beam-type collision is very informative for phosphoproteomics and peptide quantification that uses low-mass isobaric tag reporter ions (iTRAQ and TMT) [60, 61]. Formerly denominated higher energy C-trap dissociation in 2007 [62] HCD is a CID beam type incorporated to the Orbitrap® mass spectrometers. It is identical to the fragmentations observed in QqQ and Q-ToF and is part of the architecture of the LTQ-Orbitrap mass spectrometers. HCD is performed outside the LTQ.

Alternative methods are *electron capture dissociation* (ECD), *electron transfer dissociation* (ETD), and *proton transfer charge reduction* (PTR). Introduced in 1998 [63], ECD uses low-energy electrons to react and be incorporated into protonated peptides in the magnetic field of an FT-ICR-MS producing cations with an additional electron. The odd electron rearranges quickly in the polypeptide chain and subsequently dissociates fragmenting the peptide. Advantages of this fragmentation technique for peptide and protein analysis are random cleavage—with the exception of N-terminal proline—yielding higher coverage than CID. Also labile PTMs are not cleaved or degraded during the fragmentation process. When multi-charged ions are irradiated with low-energy electrons they form new ions with reduced charge that dissociate via radical-directed routes. Main products of the ECD reaction for protonated peptides—$[M+nH]^{n+}$—are precursor ions with reduced charge—$[M+nH]^{(n-1)+\bullet}$—that capture one electron but do not dissociate. The polypeptide chain is most cleaved at the N–Cα bond giving rise to the ion series c and z.

Developed in 2004, ETD [64] is a technique that transfers electrons from an anionic radical (e.g., anthracene, fluoranthene) to a protonated peptide. As with ECD, fragmentation produces the same ions c and z. ETD proceeds in steps: first, the cationic analyte and the anionic reagent are generated; then, both ions are transferred at the same time to the same site; and finally the fragment ions are analyzed. The multi-charged peptide submitted to ETD is produced by ESI; the anionic radical is formed by chemical ionization (CI) through the impact of electrons on, e.g., fluoranthene, the mostly used reagent. Electron transfer from the anion to the peptides occurs in the LTQ where the multi-charged peptides are stored. ETD fragmentation mechanism is similar to ECD; however, fragmentation efficiency is lower especially for double-charged peptides. ETD has been successfully employed in *top-down* experiments as well as in glycol- and phosphoproteomics.

ETD in combination with PTR has been successfully used in experiments of top-down proteomics, allowing the sequence analysis of mixtures of larger peptides or proteins even when highly modified. PTR along with ETD serves to simplify the spectrum reducing the complexity of spectra generated from these highly charged species. The c- and z-type ion products are present predominantly as one charge state [65, 66].

6 Liquid Chromatography

The theory of chromatography developed by Martin and Synge [67] models the chromatographic process as a consequence of the partitioning equilibrium (partition coefficient of molecules) between two phases in each theoretical plate. In chromatography, analytes dissolved in a mobile phase flow through the column and interact with the stationary phase.

Highly complex samples pose a special problem to proteomics: identification in space and time of its components is a requirement for further understanding of their involvement on biology. To best address this problem shotgun proteomics focuses in three approaches and techniques which have become the golden standards in recent decades: reverse-phase (RP) chromatography, MudPIT, and the use of longer columns, temperature, smaller particles, and elution/gradient times.

Most used chromatographic techniques in shotgun proteomics are ion exchange (IXC), RP, and a combination of both MudPIT [68]. IXC separates analytes based on their surface charge differences; reverse phase separates analytes based on differences in their hydrophobicity.

Separation uses HPLC (high performance) and UHPLC (ultrahigh performance) liquid chromatographies. High- and ultrahigh-pressure pumps favor the use of capillary columns of 50–100 μm I.D. filled with 1.7–5 μm beads with nanoflow rates of up to 300 nL.

6.1 Reverse Phase

This term means that the stationary phase is less polar than the mobile. It is the universal technique used for the separation of peptides and proteins in proteomics. Separation depends on the hydrophobic interaction of molecules present in the mobile phase with immobilized hydrophobic ligands in the stationary phase.

Elution is performed with organic solvents as mobile phases—mostly acetonitrile, and methanol and isopropanol—whose concentration is made to vary to favor desorption of the solute from the stationary phase. Important parameters to consider are matrix composition, bead and pore size, and type of immobilized ligand which is normally linear hydrocarbon, e.g., two-carbon ligand (C2, ethyl), octyl (C8), and octadecyl (C18). Pore sizes vary from 100 Å for peptides to 300 Å and over for larger peptides and proteins. Use of long columns packed with <2 μm beads and high temperature and pressures (UHPLC) increases peak capacity and resolution allowing the use of longer gradients [15, 49].

6.2 Multidimensional Protein Identification Technology

Protein abundance in a biological sample can vary by several orders of magnitude, and after enzymatic hydrolysis the number of peptide species largely increases its complexity. The introduction of orthogonal techniques helped to approach this problem.

Strong cation-exchange (SCX) chromatography followed by reverse phase is the second most used methodology. Samples applied to SCX columns are eluted online with the mass spectrometer in a series of increasing salt steps. Each sample is individually transferred online to the reverse column, and the mobile phase is directly injected into the ESI source. Combination of these techniques under the name of MudPIT [15, 49], together with longer columns and gradients at fixed temperatures, still is one of the methods of choice for the analysis of very complex peptide mixtures.

6.3 Longer Columns and Elution Times

Ideally, it is desirable to analyze a whole proteome in a single shotgun run. Recently, it was demonstrated that a combination of longer columns, smaller bead sizes, and long gradients under temperature control, coupled to a quadrupole Orbitrap® platform, reached a large dynamic range of protein identification using minute amounts of sample. Lys-C digests were fractionated at 35 °C on 50 cm columns (75 μm I.D.), 1.8 μm C_{18} beads, and a flow rate of 250 nl/min for 240 min (4-h gradient). More than 4,000 proteins were identified in a single run accounting for over 95 % of the yeast genome [69, 70]. One-hour yeast proteome analysis identified 4,000 proteins employing 75–360 μm inner–outer diameter, 35 cm long column packed with 1.7 μm C18 beads, 60 °C, and 1.3-h acetonitrile gradient [54].

7 Inferring Peptide Sequences

Bioinformatics is a cornerstone for proteomics once this field produces a lot of data with different complexities and aims. In a typical shotgun experiment mass spectrometers execute a survey or a precursor scan (MS^1 spectra) followed by isolation, fragmentation, and detection of the fragment ions (MS^2 spectra). This acquisition mode is supervised by a number of parameters previously set (*data-dependent acquisition*—DDA). In the beginning, identification of peptides required an expert; MS^2 spectra were manually interpreted. With the advent of computational tools this task became easier and widespread, except for *de novo* sequencing.

Interpretation of tandem mass spectra for peptide identification may be dependent or independent of databases. For shotgun proteomics, database-dependent mode is largely used and the strategy applied is called *unprocessed MS^2 data* search or *peptide spectrum match* (PSM) search. Among the tools that employ this strategy the most commons are SEQUEST [13], ProLuCID [71], Mascot [72], MaxQuant [73], and others [74–76].

8 Quantitative Proteomics

Descriptive proteome, bound only to identification, gives extremely important qualitative information on proteins expressed in a sample necessary to understand and argue biological systems. However, it is quantitative information on protein abundance that provides precise and richer results and characterizes proteomics analysis.

There are two types of quantitative proteomics: absolute quantification, when the amount of the protein is determined by its concentration or copy number in the sample, and relative quantification defined as the rate of a protein amount present in different samples. Quantification depending on the use of stable isotopes is label dependent and label free, respectively, as well as targeted and nontargeted. Based on stable isotopes proteins are chemically, metabolically, or enzymatically labeled with molecules that have a combination of light and heavy isotopes of ^{15}N, ^{13}C, ^{18}O, and ^{2}H in their composition. Labeled samples are mixed, and peptides labeled with different isotopes are identified as pairs in the MS spectra or as reporter ions in the MS^2. Examples are chemical isotopic labeling with *isotope-coded affinity tag* (ICAT), *isotope-coded protein label* (ICPL), and *dimethyl labeling*; isobaric labeling with *isobaric tags for relative and absolute quantification* (iTRAQ) and *tandem mass tags* (TMT); enzymatic labeling by $H_2{}^{18}O$ during protein digestion; and metabolic light and heavy substrates or amino acids supplemented in culture media. In label-free quantification two or more samples are compared based on the normalized ion intensity of identical peptides or on the spectral counts acquired for each protein.

In targeted MS a standard internal specific peptide (*proteotypic*) labeled with heavy isotopes is used for each protein to be quantified. Both the sample and the heavy peptide internal standard co-elute and are detected in the MS spectra, selected, and fragmented. Because the amount of the internal standard is known, the sample peptide can be quantified.

8.1 Stable Isotopes

8.1.1 Chemical Labeling: ICAT, ICPL, and Dimethyl

Proteomes are labeled with light and heavy isotopes in different reactive sites of peptides and proteins, e.g., sulfhydryl and all free amino groups. Quantification is performed determining the ratio between the intensity of isotope ions in MS or integrating their chromatographic areas [24].

ICAT

The reagent possesses a reactive group specific to the sulfhydryl lateral chain of cysteine, a connecting group that may have eight ^{1}H or ^{2}H (heavy isotope) and biotin for affinity isolation. The technique is based on the labeling of cysteine residues of control and sample proteins, respectively, with the light and heavy reagents. Samples are combined and trypsin digested, and labeled peptides are isolated in an avidin column and analyzed by mass spectrometry. Peptide pairs with 8 Da differences in MS are quantified by their intensity ratios. Advantage of the ICAT technique is the enrichment in peptides that contain cysteine reducing the complexity of the mixture. Disadvantages are differences in retention times of light and heavy peptides during reverse-phase chromatography; heavier elutes before the lighter. Also, the hydrophobic marker creates problems in the elution of peptides from the reverse-phase column and in interpretation of MS^2 spectra because many fragments come from the marker instead of the polypeptide chain. To avoid these problems several alternatives were developed as photo-cleavage [77], a new reagent (cICAT) cleavable by acid [78]. However, the very nature of the methodology based on cysteine labeling is the main problem because proteins/peptides that do not have cysteine are not labeled, detected, and quantified.

ICPL

Originally developed in 2005 [79] the technique is based in the isotopic labeling of all free amino groups, N-terminal, and lysine ε-amino. Sample proteins are labeled with Nic-NHS (d0- or d4-*N*-nicotinoyloxysuccinimide, duplex; or 4-plex with ^{13}C), combined, fractionated, and digested. Peptides are analyzed after reverse-phase chromatography by mass spectrometry; peptide ions appear as pairs in the mass spectrum.

Problems of this methodology are the presence of big peptides, dependence of the labeling upon peptide composition, and four times increase in the complexity of the spectrum by the 4-plex reagent.

Dimethyl Labeling

The main advantage is low cost. It was developed in 2003 [80] and improved in 2008 [81]. It uses formaldehyde to label the N-terminal and ε-amino groups of lysine through reductive amination, proline excepted. Reduction of the Schiff base formed with formaldehyde by cyanoborohydride increases mass by 28 Da for each primary amine, 32 Da if deuterated formaldehyde is used, or 36 Da if ^{13}C and deuterated formaldehyde plus deuterated borohydride are the reagents.

Disadvantages of this technique are small shifts in retention time during chromatography due to deuterium incorporation and superimposition of isotopic envelopes.

8.1.2 Chemical Isobaric Labeling: iTRAQ and TMT

iTRAQ

It is the most used and reliable labeling technique for protein quantification developed to analyze multiple samples bearing labeled peptides in Lys ε-amino and N-terminal groups [27]. The reagent contains a reporter group (*N*-methylpiperazine) and balance (carbonyl) and reactive groups (NHS ester). Each plex reagent has the same mass achieved by a combination of ^{13}C, ^{15}N, and ^{18}O in the reporter and balance groups. Derivatized peptides have similar retention times in liquid chromatography and appear as a single isobaric peak in MS; after fragmentation they produce different reporter ions, each one belonging to a labeled sample. Multiplex (4- and 8-plex) isobaric markers are commercially available to label different samples; in MS^2 their reporter ions fall in the low region of the m/z spectrum from 114 to 117 (4-plex) and 113 to 121 (8-plex, except 120). Absolute quantification is also possible.

TMT

Developed in 2003 [82] TMT exits as 2- and 6-plex versions. It also reacts with N-terminal and Lys ε-amino groups and has reporter groups that appear in the MS in the m/z region 126–131 for the 6-plex. TMT has the same characteristics and uses an analytical approach similar to iTRAQ.

Advantages of the labeling methods are the following: (1) multiplex labeling; (2) co-elution of labeled and non-labeled peptides; and (3) complexity of the MS spectra does not change because labeled peptides are isobaric.

Disadvantages are the following: (1) low mass (<150 Da) of the reporter ions which did not allow initially the use of ion traps surpassed today by the use of pulsed Q dissociation (PQD), HCD, and ETD; (2) lower number of peptide sequence/spectrum match (PSM) and of identified peptides and proteins compared to a non-labeled sample [83, 84], a problem that worsens in the order iTRAQ 4-plex, TMT 6-plex, and iTRAQ 8-plex; it also hampers the identification of phosphopeptides due to an increase in their total charges (charge effect) overcome by vaporizing 5 % ammonia to decrease peptide charge; and (3) presence of contaminating ions close to the isobaric in the MS spectra; they are also isolated and fragmented compromising accuracy of quantification. Two strategies were developed to solve this problem: one based on a further fragmentation by HCD of the more intense ion of MS^2 to obtain an MS^3 spectrum; MS^2 gives the peptide sequence, whereas MS^3 collects information on the reporter ions [85]. Problems are the availability of software to jointly analyze the identification and quantification data. The other strategy is based on the impairment of isolated precursor and contaminant charges by proton transfer increasing mass difference between both. This problem is solved by fragmentation of the precursor by HCD and CID, accumulation of fragments in the C-trap, and detection. The fragments contain information on reporter ions and peptide sequence [86]. Both strategies decrease the number of identifications.

8.1.3 Enzymatic Labeling

It is based on the incorporation of ^{18}O in peptides by action of hydrolytic enzymes (trypsin, Lys-C, Glu-C, etc.) on proteins. ^{18}O isotopes are introduced at the C-termini of peptides when the digestion is carried out in $H_2{}^{18}O$. The result is a mass increase of 2 or 4 Da for partial or total labeling, respectively. One big problem with this methodology is different labeling efficiency due to isotopes 18 and 16 exchange in downstream procedures with normal water, incomplete enzyme inactivation, and peptide labeling. Labeling inefficiency introduces an extra difficulty in the analysis of the isotopic envelopes of labeled and non-labeled peptides. To solve these problems, the use of immobilized trypsin, addition of a boiling step after digestion, and development of software to correct quantification were proposed [87].

8.1.4 Metabolic Labeling

This technique is also known as in vivo labeling. The label is introduced in proteins in culture media during cell division and growth. This is an advantageous characteristic that provides high accuracy and precision in quantification; labeling occurs at the beginning, and because there are no further labeling steps manipulation errors are reduced.

The first metabolic label quantification employed ^{15}N. It is useful for plants and bacteria. Two problems are the ability to label only two samples and complicated quantification because the number of incorporated ^{15}N is unpredictable and the peptide exact mass is unknown [88–90].

SILAC

This technique incorporates isotopic labeled amino acids in deficient cell culture media. They are incorporated in all cell proteins during growth. The first approach used tri-deuterated leucine (heavy) and leucine (light) [25]. Thereafter, other amino acids with different isotope compositions like $^{13}C6$-Arg were introduced [91].

In a typical metabolic labeling experiment one cell population grows in a culture medium containing light and the other in heavy-amino acid medium. Generally, heavy amino acids are lysine and arginine bearing ^{2}H, ^{13}C, or ^{15}N in lieu of H, ^{12}C, and ^{14}N. Normally, five cell generations are sufficient to change normal to heavy amino acids. Other strategies based in the original SILAC were devised: multiplex SILAC to compare three different cell cultures in the presence of an isotopic combination of the same amino acid—$^{12}C_6{}^{14}N_4$-Arg (0 Da, *light*), $^{13}C_6{}^{14}N_4$-Arg (6 Da, *middle*), and e $^{13}C_6{}^{15}N_4$-Arg (10 Da, *heavy*) [92] or double-triple labeling with Lys0-Arg0 (*light*), Lys4-Arg6 (*middle*), and e Lys8-Arg10 (*heavy*) [93]; *pulsed* SILAC for cells grown in light SILAC followed by transfer to a heavy medium for a certain time to show the rate of protein synthesis and degradation [94, 95] as well as the *super*-SILAC developed for the quantification of proteins in animal/human tissues based in the comparison of one set of proteins with a set grown in heavy SILAC as an internal standard. Protein ratios are calculated for quantification [96].

SILAC has limitations as the conversion of arginine to proline during cell division complicates quantification. Although used to label multicellular organisms (rat, flies, *Caenorhabditis elegans*, etc.) [97] it is mostly used in cultures of mammal cell that need essential amino acids.

8.2 Label Free

As the title suggests, this method quantifies proteins or peptides without any additional labeling step of any kind. Quantification is already present in the collected LC-MS/MS data, an attractive simple feature for its simplicity, cost, and reproducibility, deleting steps and complexity of multiplex experiments. Label-free methods accept larger dynamic range differences when compared to isotope labeling. However, the quality of the quantification is strongly dependent on the reproducibility of the LC-MS/MS data and on the bioinformatics tools for processing. Label-free quantifications are of two types: measurement of peak areas or signal intensity of the precursor ion and spectral counts of the peptide in an MS^2 experiment.

8.2.1 Extracted Ion Chromatogram: XIC

In LC-MS/MS experiments quantification is obtained by integrating areas of the ion peaks of any peptide along the chromatographic profile. The correlation between signal and peptide concentration is linear for four orders of magnitude [98, 99]. Comparative experiments between samples are made integrating peak areas of peptides between different LC-MS/MS runs of different samples. Abundances are estimated measuring areas for all, or part, of the peptides that belong to a protein. This quantification relies on the confidence of sample preparation, reproducibility of mass spectra, and ionization efficiency. It is necessary to align and perform peak paring, normalize the data, and do statistical analysis for high accuracy and robustness. The technique also requires a perfect balance between acquisition of MS^1 and MS^2 spectra for a precise quantification and, at the same time, maximizing of the number of identifications. One of the strategies to overcome this problem, especially in ToF instruments, is alternate screening during high- and low-energy collisions without precursor isolation with altogether fragmentation of all co-eluting ions. This method is denominated data-independent acquisition (DIA).

8.2.2 Spectral Count

Spectral count approaches are based in observations that the number of PSMs, number of peptides, and sequence coverage for a specific protein correlate with the protein amount in an LC-MS/MS experiment. Basically, quantification depends on the comparison of the spectral counts for each protein in different samples; this information is generated during search in data banks using conventional search tools. Because of its simplicity, quantitative spectral count approaches are becoming popular and being diversified by the use of different normalizations and statistical tools to increase dynamic range and quantification accuracy.

However, one must consider that the technique is influenced by (1) experimental parameters as the instrument reproducibility and the chromatographic setup, semi-random nature of data-dependent acquisition, ion dynamic exclusion of precursor ion, width of the chromatographic peak, and scan speed of the instrument; (2) protein-specific parameters as the size of peptides, saturation due to high-abundance proteins, number of generated spectra for the less abundant proteins; and (3) parameters used for protein identification like false discovery rates (FDR) that limit the number of most confident PSM spectra and peptides that are shared by protein isoforms [100].

8.3 Target Proteomics: Selected Reaction Monitoring or Multiple Reaction Monitoring

The goal of target proteomics is to selectively monitor specific proteins in a complex biological sample. QqQ mass spectrometers are the instruments of choice because they select precursor ions in Q1 and its fragments in Q3; Q2 is used to fragment the peptides. Quantification is absolute if stable isotope dilution (SID) is used.

Selected reaction monitoring (SRM)/multiple reaction monitoring (MRM) is a mass spectrometry technique applied for detection and quantification of predetermined specific analytes with known fragmentation properties in complex samples. It explores on-time transitions—m/z—of the pair precursor/fragments considering chromatographic retention time and signal intensity in the mass spectrometer, the coordinates for a specific transition. These two levels of strict selection result in a high selectivity once co-eluting ions are efficiently filtered. The selective nature of this method increases sensitivity by one to two orders of magnitude and the dynamic range of detection in up to five orders as compared to conventional proteomics techniques.

A set of peptides must be selected for each protein included in the analysis. These target peptides, called proteotypic—normally from tryptic hydrolysis—need to satisfy very stringent criteria: have an amino acid sequence uniquely associated to the protein and be consistently observed in LC-MS/MS experiments. Peptide singularity is essential to guarantee the identification of only one target protein. Additionally it is important to select peptides to differentiate isoforms occurring in alternative splicing or single-nucleotide polymorphisms (SNP). Peptide characteristics to be considered are recovery after sample preparation, presence of missing cleavages, good ionization efficiency, m/z inside the mass spectrometer range, as well as hydrophilicity and hydrophobicity. A common practice for peptide selection is the use of *PeptideAtlas*, *GBMdb*, and *PRIDE* databases. Tools for selection of peptides are *ESP Predictor*,

PeptideSieve, *PepFly*, *STEPP*, etc. However, no tool can predict the best peptide set for SRM/MRM because of factors that influence this decision: presence of amino acids chemically modified or with PTMs. It is advisable to select multiple peptides per protein [101].

Strategies for absolute quantification are label-free or isotope-labeled standards. Whereas label-free methodologies are of low cost and easy to implement those that employ SID are more accurate and allow precise quantitative measurements.

SID uses an isotopologue of the target peptide as an internal standard. Because mass spectrometers are not inherently quantitative a known quantity of stable standard isotopologue is added to the mixture. Physical-chemical properties of the standard and target peptide are identical resulting in equal ionization efficiency. Thus it is possible to correlate intensity of the signal of the internal standard added in known amounts to the quantity of the target peptide. A technique based in SID-MS for absolute quantification of proteins, AQUA, was published in 2003 [102, 103].

9 Concluding Remarks

We present a brief review for researchers that eventually need to collect proteomics data and understand the general terms and principles that govern shotgun proteomics. Thus, we approached and structured the subjects according to our biased experience and background covering tandem mass spectrometry and the modes used to acquire mass spectra and discussed mass analyzers and scanning/ion beam as well as trapping mass spectrometers, peptide fragmentation, liquid chromatography, and peptide identification. We hope that our words will ease other researchers to perform and interpret their experiments.

Acknowledgements

Conselho Nacional de Pesquisas (CNPq), Brazil, grant # 308819/2011-0, and Fundação de Amparo à Pesquisa do Estado do Rio de Janeiro (FAPERJ), Brazil, grant # E-26/110.138/2013. We thank Dr Magno Junqueira, Proteomics Unit, UFRJ, for critically reading the manuscript.

References

1. Hancock W, LaBaer J, Marko-Varga GA (2011) Journal of proteome research—10th anniversary. J Proteome Res 10:1–2
2. Wasinger VC, Cordwell SJ, Cerpa-Poljak A et al (1995) Progress with gene-product mapping of the Mollicutes: *Mycoplasma genitalium*. Electrophoresis 16:1090–1094
3. Yates JR (2013) The revolution and evolution of shotgun proteomics for large-scale proteome analysis. J Am Chem Soc 135:1629–1640

4. O'Farrell PH (1975) High resolution two-dimensional electrophoresis of proteins. J Biol Chem 250:4007–4021
5. Karas M, Hillenkamp F (1988) Laser desorption ionization of proteins with molecular masses exceeding 10,000 daltons. Anal Chem 60:2299–2301
6. Tanaka K, Waki H, Ido Y et al (1988) Protein and polymer analyses up to *m/z* 100 000 by laser ionization time?: of? flight mass spectrometry. Rapid Commun Mass Spectrom 2: 151–153
7. Fenn JB, Mann M, Meng CK et al (1989) Electrospray ionization for mass spectrometry of large biomolecules. Science 246:64–71
8. Matsudaira P (1987) Sequence from picomole quantities of proteins electroblotted onto polyvinylidene difluoride membranes. J Biol Chem 262:10035–10038
9. Mann M, Højrup P, Roepstorff P (1993) Use of mass spectrometric molecular weight information to identify proteins in sequence databases. Biol Mass Spectrom 22:338–345
10. Henzel WJ, Billeci TM, Stults JT et al (1993) Identifying proteins from two-dimensional gels by molecular mass searching of peptide fragments in protein sequence databases. Proc Natl Acad Sci 90:5011–5015
11. Yates JR, Speicher S, Griffin PR et al (1993) Peptide mass maps: a highly informative approach to protein identification. Anal Biochem 214:397–408
12. James P, Quadroni M, Carafoli E et al (1993) Protein identification by mass profile fingerprinting. Biochem Biophys Res Commun 195:58–64
13. Eng JK, McCormack AL, Yates Iii JR (1994) An approach to correlate tandem mass spectral data of peptides with amino acid sequences in a protein database. J Am Soc Mass Spectrom 5:976–989
14. Patterson SD, Aebersold RH (2003) Proteomics: the first decade and beyond. Nat Gen 33:311–323
15. Link AJ, Eng J, Schieltz DM et al (1999) Direct analysis of protein complexes using mass spectrometry. Nat Biotechnol 17:676–682
16. Wolters DA, Washburn MP, Yates JR (2001) An automated multidimensional protein identification technology for shotgun proteomics. Anal Chem 73:5683–5690
17. Larsen MR, Roepstorff P (2000) Mass spectrometric identification of proteins and characterization of their post-translational modifications in proteome analysis. Fresenius J Anal Chem 366:677–690
18. Harvey DJ (2001) Identification of protein-bound carbohydrates by mass spectrometry. Proteomics 1:311–328
19. MacCoss MJ, McDonald WH, Saraf A et al (2002) Shotgun identification of protein modifications from protein complexes and lens tissue. Proc Natl Acad Sci U S A 99: 7900–7905
20. Mann M, Jensen ON (2003) Proteomic analysis of post-translational modifications. Nat Biotechnol 21:255–261
21. Cournoyer JJ, Pittman JL, Ivleva VB et al (2005) Deamidation: differentiation of aspartyl from isoaspartyl products in peptides by electron capture dissociation. Protein Sci 14: 452–463
22. Edelson-Averbukh M, Pipkorn R, Lehmann WD (2007) Analysis of protein phosphorylation in the regions of consecutive serine/threonine residues by negative ion electrospray collision-induced dissociation. Approach to pinpointing of phosphorylation sites. Anal Chem 79:3476–3486
23. Harvey DJ (2009) Analysis of carbohydrates and glycoconjugates by matrix-assisted laser desorption/ionization mass spectrometry: an update for 2003–2004. Mass Spectrom Rev 28:273–361
24. Gygi SP, Rist B, Gerber SA et al (1999) Quantitative analysis of complex protein mixtures using isotope-coded affinity tags. Nat Biotech 17:994–999
25. Ong SE, Blagoev B, Kratchmarova I et al (2002) Stable isotope labeling by amino acids in cell culture, SILAC, as a simple and accurate approach to expression proteomics. Mol Cell Proteomics MCP 1:376–386
26. Zhu H, Pan S, Gu S et al (2002) Amino acid residue specific stable isotope labeling for quantitative proteomics. Rapid Commun Mass Spectrom RCM 16:2115–2123
27. Ross PL, Huang YN, Marchese JN et al (2004) Multiplexed protein quantitation in Saccharomyces cerevisiae using amine-reactive isobaric tagging reagents. Mol Cell Proteomics MCP 3:1154–1169
28. Ishihama Y, Oda Y, Tabata T et al (2005) Exponentially modified protein abundance index (emPAI) for estimation of absolute protein amount in proteomics by the number of sequenced peptides per protein. Mol Cell Proteomics MCP 4:1265–1272
29. Pratt JM, Simpson DM, Doherty MK et al (2006) Multiplexed absolute quantification for proteomics using concatenated signature peptides encoded by QconCAT genes. Nat Protoc 1:1029–1043

30. Wang G, Wu WW, Zeng W et al (2006) Label-free protein quantification using LC-coupled ion trap or FT mass spectrometry: reproducibility, linearity, and application with complex proteomes. J Proteome Res 5:1214–1223
31. Zhang Y, Fonslow BR, Shan B et al (2013) Protein analysis by shotgun/bottom-up proteomics. Chem Rev 113:2343–2394
32. Glish GL, Vachet RW (2003) The basics of mass spectrometry in the twenty-first century. Nat Rev Drug Discov 2:140–150
33. Gelpi E (2008) From large analogical instruments to small digital black boxes: 40 years of progress in mass spectrometry and its role in proteomics. Part I 1965-1984. J Mass Spectrom JMS 43:419–435
34. Gelpi E (2009) From large analogical instruments to small digital black boxes: 40 years of progress in mass spectrometry and its role in proteomics. Part II 1985–2000. J Mass Spectrom JMS 44:1137–1161
35. Thelen JJ, Miernyk JA (2012) The proteomic future: where mass spectrometry should be taking us. Biochem J 444:169–181
36. Yates JR 3rd, Washburn MP (2013) Quantitative proteomics. Anal Chem 85:8881–8881
37. Domon B, Aebersold R (2006) Mass spectrometry and protein analysis. Science 312: 212–217
38. Spengler B (1997) Post-source decay analysis in matrix-assisted laser desorption/ionization mass spectrometry of biomolecules. J Mass Spectrom 32:1019–1036
39. Spengler B, Kirsch D, Kaufmann R et al (1992) Peptide sequencing by matrix-assisted laser-desorption mass spectrometry. Rapid Commun Mass Spectrom 6:105–108
40. Brown RS, Lennon JJ (1995) Sequence-specific fragmentation of matrix-assisted laser-desorbed protein/peptide ions. Anal Chem 67:3990–3999
41. Medzihradszky KF, Campbell JM, Baldwin MA et al (2000) The characteristics of peptide collision-induced dissociation using a high-performance MALDI-TOF/TOF tandem mass spectrometer. Anal Chem 72:552–558
42. Yost RA, Enke CG (1979) Triple quadrupole mass spectrometry for direct mixture analysis and structure elucidation. Anal Chem 51: 1251–1264
43. Kanu AB, Dwivedi P, Tam M et al (2008) Ion mobility-mass spectrometry. J Mass Spectrom JMS 43:1–22
44. Cooks RG, Glish GL, Mc Luckey SA et al (1991) Ion trap mass spectrometry. Chem Eng News 69:12, Medium:–X
45. March RE (1992) Ion trap mass spectrometry. Int J Mass Spectrom Ion Process 118–119:71–135
46. Marshall AG, Hendrickson CL, Jackson GS (1998) Fourier transform ion cyclotron resonance mass spectrometry: a primer. Mass Spectrom Rev 17:1–35
47. Schaub TM, Hendrickson CL, Horning S et al (2008) High-performance mass spectrometry: Fourier transform ion cyclotron resonance at 14.5 Tesla. Anal Chem 80: 3985–3990
48. Schwartz JC, Senko MW, Syka JEP (2002) A two-dimensional quadrupole ion trap mass spectrometer. J Am Soc Mass Spectrom 13: 659–669
49. Yates JR, Ruse CI, Nakorchevsky A (2009) Proteomics by mass spectrometry: approaches, advances, and applications. Annu Rev Biomed Eng 11:49–79
50. Hager JW (2002) A new linear ion trap mass spectrometer. Rapid Commun Mass Spectrom 16:512–526
51. Makarov A (2000) Electrostatic axially harmonic orbital trapping: a high-performance technique of mass analysis. Anal Chem 72: 1156–1162
52. Makarov A, Denisov E, Kholomeev A et al (2006) Performance evaluation of a hybrid linear ion trap/orbitrap mass spectrometer. Anal Chem 78:2113–2120
53. Hu Q, Noll RJ, Li H et al (2005) The Orbitrap: a new mass spectrometer. J Mass Spectrom JMS 40:430–443
54. Hebert AS, Richards AL, Bailey DJ et al (2013) The one hour yeast proteome. Mol Cell Proteomics MCP 13(1):339–47
55. Roepstorff P, Fohlman J (1984) Proposal for a common nomenclature for sequence ions in mass spectra of peptides. Biomed Mass Spectrom 11(11):601
56. Biemann K (1988) Contributions of mass spectrometry to peptide and protein structure. Biomed Environ Mass spectrom 16: 99–111
57. Jones AW, Cooper HJ (2011) Dissociation techniques in mass spectrometry-based proteomics. Analyst 136:3419–3429
58. Paizs B, Suhai S (2005) Fragmentation pathways of protonated peptides. Mass Spectrom Rev 24:508–548
59. Ting L, Cowley MJ, Hoon SL et al (2009) Normalization and statistical analysis of quantitative proteomics data generated by metabolic labeling. Mol Cell Proteomics 8: 2227–2242

60. McAlister GC, Phanstiel D, Wenger CD et al (2010) Analysis of tandem mass spectra by FTMS for improved large-scale proteomics with superior protein quantification. Anal Chem 82:316–322
61. Pichler P, Kocher T, Holzmann J et al (2011) Improved precision of iTRAQ and TMT quantification by an axial extraction field in an Orbitrap HCD cell. Anal Chem 83: 1469–1474
62. Olsen JV, Macek B, Lange O et al (2007) Higher-energy C-trap dissociation for peptide modification analysis. Nat Methods 4: 709–712
63. Zubarev RA, Horn DM, Fridriksson EK et al (2000) Electron capture dissociation for structural characterization of multiply charged protein cations. Anal Chem 72:563–573
64. Syka JE, Coon JJ, Schroeder MJ et al (2004) Peptide and protein sequence analysis by electron transfer dissociation mass spectrometry. Proc Natl Acad Sci U S A 101:9528–9533
65. Coon JJ, Ueberheide B, Syka JE et al (2005) Protein identification using sequential ion/ion reactions and tandem mass spectrometry. Proc Natl Acad Sci U S A 102:9463–9468
66. Mikesh LM, Ueberheide B, Chi A et al (2006) The utility of ETD mass spectrometry in proteomic analysis. Biochim Biophys Acta 1764:1811–1822
67. Martin AJ, Synge RL (1941) Separation of the higher monoamino-acids by counter-current liquid-liquid extraction: the amino-acid composition of wool. Biochem J 35:91–121
68. Washburn MP, Wolters D, Yates JR 3rd (2001) Large-scale analysis of the yeast proteome by multidimensional protein identification technology. Nat Biotechnol 19:242–247
69. Thakur SS, Geiger T, Chatterjee B et al (2011) Deep and highly sensitive proteome coverage by LC-MS/MS without prefractionation. Mol Cell Proteom MCP 10:M110–M003699
70. Nagaraj N, Kulak NA, Cox J et al (2012) System-wide perturbation analysis with nearly complete coverage of the yeast proteome by single-shot ultra HPLC runs on a bench top Orbitrap. Mol Cell Proteom MCP 11: M111–M013722
71. Xu T, Venable JD, Park SK et al. (2006) ProLuCID: a fast and sensitive tandem mass spectra-based protein identification program. Mol Cell Pro 5(10 suppl):S174
72. Perkins DN, Pappin DJ, Creasy DM et al (1999) Probability-based protein identification by searching sequence databases using mass spectrometry data. Electrophoresis 20: 3551–3567
73. Cox J, Mann M (2008) MaxQuant enables high peptide identification rates, individualized p.p.b.-range mass accuracies and proteome-wide protein quantification. Nat Biotechnol 26:1367–1372
74. Colinge J, Bennett KL (2007) Introduction to computational proteomics. PLoS Comput Biol 3
75. Deutsch EW, Lam H, Aebersold R (2008) Data analysis and bioinformatics tools for tandem mass spectrometry in proteomics. Phys Genomic 33:18–25
76. Eng JK, Searle BC, Clauser KR et al (2011) A face in the crowd: recognizing peptides through database search. Mol Cell proteom MCP 10:R111–R009522
77. Zhou H, Ranish JA, Watts JD et al (2002) Quantitative proteome analysis by solid-phase isotope tagging and mass spectrometry. Nat Biotechnol 20:512–515
78. Hansen KC, Schmitt-Ulms G, Chalkley RJ et al (2003) Mass spectrometric analysis of protein mixtures at low levels using cleavable 13C-isotope-coded affinity tag and multidimensional chromatography. Mol Cell Proteom MCP 2:299–314
79. Schmidt A, Kellermann J, Lottspeich F (2005) A novel strategy for quantitative proteomics using isotope-coded protein labels. Proteomics 5:4–15
80. Hsu JL, Huang SY, Chow NH et al (2003) Stable-isotope dimethyl labeling for quantitative proteomics. Anal Chem 75:6843–6852
81. Boersema PJ, Aye TT, van Veen TA et al (2008) Triplex protein quantification based on stable isotope labeling by peptide dimethylation applied to cell and tissue lysates. Proteomics 8:4624–4632
82. Thompson A, Schafer J, Kuhn K et al (2003) Tandem mass tags: a novel quantification strategy for comparative analysis of complex protein mixtures by MS/MS. Anal Chem 75: 1895–1904
83. Thingholm TE, Palmisano G, Kjeldsen F et al (2010) Undesirable charge-enhancement of isobaric tagged phosphopeptides leads to reduced identification efficiency. J Proteome Res 9:4045–4052
84. Pichler P, Kocher T, Holzmann J et al (2010) Peptide labeling with isobaric tags yields higher identification rates using iTRAQ 4-plex compared to TMT 6-plex and iTRAQ

8-plex on LTQ Orbitrap. Anal Chem 82: 6549–6558
85. Ting L, Rad R, Gygi SP et al (2011) MS3 eliminates ratio distortion in isobaric multiplexed quantitative proteomics. Nat Methods 8:937–940
86. Wenger CD, Lee MV, Hebert AS et al (2011) Gas-phase purification enables accurate, multiplexed proteome quantification with isobaric tagging. Nat Methods 8:933–935
87. Miyagi M, Rao KC (2007) Proteolytic 18O-labeling strategies for quantitative proteomics. Mass Spectrom Rev 26:121–136
88. Oda Y, Huang K, Cross FR et al (1999) Accurate quantitation of protein expression and site-specific phosphorylation. Proc Natl Acad Sci U S A 96:6591–6596
89. Liao L, McClatchy DB, Park SK et al (2008) Quantitative analysis of brain nuclear phosphoproteins identifies developmentally regulated phosphorylation events. J Proteome Res 7:4743–4755
90. McClatchy DB, Liao L, Park SK et al (2007) Quantification of the synaptosomal proteome of the rat cerebellum during post-natal development. Genome Res 17:1378–1388
91. Ong SE, Kratchmarova I, Mann M (2003) Properties of 13C-substituted arginine in stable isotope labeling by amino acids in cell culture (SILAC). J Proteome Res 2:173–181
92. Blagoev B, Ong SE, Kratchmarova I et al (2004) Temporal analysis of phosphotyrosine-dependent signaling networks by quantitative proteomics. Nat Biotechnol 22:1139–1145
93. Olsen JV, Blagoev B, Gnad F et al (2006) Global, in vivo, and site-specific phosphorylation dynamics in signaling networks. Cell 127:635–648
94. Schwanhausser B, Gossen M, Dittmar G et al (2009) Global analysis of cellular protein translation by pulsed SILAC. Proteomics 9: 205–209
95. Cambridge SB, Gnad F, Nguyen C et al (2011) Systems-wide proteomic analysis in mammalian cells reveals conserved, functional protein turnover. J Proteome Res 10: 5275–5284
96. Geiger T, Cox J, Ostasiewicz P et al (2010) Super-SILAC mix for quantitative proteomics of human tumor tissue. Nat Methods 7: 383–385
97. Ong SE (2012) The expanding field of SILAC. Anal Bioanalytical Chem 404:967–976
98. Bondarenko PV, Chelius D, Shaler TA (2002) Identification and relative quantitation of protein mixtures by enzymatic digestion followed by capillary reversed-phase liquid chromatography-tandem mass spectrometry. Anal Chem 74:4741–4749
99. Chelius D, Bondarenko PV (2002) Quantitative profiling of proteins in complex mixtures using liquid chromatography and mass spectrometry. J Proteome Res 1:317–323
100. Bantscheff M, Lemeer S, Savitski MM et al (2012) Quantitative mass spectrometry in proteomics: critical review update from 2007 to the present. Anal Bioanal Chem 404: 939–965
101. Gallien S, Duriez E, Domon B (2011) Selected reaction monitoring applied to proteomics. J Mass Spectrom JMS 46:298–312
102. Gerber SA, Rush J, Stemman O et al (2003) Absolute quantification of proteins and phosphoproteins from cell lysates by tandem MS. Proc Natl Acad Sci U S A 100:6940–6945
103. Kirkpatrick DS, Gerber SA, Gygi SP (2005) The absolute quantification strategy: a general procedure for the quantification of proteins and post-translational modifications. Methods 35:265–273

Part II

Sample Pre-fractionation

Chapter 2

LC-MALDI-TOF/TOF for Shotgun Proteomics

Patricia Fernández-Puente, Jesús Mateos, Francisco J. Blanco, and Cristina Ruiz-Romero

Abstract

Automated liquid chromatography tandem mass spectrometry (LC-MS/MS) is a well-established technique for identification of components from complex mixtures in shotgun proteomics experiments. Approaches involving the use of matrix-assisted laser desorption/ionization mass spectrometry (LC-MALDI-MS/MS) comprise the preparation of protein extracts, their enzymatic digestion, the separation of the resulting peptides by nanoscale liquid chromatography coupled to a collector that deposits the microfractions onto a MALDI plate, and finally the mass spectrometry analysis of the fractions. Using an LC-MALDI strategy, the rate of the collection of MS/MS data is decoupled from the chromatographic separation, thus allowing high-quality data which are often complementary to electrospray ionization (LC-ESI-MS/MS) techniques.

Key words Mass spectrometry, MALDI, Liquid chromatography, Proteomics, Proteins, Peptides

1 Introduction

Mass spectrometry has evolved into an indispensable tool for proteomics research due to its ability to identify and quantify thousands of proteins from biological samples. Usually, this technique is coupled to liquid chromatography (LC-MS/MS), which resolves the peptides prior to mass spectrometry analysis, thus dramatically increasing the number of protein identifications and sequence coverage in complex samples.

Although electrospray ionization (ESI) has become the preferred method for coupling the liquid chromatography to the mass spectrometer [1], many recent studies have focused on utilizing a MALDI source as an interface for chromatographic separations, an option that is particularly attractive with the availability of MALDI-TOF/TOF for MS/MS analysis [2]. Using an LC-MALDI strategy, the collection of MS/MS data can be decoupled from the chromatographic separation, allowing more time to acquire spectra without missing components as they elute from the LC column.

Daniel Martins-de-Souza (ed.), *Shotgun Proteomics: Methods and Protocols*, Methods in Molecular Biology, vol. 1156, DOI 10.1007/978-1-4939-0685-7_2, © Springer Science+Business Media New York 2014

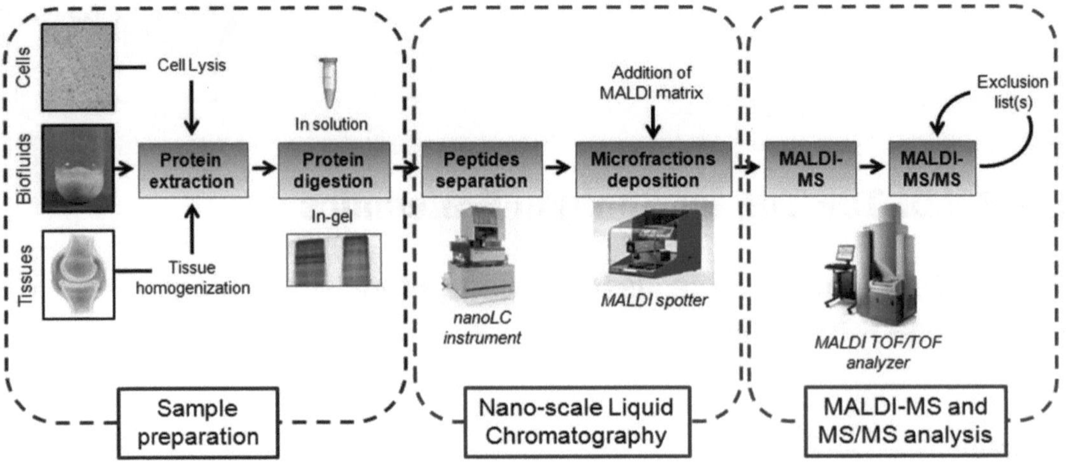

Fig. 1 Workflow of a typical LC-MALDI approach, which involves three independent steps: the preparation of the sample, separation of the peptides by nanoliquid chromatography, and analysis of the correspondent LC fractions by MALDI mass spectrometry (MS or MS/MS)

Furthermore, LC-MALDI-based strategies allow reanalyzing the samples using optimized MS/MS parameters after the initial interpretation is complete, for example by performing *inclusion list*-driven peptide identifications [3]: as the MALDI plates with the nLC fractions can be stored for several days and reanalyzed if necessary, it is then possible to acquire fragmentation ion spectra for all MS parent ions that are of interest. In addition, performing *exclusion list*-driven peptide identifications eliminates redundant MS/MS acquisitions in subsequent runs, thus allowing identification of a larger number of peptide identifications in the cumulative dataset. Finally, spectra are easier to interpret and compare because mostly singly charged ions are produced and detected [4].

Considering the advantages described above, LC-MALDI is particularly useful when greater coverage of proteins is desired or when sample complexity requires the acquisition of many MS/MS spectra and also when isobaric peptides are present [5]. Quantitative proteomics with various isotope labeling strategies (SILAC, iTRAQ) can be performed using this technique [6]. Additional benefits of this strategy include its compatibility with a wide selection of chromatographic mobile phases and its complementarity with LC-ESI approaches [2]. Combining MALDI and ESI strategies to analyze the same sample has been shown to improve the number of proteins identified and the confidence of the identifications [7], probably because each of them tend to favor the ionization of different sets of peptides [8].

In this chapter, we describe the different steps of an off-line LC-MALDI workflow (Fig. 1). In contrast to online procedures, this technique is especially attractive because each step can be performed at its individual optimum, as chromatographic separation and MS or MS/MS analysis can be optimized independently by individual adjustment of operating parameters. Procedures for sample preparation, digestion of proteins, separation of the correspondent peptides by nanoscale liquid chromatography coupled to the deposition of its microfractions onto a MALDI plate, and finally MS or MS/MS analysis of the fractions are illustrated.

2 Materials

Prepare all solutions using mass spectrometry (MS)-grade water and analytical grade reagents. Prepare and store all reagents at room temperature (unless indicated otherwise). Diligently follow all waste disposal regulations when disposing waste materials.

2.1 Preparation of Protein Extracts

1. Urea lysis buffer: 30 mM Tris 8 M urea, 2 M thiourea, 4 % 3-[(3-cholamidopropyl)dimethylammonio]-1-propanesulfonate (CHAPS). Dissolve 7.27 mg of Trizma base (≥99 %), 0.96 g urea for electrophoresis (≥99 %), 0.304 g thiourea (≥99 %), and 80 mg of CHAPS in 2 mL of water (*see* **Note 1**). Make 100 or 200 μl aliquots, and store them at −20 °C.
2. Sonication bath.
3. Refrigerated microcentrifuge.
4. For tissue extractions: Tissue homogenizer (e.g., Mixer Mill MM 200, Retsch, Haan, Germany, and zirconium oxide grinding balls).
5. Protein quantification kit.
6. Protein loBind tubes (Eppendorf).

2.2 Protein Digestion

1. 25 mM Ammonium bicarbonate (AmBi, ≥99 %) in water: Dissolve 19 mg in 10 mL of HPLC water (*see* **Note 2**).
2. 500 mM DL-dithiothreitol (DTT) stock (≥99 %) in 25 mM AmBi: Dissolve 77.1 mg DTT in 1 mL of 25 mM AmBi. Store in aliquots at −20 °C.
3. 100 mM Iodoacetamide (IAA) solution (≥99 %): Dissolve 18.5 mg IAA in 1 mL of 25 mM AmBi (*see* **Note 2**).
4. 0.5 % Trifluoroacetic acid (TFA) solution: Add 5 μL of 99 % TFA to 995 μL of water.
5. 0.1 % TFA solution: Add 200 μL of 0.5% TFA solution to 800 μL of water.
6. 0.1 % TFA, 80 % acetonitrile (AcN) solution: Add 800 μL of AcN (HPLC grade AcN) to 200 μL of 0.5 % TFA solution.

7. 50 Mm Acetic acid: Add 1.42 μL of acetic acid glacial (100 % anhydrous) to 498.58 μL of H_2O.
8. Sequencing-grade trypsin: 100 μg in 100 μL of 50 Mm acetic acid glacial, 1 μg/μL stock (*see* **Note 3**).
9. Refrigerated microcentrifuge.
10. Sonication bath.
11. Thermoblock.
12. Vacuum centrifuge.
13. NuTips NT2 C18 (Glygen Corp, Columbia, USA).
14. Binding (0.1 % TFA) and elution (0.1 % TFA in 80 % AcN) solutions for the cleaning of samples using NuTips.

2.3 Nanoscale Liquid Chromatography and Deposition of Microfractions

1. Buffer A: 0.1 % TFA, 2 % AcN in LC-MS water. Add about 98 mL water to a 100 mL graduated cylinder, and make up to 100 mL with AcN. Transfer to a bottle, and add 100 μL of TFA (*see* **Note 4**).
2. Buffer B: 0.1 % TFA in 80 % AcN. Add about 80 mL acetonitrile to a 100 mL graduated cylinder, and make up to 100 mL with water. Transfer to a bottle, and add 100 μL of TFA (*see* **Note 4**).
3. Sonicate the buffers for15 min to degas the solution in a sonication bath (*see* **Note 5**).
4. Trap column (0.5 × 2 mm, Michrom Bioresources).
5. C18 column (Integrafit C18, Proteopep™ II, 75 μm id, 15 cm, 5 μm, 300 Å, New Objective).
6. NanoLC system (Tempo Eksigent, ABSciex).
7. Angiotensin peptide internal standard (≥93 % [HPLC]): Add 1 mL of water to the vial to make a 1 mg/mL stock. Then, make a 1/100 dilution to have a total concentration of 0.01 μg/μL (10^4 fmol/μL), distribute in aliquots, and store at −20 °C. Each aliquot can be used several times.
8. Matrix solution: 3 mg/mL α-cyano-4-hydroxycinnamic acid (α-CHCA) in 0.1 % TFA, 70 % AcN. Weigh 3 mg of matrix in a microfuge tube; add 700 μL of acetonitrile, 200 μL of 0.5 % TFA, and 100 μL of water.
9. Matrix solution with an internal standard: Add 1 μL of 10^4 fmol/μL of angiotensin internal standard to 1 mL of matrix solution. The final concentration of angiotensin in the matrix solution is 10 fmol/μL.
10. Sun Collect MALDI Spotter/Micro Collector (SunChrom Wissenschaftliche Geräte GmbH).
11. Opti-TOF™ LC MALDI Plate Inserts (123 × 81 mm, ABSciex).

2.4 Mass Spectrometry

1. 4700 Proteomics Analyzer Calibration Mixture (ABSciex).
2. 4800 MALDI-TOF/TOF system (ABSciex).
3. 4000 Series explorer software (ABSciex).
4. Opti-TOF™ LC MALDI Plate Inserts (123 × 81 mm, ABSciex).
5. Opti-TOF™ Magnetic Holder (ABSciex).

3 Methods

Carry out all procedures at room temperature unless otherwise specified.

3.1 Preparation of Protein Extracts

1. For cellular extracts, harvest cells and collect by centrifugation (1,200 × *g*, 4 °C, 10 min). For tissue extracts, disrupt the snap-frozen tissue biopsy using a tissue homogenizer and proceed to **step 4**. For biofluid extracts, proceed directly to protein quantification (**step 8**) (*see* **Note 6**).
2. Resuspend and wash the pellet carefully with PBS or saline.
3. Centrifuge again, and aspirate any remaining liquid. Repeat **steps 2–3** twice (*see* **Note 7**).
4. Solubilize pellet (approx. 1:10 w/v) in urea lysis buffer. Vortex thoroughly.
5. Sonicate 3 × 20 s, and burst on ice, with a 1-min chilling interval.
6. Centrifuge the lysate at 14,000 × *g* in a precooled centrifuge for 10 min.
7. Immediately transfer the supernatant to a fresh centrifuge tube, and discard the pellet.
8. Proceed to protein quantification. Protein samples can also be divided into aliquots and stored at −80 °C.

3.2 Protein Digestion

1. Take the amount of cellular extract corresponding to 5–10 μg of protein.
2. Add 25 mM Ammonium bicarbonate (AmBi) to the extract. A minimum final volume of 10 μL is required for the digestion.
3. Check pH to be 8. This is essential to preserve the enzymatic activity.
4. Add the reducing agent DTT to a concentration of 10 mM and incubate for 1 h at 37 °C.
5. Add IAA 100 mM to a final concentration of 50 mM, and incubate the resultant mixture for 45 min at room temperature in darkness.
6. Dilute fourfold with 25 mM AmBi to reduce urea concentration under 1 M.

7. Add trypsin at a 1:20–1:50 protease-to-protein ratio, and incubate overnight at 37 °C.
8. Stop the enzymatic reaction by acidifying with 5 μL of 0.5 % TFA. Check the pH and add more, if necessary, until the solution is acid.
9. Perform a cleanup to eliminate the salts or the impurities than can interfere with the LC-MS/MS analysis using NuTips. To conditionate the column, attach the NuTip to a micropipette, aspirate/expel 10–50 μL of the appropriate elution solution five times, and wash five times. Then, bind the sample by aspirating/expelling it (10–50 μL) 20–50 times (or higher, depending on sample volume and concentration of proteins) to allow peptides adsorbing to the RPC material. Wash the sample by aspirating/expelling ten times with 25–50 μL of binding solution, discarding the expelled solution each time. Finally, elute the cleaned sample by aspirating/expelling ten times with 10–30 μL of elution solution, collecting the eluate in a suitable clean tube (*see* **Note 8**). Repeat with a fresh portion of elution solution if you want to be sure of releasing and collecting all of the adsorbed peptide.
10. Evaporate the peptide samples to dryness and store at −20 °C until LC-MS/MS analysis.

3.3 Liquid Chromatography Coupled to Microfraction Spotting

1. Ensure that the reservoirs for the mobile phase are not empty. In case they need to be filled, close their valves by turning to a perpendicular position.
2. Sonicate the mobile phase for 15 min. Fill the mobile-phase reservoir slowly with a pipette to avoid air bubbles.
3. Open the valve on each reservoir by rotating to a vertical orientation.
4. Prime all pumps by inserting the priming toll down through the reservoir and gently pushing down the check valve (*see* **Note 9**).
5. Before purging and flushing the system, disconnect the capillaries (the microtight fittings) from the outlets of channels 1 and 2.
6. Select System-Mobile Phases from the main window.
7. Select the desired mobile phase from the drop menu.
8. Type in the composition of the mobile phase.
9. Click the more button to reveal the purge/flush settings.
10. Type 10 purge cycles, and select both A and B sides.
11. Click the Purge Now button to purge the system from the pump to the purge valve.
12. Monitor the purge output, and measure the volume. Each pump has an internal volume of approximately 600 μL (*see* **Note 10**).

Table 1
Typical nanoLC gradient separation

Time (min)	Mobile phase A (%)	Mobile phase B (%)
0	95	5
3	95	5
120	50	50
122	5	95
132	5	95
134	95	5
150	95	5

13. Type in the desired flush volume and total flow rate (*see* **Note 11**).
14. Click Flush Now button to flush the system from the purge valve to the mixing tee.
15. After the flush sequence ends, click OK to close the window.
16. Repeat these steps for channel 2.
17. Reconnect the microtight fittings to the outlets of the mixing tees.
18. Perform the equilibration of the column, and record the pressure of channel 2 line A and B at a flow of 350 nL/min changing the mobile phase from 50 to 95 % of buffer A, which are the initial conditions of the gradient (*see* **Note 12**).
19. Dissolve the dried peptide fractions in buffer A.
20. Inject 5 μL of the samples (2 μg of total protein on column) on a trap column at a flow rate of 15 μL/min by channel 1 (Tempo nanoLC, ABSciex) to remove any highly polar impurities for 15 min.
21. Load the desalted peptide mixture to a C18 column at a constant flow rate of 350 nL/min by channel 2 to perform the peptide separation.
22. Separate the peptides using a linearly increasing concentration of acetonitrile in buffer B given in Table 1: The gradient starts with 5 % buffer B for 3 min and rises to 50 % buffer B for 120 min, then finalize with 95 % buffer B for 10 min, and equilibrate the nanoLC system with 95 % buffer A for 15 min. In the case of standards or pre-fractionated protein digests from highly complex samples as biofluids or tissues, the gradient can be shortened by increasing the slope of acetonitrile gradient used in nanoLC system.

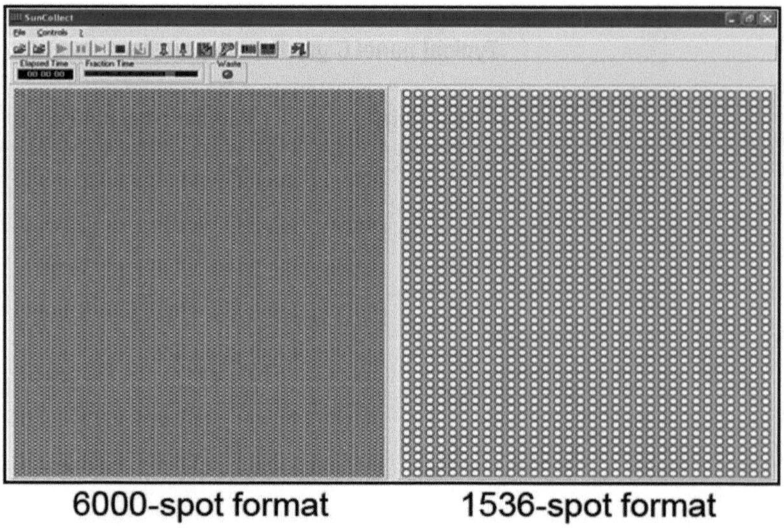

Fig. 2 Examples of different spot density layouts that can be employed for LC-MALDI experiments. A 6,000-spot plate (*left*) and a 1,536-spot plate (*right*) are shown, although the design can be adapted to the requirements of each experiment

23. Collect the reversed-phase fractions with a Sun Collect MALDI Spotter/Microfraction Collector, and, using this system, mix them with the matrix α-cyano-4-hydroxycinnamic acid (3 mg/mL in 0.1 % TFA, 70 % AcN) spiked with an internal standard (angiotensin, *see* Subheading 2.3, **item 11**) with a concentration of 3 fmol/spot at a flow rate of 1.2 μL/min.
24. Finally, spot the microfractions each for 15 s onto a MALDI plate for MS analysis. The spotter software allows the design of different plate formats, depending on the volume and number of spots that are fitted on the plate, as shown in Fig. 2. This design should be loaded on the mass spectrometer to permit the unsupervised analysis of the plate (*see* **Note 13**).
25. Clean the plate with dry air before loading in the mass spectrometry.
26. Clean the syringe of the spotter with isopropanol when the LC separation is finished to avoid crystallization of the matrix in the capillaries.

3.4 Mass Spectrometry Analysis

1. Spot the calibration mixture in the calibrant positions of the MALDI plate, and ensure that the samples are totally dry.
2. Load the sample plate: Insert the plate in the magnetic holder, and deposit it on the "load position" in the loading chamber of the 4800 TOF/TOF. Close the lid and slightly press down. Press the "Load plate" icon on the 4000 series explorer.

A small window indicating pressure in both loading chamber and source 1 chamber will appear. The loading cycle will end in about 2 min.

3. Turn the laser and high voltage on, by opening the proper MS acquisition and processing methods and shooting manually over a calibrant position at low laser output (i.e., 3,200). Check signal intensity and resolution, and wait for 5 min for stabilization.
4. Align the sample plate: Press "Align sample plate" icon on the software, and follow the instructions given by the software. Using the joystick, center the "alignment positions" -typically the four external corners of the set of chromatograms-. Once centered, for each position press the "1" button on the joystick and move to the next position.
5. Plate model and default calibration (*see* **Note 14**): Open the processing method used for the calibration, and press the "set as active" icon. Shoot manually over a calibrant position at low laser output (i.e., 3,200, 1,000 total shots), and check signal intensity and resolution along the mass range. If the final spectra do not pass the internal calibration specifications perform manual tuning. If it does, select all the calibrant positions on the spot set manager and send an automatic job using the proper acquisition and processing methods for plate model and default calibration using fixed parameters (i.e., 3,200 laser output, 1,000 total shots). If the calibration is successful, move to the next step. If not, repeat slightly increasing laser output (i.e., 3,400 max) or perform tuning of the reflector operating mode.
6. Manually refine the chromatograms: In order to reduce the total time of the automatic analysis, it is advisable to manually limit the positions where peptides have eluted in each chromatogram. Typically, there are no sample peptides in the initial and final positions, but this should be manually checked for each individual chromatogram.
7. Perform MS analysis: Select the positions to analyze in the spot set manager of the software, and send an automatic job with the proper acquisition and processing methods. During the job, check the signal intensity and the resolution of the sample peptides in the individual spectra (*see* **Note 15**). Once the MS analysis is concluded, proceed to MS/MS analysis (*see* **Note 16**).
8. Initialize MS/MS parameters: Open the proper MS/MS acquisition and processing methods. Manually shoot over a calibrant to turn CID system on and adjust high voltage. Wait for 5 min for stabilization.
9. MS/MS calibration: For MS/MS calibration, default calibration of the plate is not necessary in the 4800 TOF/TOF system. However, it is strongly recommended to check manually the fragmentation of the Glu-Fib calibrant using a Glu-Fib

calibration method. Optionally, fragmentation of some sample peptides (randomly chosen all over the chromatogram) with different signal-to-noise ratio can be checked in order to refine acquisition parameters (laser output, total number of shots, and TIS resolution). Perform manual tuning of the MS/MS operating mode if necessary.

10. Perform MS/MS analysis (*see* **Note 16**): Select the sample positions in the spot set manager based on the automatic MS run, and send an automatic MS/MS run using the proper interpretation method. This should include (a) signal-to-noise threshold (typically >80); (b) adduct exclusion list; (c) mass inclusion list (if necessary); (d) mass exclusion list (typically, trypsin self-digestion peptides, known contaminants and peptides used for internal MS calibration); and (e) MS/MS acquisition (typically 4,200 laser output, 2,000 total shots CID on, and 300 TIS resolution) and processing modes (typically default calibration).
11. Optionally, perform a second MS/MS analysis using the masses selected in the first run as an *exclusion list* in order to identify low-abundance peptides. The exclusion list eliminates redundant MS/MS acquisitions in subsequent runs, thus reducing MALDI sample depletion and allowing identification of a larger number of peptide identifications in the cumulative dataset.
12. To obtain the exclusion list for each individual chromatogram select the MS/MS job run of the chromatogram in the spot manager and view in Peak Explorer.
13. Copy to the clipboard the precursor masses list.
14. In an Excel datasheet, make a four-column list as follows: precursor masses, m/z tolerance (typically 0.2), retention time (in which the mass should be excluded), and retention time tolerance (typically 2 min).
15. Copy the four-column list and paste in the exclusion list window of the interpretation method in the 4000 Series Explorer software.
16. (Optional) Decrease signal-to-noise threshold (i.e., >30).
17. Rename the interpretation method for each chromatogram.
18. Launch a second MS/MS job run using the correspondent exclusion list for each chromatogram.

4 Notes

1. Urea solutions must not be heated above 30 °C, as this may produce protein carbamylation. Thiourea is insoluble in water; therefore, dissolve first the urea, then the thiourea, and finally the rest of the reagents.

2. The AmBi and IAA solutions must be prepared fresh each time.
3. Trypsin stock (1 μg/μL) can be kept frozen in aliquots at −20 °C.
4. Cell pellets can also be stored at this point at −80 °C.
5. Use the same graduate cylinder each time to prevent contaminations.
6. Some biofluids may require cleaning or purification steps prior to the quantification. Protocols must be optimized for each type of sample.
7. It is advisable to pre-wash collection tubes to eliminate leachable contaminants.
8. The cap should be a little open to prevent heat.
9. A complete system purge and flush are critical to get maximum performance. Purging is necessary to flush out any air, shipping fluid, and/or gas bubbles that may disrupt flow. Purging is automatic and rapidly replaces the solvent between the pumps and the purge values. Flushing is the replacement of solvent from the purge valve through the flowmeters and out to the sample injector.
10. If you only get drops coming from the purge line you need to purge again the system several times because air is in the system. If liquid is not coming through the purge line the filter of the pump is blocked and you need to replace it.
11. We use a volume of 500 μL and a total flow rate of 15 μL/min for channel 1 and a volume of 10 μL and a total flow rate of 500 nL/min for channel 2.
12. We record the pressure of the column and trapping to check the stability of the system during different days. It can give us an idea if the trap or the column is blocked or if there is any problem in the system by comparing the values of the pressure using the same percentage of mobile phase in channels 1 and 2.
13. We usually employ a 1,536-spot format in our analysis, but other formats are possible if smaller volumes are spotted.
14. Default calibration must be done on each plate when working with the reflector mode after the alignment.
15. In case an internal calibration processing method is being used, the internal calibrant signal should also be checked during the MS analysis.
16. Remember to check at any time signal intensity and resolution in both MS and MS/MS modes and perform manual tuning of the operating modes when needed.

Acknowledgement

Our work is supported by grants from Fondo Investigación Sanitaria-Spain (CP09/00114, PI11/02397, PI12/00329) and Secretaría I + D + I Xunta de Galicia (10CSA916058PR). J. Mateos (CA11/00050) and P. Fernández-Puente (CA09/00458) are supported by Fondo Investigación Sanitaria-Spain. C. Ruiz-Romero is supported by the Miguel Servet program from Fondo Investigación Sanitaria-Spain (CP09/00114).

References

1. Wolters DA, Washburn MP, Yates JR 3rd (2001) An automated multidimensional protein identification technology for shotgun proteomics. Anal Chem 73:5683–5690
2. Yang Y, Zhang S, Howe K, Wilson DB, Moser F, Irwin D, Thannhauser TW (2007) A comparison of nLC-ESI-MS/MS and nLC-MALDI-MS/MS for GeLC-based protein identification and iTRAQ-based shotgun quantitative proteomics. J Biomol Tech 18:226–237
3. Mueller DR, Voshol H, Waldt A, Wiedmann B, Van Oostrum J (2007) LC-MALDI MS and MS/MS–an efficient tool in proteome analysis. Subcell Biochem 43:355–380
4. Mateos J, Lourido L, Fernandez-Puente P, Calamia V, Fernandez-Lopez C, Oreiro N, Ruiz-Romero C, Blanco FJ (2012) Differential protein profiling of synovial fluid from rheumatoid arthritis and osteoarthritis patients using LC-MALDI TOF/TOF. J Proteome 75: 2869–2878
5. Fernandez-Puente P, Mateos J, Fernandez-Costa C, Oreiro N, Fernandez-Lopez C, Ruiz-Romero C, Blanco FJ (2011) Identification of a panel of novel serum osteoarthritis biomarkers. J Proteome Res 10:5095–5101
6. Calamia V, Rocha B, Mateos J, Fernandez-Puente P, Ruiz-Romero C, Blanco FJ (2011) Metabolic labeling of chondrocytes for the quantitative analysis of the interleukin-1-beta-mediated modulation of their intracellular and extracellular proteomes. J Proteome Res 10:3701–3711
7. Bodnar WM, Blackburn RK, Krise JM, Moseley MA (2003) Exploiting the complementary nature of LC/MALDI/MS/MS and LC/ESI/MS/MS for increased proteome coverage. J Am Soc Mass Spectrom 14:971–979
8. Zhen Y, Xu N, Richardson B, Becklin R, Savage JR, Blake K, Peltier JM (2004) Development of an LC-MALDI method for the analysis of protein complexes. J Am Soc Mass Spectrom 15: 803–822

Chapter 3

Fully Automatable Multidimensional Reversed-Phase Liquid Chromatography with Online Tandem Mass Spectrometry

Maggie P.Y. Lam, Chun Hin Law, Quan Quan, Yun Zhao, and Ivan K. Chu

Abstract

Liquid chromatography (LC) is essential for sample fractionation in shotgun proteomics applications. With suitable design, common LC separation chemistries, including reversed-phase (RP) and strong cation exchange (SCX) mode, can be combined in online multidimensional LC to greatly enhance the overall separation power and, thus, proteome coverage. This protocol describes the design and assembly of a flexible online multidimensional RP-SCX-RP LC system that is compatible with deep proteome profiling on common shotgun proteomics platforms.

Key words Multidimensional liquid chromatography, Peptide separation, RP-RP, SCX-RP

1 Introduction

An integral goal of shotgun proteomics is to interrogate the expression of every protein in a system and discover its associated biology. Because of the extreme complexity of proteomes and the finite sensitivity of mass spectrometers, current shotgun experiments tend to oversample abundant proteins (e.g., housekeeping proteins) while disfavoring the detection of low-copy proteins (e.g., transcription factors), thereby limiting the comprehensiveness of proteomics studies and creating the imperative to improve proteome coverage.

Separating the components of a sample based on their biological (e.g., subcellular localization) or chemical (e.g., hydrophobicity) properties is a highly effective method for improving the coverage of any shotgun proteomics platform. Better separation affords a simpler sample mixture, which minimizes signal suppression of low-abundance species and favors their detection by the mass spectrometer.

The authors Maggie P.Y. Lam, Chun Hin Law, and Quan Quan have contributed equally to this work.

Daniel Martins-de-Souza (ed.), *Shotgun Proteomics: Methods and Protocols*, Methods in Molecular Biology, vol. 1156, DOI 10.1007/978-1-4939-0685-7_3, © Springer Science+Business Media New York 2014

Multidimensional liquid chromatography (MDLC) is an attractive separation strategy in which each fraction from one LC column is further separated by another column, thereby combining the separation power of two (or more) LC chemistries. Two prominent MDLC implementations are strong cation exchange/reversed-phase (SCX-RP) [1] and RP-RP, the latter employing two RP columns operated at different values of pH to achieve orthogonal chromatography [2]. MDLC is simple to perform off-line—each fraction is manually collected from one LC column and re-injected into a second column [3]—but such implementation introduces additional labor costs and sample loss. The preferred method is to combine LC columns online [4–11]: the LC flow path is built to divert the analytes between columns automatically and with minimal loss. This process adds complications related to solvent incompatibility between the LC columns, but they can be overcome with judicious fluidics designs.

Here, we provide a detailed protocol for the construction of an online MDLC implementation that is simple, flexible, and scalable; it is based on our previously reported [4] procedures (Fig. 1). This MDLC system features a dual-trap column design, which reduces the idle time in fraction elution, trapping, and phase focusing between successive fractions with significant improvement in throughput. It requires only one integrated dual-gradient LC system, two ten-port switch valves, and common capillary connections and can be assembled readily in a proteomics laboratory with almost any mass spectrometer (Fig. 2). The fully automatable design can be configured to operate in two-dimensional RP-RP separation mode for routine analyses at low-microgram levels or in three-dimensional RP-SCX-RP separation mode when deeper proteome coverage is desired. Therefore, the total run time per analysis ranges from 20 to 50 h. In our laboratory, we have used multidimensional RP-based LC systems as the preferred protein discovery and identification technology for mammalian, yeast, and plant cell/tissue analyses [4–8, 11] as well as for iTRAQ quantitative applications (Fig. 3).

2 Materials

All solutions were prepared using high-performance liquid chromatography-grade water. The described columns were packed in-house using an ultrahigh-pressure syringe pump (Teledyne Isco). A pressure injection cell could also be used for column packing or prepacked columns could be purchased.

2.1 Protein Extraction and Digestion

1. Cell culture medium: F-12K medium supplemented with 15 % (v/v) heat-inactivated horse serum, 2.5 % (v/v) fetal bovine serum (FBS), penicillin (100 U/mL), and streptomycin (100 μg/mL).

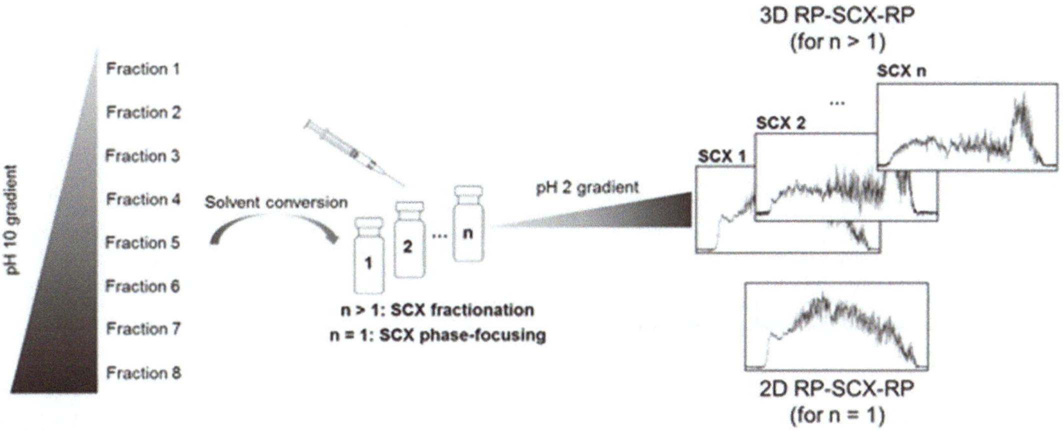

Fig. 1 Workflow of the online MDLC RP-SCX-RP system for proteomics analysis. The complete analysis of a single sample comprised eight fractions of the first-dimension separation, each further refocused (for 2D RP-SCX-RP system) or subfractioned (for 3D RP-SCX-RP system with n plugs of salt solution of increasing strength) in the SCX-RP trap column prior to eluting all peptides for further separation on the conventional low-pH RP separation

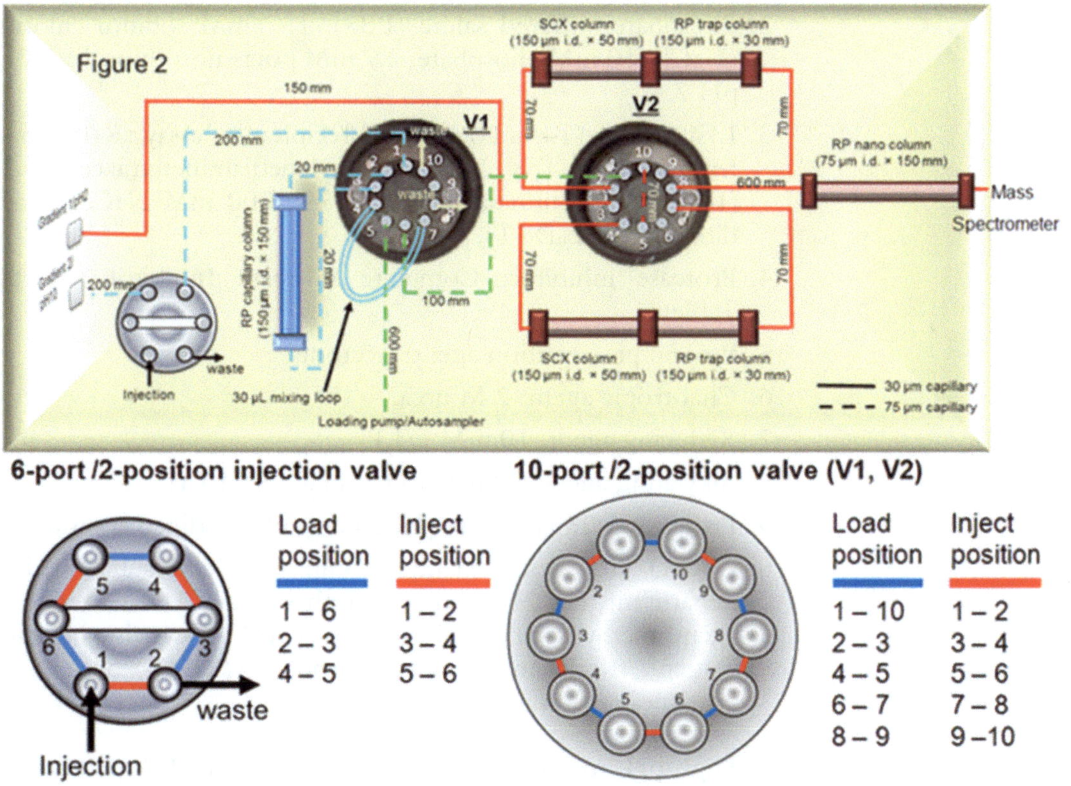

Fig. 2 Schematic representation of MDLC RP-SCX-RP with valve configuration at each step during the gradient profile, the pH 10, SCX, and pH 2 mobile-phase flow are indicated with *blue*, *green*, and *red lines*, respectively. The flow and valve positions are depicted (*low panel, left*) to illustrate the six-port valve for the sample injection in the first dimension; (*low panel, right*) the two ten-port valves (V1 and V2) were set up identically; thus, only one is being shown. The eluent from the first-dimension RP separation was injected onto the SCX-RP trap columns via a 30-µL mixing loop connected to ports 4 and 7 of V1 to resolve the issue of solvent incompatibility between the high-pH RP dimension and the SCX dimension [5]

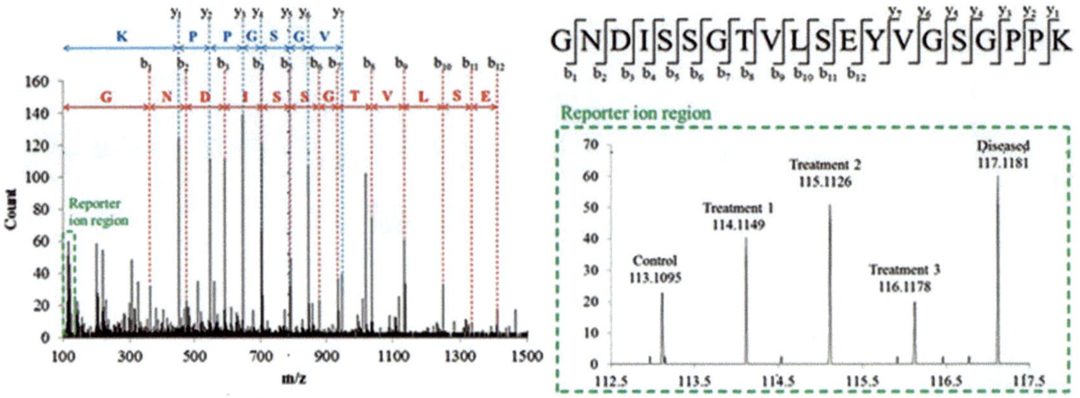

Fig. 3 Representative spectra showing full MS/MS spectrum and the reporter ion region (the inset) for tryptic peptide of GNDISSGTVLSEYVGSGPPK with 5-plex iTRAQ tagged of a protein that is differentially expressed in Parkinson's disease sample versus normal control of the PC12 lysate. The labeled peaks show b and y product ions that have been matched to the identified peptide sequence, which is uniquely matched to the phosphatidylethanolamine-binding protein 1 in the protein sequence database

2. Phosphate-buffered saline (PBS): 137 mM sodium chloride (NaCl), 10 mM phosphate, 2.7 mM potassium chloride (KCl), pH 7.4.
3. Lysis buffer (1× radioimmunoprecipitation assay (RIPA) buffer): 50 mM Tris, 1.0 mM ethylenediaminetetraacetic acid (EDTA) disodium salt, 1 % NP-40, 500 mM NaCl, 1 mM dithiothreitol (DTT).
4. Protease inhibitor: Complete Protease Inhibitor Cocktail Tablets.
5. Protein precipitation agent: Acetone.
6. Chaotropic agent: 2 M urea.
7. Reducing agent: 10 mM DTT.
8. Alkylating agent: 20 mM iodoacetamide (IAA).
9. Protease: 1:25 (w/w) sequencing-grade modified trypsin.
10. iTRAQ kit, which includes sodium dodecyl sulfate (SDS) denaturant, tris-(2-carboxyethyl)phosphine (TCEP) reducing agent, methyl methanethiosulfonate (MMTS) cysteine blocking reagent, triethylammonium bicarbonate (TEAB) buffer, isopropanol (AB SCIEX).

2.2 LC Mobile Phases

1. High-pH solvent A: 20 mM NH_4OAc in 2 % acetonitrile (MeCN); ca. pH 10.
2. High-pH solvent B: 20 mM NH_4OAc in 90 % MeCN; ca. pH 10.
3. Salt buffer for SCX fractionation: 20 mM, 100 mM, 1,000 mM ammonium acetate (NH_4OAc).

4. Low-pH solvent A: 2 % MeCN in 0.5 % formic acid; ca. pH 2.
5. Low-pH solvent B: 98 % MeCN in 0.5 % formic acid; ca. pH 2.

2.3 Columns and Fittings

1. First-dimension RP column: 150 μm [internal diameter (i.d.)] × 150 mm (length) column packed with Jupiter C18 (3-μm particle, 300-Å pore) beads (Phenomenex).
2. Second-dimension RP column: 75 μm (i.d.) × 150 mm (length) column packed with Jupiter C18 (3-μm particle, 300-Å pore) beads (Phenomenex).
3. SCX-RP trap columns: Two biphasic trap columns (i.d.: 150 μm), each containing a 30-mm length of Jupiter C18 beads following a 50-mm length of PolySULFOETHYL A™ (5-μm particle, 300-Å pore) SCX material (PolyLC).
4. Polyimide-coated fused-silica capillaries: 363 μm [outer diameter (o.d.)] and 30 μm (i.d.), 363 μm (o.d.), and 75 μm (i.d.) (Polymicro Technologies).
5. Nuts, ferrules, unions, and screens for connections (VICI).
6. PEEK tubing: 0.0305 in. (o.d.) × 0.015 in. (i.d.) (IDEX).
7. Two ten-port/two-position nano-bore automatic switch valves (*see* **Note 1**) and a six-port/two-position sample injection valve.

2.4 Instrumentation

1. LC system: Eksigent NanoLC Ultra 2D Plus system with minor modifications on the trigger wiring (*see* **Note 1a**). Other capillary or nano-flow LC system(s), either in an integrated four-channel system or in conjunction with multiple binary systems, could be assembled in-house to deliver the two gradients and one isocratic flow (*see* **Note 1b**).
2. Mass spectrometer: AB SCIEX TripleTOF™ 5600 system (*see* **Note 2**).
3. Instrumentation control software: Analyst TF 1.5.1 (AB SCIEX).
4. Data analysis software: PeakView 1.1 and ProteinPilot 4.5 (AB SCIEX).

3 Methods

The system contains a first-dimension RP column, which separates peptides in terms of their hydrophobicity at pH 10, and a second (last)-dimension RP column, which separates peptides in terms of their hydrophobicity at pH 2. The system can freely operate in two-dimensional mode or in three-dimensional separation mode with the addition of an SCX separation column, which fractionates peptides based on their charge, between the two RP columns.

The flow features a double-trap column design to minimize waiting time. While the analytical gradient resolves and separates the peptides held in one trap column, the other trap column is free to hold and recondition for the eluent from the next fraction of the first-dimension RP column. All procedures are conducted at ambient temperature unless otherwise specified.

3.1 Sample Preparation

1. Protein lysates should be prepared according to the type of cells, tissue, or organism of interest. We used cultured pheochromocytoma cells (PC12) from rats [12].
2. Precipitate the protein lysate in 4× acetone at −20 °C for 2 h [13]. Redissolve the proteins in 2 M urea. Quantify the proteins using a Bradford assay.
3. Treat the proteins with the reducing agent for 30 min at 60 °C and then with the alkylating reagent for 60 min in the dark. Digest with trypsin at 37 °C overnight. The resulting peptide mixture is ready for analysis.
4. The sample for iTRAQ analysis is prepared with commercial iTRAQ reagent. Briefly, the iTRAQ protein digest is prepared through the addition of SDS, TCEP, and MMTS for cysteine blocking, following the instructions provided with the iTRAQ Reagents kit. Samples are proteolyzed with 1:33 sequencing-grade trypsin at 37 °C for 12–16 h. Each trypsinized sample is derivatized with a corresponding iTRAQ label at room temperature for 2 h, following the manufacturer's instructions.

3.2 MDLC Flow Path Assembly

1. Assemble the system as displayed in Fig. 2.
2. There are three solvent inlets into the system: gradient 1 (low-pH solvents A and B), gradient 2 (high-pH solvents A and B), and an isocratic flow (solvent A, pH 2) from the loading pump. The central part of the system consists of two ten-port/two-position automatic switch valves: V1 is connected to the first-dimension RP column, a 30-μL solvent mixing loop, and receives solvent flow directly from the isocratic flow and the pH-10 gradient 2; V2 houses the two SCX-RP biphasic trap columns and is connected to the second-dimension RP column.

3.3 Two-Dimensional RP-RP Separation

1. Turn on the pump(s), and reset all valve positions to "inject" (1–2).
2. Connect a six-port/two-position sample injection valve with an injection loop of appropriate size for the sample volume. Connect the sample injection valve to V1 at the position displayed in Fig. 2 (*see* **Note 3**).
3. Condition the first-dimension column with 80 % high-pH solvent B from gradient 2 for at least 20 min at 1 μL/min (20 μL), followed by 5 % high-pH solvent B from gradient 2 for at least 30 min at 1 μL/min (30 μL).

Table 1
Mobile-phase gradient profile of the final dimension LC pump in the 2D and 3D RP-SCX-RP

Accumulation time (min)	% of low-pH solvent B
0	5
1	8 (fraction dependent, *see* Tables 2 and 3)
61	20 (fraction dependent, *see* Tables 2 and 3)
66	20 (fraction dependent, *see* Tables 2 and 3)
76	50 (or 80 for the last fraction of the experiment)
81	50 (or 80 for the last fraction of the experiment)
86	5
110	5

4. Using a syringe port, inject samples (15–30 μg) into the injection loop. Switch the sample injection valve position to "inject" (1–2) and load the sample with 5 % high-pH solvent B from gradient 2 at 1 μL/min until at least twice the loop volume has elapsed. Switch the sample injection valve back to the "load" (1–6) position.
5. Start the isocratic flow at 3 μL/min. Program the LC solvent flows and automatic switch valves to perform the following steps.
6. Condition the first biphasic trap column and second-dimension RP column with 80 % low-pH solvent B from gradient 1 at 300 nL/min for 20 min.
7. Switch V2 to the "load" (1–10) position to condition the other biphasic trap column with 80 % low-pH solvent B from gradient 1 at 300 nL/min for 20 min, followed by 5 % low-pH solvent B from gradient 1 at 300 nL/min for 20 min.
8. Program the LC solvent flow and automatic switch valves to go through the following steps to deliver the analytical gradient; start data acquisition at the mass spectrometer. Tables 1 and 2 summarize the gradient profiles of the first- and second-dimension LC pumps.
9. Switch V1 to the "load" (1–10) position. Allow the 5 % high-pH solvent B flow from gradient 2 to stabilize at 1 μL/min for 1 min while bypassing the first-dimension RP column. Then switch V1 to the "inject" (1–2) position, and ramp gradient 2 from 5 to 12 % high-pH solvent B for 5 min to elute the first peptide fraction into the solvent mixing loop (*see* **Note 4**). Maintain the isocratic flow of the low-pH solvent A at 3 μL/min and the gradient 1 of the 5 % low-pH solvent B at 300 nL/min throughout the process, unless otherwise specified.

Table 2
Mobile-phase gradient profile of the first- and second-dimension LC pumps for 2D RP-SCX-RP

	pH 10 RP fractionation (% of high-pH solvent B)	pH 2 gradient (% of low-pH solvent B)
Fr. 1	5–12	8–20
Fr. 2	12–14	8–25
Fr. 3	14–16	10–25
Fr. 4	16–19	10–28
Fr. 5	19–22	12–30
Fr. 6	22–26	12–32
Fr. 7	26–30	15–35
Fr. 8	30–45	20–40

10. Switch V1 to the "load" (1–10) position to bypass the first-dimension RP column. The isocratic flow will now push the first fraction, eluted from the first-dimension column, away from the mixing loop into trap column 1 for 20 min.
11. Stop the flow at gradient 2, and switch V1 to the "inject" (1–2) position. Instruct the autosampler to inject a salt plug (8 μL) of 1,000 mM ammonium acetate (NH_4OAc), which will be carried by the isocratic flow into the trap column for 30 min. The salt plug will dislodge all peptides lodged in the SCX phase of the trap column into the RP phase of the trap column.
12. Switch V2 to the "inject' (1–2) position, and repeat **steps 9–11**, except using the next fraction solvent composition from gradient 2 at **step 9**. Initiate the second-dimension analytical gradient at gradient 1 (*see* **Note 5**).
13. Switch V2 to the "load" (1–10) position. Repeat **steps 9–13** until all fractions are completed. Perform protein database search and other data analyses as described elsewhere.

3.4 Three-Dimensional RP-SCX-RP Separation

For highly complex samples, such as a mammalian cell digest, the two-dimensional RP-SCX-RP separation might not provide sufficient peak capacity to minimize the sample complexity for MS analysis. Therefore, three-dimensional RP-SCX-RP separation might be required to overcome this problem (*see* **Note 6**).

1. Follow **steps 1–7** under Subheading 3.3.
2. Program the LC solvent flow and automatic switch valves to pass through the following steps to deliver the analytical gradient; start data acquisition at the mass spectrometer. Tables 1 and 3 summarize the gradient profiles of the first-, second-, and third-dimension LC pumps.

Table 3
The gradient used for 3D RP-SCX-RP in the first, second, and final dimension

	pH-10 RP fractionation (% of high-pH solvent B)	SCX fractionation (concentration of NH_4OAc) (mM)	pH-2 gradient (% of low-pH solvent B)
Fr. 1	5–12	20	5–20
Fr. 2	12–14		8–20
Fr. 3	5–12	100	5–20
Fr. 4	12–14		8–20
Fr. 5	5–12	1,000	5–20
Fr. 6	12–14		8–20
Fr. 7	14–16	20	8–22
Fr. 8	16–19		10–25
Fr. 9	14–16	100	8–22
Fr. 10	16–19		10–25
Fr. 11	14–16	1,000	8–22
Fr. 12	16–19		10–25
Fr. 13	19–22	20	12–28
Fr. 14	22–26		15–32
Fr. 15	19–22	100	12–28
Fr. 16	22–26		15–32
Fr. 17	19–22	1,000	12–28
Fr. 18	22–26		15–32
Fr. 19	26–30	20	20–35
Fr. 20	30–45		20–35
Fr. 21	26–30	100	20–35
Fr. 22	30–45		20–35
Fr. 23	26–30	1,000	20–35
Fr. 24	30–45		20–35

3. Switch V1 to the "load" (1–10) position. Allow the 5 % high-pH solvent B flow from gradient 1 to stabilize at 1 μL/min for 1 min while bypassing the first-dimension RP column. Then switch V1 to the "inject" (1–2) position, and ramp gradient 2 from 5 to 12 % high-pH solvent B for 5 min to elute the first peptide fraction into the solvent mixing loop (*see* **Note 4**). Maintain the isocratic flow of the low-pH solvent A at 3 μL/min and the gradient 1 of the 5 % low-pH solvent B at 300 nL/min throughout the process, unless otherwise specified.
4. Switch V1 to the "load" (1–10) position to bypass the first-dimension RP column. The isocratic flow will now push the first fraction, eluted from the first-dimension column, away from the mixing loop and into trap column 1 for 20 min.

5. Stop the flow at gradient 2, and switch V1 to the "inject" (1–2) position. Instruct the autosampler to inject a salt plug (3 μL) of 20 mM NH_4OAc, which will be carried by the isocratic flow into the trap column for 30 min. The salt plug will elute the first SCX fraction into the RP phase of the trap column.
6. Switch V2 to the "inject" (1–2) position, and repeat **steps 3–5**, except using the next fraction solvent composition from gradient 1 at **step 3**. Initiate the third-dimension analytical gradient at gradient 1 (*see* **Note 5**).
7. Switch V2 to the "load" (1–10) position.
8. Stop the flow at gradient 2, and switch V1 to the "inject" (1–2) position. Instruct the autosampler to inject a salt plug (3 μL) of 100 mM NH_4OAc, which will be carried by the isocratic flow into the trap column for 30 min. Initiate the third-dimension analytical gradient at gradient 1 in the other trap column.
9. Switch V2 to the "inject" (1–2) position, and repeat **step 8** for SCX separation in the second trap column.
10. Repeat **steps 6–9**, except for instructing the autosampler to inject 8 μL of 1,000 mM NH_4OAc for the third SCX fraction.
11. Switch V1 to the "load" (1–10) position, and repeat from **step 3**, except use the next fraction solvent composition until all first-dimension fractions are completed. Perform a protein database search and other data analyses as described elsewhere (http://appliedbiosystems.cnpg.com/Video/flatFiles/20/index.aspx). MGF file was first generated from the data acquired using PeakView 1.1. The MGF file was searched against appropriate database using Paragon algorithm with ProteinPilot 4.5. For iTRAQ data analysis, bias and background correction were applied in the quantitation using ProteinPilot 4.5.

4 Notes

1. **a.** The Eksigent NanoLC Ultra 2D Plus system we have used contains four solvent channels that are capable of delivering two gradients (gradients 1 and 2) simultaneously, plus a loading pump connected to an autosampler, capable of isocratic micro-flow delivery. Modifications were made to the trigger valves 1 and 2 synchronously. Originally connected to the internal valve control circuit, pins 5 and 6 on the digital I/O port of the internal microelectric actuator designated for V1 were connected to O4 and O5 of gradient 2 directly.

b. As alternatives to the Eksigent NanoLC Ultra 2D Plus system, Agilent capillary or nano-flow LC system(s) (Agilent 1100 or later model, Agilent Technologies, Wilmington, DE) could be used. The different components in the system were synchronized with a customized hardware profile in the Analyst software. One capillary pump, one nano pump (or another capillary pump), and one autosampler were connected to the controlling computer via RS-232 cables. The switching valves were controlled via the serial ports on the computer. (#Supplementary Information: Detailed installation protocol and LC method are available free of charge via the Internet at http://web.hku.hk/~ivankchu/ichu.html/RP_MDLC_shotgun_Proteomics. Other information is also available upon request.)

2. Any mass spectrometer suitable for shotgun proteomics, such as Q-ToF or Orbitrap, may be used. In our laboratory, we operated a TripleTOF™ 5600 system (AB SCIEX, Concord, ON) fitted with a Nanospray III source (AB SCIEX) under the following settings: ion spray voltage, 2.5 kV; curtain gas, 30 psi; nebulizer gas, 15 psi; interface heater temperature, 150 °C; information-dependent acquisition mode; and 20-s dynamic exclusion. MS1 scans were acquired over 250 ms in the range *m/z* 100 1,250, followed by MS2 scans of the 20 most abundant peaks exceeding 125 cps and carrying charge of +2 to +5 in the range *m/z* 100–1,500.

3. At this point it is an option to disconnect the first-dimension column outlet from the 30-μL solvent mixing loop until sample injection is complete. This process will prevent any non-binding analytes from entering the downstream system and decrease the conditioning time. The sample injection valve may also be disconnected from the system after sample injection but prior to gradient separation.

4. The 30-μL solvent mixing loop is necessary to alleviate solvent incompatibility between the different LC dimensions.

5. We used an optimized last-dimension analytical gradient for 60–90 min, as listed in Tables 1–3. Because RP-RP is not perfectly orthogonal, late fractions from the first dimension contained peptides that were also more hydrophobic at pH 2; therefore, we used a more organic gradient to maximize efficiency.

6. The choice of two- or three-dimensional separation must be made to balance between analytical time and proteome depth. The three-dimensional separation yielded 20–34 % more proteins with the extended sequence coverage relative to the performance of the two-dimensional separation. The analytical time was, however, ~3 times longer than that when using the two-dimensional separation. The performance of the 3D system

has been benchmarked through analysis of the total lysate of *Saccharomyces cerevisiae* lysates at low-microgram levels. Over 50 % of the yeast proteome was identified; the estimated dynamic range of the 3D-LC platform spans over five orders of magnitude, from approximately 40 to 10^6 protein copies per cell (S. Figure 1).

Acknowledgment

This study was supported by the Hong Kong Research Grants Council (HKU7018/10P and HKU7016/11P). We thank AB SCIEX (Hong Kong) Limited for loan of the Eksigent NanoLC ultra 2D Plus system; Dr. Ricky P. W. Kong for performing some preliminary experiments and many helpful discussions; Dr. Tina Settineri, Eksigent (Dublin, CA, USA), Ms. Yuko P. Y. Lam, and Mr W.-Y. Yuen for their comments and suggestions; and the school of biological Science, HKU, for access to the 5600 mass spectrometer.

Supporting Information Available

S Figure 1 and detailed installation protocol. This material is available free of charge via the Internet at http://web.hku.hk/~ivankchu/ichu.html/RP_MDLC_shorgun_Proteomics. Other information is also available upon request.

References

1. Gygi SP, Rist B, Griffin TJ et al (2002) Proteome analysis of low-abundance proteins using multidimensional chromatography and isotope-coded affinity tags. J Proteome Res 1:47–54
2. Gilar M, Olivova P, Daly AE et al (2005) Two-dimensional separation of peptides using RP-RP-HPLC system with different pH in first and second separation dimensions. J Sep Sci 28:1694–1703
3. Nagaraj N, Wisniewski JR, Geiger T et al (2011) Deep proteome and transcriptome mapping of a human cancer cell line. Mol Syst Biol 7:548
4. Kong RP, Siu SO, Lee SS et al (2011) Development of online high-/low-pH reversed-phase-reversed-phase two-dimensional liquid chromatography for shotgun proteomics: a reversed-phase-strong cation exchange-reversed-phase approach. J Chromatogr A 1218: 3681–3688
5. Siu SO, Lam MP, Lau E et al (2011) Fully automatable two-dimensional reversed-phase capillary liquid chromatography with online tandem mass spectrometry for shotgun proteomics. Proteomics 11:2308–2319
6. Siu SO, Lam MP, Lau E et al (2010) N-linked glycoprotein analysis using dual-extraction ultrahigh-performance liquid chromatography and electrospray tandem mass spectrometry. Methods Mol Biol 600:133–143
7. Lam MP, Siu SO, Lau E et al (2010) Online coupling of reverse-phase and hydrophilic interaction liquid chromatography for protein and glycoprotein characterization. Anal Bioanal Chem 398:791–804
8. Zhao Y, Kong RP, Li G et al (2012) Fully automatable two-dimensional hydrophilic interaction liquid chromatography-reversed phase liquid chromatography with online tandem mass spectrometry for shotgun proteomics. J Sep Sci 35:1755–1763

9. Wolters DA, Washburn MP, Yates JR (2001) An automated multidimensional protein identification technology for shotgun proteomics. Anal Chem 73:5683–5690
10. Washburn MP, Wolters D, Yates JR 3rd (2001) Large-scale analysis of the yeast proteome by multidimensional protein identification technology. Nat Biotechnol 19:242–247
11. Zhao Y, Law CH, Kong Ricky PW et al (2013) Fully automatable multidimensional liquid chromatography system for shotgun proteomics in 61st ASMS Conference on Mass Spectrometry and Allied Topics, Minneapolis, US
12. Zhang ZJ, Cheang LC, Wang MW et al (2012) Ethanolic extract of fructus Alpinia oxyphylla protects against 6-hydroxydopamine-induced damage of PC12 cells in vitro and dopaminergic neurons in zebrafish. Cell Mol Neurobiol 32:27–40
13. Lin Y, Liu H, Liu Z et al (2012) Shotgun analysis of membrane proteomes using a novel combinative strategy of solution-based sample preparation coupled with liquid chromatography-tandem mass spectrometry. J Chromatogr B 901:18–24

Chapter 4

GeLC-MS/MS Analysis of Complex Protein Mixtures

Monika Dzieciatkowska*, Ryan Hill*, and Kirk C. Hansen

Abstract

Discovery-based proteomics has found its place in nearly every facet of biological research. A key objective of this approach is to maximize sequence coverage for proteins across a wide concentration range. Fractionating samples at the protein level is one of the most common ways to circumvent challenges due to sample complexity and improve proteome coverage. Of the available methods, one-dimensional sodium dodecyl sulfate-polyacrylamide gel electrophoresis followed by liquid chromatography-tandem mass spectrometry (GeLC-MS/MS) is a robust and reproducible method for qualitative and quantitative proteomic analysis. Here we describe a general GeLC-MS/MS protocol and include technical advice and outline caveats to increase the probability of a successful analysis.

Key words GeLC-MS/MS, SDS-PAGE, Protein fractionation, Protein digestion, Liquid chromatography, Mass spectrometry, Protein identification

1 Introduction

MS-based proteomic methods are currently unparalleled in their ability to rapidly characterize protein mixtures from biological samples. MS-based methods, however, often fail to reach the sensitivity of immunohistochemical methods that benefit from massive amplification of signal. Identification of low-abundance proteins to obtain as global of an analysis as possible is an ongoing challenge in the field [1]. While there is a substantial effort to improve proteome coverage during data acquisition steps, some of the most straightforward and effective methods are at the sample preparation phase of analysis. Broadly speaking, there are four levels of preparative fractionation methods: (1) cellular and (2) subcellular fractionation, followed by (3) protein- and (4) peptide-level fractionation. Popular examples of each include cell sorting, differential centrifugation, one-dimensional sodium dodecyl sulfate-polyacrylamide gel electrophoresis (1D

* Monika Dzieciatkowska and Ryan Hill have contributed equally to this work.

Daniel Martins-de-Souza (ed.), *Shotgun Proteomics: Methods and Protocols*, Methods in Molecular Biology, vol. 1156, DOI 10.1007/978-1-4939-0685-7_4, © Springer Science+Business Media New York 2014

SDS-PAGE), and orthogonal chromatography of digested peptides, respectively. The first two methods are highly dependent on the sample type, are compatible with the protein- and peptide-level fractionation methods, and have been reviewed elsewhere [2]. Peptide-level fractionation has been a popular approach for improved sequence and proteome coverage and has gained widespread popularity with the advent of the online tandem SCX-RP mud-pit approach [3] as discussed in Chapter 3. While these methods have proven successful on many levels, there are some distinct advantages to fractionating at the protein level. For example, if samples contain a wide range of protein abundances (wide dynamic range), peptides from the abundant proteins tend to dominate the MS acquisition time and hence impede the identification of lower level proteins. Protein-level separation allows for isolation of abundant proteins for improved low-level protein coverage. In addition, there is the potential to obtain isoform information based on the physical principle of the separation and gel migration by PAGE, in this example, to resolve isoforms that differ by molecular weight.

Considerations for selecting a protein separation method depend on many factors including resolving power, concentration constraints, reagent compatibility, and post-separation sample compatibility to name a few. The 1D SDS-PAGE method is a logical choice for a wide range of protein samples due to high resolving power, protein capacities in excess of sample loads for modern nLC-MS/MS instruments, compatibility with many detergents and chaotropes used in sample extraction, cost, and availability. For many years 2D-PAGE was the method of choice for protein-level fractionation prior to MS analysis. The approach has the capacity for high resolution based on two orthogonal physical properties of a protein and the ability to focus identification efforts (albeit only for more abundant proteins). However, the method is poorly suited for more global protein identification. While it is always possible to increase proteome overage by performing higher dimensional fractionation, there is a trade-off in sample loss and instrument time required for sample analysis. Many proteomic projects are constrained by limited sample quantities and instrument availability. The combination of protein-level separation by 1D-SDS-PAGE followed by RP LC-MS/MS analysis of digests from all bands, referred to as GeLC-MS/MS, offers a powerful analytical approach that balances real-world constraints with obtaining optimal proteome coverage.

The protocol described here is based on the original in-gel digestion approach presented by Rosenfeld et al. with subsequent modifications [4, 5] (*see* Fig. 1). Proteins are separated by SDS-PAGE, and the entire gel lanes are excised and subdivided into bands. The proteins in these gel sections are subsequently

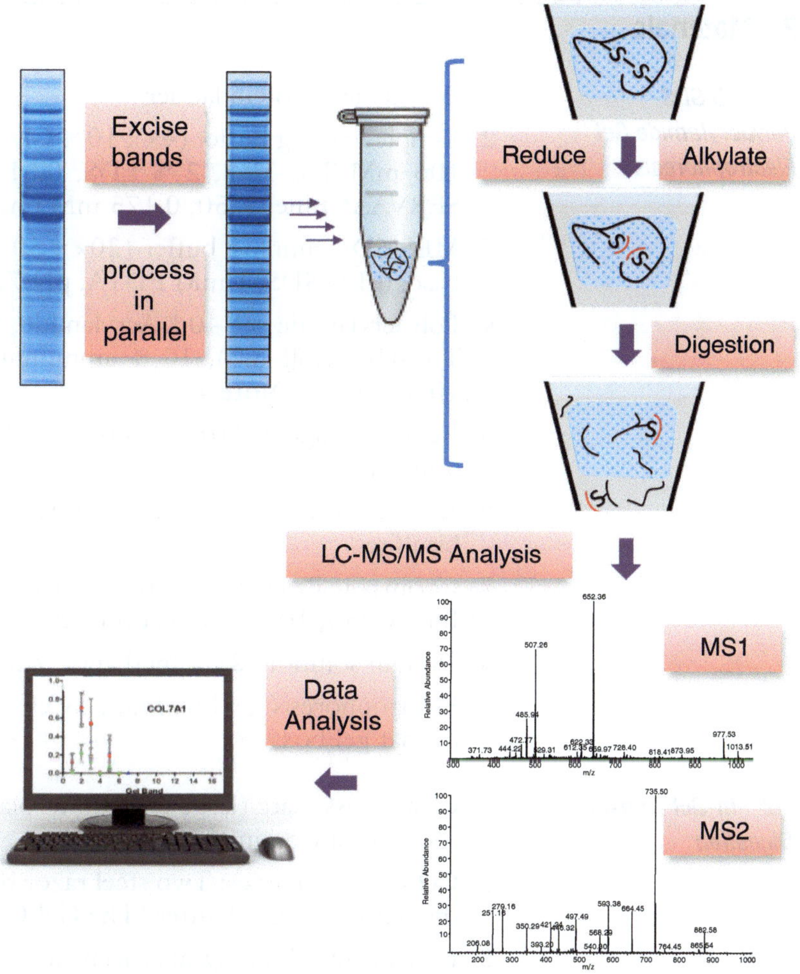

Fig. 1 A schematic overview of the in-gel digestion procedure. Protein sample is separated by 1D SDS-PAGE, and the lane is cut into bands for parallel processing. Reduction and alkylation are performed prior to enzymatic digestion of proteins. Peptides are extracted for LC-MS/MS analysis followed by protein sequence database searches

digested in-gel, and the extracted peptides are subjected to nanoflow reversed-phase LC-MS/MS analysis to obtain peptide sequence identifications that can be mapped to proteins in a sequence database [6, 7]. The database search results can be queried by band to yield information regarding intact MW and search for evidence of protein isoforms or combined to yield the total sample identifications and relative quantification per sample. Here we present the steps for performing GeLC-MS/MS analysis and give suggestions to optimize peptide identification results using this method.

2 Materials

2.1 1D SDS-Polyacrylamide Gel Electrophoresis

1. Pre-stained protein ladder.
2. Sample loading buffer (4×): 10 % Glycerol, 141 mM Tris base, 106 mM Tris–HCl, 2 % LDS, 0.51 mM EDTA, 0.22 mM SERVAÂ® Blue G250, 0.175 mM phenol red, pH 8.5.
3. MES SDS running buffer (20×): 50 mM MES, 50 mM Tris base, 0.1 % SDS, 1 mM EDTA, pH 7.3.
4. Polyacrylamide gel: 40 % Acrylamide, 1 % bisacrylamide, 1.5 M Tris–HCl (pH 8.7), 10 % ammonium persulfate, TEMED, 10 % SDS (*see* **Note 1**).
5. Reducing agent (10×): 500 mM Dithiothreitol or 2.5 % β-mercaptoethanol.
6. Antioxidant: 10 % *N*,*N*-dimethylformamide, 15 % sodium bisulfite.
7. Coomassie Brilliant Blue-based stain: 0.1 % Coomassie Brilliant Blue G250, 10 % glacial acetic acid, 40 % methanol.
8. Destain solution: 25 % methanol, 7.5 % glacial acetic acid.
9. Electrophoresis gel box (e.g., XCell SureLock or other small-format gel box).

2.2 In-Gel Trypsin Digestion

1. Clean glass plate (large enough to place the entire gel on and room for a working area, 8″ × 8″).
2. Gel-cutting devices: Two steel razor blades, surgical scalpel, or the MEG-1.5 Gel Cutter (The Gel Company).
3. Low-protein-binding microcentrifuge tubes (0.65 or 1.5 mL).
4. Gel-loading pipette tips.
5. Autosampler vials with perforated caps.
6. SpeedVac concentrator.
7. 25 mM ammonium bicarbonate: Dissolve 100 mg ammonium bicarbonate in 50 mL double-distilled water.
8. Destain solution: 25 mM ammonium bicarbonate/50 % acetonitrile (ACN). Weigh 100 mg ammonium bicarbonate, and prepare a 50 mL solution.
9. Extraction solution: 1 % formic acid.
10. Tris[2-carboxyethyl]phosphine (TCEP)-HCl stock solution: 5 mM TCEP in 25 mM ammonium bicarbonate.
11. Iodoacetamide (IAM) stock solution: 20 mM in 25 mM ammonium bicarbonate (always prepare fresh, light sensitive).
12. 10 ng/μl Trypsin, sequencing grade (use 25 mM ice-cold ammonium bicarbonate to dilute stock trypsin solution, immediately before adding to gel pieces).

2.3 Liquid Chromatography-Tandem MS Analysis

1. Formic acid (LC/MS grade).
2. Solvent A: Double-distilled water from an all glass still, with 0.1 % formic acid. Add acid using a glass syringe.
3. Solvent B: ACN (LC/MS grade) with 0.1 % formic acid. Add acid using a glass syringe.
4. Trap column: ZORBAX 300SB-C18, dimensions 5 × 0.3 mm, 5 µm (Agilent Technologies).
5. Analytical column: Self-packed, KaSil fritted 100 µm i.d. × 150 mm fused silica capillary packed with Synergi C_{18} resin (Phenomenex; Torrance, CA), 15 µm tip ID (New Objective).
6. High-performance LC system, capable of nanoliter flow rates, with a chilled autosampler (e.g., Agilent 1260 nano system).
7. Mass spectrometer with tandem MS capabilities (e.g., LTQ Orbitrap Velos, Thermo Fisher Scientific).
8. Peak list extraction software (e.g., manufacture-supplied extractor, Mascot distiller, ProteoWizard [8])
9. MS/MS data analysis software (e.g., ProteinProspector or Mascot).
10. Optional: Software for differential analysis (spreadsheet program with statistical analysis package or dedicated programs for this purpose, e.g., Scaffold, Progenesis LC-MS)

3 Methods

3.1 1D SDS-PAGE

SDS-PAGE offers size-based separation and provides a convenient method for sample cleanup prior to trypsin digestion. Efficient protein extraction and isolation prior to GeLC-MS/MS are critical for obtaining an accurate representation of the proteome under study. Proteins can be prepared from in vivo (tissue or bodily fluids) and in vitro (cell culture or immunoprecipitations) sources. While preparation methods are sample specific, protocols commonly call for one or more of the following steps: (1) mechanical lysis of tissues or cells, or the use of alternative homogenization techniques, (2) solubilization in buffer (some classes of protein require strong detergents or chaotropes for solubilization), (3) subcellular fractionation approaches [9], and (4) protein complex isolation. The most basic form of protein isolation is by direct solubilization in SDS-PAGE sample buffer with heat [10]. However, if this is the method of choice, it is important to remove salts and incompatible detergents and chaotropes (such as guanidine hydrochloride) in order to avoid streaking and aberrant protein migration. This can readily be performed by precipitation of proteins prior to solubilization in sample buffer [11].

1. Add LDS sample buffer (4×) to the protein sample (with or without reducing agent), and heat at 70 °C for 10 min. Centrifuge the heated samples at 2,400 × *g* for 30 s to bring down insoluble material (*see* **Notes 2–4**).
2. Obtain 1.5 mm thick precast Bis-Tris 4–12 % gradient gel (*see* **Notes 5** and **6**).
3. Remove the Bis-Tris 4–12 % gradient gel from the storage pouch, and pull the plastic strip from the bottom of the cassette.
4. Place gel into mini gel box electrophoresis system with the gel comb opening facing towards the inside of the buffer tank.
5. Dilute MES SDS running buffer (10×) to a 1× solution using ddH_2O. Make up to 1 L (*see* **Note 7**).
6. Fill the inner tank with the 1× running buffer. The buffer level must exceed the level of the wells.
7. Add 500 μL of antioxidant in the inner tank.
8. Gently pull the comb out of the cassette, and rinse the sample wells thoroughly with 1× running buffer to remove air bubbles and displace any storage buffer.
9. Apply protein sample up to 40 μg protein concentration per gel (*see* **Notes 8–10**).
10. Load 5–8 μL of the marker into a sample well of the gel to be run.
11. Attach the electrophoresis gel box lid, and connect the electrode cords to the power supply [red to (+) jack, black to (−) jack].
12. Set power supply for 180 V constant voltage and 35 min run time. Start the run. The bromophenol blue tracking dye elutes from the gel at approximately 30–35 min.
13. After electrophoresis is complete, shut off the power, disconnect electrodes, and remove gel(s) from electrophoresis gel box.
14. Insert the gel knife into the gap between the cassette's two plates. Twist to open gel cassette (*see* **Note 15**).
15. Carefully remove and discard the top plate, allowing the gel to remain on the bottom plate.
16. Let the gel fall off into 100 mL of ddH_2O in a clean staining container.
17. Rinse the gel three times for 10 min with ddH_2O (*see* **Note 11**).
18. Discard water from gel, and add 50 mL of Coomassie stain solution or sufficient volume to completely cover the gel (*see* **Note 12**).
19. Stain for 1 h at room temperature with gentle shaking. Bands will begin to develop within minutes. After staining is complete, discard the stain solution.

20. Rinse gel with 100 mL ddH_2O to remove residual stain from container. Add 100 mL of destaining solution and microwave gel on high power for an additional 30 s. Gently shake the gel at room temperature for at least 1 h. Gel will have a clear background after 1 h in destain solution. The gel can be stored at 4 °C until processing for in-gel digestion (*see* **Notes 13** and **14**).

3.2 Band Excision

1. Place the gel on a clean glass plate. Cover the gel with just enough ddH_2O water to prevent dehydration during the slicing process (*see* **Note 16**).
2. Cut the gel lane using (new, if possible) scalpel, razor blade, or OneTouch GridCutter.
3. Cut each of the excised bands into 1–2 mm cubes, and transfer these cubes to a 0.65 mL siliconized microcentrifuge tube (*see* **Note 17**).

3.3 In-Gel Digestion

4. Add ~100 μL (or enough to cover) of 25 mM ammonium bicarbonate/50 % ACN and vortex for 10 min. Using gel-loading pipet tip, extract the supernatant and discard. The procedure should be repeated until the stain is completely removed. Two additional washes should be sufficient for moderately intense bands.
5. Add 100 μL of 5 mM TCEP, and incubate for 30 min at 56 °C. Spin. Discard all the liquid afterwards (*see* **Note 18**).
6. Allow samples to cool to room temperature.
7. Add 100 μL of 20 mM IAM, and incubate the gel pieces in the dark for 45 min at room temperature. Spin. Discard the liquid afterwards (*see* **Note 19**).
8. Wash the gel pieces with 100 μL of 25 mM ammonium bicarbonate, vortex for 10 min, and spin. Discard the liquid afterwards.
9. Wash the gel pieces with ~100 μL (or enough to cover) of 25 mM ammonium bicarbonate in 50 % ACN, vortex for 10 min, and spin. Discard the liquid (*see* **Note 20**).
10. Dehydrate the gel pieces in 100 % ACN for 10 min, spin, and discard the liquid afterwards.
11. Dry the sample in a SpeedVac for 10 min. The gel pieces are now ready for tryptic digestion.
12. Just before use, dilute or reconstitute trypsin with 50 mM ice-cold ammonium bicarbonate to give final concentration of the 10 ng/μL (*see* **Note 21**).
13. Add trypsin solution to just cover the gel pieces (*see* **Note 22**).
14. Rehydrate dried gel pieces on ice or at 4 °C for 30–45 min [12].
15. Check that the gel pieces are covered with trypsin solution.

16. Add 25 mM ammonium bicarbonate as needed to cover the gel pieces.
17. Spin briefly, and incubate at 37 °C for 4 h–overnight (*see* **Note 23**).
18. Stop digestion by adding 20 μL of 5 % formic acid.
19. Vortex for 15–20 min, spin, and transfer the digest solution (aqueous extraction) into a clean autosampler vial appropriate for LC/MS-MS (*see* **Note 24**).
20. To the gel pieces, add 30 μL (enough to cover) of 1 % formic acid, vortex for 15–20 min, spin, and transfer solution to the tube used above. Repeat this step once (*see* **Note 25**).
21. Concentrate peptide extracts using a SpeedVac concentrator to a volume that is slightly larger than will be used for injection during LC-MS/MS analysis (*see* **Notes 26** and **27**).
22. Store the vial with the extracted peptides at −20 °C if the samples will not be run on the same day.

3.4 Mass Spectrometry Analysis

A high-pressure liquid chromatography system running at nanoflow rates should be used for peptide fractionation prior to mass spectrometry analysis (*see* **Note 28**). Reverse-phase liquid chromatography columns (15 cm × 100 μm ID) packed in-house with C18 resins (4 μm, 100 Å beads, Phenomenex, Torrance, CA) or commercial columns can be used. An LTQ Orbitrap Velos™ mass spectrometer (Thermo Fisher Scientific) or equivalent rapid scanning mass spectrometer may be used. Specific values in the methods below will be laboratory dependent and are provided simply as a benchmark.

1. Prior to analysis, calibrate the instrument using a standard calibration mixture according to a validated SOP or the manufacturer's instructions.
2. Perform a QC run to evaluate instrument performance with LC-MS/MS analyses of 1–20 fmol of a protein digest standard. Ideally the standard protein will not be present in the sample of interest. We commonly use yeast alcohol dehydrogenase digests for this purpose.
3. Load 1–10 μL of each peptide sample using the autosampler. Optional: Desalt on a trapping column with a flow rate of 10 μL/min for 3 min (*see* **Note 29**).
4. Program the switching valve to place the trapping column online with the analytical column.
5. Separate the peptides by reverse-phase LC with a 45- or a 75-min linear gradient (60–90-min total run time) from 5 % ACN to 35 % ACN with 0.1 % FA at 300 nL/min flow rate (*see* **Note 30**).
6. Acquire MS/MS in a data-dependent mode, in which MS/MS fragmentation is performed on the most intense ions of every full

MS scan (10–20 is typical). Perform full MS scan (*m/z* 400–2,000) in the Orbitrap with 60,000 resolution (*see* **Note 31**).

7. Run standard protein digest at routine intervals and at the end of the analysis to evaluate instrument performance.

3.5 Data Analysis

Several computational algorithms have been developed to match peptide fragmentation spectra to peptides for protein identification, as the complexity of tandem MS/MS data files precludes comprehensive manual interpretation of all spectra. Software packages and associated algorithms, such as Mascot [13], X!Tandem [14], and ProteinProspector [15–17], may be used to search a given sequence database for peptides with theoretical fragmentation spectra best matching the observed spectra and subsequent assignment of the matched peptides to the corresponding protein/s. An important component of the data analysis process is evaluation of the false discovery rate for peptide- and protein-level identifications; this is performed by including appropriate decoy sequences in the queried protein database. Results from individual gel bands can be combined and a statistical comparative analysis performed across different samples. Additional information regarding protein isoforms that resolve by gel band can be obtained by comparing identification results across samples by gel band.

4 Notes

1. Adjusting the ratio of acrylamide to cross-linker (bisacrylamide) allows for tailoring the gel to the specific sample under consideration. In general, low-percent acrylamide gels will separate high-molecule proteins more efficiently and high-percent acrylamide gels will separate low molecular proteins more efficiently. For GeLC-MS/MS applications we suggest the use of commercially available gels as they typically yield more reproducible results.
2. Heating the sample at 100 °C in SDS containing buffer can result in proteolysis. It is recommend to heat samples for 5–10 min for denaturing electrophoresis at 70 °C for reduced and 90 °C for non-reduced samples for optimal results [18].
3. SDS precipitates at 4 °C. Therefore, LDS sample buffer needs to be warmed prior to use if stored at 4 °C for increased shelf life.
4. Reducing agents such as TCEP and DTT are used to reduce disulfide bonds permitting more complete protein unfolding/denaturing prior to electrophoresis.
5. Free acrylamide may react with primary amines and free thiols on proteins during polyacrylamide gel electrophoresis [19]. We recommend using commercial precast gels for GeLC-MS/MS experiments due to our experiences with reproducibility of

hand-cast gels. If hand-cast gels will be used we recommend using fresh reagents and allowing overnight polymerization. Precast gels are ready to use and offer greater convenience, more stringent quality control, and higher reproducibility than hand-cast gels in general.

6. Use gradient gels to separate samples containing a broad range of molecular weights. Gradient gels allow resolution of both high- and low-molecular-weight bands on the same gel.
7. Other buffer systems such us MOPS may also produce comparable results.
8. Overloading will result in poor to no resolution of protein bands.
9. High salt concentrations result in increased conductivity that affects protein migration and can result in gel artifacts in adjacent lanes containing samples with lower salt concentrations. Precipitate protein, and bring the pellet up in LDS sample loading buffer or perform dialysis (micro) using a lower salt buffer prior to electrophoresis.
10. Samples solubilized in guanidine-HCl have high ionic strength and produce increased conductivity similar to high salt concentrations. In addition, guanidine precipitates in the presence of SDS leading to various types of gel artifacts. If possible, change the solubilization agent by dialysis prior to electrophoresis or precipitate with cold ethanol (1:9 ratio: sample to ethanol) prior to electrophoresis [20].
11. The washing step is necessary to remove residual SDS, which interferes with dye binding.
12. A colloidal Coomassie (G250)-based stain is recommended for visualizing proteins. In GeLC-MS/MS experiments staining usually serves two purposes: location of intense protein bands to direct band cutting and to provide a global overview of gross differences by visual comparison of samples (usually this is of limited utility). If silver staining is used make sure that the protocol is compatible with mass spectrometry analysis. Glutaraldehyde, used in some protocols as a sensitizer, cross-links proteins which results in decreased trypsin digestion and protein extraction from the gel.
13. Microwaving the gels speeds up both the staining and destaining process and can be circumvented with longer incubation times [21]. Commercially available stains are available that do not require the addition of heat or destain solution to visualize gel bands.
14. Gel migration is a very crude method for estimating molecular weight of a protein. Proteins migrate at different rates based on amino acid composition and presence of modifications. Acidic proteins repel SDS and thus will migrate slower towards

the anode, whereas basic proteins bind a higher concentration of SDS and tend to migrate faster than most proteins that are used as molecular weight markers [22].

15. Reduce exposure of the gel, gel equipment, and supplies to primary sources of keratins, such as skin, hair, and clothing. Reduce the amount of exposure to dust and particulates, both of which can be a rich source of keratins. Perform as much work as possible in an area cleaned of dust, ideally in laminar flow hood. Wear powder-free gloves and sleeve protectors or a lab coat; wipe down ALL surfaces with 50 % ethanol solution, moistened lint-free cloth, or tissue, including glass plate used for cutting, the SpeedVac and centrifuge, racks, and tubes.
16. Always use siliconized or low-binding polypropylene tubes and low-retention tips (important once peptides have been generated) to minimize sample loss due to adsorption to tube or tip surfaces.
17. Dithiothreitol (DTT) is a commonly used alternative for reduction of disulfide bonds. TCEP offers several advantages including greater resistance to oxidation; it is a stronger reductant and is not prone to side reactions with peptide functional groups. Check the pH of your TCEP stock solution as it may be acidic when brought up in solution depending on buffering conditions and needs to be brought to neutral pH prior to addition to the protein sample.
18. The most commonly used agents for alkylation of cysteine thiols in protein samples prior to digestion are IAM followed by iodoacetic acid (IAA).
19. Sufficient washing of the gel pieces after alkylation is a critical step of in-gel digestion processing. Insufficient washing leaves residual alkylating agents in the gel. If present during digestion, the N-terminus of peptides is commonly alkylated which has two negative effects: (1) It splits the population of your peptides which decreases the probability of identification and (2) if not accounted for in database searches increases the odds of false-positive identifications [23].
20. Modified trypsin is preferred for protein digestion as it is less susceptible to autolysis.
21. Although trypsin is the most common enzyme of choice for proteomic methods, additional enzymes such as Asp-N, chymotrypsin, or Glu-C can be used to improve sequence coverage [24, 25].
22. 37 °C has been suggested as the optimal temperature and is the temperature most commonly used for overnight in-gel- and in-solution-based tryptic digestion.
23. This is a potential point of contamination; all plastic vials have the potential to contain polymer. Confirm that the

batch of autosampler vials has been used without incident prior to usage here.

24. A potential disadvantage of the GeLC-MS/MS approach when compared to tandem LC/LC-MS/MS approaches is the loss of peptides due to extraction efficiency from the gel. Speicher et al. have explored this issue using isotopically labeled protein and scintillation counting to determine sample losses and extraction efficiencies [26]. Their findings indicate that approximately 15–20 % of sample is lost during the entire in-gel digestion procedure. Sample is lost during destaining steps (~4 %) and to the digestion tube (~6 %), pipette tip (~1 %), and the gel (7 %). Presumably in-solution digestion methods would be subject to losses due to the digestion tube and pipette tip/s. The overall recovery of peptides was approximately 78 % in the first round of extraction and an additional 6 and 1 % in subsequent rounds.
25. Generate an autosampler vial that contains a known amount of liquid for reference. The outer tube should be removed if it is difficult to visualize liquid level.
26. If possible do not let the peptide extract dry completely; resolubilization is a source of potential sample loss. If sample volume needs to be added use buffer A from the LC system.
27. Many investigators will need to utilize a core facility or work with a collaborator to have peptide digests analyzed by mass spectrometry. Important aspects of this analysis to consider and discuss with a facility or a collaborator include the following: (1) wait time between digestion and analysis: ideally digests will be run within the first 48 h of digestion; (2) complexity of sample and objectives of the analysis; these parameters will dictate the length of LC runs, the optimal mass spectrometer, and acquisition parameters to use.
28. For differential analysis we prefer to run band digests in groups of five from one lane (sample A, bands 1–5) followed by a blank run and the same five bands from the sample/s to be compared (sample B, C, etc.) before moving to the next five bands (sample A, 6–10). This serves to minimize artifacts due to variations in instrument performance over time while keeping blank runs to a minimum. The blank runs should be analyzed for QC purposes.
29. Longer columns improve chromatographic separation and generally increase peptide identifications; further improvements are observed with longer gradients [27].
30. The type of fragmentation used will depend on experimental considerations. Combinations of ETD and HCD have proven useful for improving protein coverage, and ETD has some advantages for mapping posttranslational modifications [28].

References

1. Zubarev RA (2013) The challenge of the proteome dynamic range and its implications for in-depth proteomics. Proteomics 13:723–726
2. Michelsen U, von Hagen J (2009) Isolation of subcellular organelles and structures. Methods Enzymol 463:305–328
3. Washburn MP, Wolters D, Yates JR 3rd (2001) Large-scale analysis of the yeast proteome by multidimensional protein identification technology. Nat Biotechnol 19:242–247
4. Rosenfeld J, Capdevielle J, Guillemot JC et al (1992) In-gel digestion of proteins for internal sequence analysis after one- or two-dimensional gel electrophoresis. Anal Biochem 203:173–179
5. Hellman U, Wernstedt C, Góñez J et al (1995) Improvement of an "In-Gel" digestion procedure for the micropreparation of internal protein fragments for amino acid sequencing. Anal Biochem 224:451–455
6. Baldwin MA (2004) Protein identification by mass spectrometry. Mol Cell Proteomics 3:1–9
7. Chalkley RJ, Hansen KC, Baldwin MA (2005) Bioinformatic methods to exploit mass spectrometric data for proteomic applications. Methods Enzymol 402:289–312
8. Kessner D, Chambers M, Burke R et al (2008) ProteoWizard: open source software for rapid proteomics tools development. Bioinformatics (Oxford, England) 24:2534–2536
9. Zhang Y, Fonslow BR, Shan B et al (2013) Protein analysis by shotgun/bottom-up proteomics. Chem Rev 113:2343–2394
10. Ericsson C, Nistér M (2011) Protein extraction from solid tissue. Methods Mol Biol (Clifton, NJ) 675:307–312
11. Jiang L, He L, Fountoulakis M (2004) Comparison of protein precipitation methods for sample preparation prior to proteomic analysis. J Chromatogr A 1023:317–320
12. Havlis J, Thomas H, Sebela M et al (2003) Fast-response proteomics by accelerated in-gel digestion of proteins. Anal Chem 75: 1300–1306
13. Perkins DN, Pappin DJ, Creasy DM et al (1999) Probability-based protein identification by searching sequence databases using mass spectrometry data. Electrophoresis 20: 3551–3567
14. Craig R, Beavis RC (2004) TANDEM: matching proteins with tandem mass spectra. Bioinformatics (Oxford, England) 20: 1466–1467
15. Clauser KR, Baker P, Burlingame AL (1999) Role of accurate mass measurement (+/– 10 ppm) in protein identification strategies employing MS or MS/MS and database searching. Anal Chem 71:2871–2882
16. Chalkley RJ, Baker PR, Hansen KC et al (2005) Comprehensive analysis of a multidimensional liquid chromatography mass spectrometry dataset acquired on a quadrupole selecting, quadrupole collision cell, time-of-flight mass spectrometer: I. How much of the data is theoretically interpretable by search engines? Mol Cell Proteomics MCP 4:1189–1193
17. Chalkley RJ, Baker PR, Huang L et al (2005) Comprehensive analysis of a multidimensional liquid chromatography mass spectrometry dataset acquired on a quadrupole selecting, quadrupole collision cell, time-of-flight mass spectrometer: II. New developments in protein prospector allow for reliable and comprehensive automatic analysis of large datasets. Mol Cell Proteomics MCP 4:1194–1204
18. Kubo K (1995) Effect of incubation of solutions of proteins containing dodecyl sulfate on the cleavage of peptide bonds by boiling. Anal Biochem 225:351–353
19. Chiari M, Righetti PG, Negri A et al (1992) Preincubation with cysteine prevents modification of sulfhydryl groups in proteins by unreacted acrylamide in a gel. Electrophoresis 13: 882–884
20. Pepinsky RB (1991) Selective precipitation of proteins from guanidine hydrochloride-containing solutions with ethanol. Anal biochem 195:177–181
21. Lill JR, Nesatyy VJ (2012) Microwave-assisted protein staining, destaining, and in-gel/in-solution digestion of proteins. Methods Mol Biol (Clifton, NJ) 869:521–532
22. Shi Y, Mowery RA, Ashley J et al (2012) Abnormal SDS-PAGE migration of cytosolic proteins can identify domains and mechanisms that control surfactant binding. Protein Sci 21:1197–1209
23. Woods AG, Sokolowska I, Darie CC (2012) Identification of consistent alkylation of cysteine-less peptides in a proteomics experiment. Biochem Biophys Res Commun 419: 305–308
24. Choudhary G, Wu S-L, Shieh P et al (2003) Multiple enzymatic digestion for enhanced sequence coverage of proteins in complex proteomic mixtures using capillary LC with Ion trap MS/MS. J Proteome Res 2:59–67
25. Swaney DL, Wenger CD, Coon JJ (2010) Value of using multiple proteases for large-scale mass spectrometry-based proteomics. J Proteome Res 9:1323–1329

26. Speicher K, Kolbas O, Harper S et al (2000) Systematic analysis of peptide recoveries from in-gel digestions for protein identifications in proteome studies. J Biomol Tech JBT 11: 74–86
27. Hsieh EJ, Bereman MS, Durand S et al (2013) Effects of column and gradient lengths on peak capacity and peptide identification in nanoflow LC-MS/MS of complex proteomic samples. J Am Soc Mass Spectrom 24:148–153
28. Swaney DL, McAlister GC, Coon JJ (2008) Decision tree-driven tandem mass spectrometry for shotgun proteomics. Nat Methods 5:959–964

Chapter 5

IPG Strip-Based Peptide Fractionation for Shotgun Proteomics

Murat Eravci, Christian Sommer, and Matthias Selbach

Abstract

Efficient fractionation of peptides is an essential prerequisite for comprehensive analysis of complex protein mixtures by shotgun mass spectrometry. The separation of peptides by isoelectric focusing is particularly attractive due to its orthogonality to reverse-phase HPLC. Here, we present a protocol for in-gel peptide isoelectric focusing using immobilized pH gradient strips. The method shows high resolving power for up to 1 mg of sample and is highly reproducible.

Key words Shotgun proteomics, Peptide fractionation, Isoelectric focusing (IEF), Immobilized pH gradient (IPG)

1 Introduction

In order to decrease the complexity of protein mixtures for enhancing their analysis by liquid chromatography-mass spectrometry (LC-MS), proteins or peptides are usually prefractionated prior to RP-HPLC. Most commonly, proteins are separated by SDS-PAGE and digested "in-gel" with trypsin [1] (Chapter 3). Alternatively, peptides generated by "in-solution" digestion can be separated by ion exchange chromatography [2].

Separating peptides by isoelectric focusing (IEF) using immobilized pH gradient (IPG) gel strips prior to LC-MS analysis has several advantages such as high loading capacity, high resolving power, and high reproducibility [3–5]. The use of the p*I* as a validation and filtering criteria to identify false positives [4, 6] and posttranslational modifications (PTM) [7] is also an advantage of IEF-based prefractionation. A popular platform for peptide IEF is the OFFGEL system (Agilent Technologies) [8] which enables the recovery of peptides after IEF without any need of extraction. However, Hubner et al. [9] reported that focusing efficiency decreases significantly when increasing the amount of sample loaded: With 50 μg of sample more than 80 % of peptides were

Daniel Martins-de-Souza (ed.), *Shotgun Proteomics: Methods and Protocols*, Methods in Molecular Biology, vol. 1156, DOI 10.1007/978-1-4939-0685-7_5, © Springer Science+Business Media New York 2014

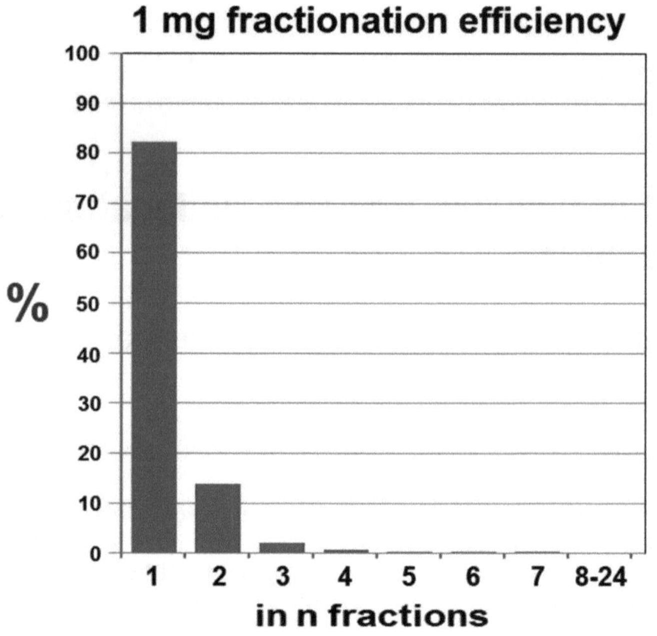

Fig. 1 Fractionation efficiency of the prefractionation of 1 mg tryptic peptides from HEK293T cells after isoelectric focusing on a 24 cm pH 3–10 IPG gel followed by an extraction of 24 gel pieces. Fractionation efficiency is the number of peptides identified in *n* fractions (in %)

found in a single fraction. However, the focusing efficiency dropped to less than 40 % when 500 μg were loaded. Thus, the OFFGEL system does not take advantage of the intrinsically high loading capacity of IPG gel [10].

Here, we describe a protocol for "in-gel" peptide IEF using IPG strips. Using this protocol we obtained fractionation efficiency of more than 80 % (peptides detected in only one of the 24 fractions) when loading 1 mg of peptides on a 24 cm IPG strip (Fig. 1). We also found that this prefractionation procedure was highly reproducible with regard to the number of proteins identified and focusing efficiency in three independent experiments (about 1 month difference between each experiment): IEF of 250 μg aliquots of the same sample on a 7 cm IPG strip with a linear gradient from pH 3–10 leads to the identification of 8,372 proteins in all three experiments, from which 7,334 (approx. 88 %) were identical in all experiments (Fig. 2), with a mean focusing efficiency of 80 % (Fig. 3). In these latter experiments we used six fractions only. This is advantageous in order to reduce the number of samples when LC-MS measurement time is limited. The method presented here is an attractive alternative to other prefractionation approaches, especially when the sample amount is not limited and large loading capacity is desirable. For example, this is typically the case for proteome profiling experiments in tissue culture cells.

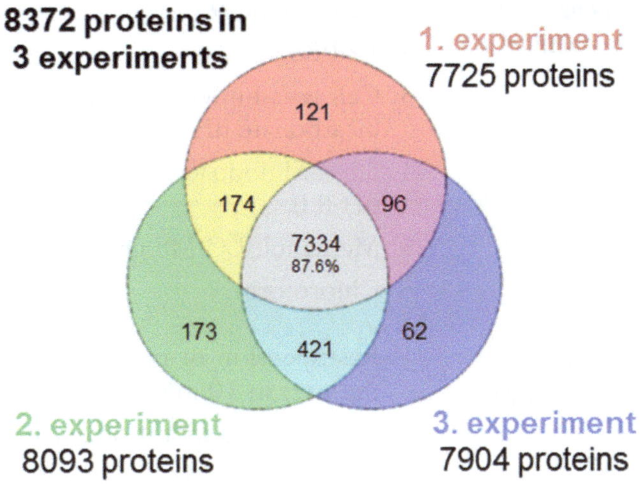

Fig. 2 Venn diagram of protein identifications from three repeated experiments. In each experiment 250 μg tryptic peptides from HEK293T cells were focused on a 7 cm pH 3–10 IPG gel followed by an extraction of six gel pieces

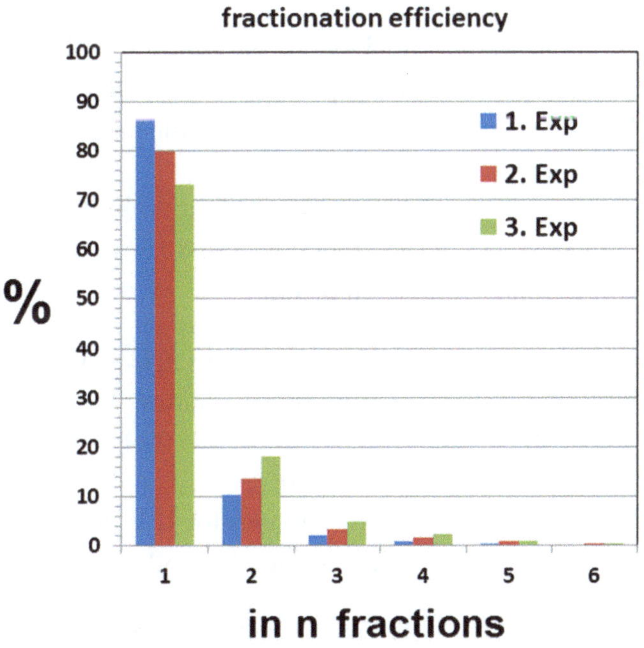

Fig. 3 Comparison of the fractionation efficiencies of three repeated experiments. In each experiment 250 μg tryptic peptides from HEK293T cells were focused on a 7 cm pH 3–10 IPG gel followed by an extraction of six gel pieces

2 Materials

All buffers and solutions were prepared using MS-grade water and MS-grade solvents.

2.1 Components for Sample Preparation

1. Cells: HEK293 cells.
2. 1× PBS pH 7.4.
3. Cell lysis buffer: 2 % SDS, 0.1 M DTT, 50 mM ammonium bicarbonate pH 8.0, complete protease inhibitor cocktail.
4. 55 mM iodoacetamide in 50 mM ammonium bicarbonate pH 8.0.
5. Methanol, LC-MS grade.
6. Chloroform.

2.2 Components for In-Solution Digestion

1. Resuspension of precipitated protein pellets: 6 M urea/2 M thiourea in 10 mM HEPES (pH 8.0).
2. Protein quantification: AmidoBlack 10B with BSA as standard.
3. Digestion enzymes: Lys-C and sequencing-grade modified trypsin.
4. Buffer for dilution of urea concentration: 50 mM ammonium bicarbonate pH 8.0.

2.3 Components for Solid-Phase Extraction (Desalting)

1. 3 ml C_{18}-SD Empore Disk Cartridges.
2. Empore Extraction Disk, C_{18}, 47 mm.
3. Equilibration buffer and washing buffer: 5 % acetonitrile, 0.1 % formic acid.
4. Sample acidification buffer: 10 % trifluoroacetic acid (TFA).
5. Elution buffer: 80 % acetonitrile.
6. Methanol.

2.4 Components for the Isoelectric Focusing of Peptides

1. IPG strips 7 cm pH 3–10 (*see* **Note 1**).
2. Peptide rehydration buffer: 10 % glycerol, 1 % ampholytes 3–10, and traces of bromophenol blue.
3. Parafilm.
4. IEF100 unit for IEF (Serva Electrophoresis).
5. Electrode filter pads.

2.5 Peptide Extraction Buffers

1. First extraction buffer: 0.1 % TFA, 99.9 % water.
2. Second extraction buffer: 0.1 % TFA, 49.9 % acetonitrile, 50 % water.
3. Third extraction buffer: 0.1 % TFA, 99.9 % acetonitrile.

2.6 Components for LC/MS

1. EASY-nLCII system (Thermo Fisher Scientific).
2. Q Exactive mass spectrometer (Thermo Fisher Scientific).
3. ReproSil-Pur C_{18}-AQ 3-μm resin (Dr. Maisch).
4. HPLC buffer.

3 Methods

3.1 Sample Preparation

1. Wash tissue culture cells with PBS, and add 150 µl cell lysis buffer to each 10 cm dish (adjust volume according to dish size). Data presented here is for HEK293 cells at 80 % confluency. Gently scrape off cells from the bottom of the dish, and transfer the sample into Eppendorf vials.
2. Vortex and sonicate samples vigorously to shear DNA and to reduce sample viscosity (*see* **Note 2**) and undergo repeated freeze–thaw cycles (*see* **Note 3**).
3. Heat samples to 95 °C for 5 min for the reduction of disulfide bonds and inactivation of protease activity.
4. Centrifuge samples at 10,000 ×*g* for 5 min at room temperature to remove any debris. Transfer supernatants to fresh 2 ml Eppendorf vials.
5. Alkylate sulfhydryl groups by adding iodoacetamide to a final concentration of 0.25 M and incubate for 20 min at room temperature and in the dark (*see* **Note 4**).
6. For depletion of SDS according to Wessel and Fluegge [11], transfer 200 µl of the protein solution to a fresh 2 ml Eppendorf tube. Fill up with LC-MS-grade water if your sample volume is smaller.
7. Add 800 µl methanol (LC-MS grade), vortex, and centrifuge briefly (10 s, 10,000 ×*g*).
8. Add 200 µl chloroform (under fume hood), vortex, and centrifuge briefly (10 s, 10,000 ×*g*).
9. Add 600 µl of water, vortex vigorously, and centrifuge for phase separation (1 min, 10,000 ×*g*).
10. Carefully remove and discard the upper phase with a pipette. Do not disturb the interphase which contains the proteins.
11. Add 600 µl methanol to the remaining lower phase and interphase, and vortex gently.
12. Centrifuge for 5 min at 10,000 ×*g*, carefully remove the supernatant, and briefly air-dry the pellet (*see* **Note 5**).

3.2 In-Solution Digestion

1. Resuspend pellet protein precipitation in 6 M urea/2 M thiourea and 10 mM HEPES (pH 8.0).
2. Perform protein quantification using the method of preference (e.g., Bradford, AmidoBlack, BCA). This step is optional—the protein amount can also be estimated based on the number of cells used.
3. For Lys-C digestion add 1 µg Lys-C solution per 50 µg sample protein and incubate for at least 3 h at room temperature.

4. Dilute samples with 50 mM ammonium bicarbonate pH 8.0 to a urea concentration below 2 M (*see* **Note 6**).
5. For trypsin digestion add 1 μg trypsin solution per 50 μg sample protein and incubate overnight at room temperature.
6. Stop the digestion by acidification of the sample to a pH below 2.5 with 10 μl of a 10 % TFA solution. The pH can be checked by spotting 0.1 μl on pH strips.

3.3 Desalting by Solid-Phase Extraction

All the following solvents or sample volumes were forced through the filter of the extraction cartridge by centrifugation at 2,000×*g* in each step.

1. Activate the 6 ml C_{18} Empore Disk Cartridge with 500 μl of MS-grade methanol.
2. Wash the cartridge with 1 ml of MS-grade water.
3. Load the acidified sample.
4. Wash with 1 ml of 5 % acetonitrile in water.
5. Wash with 1 ml MS-grade water, and check pH of the eluate using pH strips (*see* **Note 7**).
6. Elute peptides from the Empore Disk Cartridge with 300 μl of 80 % acetonitrile.
7. Optional: Quantify peptide yield spectrophotometrically (NanoDrop, Thermo), and aliquot as desired.
8. Dry samples in a SpeedVac concentrator.

3.4 Isoelectric Focusing of Peptides

1. Resuspend peptide sample pellets with 135 μl of peptide rehydration buffer and agitate for 30 min (Eppendorf thermomixer) at room temperature and 1,400 rpm.
2. Transfer the sample into the rehydration tray and sway from side to side to achieve complete coverage of the slot surface. Gently place the IPG strip into the tray with the gel side facing the sample (*see* **Note 8**).
3. To prevent evaporation of the rehydration solution and crystallization of urea, wrap the rehydration tray with Parafilm (*see* **Note 9**).
4. Rehydrate the IPG strip overnight. After complete rehydration (*see* **Note 10**) wrap the IPG strip with Parafilm as shown in Fig. 4.
5. Transfer the wrapped IPG strip into the focusing tray of the IEF 100 with the gel side facing up.
6. Place water-soaked filter pads onto the unsealed regions at both ends of the gel (*see* **Note 11**).
7. Mount the anode (+) over the filter pad at the acidic end of the gel and the cathode (−) over the filter pad at the alkaline end of the gel.
8. Start IEF with a 10-h gradient from 0 to 2,000 V (10 kVh) followed by holding a constant voltage at 2,000 V for 10 h (20 kVh) (*see* **Note 12**).

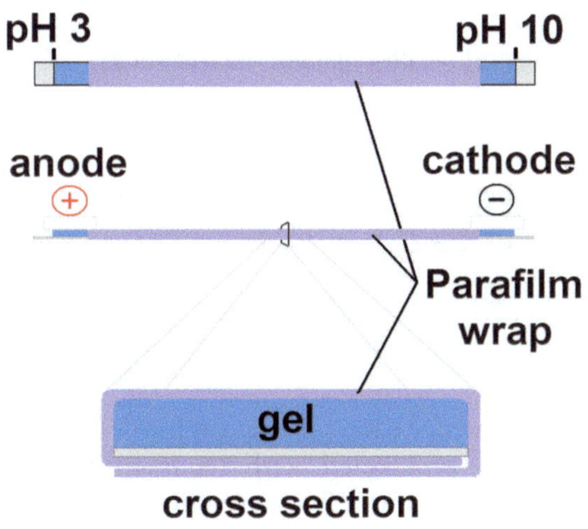

Fig. 4 Schematic diagram of the Parafilm wrapping to prevent the IPG strip from drying out during IEF, and the arrangement for isoelectric focusing

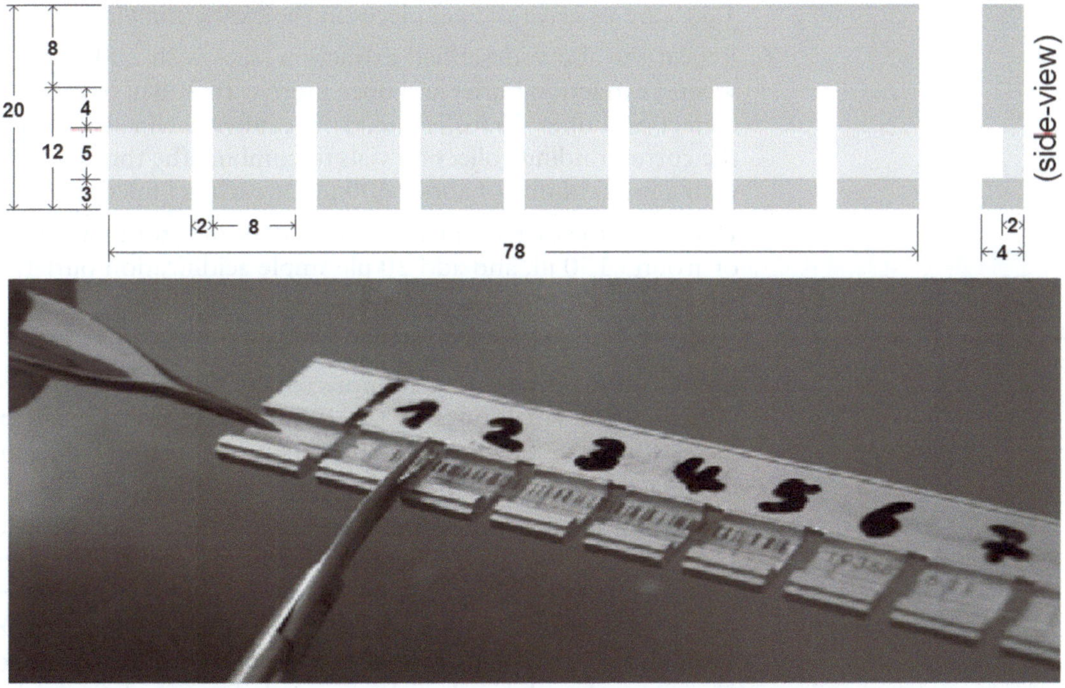

Fig. 5 The cutting comb. The use of the comb enables very fast cutting of the gel pieces and transfer into the extraction vials. This is very important to reduce the diffusion of the focused peptides

3.5 Cutting of IPG Strip and Peptide Extraction

1. After IEF has finished remove the Parafilm wrapping, place the strip on the cutting comb (Fig. 5), and cut the gel with scissors and forceps into six pieces (*see* **Note 13**). Cut the first and the last pieces to a length of 1.5 cm and the four middle pieces to a length of 1 cm (*see* **Note 14**). The theoretical pH regions of the pieces are shown in Table 1.

Table 1
Length, cutting position, and theoretical pH region of the fractions

Fraction	1	2	3	4	5	6
Length	1.5 cm	1 cm	1 cm	1 cm	1 cm	1.5 cm
Cut position	0–1.5 cm	1.5–2.5 cm	2.5–3.5 cm	3.5–4.5 cm	4.5–5.5 cm	5.5–7 cm
Theoretical pH	3–4.5	4.5–5.5	5.5–6.5	6.5–7.5	7.5–8.5	8.5–10

2. Transfer each gel section immediately into a 1.5 ml Eppendorf vial prelabeled with its fraction number and prefilled with 200 µl of the first extraction buffer (*see* **Note 15**).
3. First agitate the gel strip pieces in the first extraction buffer for 8 min on a thermomixer at 1,400 rpm and room temperature.
4. Sonicate the vials for 8 min in an ultrasonic water bath.
5. Transfer all first extracts to the corresponding collection vial. Take care to leave the gel pieces in the extraction vials.
6. Repeat the above-described extraction steps with 200 µl of the second extraction buffer and once more with 200 µl of the third extraction buffer. Transfer the extracts after each extraction to the corresponding collection vials to combine the three extracts with a final volume of about 600 µl for each gel fraction.
7. Dry down the extracts in a SpeedVac concentrator to a volume of around 150 µl, and add 20 µl sample acidification buffer.

3.6 StageTip Purification

StageTip purification is performed according to Rappsilber et al. [12]:

1. Prepare as many desalting columns (StageTips) as necessary by punching out small discs of C_{18} Empore Filter, and eject the discs (three per tip) into a P200 pipette tip. Ensure that the disc is securely wedged in the bottom of the tip.
2. Condition tips by adding 50 µl methanol to the Empore disc. Use this step to check whether the StageTips are leaky. For centrifugation (max rcf 2,000 × *g*) a standard tabletop centrifuge can be used. Punch a hole into the lid of 2 ml microfuge tube, and place the StageTips into them. Discard flow-through once it reaches the end of the tip, and always make sure that the StageTips are empty before adding the next solution.
3. Wash StageTip columns by adding 100 µl equilibration buffer by centrifugation.
4. Force the acidified peptide sample through the StageTip by centrifugation.
5. Wash the column with 120 µl equilibration buffer and washing buffer by centrifugation.

6. Elute the peptides from the C_{18} material using 50 µl elution buffer (80 % acetonitrile). Use a syringe to directly elute samples into the autosampler plate of your HPLC system.
7. Dry down the sample in a SpeedVac concentrator until around 5 µl are remaining. Fill up with buffer A of your HPLC system to the desired sample volume.

3.7 LC-MS

Peptide mixtures were separated by reversed-phase chromatography using the EASY-nLC on in-house-manufactured 20-cm fritless silica microcolumns with an inner diameter of 75 µm. Columns were packed with the ReproSil-Pur C_{18}-AQ 3-µm resin. Peptides were separated on an 8–60 % acetonitrile gradient (214 min) with 0.1 % formic acid at a flow rate of 200 nl/min. Eluting peptides were ionized online by electrospray ionization and transferred into a Q-Exactive mass spectrometer. One full scan (m/z = 300–1,700; R = 70,000; target value = 3×10^6; maximum injection time = 20 ms) was followed by ten MS/MS scans (charge >1; R = 17,500; target value = 1×10^6; isolation window = 2 m/z; maximum injection time = 60 ms). Dynamic exclusion for selected precursor ions was 30 s. This method corresponds to a previously described "fast" acquisition method [13].

4 Notes

1. In comparative experiments with IPG strips it is essential to use IPG strips which were originally from the same master gel (this is indicated by consecutive serial numbers on the IPG strips). We would not recommend the usage of IPG strips without any serial numbers or comparing IPG strips with large gaps between their serial numbers which could lead to a higher variance in comparative experiments [14].
2. Adding benzonase to cleave DNA can also help to reduce viscosity of the sample.
3. Freeze–thaw cycles are performed by repeated changes of the sample incubation between dry ice ethanol bath and 95 °C on a thermomixer.
4. Iodoacetamide is very light sensitive. If analysis of ubiquitinylation is desired, chloroacetamide should be used instead of iodoacetamide to prevent artifacts from modified lysines.
5. Do not overdry the pellet since this impedes solubilization. The pellet may be invisible.
6. Urea concentrations up to 2.5 M do not inhibit trypsin activity [15].
7. For IEF samples should have a neutral pH. For this reason it is important to check the pH before elution of the peptides with

acetonitrile. In case the pH is too low wash once more with MS-grade water.

8. Avoid trapping air bubbles between sample solution and gel. If air bubbles are present, try to remove them by lifting up the gel strip and pushing it down gently from one side with forceps. Repeat if necessary.
9. In previous experiments cover fluid (paraffin oil) was used to prevent strips from drying out. Unfortunately, even very small amounts of cover fluid interfere with MS analysis. Depletion of the cover fluid was not reproducible, and dipping in high-boiling-point petroleum ether baths in order to remove the paraffin oil [16] could lead to diffusion and loss of focused peptides. Therefore, we recommend performing rehydration and IEF without any cover fluid. To prevent the IPG strips from drying out we use Parafilm.
10. Complete rehydration can be checked by an even swelling of the gel indicated by the bromophenol blue dye in the rehydrated gel.
11. Filter pads should not be dripping wet! Excess water can be removed by blotting with paper wipes, if necessary.
12. Because of its lower cooling efficiency, it is not recommended to increase the voltage above 2,000 V when using Parafilm instead of cover fluid!
13. For 24 cm strips cut into 24 pieces (1 cm each): Using the cutting comb enables cutting and transfer into the extraction vials of 24 pieces within 3 min.
14. The 1.5 cm pieces were cut into halves because the extraction volume of 200 μl will cover up to 1 cm long gel pieces in a 1.5 ml Eppendorf vial.
15. All extraction vials were centrifuged briefly to ensure that all gel pieces are submerged in extraction buffer.

Acknowledgement

This work was supported by Deutsche Forschungsgemeinschaft (DFG) Grant SFB 958.

References

1. Lahm HW, Langen H (2000) Mass spectrometry: a tool for the identification of proteins separated by gels. Electrophoesis 21:2105–2114
2. Ishihama Y, Rappsilber J, Mann M (2006) Modular stop and go extraction tips with stacked disks for parallel and multidimensional peptide fractionation in proteomics. J Proteome Res 5:988–994
3. Cargile BJ, Bundy JL, Freeman TW, Stephenson JL Jr (2004) Gel based isoelectric

focusing of peptides and the utility of isoelectric point in protein identification. J Proteome Res 3:112–119

4. Krijgsveld J, Gauci S, Dormeyer W, Heck AJJ (2006) In-gel isoelectric focusing of peptides as a tool for improved protein identification. J Proteome Res 5:1721–1730
5. Slebos RJ, Brock JW, Winters NF, Stuart SR, Martinez MA, Li M, Chambers MC, Zimmerman LJ, Ham AJ, Tabb DL, Liebler DC (2008) Evaluation of strong cation exchange versus isoelectric focusing of peptides for multidimensional liquid chromatography-tandem mass spectrometry. J Proteome Res 7:5286–5294
6. Cargile BJ, Talley DL, Stephenson JL Jr (2004) Immobilized pH gradients as a first dimension in shotgun proteomics and analysis of the accuracy of pI predictability of peptides. Electrophoresis 25:936–945
7. Xie H, Bandhakavi S, Roe MR, Griffin TJ (2008) Preparative peptide isoelectric focusing as a tool for improving the identification of lysine-acetylated peptides from complex mixtures. J Proteome Res 5:2019–2026
8. Hörth P, Miller CA, Preckel T, Wenz C (2006) Efficient fractionation and improved protein identification by peptide OFFGEL electrophoresis. Mol Cell Proteomics 10:1968–1974
9. Hubner NC, Ren S, Mann M (2008) Peptide separation with immobilized pI strips is an attractive alternative to in-gel protein digestion for proteome analysis. Proteomics 8:4862–4872
10. Sanchez JC, Rouge V, Pisteur M, Ravier F, Tonella L, Moosmayer M, Wilkins MR, Hochstrasser D (1997) Improved and simplified in-gel sample application using reswelling of dry immobilized pH gradients. Electrophoresis 18:324–327
11. Wessel D, Fluegge UI (1984) A method for the quantitative recovery of protein in dilute solution in the presence of detergents and lipids. Anal Biochem 138:141–143
12. Rappsilber J, Mann M, Ishihama Y (2007) Protocol for micro-purification, enrichment, pre-fractionation and storage of peptides for proteomics using StageTips. Nat Protoc 2:1896–1906
13. Kelstrup CD, Young C, Lavallee R, Nielsen ML, Olsen JV (2012) Optimized fast and sensitive acquisition methods for shotgun proteomics on a quadrupole orbitrap mass spectrometer. J Proteome Res 11:3487–3497
14. Eravci M, Fuxius S, Broedel O, Weist S, Eravci S, Mansmann U, Schluter H, Tiemann J, Baumgartner A (2007) Improved comparative proteome analysis based on two-dimensional gel electrophoresis. Proteomics 7:513–523
15. Viswanatha T, Liener IE (1955) The inhibition of trypsin: III. Influence of urea. J Biol Chem 215:777–785
16. Vaezzadeh AR, Hernandez C, Vadas O, Deshusses JJ, Lescuyer P, Lisacek F, Hochstrasser DF (2008) PICarver: a software tool and strategy for peptides isoelectric focusing. J Proteome Res 7:4336–4345

Part III

In Vivo Stable Isotope Labeling

Chapter 6

SILAC Yeast: From Labeling to Comprehensive Proteome Quantification

Lyris M.F. de Godoy

Abstract

Mass spectrometry-based quantitative proteomics can identify and quantify thousands of proteins in complex mixtures, enabling characterization and comparison of cellular functional states in a proteome-wide scale. In this context, stable isotope labeling with amino acids in cell culture (SILAC) has emerged as a simple yet powerful approach, which has been applied to address different biological questions across a variety of systems, ranging from single cells to entire multicellular organisms. In this chapter, detailed instructions for SILAC labeling yeast are provided, including a series of quality checks for evaluating labeling efficiency and procedures for determining the optimal labeling parameters for a particular yeast strain. In addition, two different complete workflows for the comprehensive mass spectrometry-based SILAC quantification of close to the entire yeast proteome are described, which can be applied to assess any biological question of interest and, therefore, can be of broad use for the researchers in the field.

Key words SILAC, Yeast, Quantitative proteomics, Proteome quantification, Mass spectrometry, LC-MS/MS, Labeling, Isotope

1 Introduction

Mass spectrometry (MS)-based quantitative proteomics can identify and quantify thousands of proteins in complex mixtures, enabling in-depth characterization and comparison of cellular functional states in a proteome-wide scale. In a typical proteomics experiment, protein mixtures are digested to peptides, which are then separated by reversed-phase HPLC and analyzed online by tandem mass spectrometry (LC-MS/MS) [1]. For MS-based protein quantification a variety of methods are available, most of which (and the most accurate ones) relying on differential labeling of protein samples with stable isotopes prior to LC-MS/MS analysis. Among these, stable isotope labeling with amino acids in cell culture (SILAC) [2, 3] has emerged as a simple yet powerful approach to quantify differential changes in the proteome, which has been applied to address different biological questions across a variety of

Daniel Martins-de-Souza (ed.), *Shotgun Proteomics: Methods and Protocols*, Methods in Molecular Biology, vol. 1156, DOI 10.1007/978-1-4939-0685-7_6, © Springer Science+Business Media New York 2014

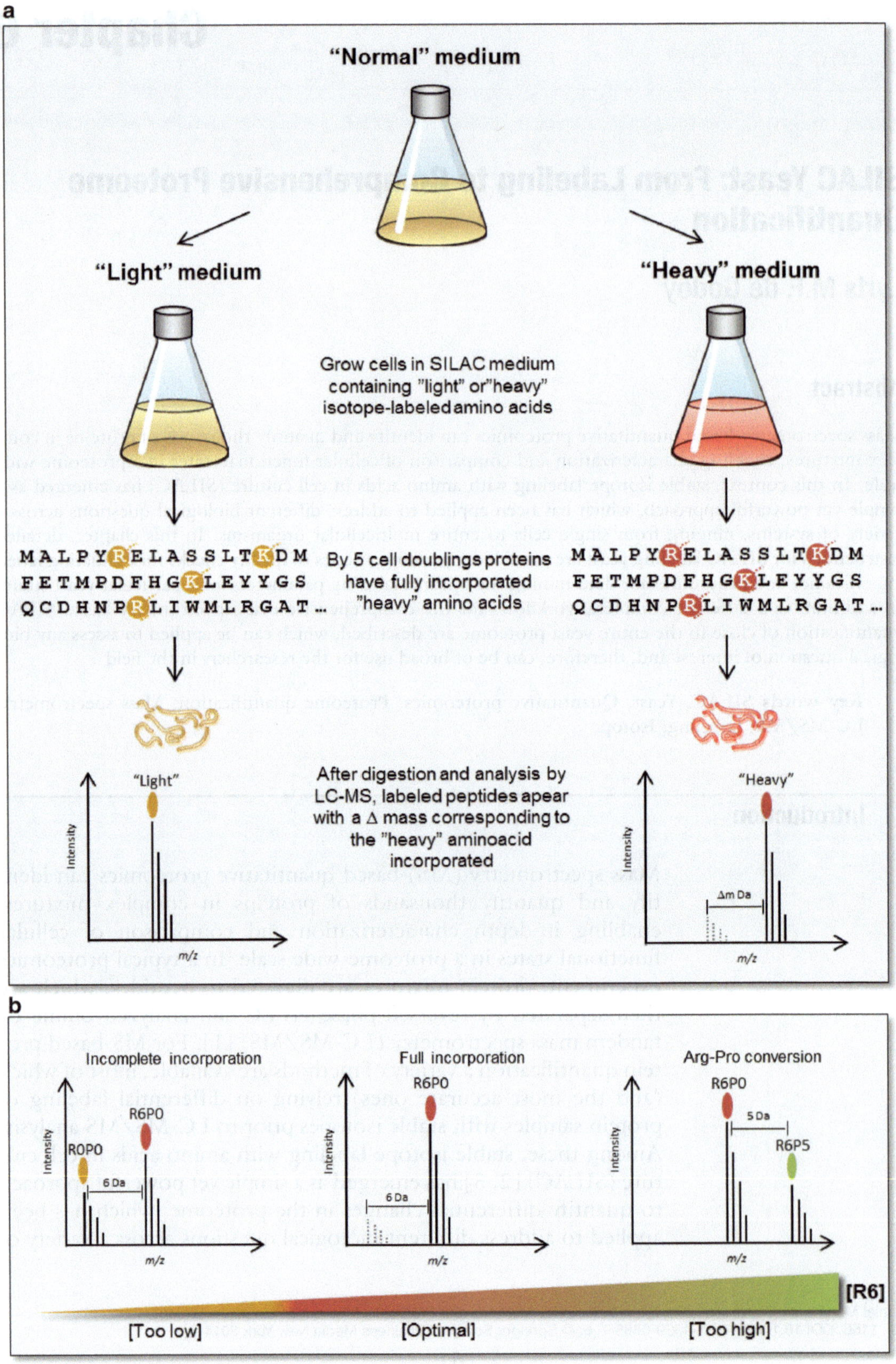
a
"Normal" medium
"Light" medium
"Heavy" medium
Grow cells in SILAC medium containing "light" or"heavy" isotope-labeled amino acids
M A L P Y R E L A S S L T K D M
F E T M P D F H G K L E Y Y G S
Q C D H N P R L I W M L R G A T ...
By 5 cell doublings proteins have fully incorporated "heavy" amino acids
M A L P Y R E L A S S L T K D M
F E T M P D F H G K L E Y Y G S
Q C D H N P R L I W M L R G A T ...
"Light"
Intensity
m/z
After digestion and analysis by LC-MS, labeled peptides apear with a Δ mass corresponding to the "heavy" aminoacid incorporated
"Heavy"
Intensity
Δm Da
m/z
b
Incomplete incorporation
R6P0
R0P0
6 Da
Intensity
m/z
Full incorporation
R6P0
6 Da
Intensity
m/z
Arg-Pro conversion
R6P0
5 Da
R6P5
Intensity
m/z
[R6]
[Too low]
[Optimal]
[Too high]

systems, ranging from single cells (e.g., mammalian and plant cell lines), yeast, bacteria, and protozoan to entire multicellular organisms (e.g., mice, fly, nematodes, and newt) [4].

For in vitro SILAC labeling, cells are grown in medium containing either "light" (normal) or "heavy" nonradioactive isotope-labeled (e.g., ^{13}C, ^{15}N, or deuterium) versions of amino acids, until the proteome is fully encoded, allowing peptides derived from each sample to be distinguished in subsequent analysis by a mass shift corresponding to the used isotope labels (Fig. 1a). In in vivo systems, labeling is achieved by feeding multicellular organisms with synthetic SILAC food or SILAC-labeled organisms, usually for multiple generations. As SILAC depends on protein synthesis for the incorporation of isotope-labeled amino acids into proteins, to efficiently label the proteome of a cell or an organism it is crucial that the amino acids used for labeling are obtained primarily (if not exclusively) by diet. Most of the higher organisms naturally present several essential amino acids, which can be explored for labeling. On the other hand, many microorganisms can synthesize most amino acids that are essential in higher eukaryotes and, therefore, might require especial manipulation prior to SILAC labeling (such as the deletion of enzymes involved in the biosynthesis of the amino acids used for labeling) to achieve full incorporation of heavy amino acids [5]. However, reliable SILAC quantification can still be performed for prototroph organisms as has been recently demonstrated for yeast [6].

In MS-based SILAC quantitative proteomics experiments, once the proteomes are labeled, cells (or protein extracts) representing the different experimental conditions are mixed in equal proportions and protein digests are analyzed by LC-MS/MS as a single sample, ensuring maximum reproducibility and accuracy for protein identification and relative quantification. SILAC quantification can be straightforwardly performed to compare two ("light" and "heavy") or three ("light", "medium heavy", and "heavy") samples (Fig. 2). In addition, SILAC can be expanded to a higher number of samples by using a common point (and label) as reference [7] or by spiking a common heavy-labeled sample, which will be used as an internal standard, into the different unlabeled samples to be quantified [8, 9].

Fig. 1 SILAC labeling. (**a**) Cells are grown in SILAC media containing either "light" or "heavy" versions of lysine and arginine, until the isotope-labeled amino acids have been fully incorporated into the proteomes. In MS spectra, "light" and "heavy" SILAC peptides can be distinguished by a mass shift corresponding to the isotope-labeled amino acid incorporated. (**b**) Too low concentrations of arginine in the labeling media can lead to incomplete incorporation (*left panel*). Optimal concentrations of arginine in labeling media should lead to full incorporation while keeping the degree of arginine-to-proline conversion not detectable or below 5 % (*middle panel*). Too high concentrations of arginine in labeling media can cause arginine-to-proline conversion, which leads to the introduction of heavy proline into the system and, in turn, results in a subpopulation of double-heavy peptides (containing both heavy arginine and heavy proline) (*right panel*)

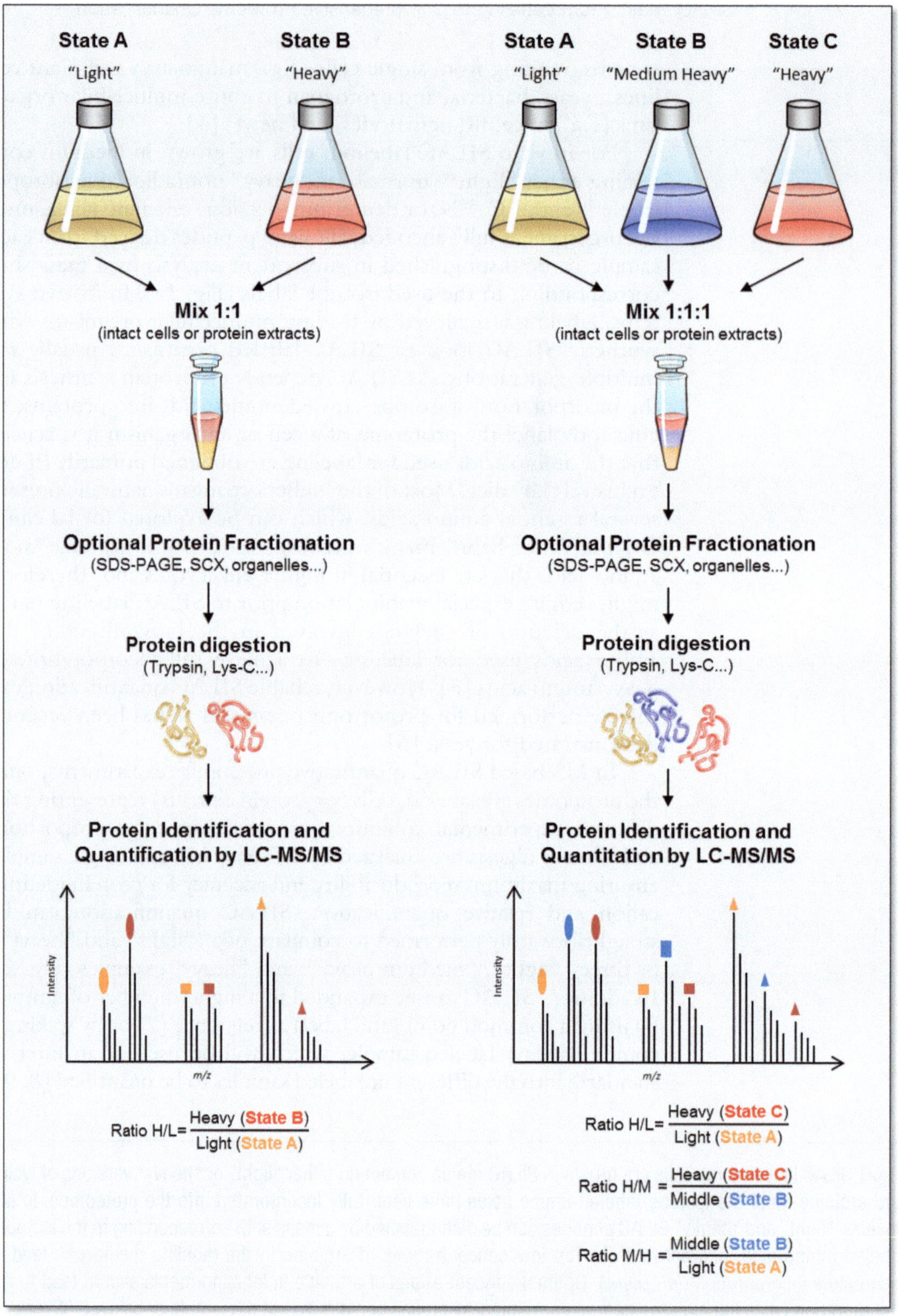

Fig. 2 Typical SILAC quantitative proteomics workflow. Cell populations are differentially labeled, mixed in equal proportions, digested to peptides, and analyzed by MS as a single sample for protein identification and quantification

The fact that a SILAC experiment differs from a conventional experiment only by introducing a labeling step previous to the subsequent experimental steps makes it versatile and compatible with several existing experimental workflows. Consequently, SILAC has been applied not only to expression proteomics experiments but also to study protein turnover [10–14] to characterize protein interactions [15–18] and posttranslational modification dynamics (e.g., methylation, phosphorylation, acetylation, ubiquitination, sumoylation) [7, 19–21] and to perform absolute protein quantification [22]. In this chapter, detailed procedures for SILAC labeling of yeast auxotroph and prototroph strains, optimization and evaluation of labeling efficiency, and sample preparation and analysis by LC-MS/MS are described. In addition, two experimental workflows to comprehensive mass spectrometry-based SILAC quantification of the yeast proteome are provided, which can be applied to assess any biological question of interest.

2 Materials

Prepare all solutions using ultrapure water with 18 MΩ cm resistance at 25 °C (MilliQ, Millipore), unless specified otherwise. All reagents should be of analytical grade or better. To prevent contamination of samples with polymers, solvents and buffers should be kept in high-quality glass bottles (Schott).

2.1 Cell Culture and SILAC Labeling

1. Yeast extract–peptone–dextrose (YPD) broth.
2. Yeast nitrogen base (YNB) w/o amino acids.
3. d(+)-Glucose monohydrate.
4. Uracil.
5. Adenine hemisulfate salt.
6. *Standard amino acids*: L-methionine, L-tryptophan, L-histidine hydrochloride, L-phenylalanine, L-leucine, L-tyrosine.
7. *"Light" SILAC amino acids*: L-arginine (Arg0, R0) hydrochloride, L-lysine (Lys0, K0).
8. *"Heavy" SILAC amino acids*: L-arginine-$^{13}C_6$, $^{15}N_4$ hydrochloride (Arg10, R10), L-lysine-$^{13}C_6$, $^{15}N_2$ hydrochloride (Lys8, K8).
9. *"Medium heavy" SILAC amino acids*: L-arginine-$^{13}C_6$ hydrochloride (Arg6, R6), L-lysine-4,4,5,5-d4 hydrochloride (Lys4, K4).
10. Lysine auxotroph yeast strain: Y15969 (BY4742, MAT α, his3Δ1, leu2Δ0, lys2Δ0, ura3Δ0, YIR034c::kanMX4) (EuroScarf).
11. Erlenmeyer culture flasks.
12. Incubator with shaker.
13. Spectrophotometer.

2.2 Cell Lysis by Cryogenic Grinding and "Classic" In-Solution Protein Digestion

1. *Cryogenic grinding lysis buffer*: 150 mM K Acetate, 2 mM Mg acetate, 1× protease inhibitor cocktail (Roche), 20 mM HEPES, pH 7.4. Prepare immediately before use.
2. Liquid N2.
3. *Denaturation buffer*: 6 M urea, 2 M thiourea in 10 mM HEPES (pH 8.0). Prepare fresh.
4. *Reduction buffer*: 100 mM Dithiothreitol (DTT) in digestion buffer (pH 8.0). Store aliquots at −20 °C.
5. *Alkylation buffer*: 550 mM Iodoacetamide (IAA) in digestion buffer (pH 8.0). Store aliquots at −20 °C, and keep the solution protected from light at all times.
6. *Digestion buffer*: 50 mM ammonium bicarbonate in water (pH 8.0). Store at room temperature.
7. Sequencing (or MS)-grade endoproteinase Lys-C.
8. Sequencing (or MS)-grade trypsin.
9. Bradford (or BCA) reagent for protein quantification.
10. Trifluoroacetic acid (TFA).
11. Milling device (MM301 Ball Mill, Retsch).

2.3 Cell Lysis in the Presence of SDS and FASP In-Solution Protein Digestion

1. Sodium dodecyl sulfate (SDS).
2. *SDS lysis buffer*: 5 % SDS, 100 mM DTT, 100 mM Tris in water (pH 7.6).
3. *UA buffer*: 8 M urea in 0.1 M Tris–HCl (pH 8.5).
4. *UB buffer*: 8 M urea in 0.1 M Tris–HCl (pH 8.0).
5. IAA solution: 5.5 mM Iodoacetamide in UA.
6. Sequencing (or MS)-grade endoproteinase Lys-C.
7. Sequencing (or MS)-grade modified trypsin (resistant to autolysis).
8. 0.5 M NaCl solution in water.
9. *Digestion buffer*: 50 mM ammonium bicarbonate in water (pH 8.0). Store at room temperature.
10. TFA.
11. 30 k Ultracentrifugal filter units.

2.4 Peptide Fractionation by IEF

1. 100 % Acetone.
2. 24 cm Immobiline DryStrip, pH 3–10 (GE Healthcare).
3. IPG Buffer, pH 3–10 (GE Healthcare).
4. 3100 OFFGEL Fractionator (Agilent Technologies).
5. 3100 OFFGEL Fractionator Kit (frames and cover seals, Agilent Technologies).
6. Immobiline DryStrip, pH 3–10, 24 cm (GE Healthcare).

7. IPG Buffer, pH 3–10 (GE Healthcare).
8. Glycerol 87 %.
9. IPG strip rehydration solution: 5 % Glycerol and IPG buffer pH 3–10 diluted 1:50 in water. Prepare immediately before use.
10. Paper electrode (GE Healthcare), cut to fit in tray lanes.
11. Mineral oil (GE Healthcare).
12. Agilent 3100 OFFGEL Fractionator (Agilent Technologies).

2.5 Purification of Peptides with RP-C18 StageTip Columns

1. C18 3 M Empore™ SPE Extraction Disks (Sigma-Supelco).
2. Methanol (MeOH).
3. Acetic acid (AcOH).
4. Acetonitrile (ACN).
5. TFA.
6. *Solution A*: 0.5 % AcOH in water. Store at room temperature.
7. *Solution B*: 0.5 % AcOH, 80 % ACN in water. Store at room temperature.
8. *Sample solution*: 1 % TFA, 5 % ACN in water. Store at room temperature.
9. Centrifugal vacuum concentrator.

2.6 LC-MS/MS

1. Nanoflow high-performance liquid chromatographer (HPLC).
2. EASY-nLC 1000 (Thermo Fisher Scientific, Odense, Denmark).
3. 15 cm fritless silica microcolumns (75 μm inner diameter), in-house manufactured.
4. 50 cm fritless silica microcolumns (75 μm inner diameter), in-house manufactured.
5. ReproSil-Pur C18-AQ 3 μm resin (Dr. Maisch GmbH).
6. 1.9 μm C18 resin (Dr. Maisch GmbH).
7. LC-MS-grade acetic acid (AcOH).
8. LC-MS-grade acetonitrile (ACN).
9. LC-MS-grade water.
10. Mobile phases for HPLC: (A) 0.5 % acetic acid in water, (B) 0.5 % acetic acid, 80 % acetonitrile (v/v) in water.
11. Linear trap quadrupole Orbitrap hybrid mass spectrometer (e.g., LTQ-Orbitrap Classic, XL, or Velos) or Q Exactive mass spectrometer (all from Thermo Scientific).
12. Nanoelectrospray ion source.
13. MaxQuant software (freeware) and desktop computer with minimal configurations required to run the software (download and instructions available at http://maxquant.org/).

3 Methods

3.1 Working with Auxotroph vs. Prototroph Strains

SILAC labeling depends on incorporation of exogenous isotope-labeled amino acids into proteins. Therefore, when planning yeast SILAC experiments for a particular strain of interest the first step is to check its genetic background regarding amino acid auxotrophy to be aware of the possibilities and limitations of that strain for metabolic labeling. Before starting a yeast SILAC experiment, review the following points carefully:

3.1.1 SILAC for Auxotroph Strains

Since yeast auxotroph strains rely entirely on exogenously provided amino acids, they allow for maximal incorporation levels of isotope-labeled amino acids and might be the best choice for most SILAC experiments.

1. In principle, any essential amino acid can be used for metabolic labeling, and a variety of them (e.g., leucine, valine, and tyrosine) have been successfully applied in SILAC experiments. However, arginine and lysine present several advantages and are by far the most commonly used in SILAC quantitative proteomics (unless the experiment requires otherwise).
2. If your strain is auxotroph for lysine, it is already prone to be SILAC labeled (single labeling, Lys-only). In this chapter, a protocol for SILAC labeling of *Saccharomyces cerevisiae* using a lysine auxotroph strain will be presented as an example.
3. To be able of performing double labeling with lysine and arginine, however, it is of course necessary to work with a double auxotroph for those amino acids.
4. If your strain of interest is not an auxotroph for lysine and/or arginine, auxotroph variants for your genetic background can be generated by standard methods.
5. In SILAC experiments using yeast auxotroph strains, incorporation levels higher than 99 % can easily be achieved.

3.1.2 nSILAC for Prototroph Strains

Although yeast prototroph strains are capable of synthesizing amino acids, they can still be SILAC labeled (at least with heavy lysine), using the recently described native SILAC (nSILAC) approach [6]. However, prototroph strains present slightly lower incorporation levels compared to those achieved for auxotroph strains.

1. In yeast prototroph cells, the presence of lysine in the medium causes downregulation of lysine metabolism enzymes, via different feedback mechanisms, favoring the uptake of exogenous lysine and allowing incorporation of heavy amino acids into the proteome.
2. However, although prototroph strains preferentially use amino acids available from the medium, there is some residual endogenous synthesis left, precluding complete labeling.

Table 1
Most commonly used isotope labels for SILAC experiments

Amino acid		Isotope label	Short names	Mass shift
Standard "light"	L-arginine L-lysine	– –	Arg0, R0 Lys0, K0	– –
SILAC "medium heavy"	L-arginine-HCL L-lysine-2HCL	$^{13}C_6$ 4,4,5,5-D_4	Arg6, R6 Lys4, K4	+6 Da +4 Da
SILAC "heavy"	L-arginine-HCL L-lysine-2HCL	$^{13}C_6{}^{15}N_4$ $^{13}C_6{}^{15}N_2$	Arg10, R10 Lys8, K8	+10 Da +8 Da

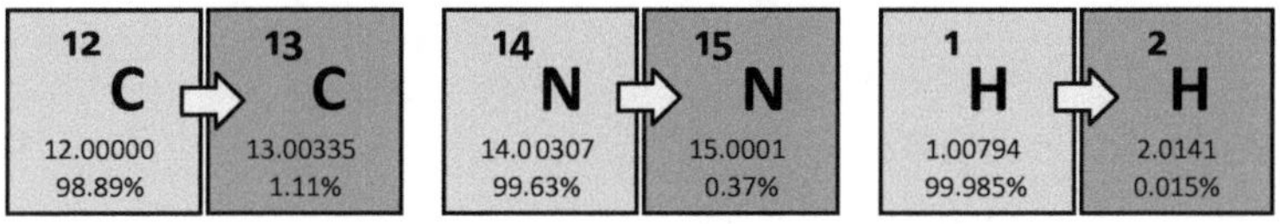

3. In nSILAC experiments using yeast prototroph strains incorporation levels around 96 % can be achieved, which is still enough for reliable quantification.

3.2 SILAC Labeling Yeast

3.2.1 Choosing the SILAC "Labeling Code"

Table 1 shows the heavy isotopes and isotope-labeled forms of lysine and arginine most commonly used in SILAC experiments.

In this scenario, proteins can be SILAC labeled only with heavy lysine (single labeling, Lys-only) or with both heavy arginine and heavy lysine (double labeling). Both labeling options lead to successful proteome-wide quantification, as long as the appropriate enzyme is used to generate peptides for mass spectrometry analysis (*see* **Note 1**). When planning a SILAC experiment, choose one of the following "labeling codes" (*see* **Note 2**):

1. For 2-plex SILAC experiments (comparing two different populations of cells), choose between:
   (a) R0K0 ("light") and R10K8 ("heavy").
   (b) R0K0 ("light") and R6K6 ("medium heavy").
   (c) R0K0 ("light") and R0K8 ("heavy").
   (d) R0K0 ("light") and R0K6 ("medium heavy").
2. For 3-plex SILAC experiments (comparing three different populations of cells), choose between:
   (a) R0K0 ("light") and R6K4 ("medium heavy") and R10K8 ("heavy").
   (b) R0K0 ("light") and R0K4 ("medium heavy") and R0K8 ("heavy").

3.2.2 Assembling Yeast SILAC Media

The yeast SILAC medium, Table 2, is prepared based on a yeast minimal dropout medium (lacking lysine and arginine), where different versions of isotope-labeled amino acids are added according to the desired labeling code.

Table 2
Composition of yeast SILAC medium

Component	Amount/liter
Yeast nitrogen base (YNB) without amino acids + ammonium sulfate	6.7 g
Dextrose	20 g
Uracil	20 mg
L-Adenine sulfate	200 mg
L-Tyrosine	100 mg
L-Histidine	10 mg
L-Leucine	60 mg
L-Methionine	10 mg
L-Phenylalanine	60 mg
L-Tryptophan	40 mg
SILAC amino acids	
L-Lysine (Lys0, Lys4, or Lys8)	30 mg[a]
L-Arginine (Arg0, Arg6, or Arg10)	5 mg[a]

[a]These concentrations of L-lysine and L-arginine should be considered only as a guideline and should be tested and optimized when performing a labeling experiment for the first time (*see* Subheading 3.3), as conditions between laboratories and for a particular strain may vary

To prevent contamination of the cell cultures, all steps involving preparation of media and other reagents as well as the manipulation of cells should be performed in sterile conditions, under a laminar flow hood.

1. Assembly of YNB "dropout" medium:
   (a) *YNB solution (100× stock)*: 67 g/L YNB, 0.55 g/L adenine, and 0.55 g/L L-tyrosine in water. Autoclave for 15 min at 121 °C, and store at 4 °C.
   (b) *20 % glucose (10× stock)*: 200 g/L glucose in water. Autoclave for 15 min at 121 °C, and store at 4 °C.
   (c) *Uracil (50× stock)*: 2.25 g/L uracil in water. Filter through a 0.2 μm filter, and store at 25 °C.
   (d) *Amino acid mix (100× stock solution)*: 1 g/L L-histidine hydrochloride, 6 g/L L-leucine, 1 g/L L-methionine, 6 g/L L-phenylalanine, and 4 g/L L-tryptophan in water. Filter through a 0.2 μm filter, and store at 4 °C. Alternatively, prepare a mix of powder amino acids and store at room temperature, protected from humidity.

(e) For 100 mL of YNB "dropout" medium, combine 10 mL 10× YNB solution, 10 mL 20 % glucose, 2 mL 50× uracil, 1 mL 100× amino acid mix, and 77 mL sterile water. Store at 4 °C.

2. *Assembly of yeast SILAC media:*

(a) *"Light" standard amino acid stocks (100 mg/mL)*: Prepare a separate stock solution for each standard amino acid by dissolving L-lysine and L-arginine hydrochloride in water separately. Filter each separate stock through a 0.2 μm filter, and store at 4 °C.

(b) *SILAC "heavy" and "medium heavy" amino acid stocks (100 mg/mL)*: Prepare a separate stock solution for each SILAC amino acid by dissolving each isotope form of L-lysine (Lys0, Lys4, or Lys8) and L-arginine hydrochloride (Arg0, Arg6, or Arg10) in sterile water. Store at 4 °C. To avoid losses, do not filter.

(c) Take a volume of YNB dropout medium sufficient to prepare the desired volume of all distinct labeling medium and split equal volumes into separate bottles of appropriate size under sterile conditions.

6. To produce distinct labeling media, add the proper amount of L-lysine (Lys0, Lys4, or Lys8) and L-arginine hydrochloride (Arg0, Arg6, or Arg10) stock solutions into each bottle of yeast dropout medium in a pairwise fashion to the desired final concentrations. Complete SILAC media can be stored at 4 °C for up to 2 months.

3.2.3 SILAC Labeling and Harvesting of Cells

Yeast cells are grown in Erlenmeyer flasks on a shaker at 30 °C with 180 rpm. If the biological experiment requires special conditions (e.g., higher/lower temperature incubations) adapt the protocol accordingly.

1. Prepare one small yeast pre-culture by growing cells in "light" SILAC medium or YPD liquid medium until they reach stationary phase (overnight culture).
2. Inoculate each "heavy" SILAC media and a fresh culture of "light" SILAC medium with cells from the pre-culture (1:10,000 dilution).
3. Grow cells until they reach mid-log phase (OD_{600} = 0.7) (*see* **Note 3**).
4. Harvest cells by centrifugation at 4,000 × *g* for 5 min at 4 °C, and discard the supernatant.
5. Wash cells two times with cold water by centrifugation for 5 min at 4,000 × *g* at 4 °C. Discard the supernatant each time. Keep samples on ice.

3.3 Optimizing SILAC Labeling Conditions

In SILAC experiments, the concentration of heavy amino acids in the media can affect the growth rate of the cell culture, efficiency of heavy amino acid incorporation, and, in case heavy arginine is used for labeling, degree of arginine-to-proline conversion (*see* **Note 4**, Fig. 1b). Another aspect to be considered is the high cost of the isotope-labeled amino acids, which can impact the viability of large-scale experiments. Therefore, before engaging a large-scale SILAC quantitative proteomics experiment, it is important to establish the optimal concentration of lysine and arginine (*see* **Note 5**) in SILAC media for successful labeling (*see* **Note 6**). This step is crucial and should always be performed when working with a particular strain for the first time before running a full SILAC experiment. Once the optimal labeling conditions for a specific system have been determined, however, they can be used for further experiments and should lead to reproducible results.

The optimal SILAC labeling conditions for a system can be determined by performing an initial small-scale titration experiment (*see* **Note 7**) and a series of quality checks. For a thorough evaluation, we suggest analyzing both "heavy" samples and 1:1 mixtures of "heavy" and "light" labeled samples under different labeling conditions, as described below:

3.3.1 Performing a Small-Scale Amino Acid Titration Experiment

1. Choose at least three different concentrations of each labeling amino acid to be tested in the titration experiment (for example, 100, 50, and 25 % of the default concentration used in regular media).
2. Following the chosen SILAC "labeling code" (*see* Subheading 3.2.1) and the desired test concentrations, prepare each distinct "heavy" and "light" SILAC media, as described in Subheading 3.2.2 (*see* **Note 8**).
3. Inoculate each "heavy" SILAC media and a fresh culture of "light" SILAC medium with cells from the pre-culture (1:10,000 dilution).
4. Grow and harvest cells as described in Subheading 3.2.3 (*see* **Note 9**).

3.3.2 Preparing and Analyzing Test Samples

1. Lyse cells by cryogenic grinding (*see* Subheading 3.4.1) or disruption in the presence of SDS (*see* Subheading 3.4.2). Alternatively, any disruption method (e.g., glass beads) and buffer routinely used for yeast can be used for this step, as long as detergents are avoided or properly removed before mass spectrometric analysis.
2. Estimate protein concentration of protein extracts by Bradford procedure or another method of choice (e.g., BCA).
3. *"Heavy-only" samples*: From each lysate of heavy labeled cells, take out a volume corresponding to 10–50 μg of total protein to proceed with the next steps.

4. *"1:1 mix" samples:* From each lysate of heavy labeled cells, take 10–50 μg of total protein and combine it with the same amount of protein from "light" cells. Alternatively, equal OD units of heavy labeled and light cells can be mixed before cell lysis.
5. Store the remaining lysates at −20 °C for future use or in case the test needs to be repeated.
6. Digest proteins as described in Subheading 3.5.1 or 3.5.2 if lysing cells by cryogenic grinding or SDS disruption, respectively. Be sure to use the appropriate enzyme compatible with the labeling amino acids (*see* **Note 1**).
7. Desalt and concentrate samples as described in Subheading 3.7.
8. Analyze samples by LC-MS/MS as described in Subheading 3.8.1 or 3.8.2.
9. Identify and quantify protein as described in Subheading 3.9.

3.3.3 Determining the Optimal Labeling Parameters

1. Perform the quality checks to assess labeling efficiency (*see* Subheading 3.10.1). When using heavy arginine for labeling, also check for the occurrence of arginine-to-proline conversion (*see* Subheading 3.10.2).
2. Select the lowest lysine concentration leading close to full incorporation without impairing cellular growth.
3. For double-labeling experiments, select the arginine concentration which provides close to full incorporation while still keeping the degree of Pro conversion at an appropriate low level and without impairing cellular growth (*see* **Note 10**).

3.4 Obtaining Protein Extracts

3.4.1 Cell Lysis and Protein Extraction by Cryogenic Grinding

Yeast cell lysis by cryogenic grinding is performed in a mixer mill according to the instructions of the manufacturer, with some modifications.

1. Resuspend cells in lysis buffer for cryogenic grinding at a density of 50 OD/mL (*see* **Note 11**).
2. Use a 1 mL pipette to mix and aspirate yeast slurry.
3. Drip cell slurry slowly into a liquid N_2-filled 50 mL conical tube, so that it freezes in small spherical pellets.
4. If storing pellets for future use, place liquid N_2-filled tubes upright in a −80 °C freezer to allow liquid N_2 to evaporate. Use preferably vented cap tubes, or pierce caps to allow venting. Frozen pellets can be stored at −80° until use. If processing samples immediately, transfer tubes to a container filled with dry ice, allow liquid N_2 to evaporate, and proceed. Keep frozen yeast pellets on dry ice at all times, and pre-chill all equipment in liquid nitrogen to prevent the cells from thawing.
5. Place the frozen yeast pellets and the grinding ball into the chamber, and screw it shut tightly. Return the sealed chamber to liquid N_2 until it is fully chilled (boiling stops). Remove the

chamber from liquid N_2, loosen it one-quarter turn, and mount it on the milling device.

6. Grind the sample using three cycles of 3-min milling at 10 Hz, intercalated by immersion in liquid N_2 (i.e., between each cycle, remove the chamber milling device, retighten the seal, and chill it completely in liquid nitrogen).
7. After the last grinding cycle, remove the chamber from the milling device, retighten the seal, and chill it completely in liquid nitrogen.
8. Using a pre-chilled metal scoop of proper size, transfer frozen cell powder from the chamber into a fresh 50 mL conical tube filled with liquid N_2 (use preferably vented cap tubes, or pierce caps to allow venting).
9. Place tubes on ice (4 °C) to allow liquid N_2 to evaporate and cell powder to thaw. Keep samples on ice for now on, until stated otherwise.
10. Transfer fully thawed lysates to 1.5 mL microfuge tubes pre-chilled on ice and centrifuge at 20,000 × *g* for 10 min at 4 °C to pellet cellular debris.
11. Recover the supernatant, avoiding both the pellet and the lipid layer at the top of the tube. Transfer clarified supernatants to fresh pre-chilled 1.5 mL microfuge tubes on ice.
12. Estimate protein concentration of protein extracts by Bradford procedure or another method of choice (e.g., BCA).
13. Frozen lysates can be stored at −80 °C until use.

3.4.2 Cell Lysis and Protein Extraction in the Presence of SDS

1. Resuspend cell pellets in SDS lysis buffer (1:10 sample-to-buffer ratio) (*see* **Note 12**).
2. Incubate samples for 3–5 min at 95 °C.
3. Sonicate lysates to shear DNA, and reduce sample viscosity.
4. Clarify the lysate by centrifugation at 16,000 × *g* for 5 min at room temperature.
5. Transfer clarified lysates to 1.5 mL microfuge tubes pre-chilled on ice.
6. Estimate protein concentration of protein extracts by Bradford procedure or another method of choice (e.g., BCA).
7. Frozen lysates can be stored at −80 °C until use.

3.5 Preparing Peptide Digests

3.5.1 "Classic" In-Solution Protein Digestion

In this procedure all steps are done at room temperature to reduce unwanted derivatization of amino acid side chains by the denaturant agents. So, never heat your sample!

1. Solubilize the proteins in urea/thiourea denaturation buffer (*see* **Note 13**).

2. Add DTT to a final concentration of 1 mM, vortex, and incubate protein samples for 30 min at room temperature to promote reduction of cysteine disulfide bonds.
3. Add IAA to a final concentration of 5.5 mM, vortex, and incubate protein samples in the dark for 20 min at room temperature to promote alkylation of cysteines.
4. Add endoproteinase Lys-C at an enzyme:substrate ratio of 1:50, and incubate for 12 h (or overnight) at room temperature (*see* **Note 14**). When SILAC labeling only with heavy lysine, DO NOT perform trypsin digestion (**step 5**) and go straight to **step 6** of this section.
5. *Only when double labeling proteins with heavy lysine and heavy arginine:* Dilute samples fourfold with digestion buffer (or enough to reduce the final concentration of urea/thiourea below 2 M). Add trypsin at an enzyme:substrate ratio of 1:50, and incubate for 12 h (or overnight) at room temperature.
6. Stop the digestion by acidifying the sample to pH < 2.5 with TFA (usually to a final concentration between 0.5 and 1 %). To facilitate pipetting and reduce precipitation, a 10 % TFA solution in LC-MS-grade water can be used. Acidified peptide digests can be stored at −20 °C until use.
7. Proceed with peptide desalting and concentration (*see* Subheading 3.7) and analysis by LC-MS/MS (*see* Subheading 3.8).

3.5.2 FASP In-Solution Digestion

FASP in-solution digestion is performed as previously described [23], with some modifications. All centrifugation steps are performed for 15 min (or until all the liquid has passed through the filter units), unless stated otherwise.

1. Dilute samples 1:10 with UA buffer to a final concentration of 0.5 % SDS.
2. Transfer diluted sample into a filter unit and centrifuge at 14,000 × *g* at room temperature.
3. Wash sample twice by adding 200 μl of UA to the filter unit and centrifuging at 14,000 × *g* at room temperature. Discard the flow-through from the collection tube.
4. Add 200 μl of 5.5 mM IAA, and mix in a thermomixer at 600 rpm for 1 min, at room temperature, protected from light. Incubate without shaking for a further 20 min at 25 °C, protected from light.
5. Centrifuge samples for at 14,000 × *g*, at room temperature, to remove the liquid.
6. Wash samples four times by adding 200 μl of UB buffer to the filter unit and centrifuging at 14,000 × *g* at room temperature. Discard the flow-through from the collection tube.

7. Add 60 μl of UB containing Lys-C (at an enzyme:substrate ratio of 1:50) to the spin column, and mix in a thermomixer at 600 rpm for 1 min at room temperature. Incubate the cartridges in a wet chamber overnight in a wet chamber at 37 °C (or room temperature). When SILAC labeling only with heavy lysine, DO NOT perform trypsin digestion (**step 8**) and go straight to **step 9** of this section.
8. *When double labeling proteins with heavy lysine and heavy arginine:* Add 40–100 μl of digestion buffer, and spin as before. Repeat this step twice. Add 120 μL digestion buffer containing trypsin (enzyme-to-protein ration 1:50), and mix at 600 rpm in thermomixer for 1 min. Incubate the cartridges overnight in a wet chamber at 37 °C (room temperature).
9. Remove old collection tubes from column and replace with fresh ones.
10. Recover peptides by centrifuging columns at 14,000 × *g* and collecting the flow-through. Place the cartridges into the collection tube, and elute the remaining peptides by adding 40–200 μl of 0.5 NaCl and centrifuging at 14,000 × *g*.
11. Stop the digestion by acidifying the sample to pH < 2.5 with TFA (usually to a final concentration between 0.5 and 1 %). To facilitate pipetting and reduce precipitation, a 10 % TFA solution in LC-MS-grade water can be used.
12. Centrifuge for 5 min at 17,000 × *g* to clear the samples from precipitates, recover, and transfer the supernatant to a fresh tube. Acidified peptide digests can be stored at –20 °C until use.

3.6 Peptide Fractionation by IEF

This section follows the Agilent 3100 OFFGEL Fractionator Kit Quick Start Guide (Fifth Edition, Agilent Technologies) with some modifications. The system was set up according to the manual of the High Res Kit, pH 3–10 (Agilent Technologies), with some modifications [24].

1. Prepare peptide IPG strip rehydration solution.
2. Prepare samples for fractionation at least in triplicate. For each replicate, dilute 75 μg (*see* **Note 15**) of peptide digests separately in rehydration solution to a final volume of 3,600 μL (for 24 cm strips, with a 24-well frames).
3. Place tray with fixed electrode side on the left and tray handle on the right.
4. For each gel strip, remove the protective plastic film in length from the strip and place it in the tray with the gel side up and the low pH extremity (anode) on the left. Pull the strip as far left as possible until it touches the left edge of the tray.
5. Keeping the right side of the frame slightly elevated, place the left side of the frame against the mechanical stop (about 2 cm,

or 1 inch, from the end of the tray) and lower the right side down on the gel, gently pressing until it snaps in.

6. Rehydrate the gel by pipetting 20 μL/well of a solution containing 5 % glycerol and ampholytes diluted 1:50, being careful not to touch the gel.
7. Gently tap the tray onto the desk to ensure that the IPG strip rehydration solution reaches the gel. Visually inspect to determine that the rehydration solution has wetted the gel surfaces completely. For better performance, be sure to also rehydrate both gel extremities. Wait for 15 min for the gels to swell, and then absorb with a pipette tip the excess of solution at the strip extremities.
8. Using tweezers, wet two paper electrode pads (for each strip) with rehydration solution or ultrapure water and place one wetted electrode pad on each protruding end of the strips, making sure that there is no gap between the pads and the frames.
9. Pipette 150 μL of sample in each of the 24 wells, being careful not to touch the gel. Place a cover seal over each frame, and press it down gently on each well to secure proper fit, being careful not to move the frame in the tray.
10. Place the tray on the OFFGEL platform. When running only one or a few IPG strips in parallel, position them in the middle of the tray.
11. Slowly pipette mineral oil onto each of the gel strip ends. For 24-well frames, pipette approximately 200 and 400 μL mineral oil onto the anode end (fixed electrode) and the cathode side (movable electrode) of the IPG strip, respectively. Wait for 1 min for the oil to slip along the frame, and then reapply an additional 200 μL mineral oil to each end of the strip. After 3 min add an additional 200 μL cover fluid onto the anode end (fixed electrode) of the strip. Pipette 2 mL mineral oil in all unused tray lanes.
12. Assemble the fixed electrode (anode) by placing the two tabs on the electrode into the slots on the left side of the tray. Rotate the fixed electrode down into position over the electrode pad and push down until the fixed electrode clicks into place. Insert the moveable electrode (cathode) and push down until it clicks in place. Electrodes must be in contact with paper electrode pads. Reposition the electrode pads if necessary.
13. Close the lid, set run temperature to 20 °C, and start the appropriate fractionation program. In this example, set up focusing conditions for 50 kVh at maximum current of 50 μA, maximum voltage of 8,000 V, and maximum power of 200 mW. Average run time should be around 30 h.
14. At the end of the fractionation (flashing blue light), stop the apparatus, and carefully remove electrodes and frame cover seals without disassembling the frame.

15. Recover each fraction by aspirating the liquid from each well and transferring into an identified 1.5 mL microfuge tube. Be careful not to touch the gel.
16. Acidify each peptide fraction by adding 3 % acetonitrile, 1 % TFA, and 0.5 % acetic acid, and proceed with peptide desalting and concentration as described in Subheading 3.7.

3.7 Desalting and Concentration of Peptides

Before analysis by LC/MS/MS, it is crucial to remove salts and other contaminants from your peptide digests that can interfere with chromatography or mass spectrometry signals. There are several ways to achieve that depending on your specific setup. Here, we describe the method for desalting, concentration, and storage of peptides for using RP-C18 StageTip columns [25, 26], which result in peptide samples suitable for direct loading into nanoscale liquid chromatography columns (even in LC setups without a pre-column):

1. *Prepare C18 StageTip desalting columns*: Place a C18 3 M Empore™ SPE Extraction Disks on a flat, clean surface such as a disposable plastic petri dish. Punch out small disks using a blunt-tipped hypodermic needle, and push the disks out of the needle into a 200 μL pipette tip by tapering a piece of fused silica or tubing fitting in the inside of the needle. Prepare as many C18 StageTip desalting columns as necessary. StageTips can be stored dry at room temperature for several months.
2. *Condition the columns with methanol*: Load the tip from the top with 50 μL 100 % methanol, and press the liquid through the disk in one of the two following ways: (1) manually, using a plastic syringe fitted with a pipette tip that was trimmed on both ends to serve as adapter between syringe and StageTip, or (2) by centrifuging the tips at approximately $3{,}000 \times g$ for 2 min at room temperature. This step is also used to ensure that the columns are not leaky (the liquid should leak without centrifugation and should take longer than a minute to pass through the column during centrifugation).
3. *Equilibrate the columns twice with "solution A"*: Load the tip from the top with 200 μL *"solution A"* (or 2× the volume of methanol used for conditioning), and press the liquid through the disk as in **step 2**. This step guarantees that any organic solvent is removed and gets the column ready for peptide binding.
4. *Prepare the peptides for binding to the column*: Acidify the peptide samples to $pH < 2.5$ by adding TFA (usually to a final concentration between 0.5 and 1 %). To facilitate pipetting and reduce precipitation, a 10 % TFA solution in LC-MS-grade water can be used. For in-solution digests, acidification is already done when the digestion is stopped, but samples can be mixed 1:1 with washing solution to assure the low pH and also to dilute urea/thiourea.

5. *Bind the peptides to the columns*: Load the tip from the top with acidified peptide samples, and allow peptide binding to the C18 material by passing the sample through the disk in one of the following ways: (1) manually, as slowly as you can, using a plastic syringe fitted with a pipette tip that was trimmed on both ends to serve as adapter between syringe and StageTip or (2) by centrifuging the tips at approximately 1,000 × *g* for 2 min at room temperature. For more efficient peptide binding, the flow rate should be smaller than in the other steps (~20 μL/min), so adjust the centrifugation conditions accordingly. Collect the flow-through, and repeat.
6. *Wash the columns with "solution A"*: Load the tip from the top with 200 μL *"solution A"* (or 2× the sample volume), and press the liquid through the disk as described in the previous step. After washing, peptides can be stored in loaded StageTips for several months.
7. *Elute the peptides from the column*: Load the tip from the top with 20–50 μL *"solution B,"* incubate at room temperature for 2 min, and elute peptides by pressing the liquid slowly through the disk as described in **step 2** and directly into the microfuge tube or the autosampler plate to be used for LC-MS/MS.
8. *Remove organic solvent from the eluates*: Dry the eluates in a vacuum centrifuge down to 2–3 μL to remove organic solvent. To avoid losing the peptides, do not dry completely!
9. For reversed-phase LC-MS/MS experiments add *"sample buffer"* to each sample to a final volume of approximately 8–10 μL and proceed with the analysis.

3.8 LC-MS/MS

Mass spectrometry technology, proteomics methods, and computational biology continue to evolve rapidly, and a thorough discussion of LC-MS/MS and data analysis is outside the scope of this chapter. Countless reports describing state-of-the-art mass spectrometry-based proteomic experiments exist in the literature, a variety of software designed to handle large-scale quantitative proteomics data are available [27], and the conditions can be adapted according to the possibilities, needs, and preferences of different laboratories. That being said, for SILAC large-scale quantitative proteomics, we recommend the following general setup:

3.8.1 LTQ-Orbitrap Instruments

1. Perform LC-MS/MS experiments on a nanoflow HPLC system connected online to a hybrid LTQ-Orbitrap (Classic, XL, or Velos) equipped with a nanoelectrospray ion source (*see* **Note 16**).
2. For reversed-phase nanoscale liquid chromatography, use fritless 15 cm silica microcolumns with 75 μm inner diameter, packed with ReproSil-Pur C18-AQ 3 μm resin (Dr. Maisch GmbH), which can be mounted in-house [28].

3. For each LC run, inject approximately 2–5 μg of desalted peptide sample into the column (*see* **Note 17**).
4. Load samples onto the analytical column in 2 % HPLC mobile phase B, and separate peptides through a 90-min gradient ranging from 5 to 40 % acetonitrile in 0.5 % acetic acid at a flow rate of 250 nL/min. Wash the column for 10 min with high concentration of organic solvent (90 % HPLC mobile phase B) and reconstitute for another10 min with HPLC mobile phase A prior to loading of the next sample (*see* **Note 18**).
5. Submit eluting peptides to electrospray ionization and transfer into the mass spectrometer.
6. Perform mass spectrometry analysis in the data-dependent mode by automatically switching between full survey scan MS and consecutive MS/MS.
7. Acquire survey full scan MS spectra in the Orbitrap section (m/z 300–2,000, resolution of $R = 60{,}000$ at m/z 400, target value = 1 × 106 in the linear ion trap). For all survey scan measurements with the Orbitrap detector, enable the lock-mass option, using ions on from ambient air (m/z 391.284286, 429.08875, and 445.120025) for internal calibration [29] to achieve overall sub-ppm mass accuracy for all detected peptides.
8. For MS/MS, select the 5–10 (Classic and XL instruments) or 20 (Velos) most intense ions in each survey scan (with ion intensity above 500 counts and a charge state ≥2) for sequential isolation to a target value of 5,000 and fragmentation in the linear ion trap by collisionally induced dissociation (CID/CAD).
9. Set up dynamic exclusion to automatically put all peaks selected for fragmentation in an exclusion list for 90 s (*see* **Note 19**).

3.8.2 Q-Exactive Instruments

1. Perform LC-MS/MS experiments on an ultrahigh-pressure nanoflow high-performance liquid chromatography (UHPLC) system connected online to a bench top Orbitrap instrument (Q-Exactive) equipped with a nanoelectrospray ion source (*see* **Note 16**).
2. For reversed-phase nanoscale liquid chromatography, use fritless 50 cm silica microcolumns with 75 μm inner diameter packed with 1.9 μm C18 resin (Dr. Maisch GmbH), which can be mounted in-house [28].
3. For each LC run, inject approximately 2–5 μg of desalted peptide sample into the column (*see* **Note 17**).
4. Load samples onto the analytical column in 2 % HPLC mobile phase B, and separate peptides through a 240-min gradient ranging from 5 to 40 % acetonitrile in 0.5 % acetic acid at a flow

rate of 250 nL/min. Wash the column with high concentration of organic solvent (90 % HPLC mobile phase B) and reconstitute with HPLC mobile phase A (0.5 % acetic acid) prior to loading of the next sample (*see* **Note 18**).

5. Submit eluting peptides to electrospray ionization and transfer into the mass spectrometer.
6. Perform mass spectrometry analysis in the data-dependent mode by automatically switching between full survey scan MS and consecutive MS/MS.
7. Acquire survey full scan MS spectra in the Orbitrap section (m/z 300–2,000, resolution of $R = 50{,}000$ at m/z 400, target value $= 1 \times 10^6$, 20 ms maximum injection time).
8. For MS/MS, select up to ten most intense ions in each survey scan (with ion intensity above 500 counts and a charge state ≥ 2) for sequential isolation with a window of 1.6 Th (resolution of $R = 17{,}500$ at m/z 200, target value of 1×10^6, 60-ms maximum injection time) and fragmentation by higher energy collisional induced dissociation (HCD) [30] with normalized collision energies of 25.
9. Set up dynamic exclusion to automatically put all peaks selected for fragmentation on an exclusion list for 40 s (*see* **Note 19**).

3.9 Protein Identification and Quantification

There are several software capable of handling high-resolution SILAC data available in the literature. We recommend performing protein identification and quantification with the MaxQuant software package [31], which is specially designed to analyze large-scale high-resolution mass spectrometric data (*see* **Note 20**). Detailed installation instructions and requirements are provided at the MaxQuant Web site (http://www.maxquant.org).

1. Download the most recent version of the software from the MaxQuant homepage. MaxQuant is under constant development, and updated versions are published on a regular basis, so make sure to always work with the latest version available.
2. Download the most recent version of the yeast ORF database in .fasta format from www.yeastgenome.org.
3. Load all .raw files and the yeast database onto MaxQuant.
4. Set up all the required parameters according to the experiment, and proceed with running the analysis. A detailed and more explicative protocol for MaxQuant software usage is described by Cox et al. [32]. In addition, we recommend registering to the online MaxQuant Google group, where practical questions regarding the usage of the software can be posted and are answered by the developers and other expert users of MaxQuant all over the world.

3.10 Evaluating the Quality of SILAC Labeling

3.10.1 Checking the Incorporation Efficiency of Isotope-Labeled Amino Acids

1. *Looking at the protein SILAC ratios*: From a labeled-only (i.e., "heavy" or "medium heavy" without mixing with their "light" counterparts) sample, calculate average protein ratio (labeled/unlabeled) from all proteins identified. A successful SILAC labeling must lead to an average incorporation efficiency of at least 95 %, represented by an average ratio of 20 (log2 = 4.3). In case of virtually complete labeling, however, this approach is not recommended since most proteins will not have a ratio (due to the absence of "light" peptides), and a 1:1 (labeled:unlabeled) mixed sample should be evaluated instead. In this case, from a 1:1 ("heavy":"light" or "medium heavy":"light") mixed sample, a successful SILAC labeling is indicated by an average protein ratio (H/L) close to 1 (log2 = 0) and no bias towards positive or negative log2 ratios as well as absence of proteins with strikingly low ratios.
2. *Looking at the SILAC state of peptides*: Determine the total number of heavy and light arginine- and lysine-containing peptides per sample. From a labeled-only sample, the percentage of "heavy" or "medium heavy" peptides should be close to 100 %. When working with 1:1 mixed samples, light and heavy peptides should represent 50 % each.

3.10.2 Checking for Arg-to-Pro Conversion (When Using Isotope-Labeled Arginine)

1. Identify all proline-containing peptides from your list and their corresponding MS spectra.
2. In the MS spectra, look for the isotope cluster representing the peptide m/z value.
3. Once it has been identified, search for an isotope cluster with a mass shift of +5 Da (m/z shift of 2.5 for a doubly charged peptide) or +6 Da (m/z shift of 3 for a doubly charged peptide) when Arg6 or Arg10 was used, respectively. The presence of such isotope clusters is indicative that arginine-to-proline conversion occurred.
4. To be confident, the spectra of several proline-containing peptides should be inspected. As a general guideline the degree of proline conversion should not exceed 5 %.

3.11 Performing Proteome-Wide SILAC Quantitative Proteomics Experiments

After having established the optimal labeling parameters one is ready to perform a proteome-wide SILAC quantitative experiment. Here, we present two typical workflows that allow for the comprehensive identification and quantification of the yeast proteome (Fig. 3); use instrumentation widely available in proteomics research laboratories and core facilities [33, 34]. They are both designed for comparison of two different conditions but can be expanded to allow comparison of three different states (*see* **Note 21**).

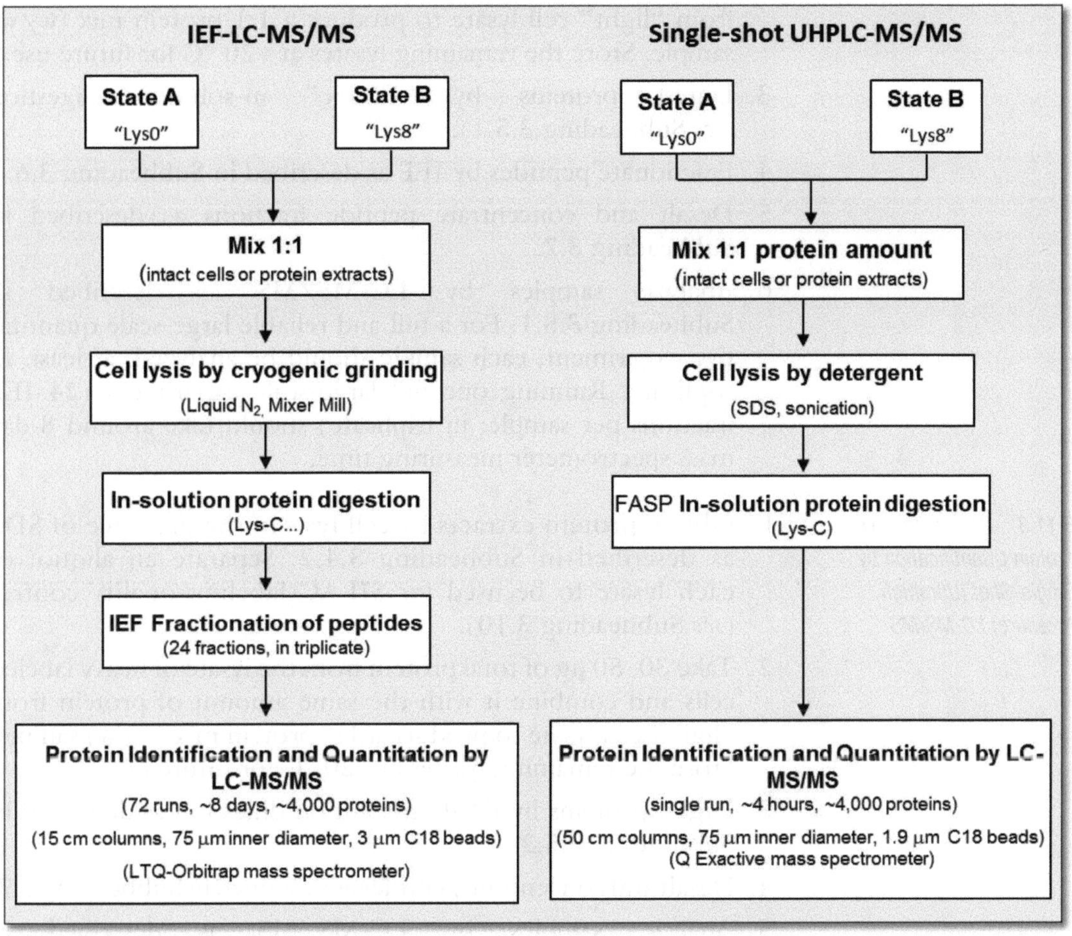

Fig. 3 Workflows for comprehensive mass spectrometry-based SILAC quantification of the yeast proteome

3.11.1 SILAC Single Labeling of Lysines and Harvesting of Cells

1. Prepare "light" and "heavy" lysine SILAC media (*see* Subheading 3.2.2) according to lysine-only labeling code (R0K0 vs. R0K8) and using the optimal amino acid concentrations (*see* Subheading 3.3).
2. Grow cells in SILAC media to allow complete labeling (*see* Subheading 3.2.3).
3. Perform the desired experimental treatment (e.g., growth factors, osmotic stress).
4. Harvest cells as described in Subheading 3.2.3.

3.11.2 Comprehensive Protein Quantification by IEF Peptide Fractionation Coupled to LC-MS/MS

1. Obtain protein extracts by cryogenic grinding (*see* Subheading 3.4.1). Separate an aliquot of each lysate to be used for SILAC labeling quality control (*see* Subheading 3.10).
2. Take 300–500 μg of total protein from the lysate of heavy labeled cells and combine it with the same amount of protein

from "light" cell lysate to produce a 1:1 protein mix (w/w) sample. Store the remaining lysates at −20 °C for future use.

3. Digest proteins by "Classic" in-solution digestion (*see* Subheading 3.5.1).
4. Fractionate peptides by IEF as described in Subheading 3.6.
5. Desalt and concentrate peptide fractions as described in Subheading 3.7.
6. Analyze samples by LC-MS/MS as described in Subheading 3.8.1. For a full and reliable large-scale quantitative experiment, each sample should be analyzed, at least, in triplicate. Running one full large-scale experiment (24 IEF fractions per sample, in triplicate) should take around 8-day mass spectrometer measuring time.

3.11.3 Comprehensive Protein Quantification by Single-Shot (Ultrahigh Pressure) LC-MS/MS

1. Obtain protein extracts by cell lysis in the presence of SDS as described in Subheading 3.4.2. Separate an aliquot of each lysate to be used for SILAC labeling quality control (*see* Subheading 3.10).
2. Take 30–50 μg of total protein from the lysate of heavy labeled cells and combine it with the same amount of protein from "light" cell lysate to produce a 1:1 protein mix (w/w) sample. Store the remaining lysates at −20 °C for future use.
3. Digest proteins by FASP in-solution digestion as described in Subheading 3.5.2.
4. Desalt and concentrate peptides as described in Subheading 3.7.
5. Analyze samples by LC-MS/MS as described in Subheading 3.8.2. For a full and reliable large-scale quantitative experiment, each sample should be analyzed, at least, in triplicate. Running one full large-scale experiment (triplicate runs) should take less than 1-day mass spectrometer measuring time.

3.11.4 Protein Identification, Quantification, and Anticipated Results

1. Perform protein identification and quantification as described in Subheading 3.9. We anticipate that both workflows described here (in a triplicate experiment) will lead to the identification and quantification of close to 4,000 proteins, meaning close to full coverage of the yeast proteome expressed in log-phase growth [33, 34].
2. Perform follow-up analysis as required and desired. A variety of tools and statistical tests exist to detect proteins as being significantly regulated as well as to assess the biological implications of the results (e.g., gene ontology, cluster analysis). One option for downstream analysis is the program Perseus. exe embedded in the MaxQuant software package.

4 Notes

1. When double labeling proteins with heavy arginine and lysine, the protein digestion is performed with trypsin, which cleaves at the C-terminal of lysine and arginine. When only heavy lysine is used, proteins can be digested with endoproteinase Lys-C, which cleaves C-terminal to lysine. Therefore, in both cases, all peptides (except for the protein C terminus) contain at least one labeled amino acid and will be available for quantitation. Double labeling with heavy arginine and lysine generates more quantifiable peptides per protein, allowing for more accurate protein quantification in comparison with lysine-only labeling. However, double labeling results in a sample twice as complex as a regular tryptic digest, hampering identification of low-abundance proteins. Therefore, for comprehensive coverage and quantification of the yeast proteome, we suggest working with lysine labeling only, which reduces sample complexity and, as a welcome bonus, avoids quantification errors due to arginine-to-proline conversion (*see* **Note 4**).
2. In classic 2-plex SILAC experiments, the isotope-labeled amino acids used should introduce into a "heavy" peptide a large enough mass difference from its "light" counterpart so that the two peak clusters can be easily distinguished in the mass spectra. In 3-plex experiments, the same rule applies in both directions (i.e., to the mass difference between the "medium heavy" peptide and its "light" and "heavy" counterparts). It is beneficial if the delta mass between light and heavy arginine and lysine differs. For example, $^{13}C_6$, $^{15}N_2$-lysine and $^{13}C_6$, $^{15}N_4$-arginine have mass differences of 8 and 10 Da compared to their light-isotope counterparts, respectively. Having different delta masses for arginine- and lysine-labeled peptides allows the determination of the C-terminal amino acid by measurement of the mass difference for a SILAC peptide pair, which helps with peptide identification.
3. This should allow for at least ten cell doublings, which are usually enough to efficiently label yeast cells. However, if incomplete labeling is observed, try to grow cells for more generations and repeat the checking.
4. Eukaryotic cells are naturally capable of converting arginine to proline. In cell culture systems, this phenomenon is frequently observed and can be enhanced by the excess of arginine in the culture medium. In the case of SILAC experiments where heavy arginine is used for labeling, the occurrence of arginine-to-proline conversion leads to the introduction of heavy proline into the system. More specifically, Pro5 (mass shift of +5 Da) and Pro6 (mass shift of +6 Da) are generated when Arg6 and

Arg10 are used for labeling, respectively. The presence of heavy proline, in turn, results in a subpopulation of double-heavy peptides (containing both heavy arginine and heavy proline), which causes erroneous quantification of proline-containing peptides and impairs accuracy of protein quantification. In the mass spectra double-heavy peptides appear as clusters with an extra mass shift and their presence is a direct indicator for arginine-to-proline conversion. The reverse metabolic conversion of proline to arginine can also occur, when cells are not provided with enough arginine, which would lead to incomplete labeling of arginine-containing peptides.

5. Ideally, one should always work with the smallest amino acid concentrations that still lead to a successful SILAC labeling while keeping the growth rate of the cell culture as close to normal as possible. To achieve successful SILAC labeling it is mandatory to eliminate all possible sources of the "light" version of labeling amino acids from the system (e.g., normal FBS must be replaced by dialyzed FBS, yeast extract/peptone must be replaced by yeast nitrogen base). This procedure can deprive cells from other essential nutrients, such as growth factors, which can impact the growth of some cell lines or organisms. To avoid unnecessary experiment costs, it is therefore advisable to first test the growth of your cells in "light" (K0/R0) SILAC medium before using the expensive heavy amino acids. It is also recommended to perform a titration experiment of L-lysine (K0) and L-arginine (R0) to empirically determine the lowest possible concentration of each required for satisfactory cell growth.
6. As a general guideline, a successful SILAC labeling must lead to an average incorporation efficiency close to full (>95%) and a negligible (<5 %) degree of arginine-to-proline conversion.
7. To access the average incorporation rate in a given experiment, complete proteome coverage is not needed and a statistically significant subset should suffice. Therefore, to check the incorporation level it is possible to simply analyze samples resulting from the in-solution digestion of a small protein amount (5–10 μg) of whole protein extracts.
8. A 100 mL culture of each "heavy" media is usually sufficient to perform a complete labeling evaluation ("heavy"-only and 1:1 mix samples, *see* Subheadings 3.3 and 3.10). A larger volume of "light" culture should be grown to allow preparation of mix samples with each tested labeling condition. Calculate the required volume of "light" culture by multiplying half the volume used for each test SILAC medium by the number of conditions tested.
9. Around 60 OD units of each labeled cells and 20 OD units of light cells per heavy condition tested are more than enough to perform a full incorporation test.

10. Sometimes, it might not be possible to optimize the arginine concentration to a level that leads to efficient labeling without impairing cellular growth. In these cases, an interesting alternative to counteract arginine-to-proline conversion is to increase the concentration of light proline in the medium. Excess of unlabeled proline, however, may reduce arginine labeling, since conversion can take place in the other direction as well (proline-to-arginine conversion), so this has to be tested for. If none of these options lead to satisfactory labeling, it might be better to switch for lysine-only labeling combined with Lys-C digestion.
11. For cryodisruption using a mixer mill, a minimum volume of 3 mL might be required (check manual of the manufacturer for your specific equipment).
12. To avoid precipitation of concentrated SDS, avoid temperatures below 15 °C and potassium salts in the buffer.
13. Keep the final volume as small as possible. When proteins are contained in large volumes, the denaturants can be added directly to the sample to avoid dilution of the denaturation buffer.
14. When SILAC labeling with heavy lysine and arginine and digesting proteins with trypsin, this step, although optional, promotes a "pre-digestion" of proteins in high concentration of denatur ant agents (which is supported by Lys-C but not by trypsin) and will knowingly improve the efficiency of trypsin digestion.
15. According to the manufacturer, up to 5 mg protein can be loaded, but in our hands separation was more efficient when loading up to 100 μg. Quality of fractionation can be evaluated by looking at the number of fractions in which each distinct peptide was found. In a good fractionation run, 95 % of the identified peptides must be present in only one or two adjacent fractions.
16. Assemble the LC system in a way the peptides are ionized as soon as they elute from the column. Any length the peptides have to travel between being eluted and ionized can cause sample diffusion and decrease impair resolution.
17. Note that, in this case, more is not always better, and higher sample amounts can impact chromatography resolution and peptide identification negatively.
18. Preferably, keep the column at a constant temperature of 35 °C in an oven regulated by a Peltier element, which can be built in-house [35] throughout the entire run.
19. This allows more unique precursor ions to be picked for MS/MS fragmentation and, consequently, increases the number of peptides identified per run. Time of exclusion can be adjusted according to the LC resolution and average peak width of a particular system.

20. Currently, MaxQuant is still restricted to the analysis of .raw data obtained from Thermo LTQ-Orbitrap or LTQ-FT-ICR instruments. However, new updates supporting other types of high-resolution mass spectrometric data should be released in the near future, so check the MaxQuant Web site on a regular basis for the most recent versions and features of the software.
21. Mixing three different labeling states will increase sample complexity and might cause the total number of identifications to decrease.

References

1. Aebersold R, Mann M (2003) Mass spectrometry-based proteomics. Nature 422(6928):198–207
2. Ong SE, Blagoev B, Kratchmarova I, Kristensen DB, Steen H, Pandey A et al (2002) Stable isotope labeling by amino acids in cell culture, SILAC, as a simple and accurate approach to expression proteomics. Mol Cell Proteomics 1(5):376–386
3. Ong SE, Kratchmarova I, Mann M (2003) Properties of 13C-substituted arginine in stable isotope labeling by amino acids in cell culture (SILAC). J Proteome Res 2(2):173–181
4. Ong SE (2012) The expanding field of SILAC. Anal Bioanal Chem 404(4):967–976
5. Gruhler A, Olsen JV, Mohammed S, Mortensen P, Faergeman NJ, Mann M et al (2005) Quantitative phosphoproteomics applied to the yeast pheromone signaling pathway. Mol Cell Proteomics 4(3):310–327
6. Fröhlich F, Christiano R, Walther TC (2013) Native SILAC: metabolic labeling of proteins in prototroph microorganisms based on lysine synthesis regulation. Mol Cell Proteomics 12(7):1995–2005
7. Olsen JV, Blagoev B, Gnad F, Macek B, Kumar C, Mortensen P et al (2006) Global, in vivo, and site-specific phosphorylation dynamics in signaling networks. Cell 127(3):635–648
8. Geiger T, Cox J, Ostasiewicz P, Wisniewski JR, Mann M (2010) Super-SILAC mix for quantitative proteomics of human tumor tissue. Nat Methods 7(5):383–385
9. Krüger M, Moser M, Ussar S, Thievessen I, Luber CA, Forner F et al (2008) SILAC mouse for quantitative proteomics uncovers kindlin-3 as an essential factor for red blood cell function. Cell 134(2):353–364
10. Pratt JM, Petty J, Riba-Garcia I, Robertson DH, Gaskell SJ, Oliver SG et al (2002) Dynamics of protein turnover, a missing dimension in proteomics. Mol Cell Proteomics 1(8):579–591
11. Doherty MK, Whitehead C, McCormack H, Gaskell SJ, Beynon RJ (2005) Proteome dynamics in complex organisms: using stable isotopes to monitor individual protein turnover rates. Proteomics 5(2):522–533
12. Doherty MK, Hammond DE, Clague MJ, Gaskell SJ, Beynon RJ (2009) Turnover of the human proteome: determination of protein intracellular stability by dynamic SILAC. J Proteome Res 8(1):104–112
13. Selbach M, Schwanhäusser B, Thierfelder N, Fang Z, Khanin R, Rajewsky N (2008) Widespread changes in protein synthesis induced by microRNAs. Nature 455(7209): 58–63
14. Schwanhäusser B, Busse D, Li N, Dittmar G, Schuchhardt J, Wolf J et al (2011) Global quantification of mammalian gene expression control. Nature 473(7347):337–342
15. Schulze WX, Mann M (2004) A novel proteomic screen for peptide-protein interactions. J Biol Chem 279(11):10756–10764
16. Blagoev B, Kratchmarova I, Ong SE, Nielsen M, Foster LJ, Mann M (2003) A proteomics strategy to elucidate functional protein-protein interactions applied to EGF signaling. Nat Biotechnol 21(3):315–318
17. Mittler G, Butter F, Mann M (2009) A SILAC-based DNA protein interaction screen that identifies candidate binding proteins to functional DNA elements. Genome Res 19(2):284–293
18. Ong SE, Li X, Schenone M, Schreiber SL, Carr SA (2012) Identifying cellular targets of small-molecule probes and drugs with biochemical enrichment and SILAC. Methods Mol Biol 803:129–140
19. Wagner SA, Beli P, Weinert BT, Nielsen ML, Cox J, Mann M et al (2011) A proteome-wide, quantitative survey of in vivo ubiquitylation sites reveals widespread regulatory roles. Mol Cell Proteomics 10(10): M111.013284
20. Choudhary C, Kumar C, Gnad F, Nielsen ML, Rehman M, Walther TC et al (2009) Lysine

acetylation targets protein complexes and co-regulates major cellular functions. Science 325(5942):834–840

21. Martin BR, Wang C, Adibekian A, Tully SE, Cravatt BF (2012) Global profiling of dynamic protein palmitoylation. Nat Methods 9(1): 84–89
22. Hanke S, Besir H, Oesterhelt D, Mann M (2008) Absolute SILAC for accurate quantitation of proteins in complex mixtures down to the attomole level. J Proteome Res 7(3): 1118–1130
23. Wiśniewski JR, Zougman A, Nagaraj N, Mann M (2009) Universal sample preparation method for proteome analysis. Nat Methods 6(5):359–362
24. Hubner NC, Ren S, Mann M (2008) Peptide separation with immobilized pI strips is an attractive alternative to in-gel protein digestion for proteome analysis. Proteomics 8(23–24): 4862–4872
25. Rappsilber J, Ishihama Y, Mann M (2003) Stop and go extraction tips for matrix-assisted laser desorption/ionization, nanoelectrospray, and LC/MS sample pretreatment in proteomics. Anal Chem 75(3):663–670
26. Rappsilber J, Mann M, Ishihama Y (2007) Protocol for micro-purification, enrichment, pre-fractionation and storage of peptides for proteomics using StageTips. Nat Protoc 2(8):1896–1906
27. Nesvizhskii AI (2010) A survey of computational methods and error rate estimation procedures for peptide and protein identification in shotgun proteomics. J Proteomics 73(11):2092–2123
28. Ishihama Y, Rappsilber J, Andersen JS, Mann M (2002) Microcolumns with self-assembled particle frits for proteomics. J Chromatogr A 979(1–2):233–239
29. Olsen JV, de Godoy LM, Li G, Macek B, Mortensen P, Pesch R et al (2005) Parts per million mass accuracy on an Orbitrap mass spectrometer via lock mass injection into a C-trap. Mol Cell Proteomics 4(12):2010–2021
30. Olsen JV, Macek B, Lange O, Makarov A, Horning S, Mann M (2007) Higher-energy C-trap dissociation for peptide modification analysis. Nat Methods 4(9):709–712
31. Cox J, Mann M (2008) MaxQuant enables high peptide identification rates, individualized p.p.b.-range mass accuracies and proteome-wide protein quantification. Nat Biotechnol 26(12):1367–1372
32. Cox J, Matic I, Hilger M, Nagaraj N, Selbach M, Olsen JV et al (2009) A practical guide to the MaxQuant computational platform for SILAC-based quantitative proteomics. Nat Protoc 4(5):698–705
33. de Godoy LM, Olsen JV, Cox J, Nielsen ML, Hubner NC, Fröhlich F et al (2008) Comprehensive mass-spectrometry-based proteome quantification of haploid versus diploid yeast. Nature 455(7217):1251–1254
34. Nagaraj N, Kulak NA, Cox J, Neuhauser N, Mayr K, Hoerning O et al (2012) System-wide perturbation analysis with nearly complete coverage of the yeast proteome by single-shot ultra HPLC runs on a bench top Orbitrap. Mol Cell Proteomics 11(3): M111.013722
35. Thakur SS, Geiger T, Chatterjee B, Bandilla P, Fröhlich F, Cox J et al (2011) Deep and highly sensitive proteome coverage by LC-MS/MS without prefractionation. Mol Cell Proteomics 10(8):M110.003699

Chapter 7

Analysis of Proteome Dynamics in Mice by Isotopic Labeling

John C. Price and Sina Ghaemmaghami

Abstract

Recent advances in mass spectrometry and in vivo isotopic labeling have enabled proteome-wide analyses of protein turnover in complex organisms. Here, we describe a protocol for analyzing protein turnover rates in mouse tissues by comprehensive ^{15}N labeling. The procedure involves the complete isotopic labeling of blue green algae (*Spirulina platensis*) with ^{15}N and utilizing it as a source of dietary nitrogen for mice. We outline a detailed protocol for in-house production of ^{15}N-labeled algae, labeling of mice, and analysis of isotope incorporation kinetics by mass spectrometry. The methodology can be adapted to analyze proteome dynamics in most murine tissues and may be particularly useful in the analysis of proteostatic disruptions in mouse models of disease.

Key words Protein turnover, Isotopic labeling, Proteome dynamics, Protein degradation, Quantitative proteomics

1 Introduction

The concept that cellular proteins are in a continual state of flux was introduced in the early 1930s [1]. However, it was not until the late 1930s that the advent of radioactive tracer techniques opened the door to reliable quantitative analyses of protein turnover [2]. These early experiments established the fundamental kinetic principles of protein turnover that endure to this day [3, 4]. Proteins are continuously synthesized and degraded in vivo. Whereas protein synthesis is typically modeled as a zero-order process, protein degradation is first order with respect to protein concentration. Thus, the expression level of a protein is determined by the ratio of its synthesis and degradation rates and the time it takes to establish the steady-state level is governed by degradation rate alone. Degradation rates are highly variable within proteins and are influenced by their chemical properties (e.g., primary sequence and higher order structure) and the cellular environment within which they are expressed (e.g., organism, tissue, organelle, and proteostatic stress).

Daniel Martins-de-Souza (ed.), *Shotgun Proteomics: Methods and Protocols*, Methods in Molecular Biology, vol. 1156, DOI 10.1007/978-1-4939-0685-7_7, © Springer Science+Business Media New York 2014

Multiple substrate-specific and nonselective degradative mechanisms operate in the cell. These include a large number of cytosolic and organelle-specific proteases, lysosomal processes such as autophagy and endosomal proteolysis, and the ubiquitin–proteasome system [5–7]. The fact that protein degradation plays a critical role in establishing and modulating protein expression levels in numerous normal and pathogenic biological processes has maintained continued interest in the development of improved technologies for its analysis.

Although the use of radioactive tracers such as ^{14}C and ^{3}H dominated early research in protein turnover, the potential utility of stable isotope incorporation (^{15}N, ^{13}C, and ^{2}H) as detected by mass spectrometry (MS) was appreciated as early as 1932 [8]. However, it was not until the beginning of the modern proteomic era that the true power of this approach was fully realized. In comparison to radioactive tracer studies, MS-based detection of stable isotopes offers two fundamental advantages. First, it allows the simultaneous detection and analysis of multiple polypeptide species within a complex mixture. Thus, in a single experiment, degradation kinetics of multiple proteins can be analyzed without biochemical purification of individual components. Second, whereas radiotracer methods can provide radioactivity measurements that are proportional to total isotope incorporation within a protein population, MS provides information on the mass distribution of the labeled species (i.e., isotopomer distribution).

Isotopomer analysis is particularly important in addressing one of the fundamental complications of radioactive tracer experiments: quantitation of the labeled precursor pool. In a typical tracer labeling experiment, isotope incorporation at steady state is influenced both by the degradation rate of the protein and labeling of the aminoacyl-tRNA precursor pool. As discussed above, the degradation kinetics of a protein typically follows simple first-order kinetics. However, patterns of precursor labeling can be substantially more complicated and are influenced by the kinetics of tracer uptake, transport, metabolism, catabolism, and recycling. The historical literature in protein turnover is replete with elegant mathematical compartment models, sophisticated pulse-chase techniques, and biochemical approaches for analysis of amino acid pools within cells and tissues—all aimed at proper quantitation and interpretation of precursor labeling. However, MS-based analysis of stable isotope incorporation typically circumvents these complications as the extent of precursor labeling at the site of synthesis can be directly determined by resolving the isotopomer distribution of newly synthesized proteins. These analyses, pioneered by the Hellerstein and Beynon laboratories [9, 10], enable the deconvolution of degradation kinetics from precursor labeling within a mass spectrum. This approach is particularly important in the analysis of isotopic labeling in complex organisms where label uptake into tissues can be slow relative to rates of protein turnover and follows complicated kinetics that are difficult to model mathematically.

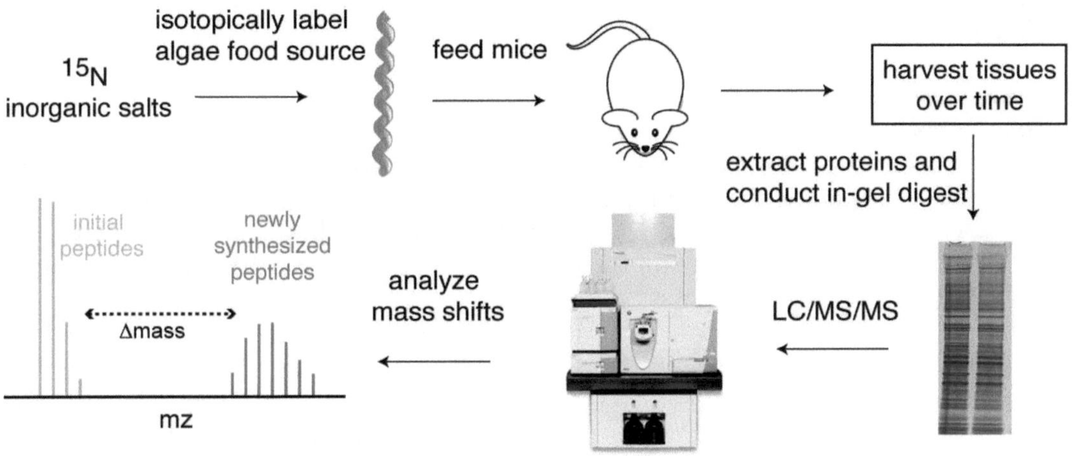

Fig. 1 Overview of experimental protocol. ^{15}N inorganic salts are used to make isotopically labeled media for growth of spirulina. Dried spirulina is used as the protein component of the mouse diet. After the initiation of the labeled diet, mice are collected over time and tissues of interest are isolated. Tissue extracts are resolved on SDS-PAGE and subjected to in-gel digests by trypsin prior to mass spectrometric analysis. Quantification of time-dependent mass shifts of peptides derived from newly synthesized proteins is used to calculate the turnover rate of the protein

From a methodological standpoint, proteome-wide analyses of turnover by mass spectrometry have been greatly propelled by the development of stable isotope labeling in cell culture (SILAC)—a set of standardized protocols, reagents, and analysis tools aimed at quantifying the incorporation of $^{15}N/^{13}C$-labeled amino acids into proteins [11]. Conducting dynamic (i.e., time-resolved) SILAC experiments is now fairly routine and has enabled global analyses of protein turnover in a number of cellular systems (e.g., [12, 13]). Cell culture experiments typically avoid the complications of precursor labeling as the introduction of excess levels of labeled amino acids in the media and the continuous growth conditions typically employed in these studies ensure complete labeling of the precursor pool at a rate that exceeds the rate of turnover for most proteins. However, the precursor problem cannot be similarly ignored in in vivo turnover experiments. Here, the utilization of fully ^{15}N-labeled isotope is particularly advantageous as it allows the direct measurement of precursor pool labeling using isotopomer analysis.

In a recent study [14], we modified a ^{15}N labeling strategy developed by Wu et al. [15] to measure in vivo turnover rates in various mouse tissues. The procedure involves the complete isotopic labeling of blue green algae (*Spirulina platensis*) with ^{15}N-labeled salts and utilizing it as a dietary source of nitrogen for mice (Fig. 1). By quantifying mass shifts and isotopomer distributions we were able to obtain turnover rates for more than 2,500 unique proteins across multiple tissues. The results highlighted a number of interesting relationships between turnover rates and protein functions. Here, we describe the protocol used in this study in some detail.

2 Materials

2.1 Algae Bioreactor and Culture Media

1. Bioreactor: 40 L pyrex glass carboy (modified after purchase with 2.5 cm diameter hose barbs at the neck, and a second horse barb to the side and two threaded 10 cm ports at the bottom, one in the center and the second offset to one side—*see* Fig. 2), 27 individual glass tubes (2.5 cm diameter, 1 m long) for the flow path, 10 m of 2.5 cm diameter tygon tubing,

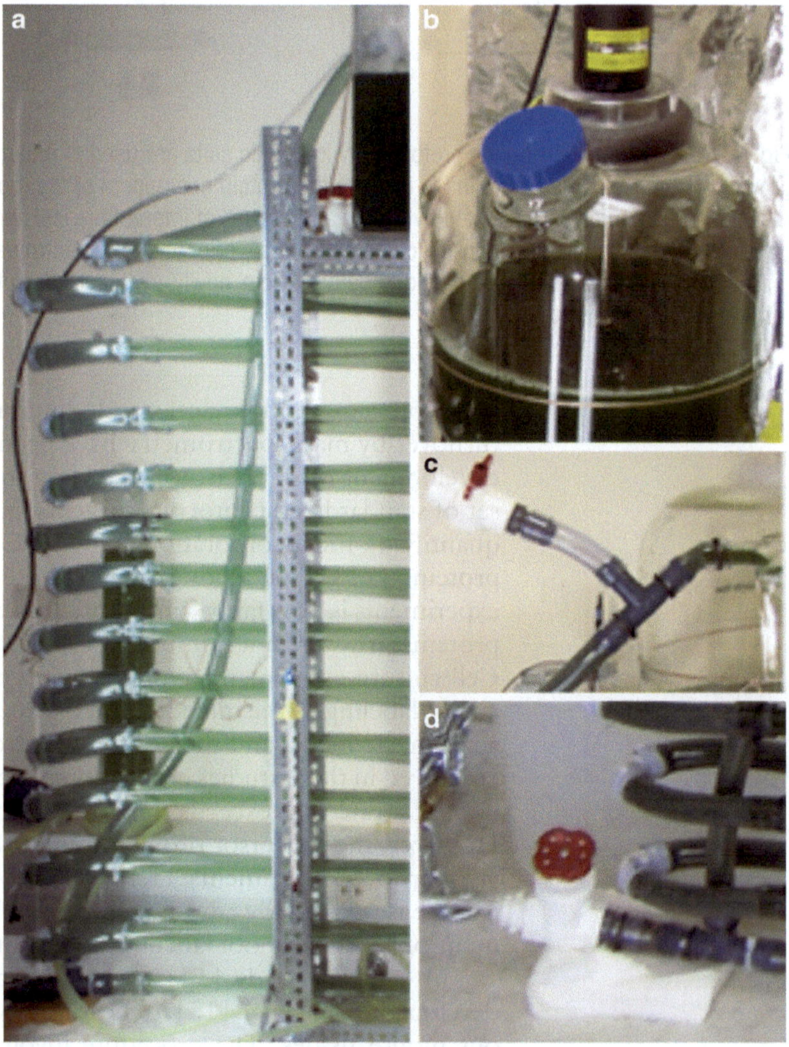

Fig. 2 Custom bioreactor for the production of ^{15}N-labeled *Spirulina*. Broth circulates steadily down toward the floor through 27 glass tubes, each measuring 1.5 m connected by 180° "u"-shaped connections (**a**). At the floor level, air is injected into the bubble lift (*panel d*, from *bottom left* to the *top* of *panel a*); this sudden increase in volume displaces broth into the modified 48-L glass carboy (**b**). The displaced broth then circulates back into the glass tubing. Adjusting the volume of air injected into the riser controls the rate of circulation. Algae can be harvested either using the port at the top of the bioreactor (**a**) or by adding a valve on the bubble lift riser (**c**)

30ea 2.5 cm diameter 90° barbed unions for the tygon tubing, 3ea 2.5 cm "t"-shaped barbed junctions, an air pump rated for operation under a 1.2 m water column, an adjustable pressure release valve, ~5 m rigid 0.5 cm gas line, a variable speed mechanical stirrer, 10ea Octron 3500K warm spectrum fluorescent lights, a handheld lux meter.

2. Spirulina starter culture: *Spirulina platensis* (LB1926) obtained from the UTEX Texas Algae Culture Collection. Aquarium air pump, tubing for air pump, 1 mL graduated glass pipets, 2 L glass graduated cylinders.
3. Metal-free broth stock solution: $NaHCO_3$ (162 mM), $NaCO_3$ (48 mM), K_2HPO_4 (3 mM), $Na^{15}NO_3$ (11.8 mM, Sigma Aldrich 364606), K_2SO_4 (5.7 mM), NaCl (17 mM), $MgSO_4$ (0.81 mM) in 16 L of H_2O.
4. Calcium stock solution: $CaCl_2$ (0.27 mM) in 1 L of H_2O.
5. Iron stock solution: $FeCl_3$-$6H_2O$ (6 μM) and 1.75 g ethylenediaminetetraacetic acid (EDTA) in 0.1 L of H_2O.
6. Trace metal mix A5 solution: H_3BO_3 (46 μM), $MnCl_2$-$4H_2O$ (9 μM), $ZnSO_4$-$7H_2O$ (0.76 μM), Na_2MoO_4-$2H_2O$ (1.6 μM), $CuSO_4$-$5H_2O$ (0.3 μM), $CoCl_2$-$6H_2O$ (0.17 μM) in 1 L of H_2O.
7. B6 stock solution: $^{15}NH_4Cl$ (0.04 μM), $Na^{15}NO_3$ (0.04 μM), $K_2Cr_2(SO_4)_4$ (3.8 nM), $NiSO_4$-$7H_2O$ (3.4 nM), Na_2WO_4-2H2O (0.1 nM) in 1 L of H_2O.

2.2 Mouse Feed

1. Unlabeled spirulina powder grown in-house or purchased from commercial vendor.
2. ^{15}N spirulina powder grown in-house as described below or purchased from commercial vendors.
3. Powdered protein-free mouse chow.
4. Food dehydrator or other low-heat oven for drying food pellets.

2.3 Tissue Preparation

1. Carbon dioxide euthanization station.
2. Dry ice.
3. Dulbecco's PBS (pH = 7.2).
4. Protease inhibitor cocktail.
5. Tissue homogenizer.
6. Surgical scissors, syringes, spatula, and forceps for isolation of tissues of interest.

2.4 Peptide Sample Preparation

1. Bicinchoninic acid assay.
2. 2 mM thick 4–12 % acrylamide gradient SDS-PAGE gels.
3. CMS-grade acetonitrile and formic acid.
4. Sequencing-grade modified trypsin.
5. 37 °C incubator, bath sonicator, vortexer, clean biohazard hood.

6. 25 mM Ammonium bicarbonate solution.
7. Low-bind Eppendorf tubes.
8. 500 mM Dithiothreitol–water solution.
9. Fresh 250 mM Iodoacetamide–water solution.
10. G250 Biosafe Coomassie stain.
11. Disposable scalpel.
12. 4× LDS gel sample loading buffer.
13. SDS-PAGE molecular weight standards.

2.5 Liquid Chromatography/Tandem Mass Spectrometry (LC/MS/MS)

1. LTQ-FT mass spectrometer interfaced with a Waters NanoAcquity ultra-performance liquid chromatography (UPLC) or an equivalent LC/MS/MS system with sufficiently high proteomic coverage and isotopic resolution (*see* below for details).
2. Software for data visualization (e.g., Xcalibur from Thermo Scientific), MS/MS sequence database search (e.g., Mascot, Protein Prospector), data export and conversion (e.g., Pava, msconvert), and analysis of isotopomer distribution and turnover (*see* below for details).

3 Methods

An overview of the experimental protocol is presented in Fig. 1. We have divided the protocol into five parts: production of isotopically labeled algae; labeling of mice and harvesting of tissues; preparation of MS samples; mass spectrometry; and data analysis. General considerations for different aspects of the protocol are presented in Subheading 4.

3.1 Production of Algae (See Note 1)

Production of labeled algae is conducted in a bioreactor that maintains a constant flow rate, removes oxygen, and provides ample light exposure. Here, we describe the design of a closed-loop bioreactor that incorporates an air lift to induce flow and mixing of the culture (Fig. 2).

3.1.1 Bioreactor Design

1. A modified 40 L glass carboy is supported on top of a 1.2 m (height) by 23 cm (width) framework. A series of interconnected 2.5 cm diameter 1 m long glass tubes are secured to the framework beneath the carboy. The carboy is modified so that the neck ends in a glass hose barb. A second glass hose barb is fused to the side close to the bottom of the carboy, and two threaded 10 cm ports are added to the bottom: one in the center and the second offset to one side (Fig. 2).

2. The modified carboy is installed upside down so that the neck acts as a funnel connecting to the glass tubing below and the ports are at the top of the bioreactor. Short lengths of tygon tubing connected to 90° PVC elbows are used to form 180° "u"-shaped junctions connecting the glass tubing, resulting in the broth exiting the neck/funnel of the carboy and zigzagging back and forth through thirty 180° turns as it descends to the floor.
3. At floor level a "t"-shaped junction is used to connect a tygon riser from floor level to the hose barb on the side of the carboy. The terminal position of the "t" is connected to an air pump by injecting air into the "t" so that it displaces broth from the riser into the top of the carboy. An adjustable pressure-release valve should be installed in the air line to allow control of the air volume injected into the riser. Controlling the air volume allows control of the rate of circulation, which is optimally ~0.1 m/s for solution traveling through the tubing. This can be assessed by timing the transit of algal particles through a section of tubing.
4. A mechanical stirrer in the carboy is necessary to ensure that the broth in the carboy is circulating and the spirulina does not deposit on the sides of the funnel. The mechanical stirrer is positioned in the central port so that the second port can be used for adding broth or harvesting algae.
5. Banks of Octron 3500K warm spectrum fluorescent lights are positioned on a horizontal rack to illuminate the glass tubes of the bioreactor. During the growth cycle, light intensity is modulated by changing the distance between the lights and the bioreactor as well as changing the number of lights in use. A handheld lux meter held at the surface of the bioreactor tubing can be used to quantify the light intensity. At initial inoculation densities of approximately $OD_{750} = 0.1$, light should be maintained at a minimal intensity of 500 lux. As the culture density increases the light intensity should be increased linearly to a maximum of 5,000 lux when the culture density exceeds $OD_{750} = 1.2$.

3.1.2 Preparation of Spirulina Growth Media

For convenience, the media should be initially formulated as five different stock solutions: metal-free broth (3×), calcium (1,000×), iron (10,000×), A5 trace metals (1,000×), and B6 trace metals (50,000×). The compositions of stock solutions are described in Subheading 2. All stocks are formulated in 18 MΩ H_2O.

The large volume of growth media needed for the algal growth can be conveniently mixed in the bioreactor. Add 1/3 volume of the metal-free media stock solution and 2/3 volume of purified water to the bioreactor. Start the air pump, begin circulating the solution within the bioreactor, and allow 1 h of mixing before adding 1 mL calcium stock per liter of media. Some precipitate may form upon addition of calcium; circulate the broth for 1 h in the bioreactor to resolubilize the particulate. Next, slowly add the iron stock (100 μL stock per L broth), A5 (1 mL per L), and B6 (20 μL per L) during continued circulation.

3.1.3 Starter Culture

Initiate a starter culture by growing the spirulina cultures in 2 L of the media in graduated cylinders. Mixing of the starter culture can be accomplished with an aquarium air pump to induce turbulence. The pump is turned on intermittently (for 2 min every hour). The starter cultures are illuminated with 500 lux from fluorescent lights.

3.1.4 Algal Growth

Use the 2 L starter cultures of spirulina to inoculate the large bioreactor at a minimum cell density of 0.1 g/L. During algal growth the broth will become increasingly basic. Algal growth is inhibited at $pH > 10$. pH monitoring can be accomplished by measuring the circulating broth directly with an electrode inserted into the top of the carboy. During the exponential phase of the growth curve, $CO_2(g)$ should be added to the broth each morning to adjust the pH. This can be accomplished by inserting a $CO_2(g)$ supply line capped with a fritted glass gas diffuser into the broth of the carboy. $CO_2(g)$ should be added until the pH reaches ~9.5.

3.1.5 Harvesting

Start harvesting the algae once the OD_{750} growth curve reaches saturation. The algae can be harvested either by collecting the congealed algal mass that accumulates on the top of the broth or diverting the broth into 1 L centrifuge bottles. The algae is then pelleted by centrifugation for 30 min at $1{,}000 \times g$. The decanted broth can be added back to the bioreactor. Resuspend the pellets in a minimal volume of 100 mM acetic acid, and pool the suspensions into a single centrifuge bottle. Re-pellet the algae by centrifugation at $1{,}000 \times g$ for 30 min. Freeze and lyophilize the washed algal pellets.

3.1.6 Analysis of Isotopic Enrichment

Measure the ^{15}N enrichment of the collected algae using the same procedure used for mouse tissues described below (*see* **Note 2**).

3.2 Feeding of Mice and Harvesting of Tissues

Mice minimally require 5 g of commercial feed per day. Our formulation consists of 20 % algae and 80 % protein-free mix. Therefore, labeling a single mouse for 1 day requires 5 g of feed comprising 1 g of spirulina and 4 g of protein-free mix (*see* **Note 3**).

1. Both control unlabeled algae and ^{15}N-labeled algae are prepared according to the same protocol. Combine lyophilized algae powder (20 % of final mass) with powdered protein-free mouse chow (Harlan Teklad, TD93328) and mix thoroughly. Form a thick paste from this powder by mixing in a minimal volume of water.

Roll out the paste into long ~3 cm thick rods. Cool the rods for 1 h at −20 °C, and then cut into 1 cm long sections. Place pellets on a drying rack in a food dehydrator (Oster) and dry overnight at ~40 °C. Dried food pellets can be stored at 4 °C until use. The dried pellets can be supplied to mice in the same manner as normal lab chow pellets.

2. Several days in advance of the labeling experiment, start feeding mice with the unlabeled algae diet using the same feeding schedule planned for the labeling time course (*see* **Note 4**).
3. Initiate the labeling time course by introducing the ^{15}N-labeled algae feed to the mouse cohort (*see* **Note 5**).
4. At each time point, anesthetize mice and harvest the desired tissues (*see* **Note 6**). For example, blood can be collected by opening the chest cavity and drawing up to 1 mL of blood directly from the heart. Collected blood should be quickly mixed with the appropriate anticoagulant buffer and stored at 4 °C. If desired, blood can be fractionated into different cellular populations prior to analysis by centrifugation. Tissues such as brain and liver can be snap frozen on dry ice and homogenized in PBS containing 1 mM PMSF using a bead beater. For more detailed descriptions of different tissue preparation protocols readers are referred to the review article by Bodzon-Kulakowska et al. [16].

3.3 Preparation of MS Samples

Below, we describe a protocol for the preparation of peptides utilizing in-gel trypsin digestion. However, it should be noted that for some applications, gel-free digests may be advantageous (*see* **Note 7**). As much as possible, work in the hood and wear gloves and long sleeves to avoid contamination.

1. Use BCA assays to measure concentrations of protein extracts and adjust the concentration to 5 mg/mL in extract buffer (10 mM Tris–Cl, pH 7.5, 1 mM EDTA, 100 mM NaCl, 1 mM PMSF).
2. To 50 μL of 5 mg/mL extract add 22.5 μL of 4× loading buffer, 1 μL of 500 mM DTT, and 1.5 μL water. Gently mix, and spin down.
3. Heat samples to 60 °C for 1 h. Let cool to room temperature.
4. Add 5 μL of 250 mM iodoacetamide (made fresh), and incubate in the dark at room temperature for 1 h.
5. Add 2.5 μL of 500 mM DTT. Heat samples to 60 °C for 1 h or boil at 95 °C for 5 min.
6. Load 35–40 μL of each sample on the SDS-PAGE (~100 μg of total protein), and run gel at the recommended setting to ensure good separation.

7. Wash gel in ddH_2O three times for at least 10 min, stain gel for 1 h in Biosafe Coomassie G-250, and destain in ddH_2O for one quick wash and then overnight.
8. Cut each lane into ten pieces with boundaries set by the MW marker with a scalpel, and place each piece in a 1.5 mL protein low-binding Eppendorf tube. Use the backside of the scalpel to "dice" the gel section into small (~1 mm^2) pieces.
9. Wash the gel pieces by adding ~100 μL (or enough to cover) of 25 mM NH_4HCO_3/50 % ACN and vortex for 10 min. Using gel loading or P20 pipet tip, extract the supernatant and discard. Repeat wash two more times. Most of the dye should wash away.
10. Completely dry the gel pieces using a speed vacuum.
11. Reconstitute trypsin (100 μg) by adding 100 μL of 50 mM NH_4HCO_3 for 10 min on ice. Once dissolved, add 550 μL of additional 50 mM NH_4HCO_3 and 750 μL 2× trypsin buffer.
12. Add the above trypsin solution to just barely cover the gel pieces. This volume will vary from sample to sample, but on average ~5–25 μL is sufficient.
13. Rehydrate the gel pieces on ice or at 4 °C for 10 min. Add 25 mm NH_4HCO_3 as needed to cover the gel pieces. Incubate at 37 °C overnight.
14. Transfer the digest solution into clean low-binding tubes. To the gel pieces add 50 % ACN/4 % formic acid (enough to cover), vortex for >20 min using a multi-vortexer, and sonicate for 5 min. Transfer the solution to the previously isolated solution. Repeat the entire process one more time. The total volumes should be 400–600 μL at this point.
15. Using a speed vacuum reduce volume to 10 μL. Add 20 μL of 4 % formic acid to get a total volume of 30 μL. Spin down at 10,000 rpm in a desktop centrifuge for 20 s. Transfer top two-thirds (~20 μL) to new 0.6 mL Eppendorf low-binding tubes. Store at −4 °C until MS analysis.

3.4 LC/MS/MS

The exact LC/MS/MS protocol used will be dependent on the available instrumentation (*see* **Note 8**). Here, we describe an LC/MS/MS protocol utilizing an LTQ-FT system (ThermoFisher).

1. Interface the LTQ-FT instrument with a NanoAcquity LC system (Waters) and a 75 μm ID, 150 mm length, 3 μm particles, 100A pore size C18 column.
2. For each run, inject 1 μL of the peptide sample with an injection time of 500 ms, and run a 60-min 3–40 % B linear gradient at 300–800 nl/min.
3. Acquire one survey scan every 3 s in FTICR at a mass resolution of 100,000 and six secondary MS/MS scans in the ion trap for each survey scan.

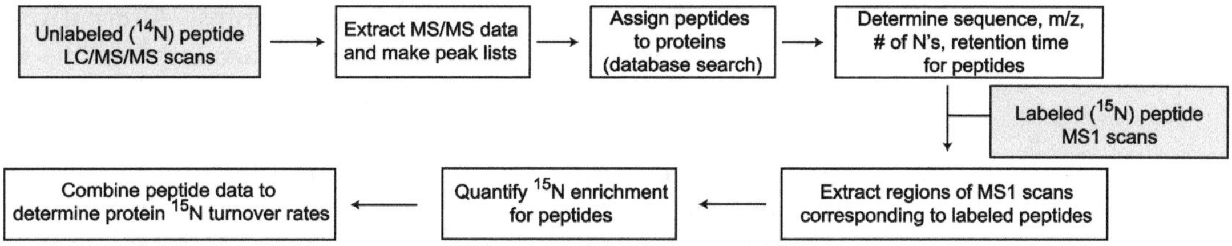

Fig. 3 Summary of data processing pipeline. Peptide-level information obtained from unlabeled MS/MS spectra is used to identify and quantify MS1 spectral regions corresponding to the labeled peptides in later isotope incorporation times. The kinetics of ^{15}N incorporation in all peptides corresponding to a given protein is used to measure the rate constant for turnover. The *gray squares* represent inputted data

3.5 Data Analysis

The general flow chart of the data analysis procedure is shown in Fig. 3. It is fairly straightforward to conduct this analysis manually for individual peptides identified in database searches of unlabeled samples using commonly available commercial and public software. However, the global analysis of partially labeled samples is considerably more complex (*see* **Note 9**). A number of software platforms have been developed in the past few years to tackle this problem [17–19]. Here, we describe the manual analysis of individual peptides in detail and provide a brief description of global analysis of proteome-wide data using the software platform developed by Guan et al. [19, 20].

3.5.1 Database Search

The first step in analysis of global turnover rates is conducting a sequence database search of the MS/MS data obtained from the unlabeled (^{14}N) control samples. The goal is to obtain the following information for each parent peptide: (1) *m/z* and retention time: these parameters will be used to identify and extract peptide MS1 scans from the labeled (^{15}N) samples for quantitation; (2) peptide sequence: this will enable the calculation of the number of nitrogens in the peptide which will be used to analyze the isotopomer distribution in the labeled samples; (3) protein assignment: to aggregate results of peptides derived from the same protein. There are multiple commercial and publicly available MS/MS search algorithms in common use. Some of the most popular platforms include MASCOT, SEQUEST, X! Tandem, and Protein Prospector. Detailed descriptions of each algorithm are beyond the scope of this review (readers are referred to [21] for a recent review). In theory, any search algorithm that can generate the above peptide-level information with justified statistical significance can be utilized. Here, we describe the use of Protein Prospector [22] and discuss some important search parameter considerations. More detailed tutorials on Protein Prospector are available online (http://prospector.ucsf.edu).

1. Extract MS/MS data from raw data, and create peak lists from each unlabeled sample. Individual MS platforms often provide their own software packages that generate MS/MS peak lists

using vendor-specific peak picking algorithms. Alternatively, two publicly available algorithms that can accomplish this are msconvert (part of the ProteoWizard package [23]) and PAVA [19] (for Thermo .Raw files only). Ensure that peaks are centroided and filtered to produce a manageable number of peaks for each scan. Protein Prospector accepts peak lists in a number of tabular formats including Mascot Generic Format (.mgf).

2. If multiple fractions (e.g., gel sections) were analyzed, combine all MS/MS data into a single zipped file.
3. Upload the MS/MS peak list to Protein Prospector (http://prospector.ucsf.edu), and conduct a database search using the appropriate search parameters: (1) Select the desired database and species. Ideally, a current protein database (e.g., Uniprot) containing decoy random sequences for calculation of false discovery rates should be used. (2) Set the mass tolerance of peptide and fragment ions depending on the mass resolution of the instrument. In our experiments we utilized an LTQ-FT instrument to collect survey scans at a mass resolution of 100,000. Accordingly, we set the parent mass tolerance of 50 ppm and fragment mass tolerance of 0.6 Da. (3) Set the enzymatic constraint depending on the utilized protease. In our analysis, we used trypsin. (4) Select the desired static and variable modifications. In our analysis, we set the fixed modifications to carbamidomethylation and the variable modifications to acetylation on the N-terminus, glutamine on peptide N-terminal to glutamic acid, methionine loss from protein N terminus, and acetylation and methionine oxidation. The maximal allowed number of variable modification peptide was set to 2. When all parameters are set, click "Start Search."
4. Once search is completed, create a peptide report for export using the "Search Compare" program in Protein Prospector. For each peptide, be sure to include the protein ID, peptide sequence, z, m/z, fraction, and retention time in the exported report.

3.5.2 Manual Measurement of Turnover Rates

1. Identify all peptides belonging to the protein(s) of interest in the peptide report file.
2. Using the m/z and retention time of unlabeled peptides, extract the corresponding MS1 spectrum from the unlabeled and labeled samples. Typically, this can be readily accomplished by vendor-supplied analysis software. For example, the Qual Browser module within the Xcalibur software platform can be used to analyze data collected in Thermo .Raw files. The extracted-ion chromatogram (XIC) can be used to isolate spectra corresponding to specific elution times. For example, unlabeled and partially labeled spectra, chromatograms, and peak intensity lists are shown in Fig. 4 for a peptide belonging to

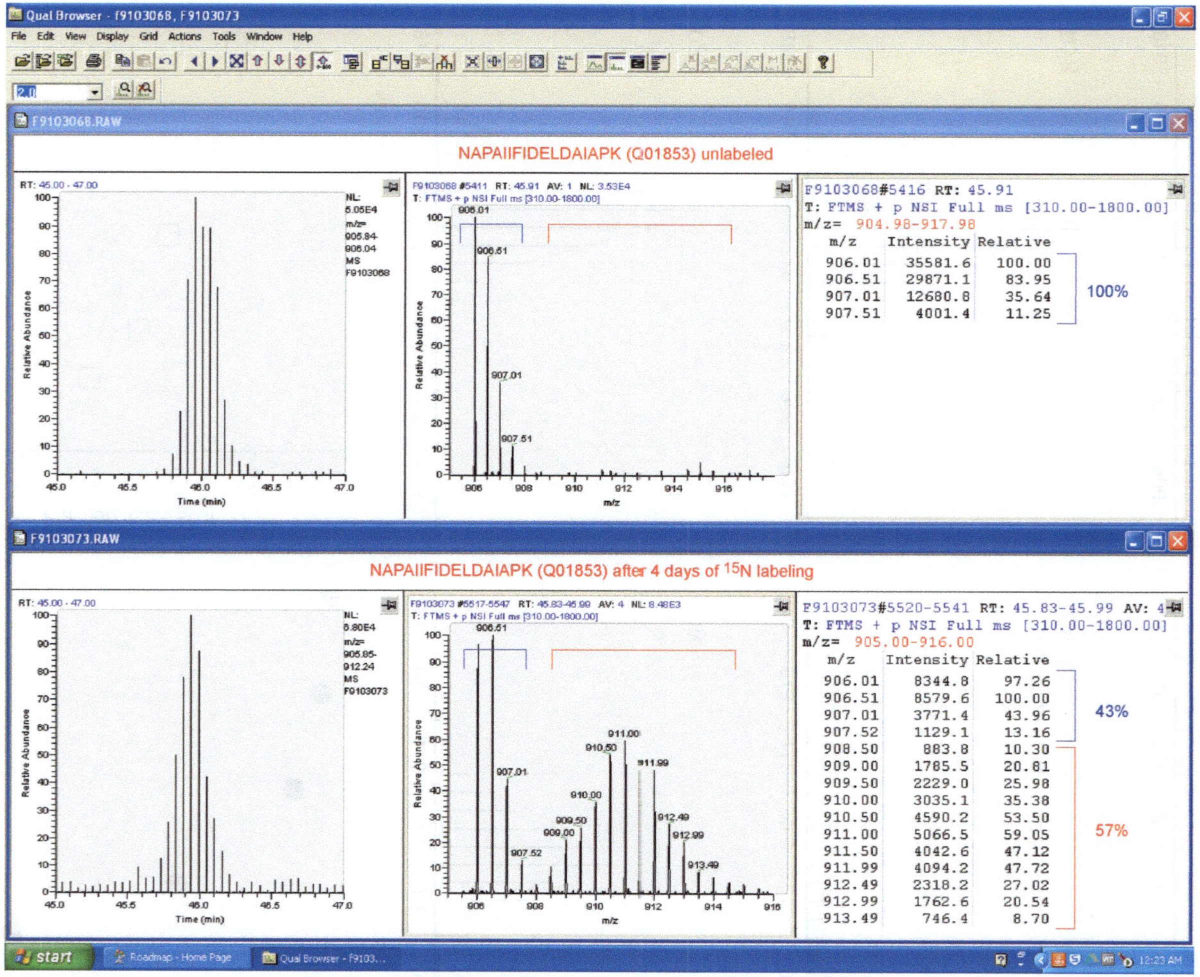

Fig. 4 Identification and quantification of ^{15}N-labeled isotopomer distribution. The chromatograms (*left*), mass spectra (*center*), and peak intensity lists (*right*) for a tryptic peptide belonging to transitional endoplasmic reticulum ATPase (Uniprot Q01853) after 0 and 4 days of labeling. The extract was obtained from brain tissue. The *blue brackets* indicate the unlabeled (*natural*) isotopomer distribution, and the *red brackets* indicate the ^{15}N-labeled isotopomer distribution. Analysis of peak intensities within each distribution indicates a fractional labeling of ~57 %. The data is visualized by Xcalibur software (Thermo Scientific)

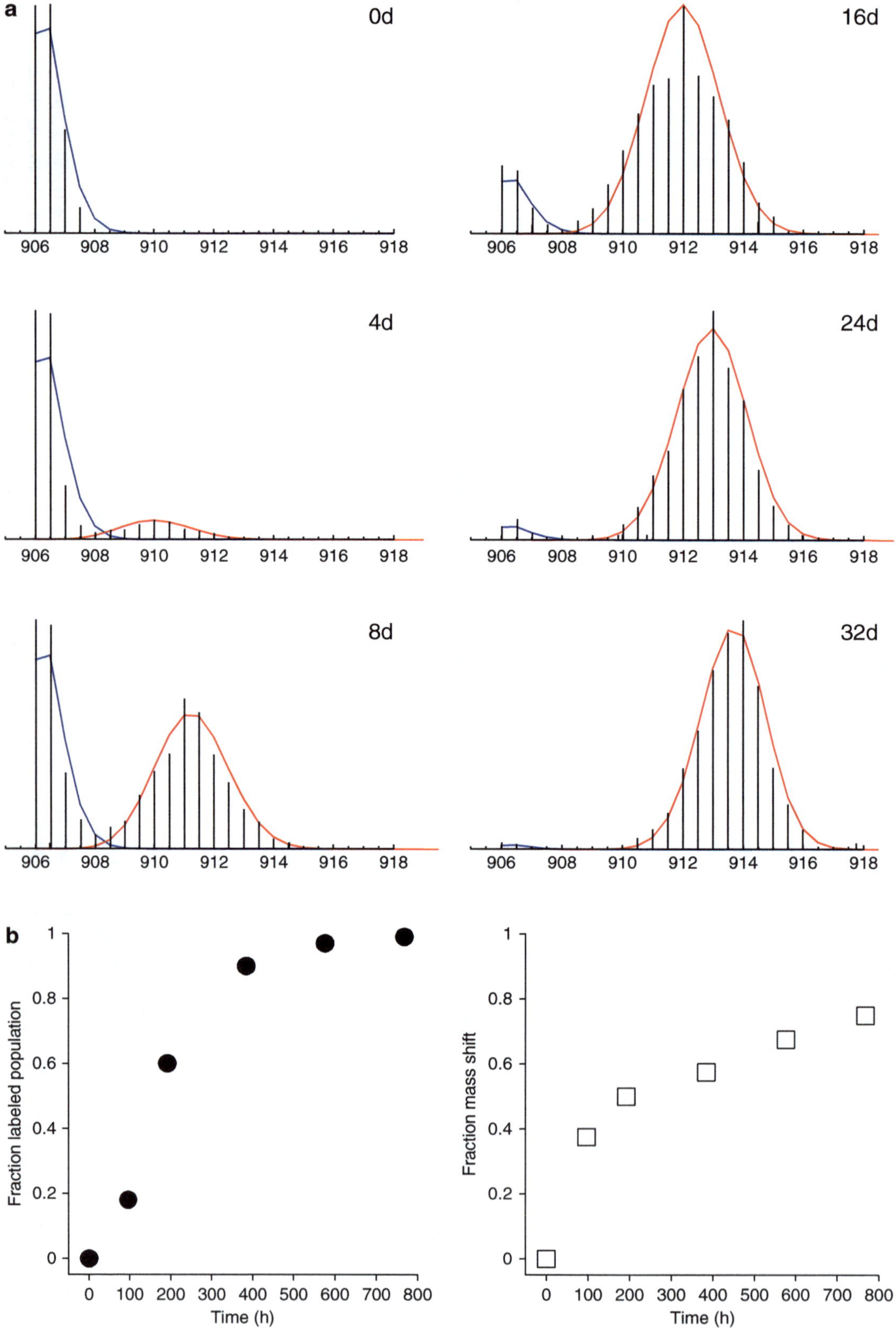

Fig. 5 Spectral shift patterns caused by time-dependent ^{15}N incorporation. (**a**) Sample spectral data showing the gradual incorporation of ^{15}N over time in the Q01853 peptide shown in Fig. 4. Changes in both the fractional labeled population (ratio of labeled and unlabeled integrated isotopomer areas) and mass shifts are evident.

transitional endoplasmic reticulum ATPase (Uniprot Q01853) isolated from mice brain after 0 and 4 days of ^{15}N labeling.

3. Export peak intensities for labeled and unlabeled spectra as text files. In Xcalibur, this can be readily done by "right-clicking" on the Spectrum List window and exporting to the clipboard.

4. Measure fractional labeled population. This can be easily accomplished by spreadsheet programs such as Excel. In comprehensive ^{15}N labeling experiments, labeled peaks are usually well resolved from the unlabeled peaks and no deconvolution is necessary to measure the ratio of the labeled and unlabeled distributions. In the example shown in Fig. 4, the fractional labeling after 4 days is ~57 % indicating that the percentage of the total protein pool has been synthesized since the introduction of the ^{15}N diet. Measurement of this parameter over time can be used to analyze the fractional turnover kinetics of the peptide.

5. Measure the mass shift. Although not necessary for evaluation of fractional protein turnover, the measurement of the mass shift of the labeled population can determine the distribution of ^{15}N incorporation ratios in newly synthesized proteins. This parameter is contingent on the fractional labeling of the precursor pool (*see* above). ^{15}N incorporation ratios can be determined by fitting the distribution of the labeled peaks with algorithms that simulate isotopomer distributions at variable ^{15}N incorporation rates. One user-friendly publicly available software that can readily accomplish this is Envelope [24]. Figure 5 shows the kinetics of fractional labeled population and mass shift for the Q01853 peptide. It can be seen that the former follows single exponential kinetics (with a first-order rate constant established by the degradation rate of the protein) whereas the latter is multiphasic (for detailed compartment modeling of mass shift kinetics refer to [20]).

6. Combine fractional labeled population data from all peptides corresponding to the protein of interest, and fit the data with a single exponential equation to determine the first-order degradation rate constant of the protein. Figure 6 shows the aggregated turnover plot for all peptides corresponding to Q01853. The data are fit to the equation $1 - e^{(-k \times t)}$ with least square fitting using the program Kleidagraph (Synergy software). The measured first-order rate constant is 0.10 day^{-1} which corresponds to a half-life of ~7 days.

Fig. 5 (continued) The *blue curves* represent fits to natural mass isotope distribution of the peptide, and the *red curves* represent fits to mass isotope distributions at the incorporation ratios plotted in (**b**). (**b**) Fractional labeled population (*circles*) and fractional mass shifts (*squares*) plotted as a function of incorporation time

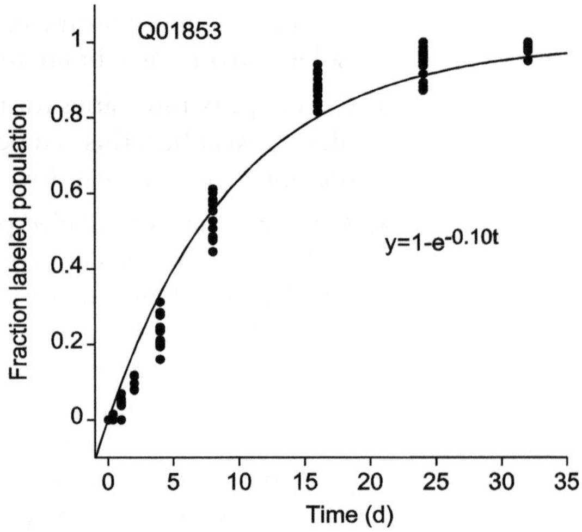

Fig. 6 Aggregated fractional labeling kinetics for all peptides mapped to Q01853. Each *circle* at a given time point represents a distinct peptide ion assigned to Q01853, and the *line* represents a single exponential fit to all data

3.5.3 Automated Proteome-Wide Analysis of Turnover Rates

A number of programs have been developed to analyze ^{15}N labeling kinetics [17–19] on proteome-wide scales. For example, the algorithm developed by Guan et al. [19] consists of a set of modules, written in Visual Basic and MATLAB that carry out the functionalities outlined in Fig. 3. As input, it requires an MS/MS peptide report generated by Protein Prospector, a set of mass spectra in .Raw format and a parameter file that associates each spectrum with a given fraction (e.g., gel segment) and time point. The program generates kinetic curves for fractional population labeling and mass shifts at the peptide and protein levels and fits the former to exponential equations to generate rate constants. The detailed descriptions and theoretical nuances of these protocols are beyond the scope of this review, and readers are referred to the corresponding manuscripts for detailed explanations of each algorithm.

4 Notes

1. Spirulina has a very high protein content (~75 % by dry weight) and has been used for decades as a source of isotopically labeled amino acids. Other algal species can potentially be used, but many are known to produce toxins (e.g., microcystin) that may confound downstream experiments. Spirulina has been tested extensively by commercial and governmental agencies, and no evidence of toxin synthesis has been observed. The utilization of ^{15}N spirulina-fed mice as a quantitative proteomic strategy was first described by Wu et al. [15]. However, it should be noted that alternative ^{15}N dietary sources have successfully

been used for labeling of mice, including the bacteria *Ralstonia eutropha* [25]. Multiple commercial sources of labeled algae are now available where large quantities of ^{15}N spirulina can be obtained for immediate use (Cambridge Isotope Labs, Sigma Isotech, and others). However, in our experience, the materials required for growing labeled spirulina in-house can be purchased for approximately 20 % of the cost of commercial algae. The decision to make or buy ^{15}N algae will thus depend on the availability of financial resources, time, labor, and a bit of engineering know-how. Our bioreactor design follows the principles outlined by Molina et al. [26]. However, it should be noted that many alternative bioreactor designs for the growth of algae have been described and some are available commercially (e.g., *see* http://www.algaelab.org/).

2. The extent of labeling in the algal feed represents the maximum achievable isotope enrichment in mouse tissues. Therefore, measurement of this incorporation plateau is important in the interpretation of in vivo labeling dynamics. The measurement of algal ^{15}N enrichment is accomplished by preparing trypsin digests of the unlabeled and ^{15}N-labeled algal extracts. Using the procedure outlined for mouse tissues, a list of tryptic peptide is compiled for an unlabeled algal sample. Using the amino acid sequences, theoretical isotopic envelopes of fully labeled species are calculated. Comparison of the peptide spectra from the labeled algal samples and the theoretical incorporation maximum allows calculation of the labeling efficiency. If the bioreactor is fully sealed, the algae should attain the same isotopic enrichment as the sodium nitrate ($^{15}NaNO_3$) used in the broth.
3. We found that mice supplied with 5 g of algae pellets per day consume the entire ration within 24 h. Although this suggests that the mice may eat more in an ad lib setting, the algae-fed mice continue to gain weight during the course of the labeling experiment, suggesting that 5 g of feed per day is sufficient to provide all the required nutrients.
4. Acclimating mice to the algae diet prior to labeling addresses two important variables in the experiment. The first is the synchronous timing of label introduction. After a day or 2 on the unlabeled diet, we observed that most mice would begin feeding on the algal diet within minutes of feed introduction. The second important variable is that proteome dynamics measurements assume a stable protein concentration during the course of the labeling experiment. Introduction of a new diet may cause significant changes in behavior and metabolism. The acclimation period is an attempt to ensure that experimental and control groups are in the state of proteome homeostasis prior to initiation of labeling.

5. Care should be taken in designing the appropriate labeling time course and the appropriate number of mice. Time points should be chosen depending on the estimated turnover rate of protein(s) of interest. For example, mouse serum albumin has a half-life of approximately 2 days (turnover rate of ~0.3 day^{-1}). Therefore, a 10-day time course consisting of multiple time points and at least two mice per time point is suitable. If resources are limited and only a single labeled time point can be collected, that time point should be chosen close to the expected half-life of the protein(s) of interest. If a broad, proteome-wide analysis of turnover is desired, multiple time points with exponentially growing spans (e.g., 0, 1, 2, 4, 8 days) is ideal. We found that a month-long time course can accurately measure the labeling kinetics of more than 90 % of observable proteins excluding the most stable proteins (e.g., myelin basic proteins, histones) that have half-lives well exceeding 30 days. It should also be noted that the time for label uptake, by an oral route, can be as long as 1 day for some tissues [14]. Thus, for analysis of fast turnover proteins (half-lives of less than 1 day), alternative routes of labeling [27] may need to be explored. The statistical power of the study should be calculated depending on the variability of rates between the experiment and control groups. The larger the difference in rates between control and experimental groups, the fewer mice will be needed for a statistically significant comparison. Also, mouse strain and sex have been observed to influence mouse weight, life-span, and cell proliferation rates. Although we have not conducted an extensive proteome-wide study of the effects of strain and sex on protein turnover, preliminary data support the assumption that the rate of protein turnover can be influenced by these variables. Care should be taken to match the strain, age, and sex of experimental and control groups.
6. Homogenization of the tissue will allow a broad view of proteome, but it also reduces the ability to describe specific protein pools within a tissue. In many cases it may be beneficial to isolate specific structures from the tissue before homogenization. The brain, for example, shows region-specific metabolism and susceptibility to disease. Therefore dissection of the brain into the different regions for comparison of protein turnover may be important in understanding disease pathology. Separation of a tissue into the various cellular components may also allow expansion of the total proteome coverage. The homogenization protocol should also take into account the subsequent peptide sample preparation approach. For example, if in-gel digests are performed salts, detergents, and buffers will typically be removed by SDS-PAGE. However, if an in-solution digest is to be performed the remaining detergents in the homogenate may adversely affect subsequent steps in the protocol.

7. One major advantage of dynamic proteomics over quantitative proteomics is that common sample preparation techniques do not influence the measurement of ^{15}N incorporation ratios. This is because the isotopomer distribution of a peptide will typically not be effected by purification yield or efficiency. Therefore, fractionation methods to enrich the proteins of interest can be applied without impacting quantitation. For small-scale experiments, in-gel digest (as described in this protocol) is a reliable and robust method for protein separation and peptide production. For larger experiments involving multiple tissues and experimental groups, gel-free or in-solution digests may be preferred. In-gel digestion requires multiple extraction and pipetting steps, whereas in-solution digestions can be conducted as "addition-only" processes that dramatically reduce the workload.
8. Four critical capabilities of the LC-MS/MS instrumentation that need to be considered are mass precision, proteome coverage, sufficient chromatographic separation of peptides, and experiment-to-experiment retention time variability. Typically, the mass resolution of the parent peptide spectra must be sufficiently high to resolve individual isotopic species. Some popular instrumental configurations that are appropriate include LTQ-FTICR (described in this protocol), LTQ-Orbitrap, Q-TOF, and TripleTOF. In order to provide sufficient peptide separation, the MS should ideally be interfaced with a nanoflow UPLC system and narrow capillary reverse-phase columns. Care should be taken to maximize the reproducibility of elution profiles from sample to sample as large deviations in retention times of peptides will significantly complicate subsequent data analysis (*see* Subheading 3.5). Injection of similar total amounts of peptides from run to run and maintaining the column in a temperature-controlled chamber will minimize retention time variation. Overloading the column will cause shifts in the retention time, while loading too little sample reduces the number of identified peptides and may increase the signal-to-noise ratio. To avoid these problems, gel staining intensities can be used to normalize sample injection volumes for samples obtained from the same gel. The buffer system and the LC solvent gradient should be optimized for the tissue of choice. For example, adipose tissue has a high percentage of hydrophobic peptides and may necessitate more extended organic washes.
9. The primary complication in the analysis of protein turnover by comprehensive ^{15}N labeling lies in the difficulty of identifying and quantifying isotopically labeled peaks corresponding to the peptide of interest. In traditional SILAC experiments, where only one or few amino acids within a given peptide have the potential of being labeled, the exact mass differences

between labeled and unlabeled peptides are predetermined and can be used as a constraint to unequivocally assign sequences to labeled spectra. In comprehensively ^{15}N-labeled samples, where multiple nitrogens within the peptide can be labeled at unknown incorporation ratios, the mass difference between labeled and unlabeled peaks is itself a variable in the experiment and the isotopic distribution pattern of labeled peaks can be complex. Hence, it is difficult to directly assign sequences to partially labeled peptides by typical database search algorithms. Instead, peptide-level information (sequence, elemental composition, *m/z*, and retention time) is gathered from MS/MS database searches of unlabeled control samples and used to extract and interpret the MS1 spectra corresponding to the isotopically labeled forms of the same peptide generated from ^{15}N-labeled experimental samples. Thus, reproducibility of peptide retention times between multiple experimental samples is an important requirement of this methodology (*see* above).

References

1. Borsook H, Keighley GL (1935) The "continuing" metabolism of nitrogen in animals. Proc R Soc Lond B 118:488
2. Schoenheimer R (1942) The dynamic state of body constituents. Harvard University Press, Boston
3. Claydon AJ, Beynon R (2012) Proteome dynamics: revisiting turnover with a global perspective. Mol Cell Proteomics 11(12):1551–1565. doi:10.1074/mcp.O112.022186
4. Hinkson IV, Elias JE (2011) The dynamic state of protein turnover: it's about time. Trends Cell Biol 21(5):293–303. doi:10.1016/j.tcb.2011.02.002
5. Hochstrasser M (1996) Ubiquitin-dependent protein degradation. Annu Rev Genet 30:405–439. doi:10.1146/annurev.genet.30.1.405
6. Klionsky DJ (2007) Autophagy: from phenomenology to molecular understanding in less than a decade. Nat Rev Mol Cell Biol 8(11):931–937. doi:10.1038/nrm2245
7. Lopez-Otin C, Overall CM (2002) Protease degradomics: a new challenge for proteomics. Nat Rev Mol Cell Biol 3(7):509–519. doi:10.1038/nrm858
8. Schoenheimer R, Rittenberg D, Foster GL et al (1938) The application of the nitrogen isotope N15 for the study of protein metabolism. Science 88(2295):599–600. doi:10.1126/science.88.2295.599
9. Hellerstein MK, Neese RA (1992) Mass isotopomer distribution analysis: a technique for measuring biosynthesis and turnover of polymers. Am J Physiol 263(5 Pt 1):E988–E1001
10. Doherty MK, Whitehead C, McCormack H et al (2005) Proteome dynamics in complex organisms: using stable isotopes to monitor individual protein turnover rates. Proteomics 5(2):522–533. doi:10.1002/pmic.200400959
11. Ong SE, Mann M (2007) Stable isotope labeling by amino acids in cell culture for quantitative proteomics. Methods Mol Biol 359:37–52. doi:10.1007/978-1-59745-255-7_3
12. Schwanhausser B, Busse D, Li N et al (2011) Global quantification of mammalian gene expression control. Nature 473(7347):337–342. doi:10.1038/nature10098
13. Cambridge SB, Gnad F, Nguyen C et al (2011) Systems-wide proteomic analysis in mammalian cells reveals conserved, functional protein turnover. J Proteome Res 10(12):5275–5284. doi:10.1021/pr101183k
14. Price JC, Guan S, Burlingame A et al (2010) Analysis of proteome dynamics in the mouse brain. Proc Natl Acad Sci U S A 107(32):14508–14513. doi:10.1073/pnas.1006551107
15. Wu CC, MacCoss MJ, Howell KE et al (2004) Metabolic labeling of mammalian organisms with stable isotopes for quantitative proteomic analysis. Anal Chem 76(17):4951–4959. doi:10.1021/ac049208j
16. Bodzon-Kulakowska A, Bierczynska-Krzysik A, Dylag T et al (2007) Methods for samples preparation in proteomic research. J Chromatogr B

Analyt Technol Biomed Life Sci 849(1–2):1–31. doi:10.1016/j.jchromb.2006.10.040

17. Hoopmann MR, Chavez JD, Bruce JE (2011) SILACtor: software to enable dynamic SILAC studies. Anal Chem 83(22):8403–8410. doi:10.1021/ac2017053
18. Hsieh EJ, Shulman NJ, Dai DF et al (2012) Topograph, a software platform for precursor enrichment corrected global protein turnover measurements. Mol Cell Proteomics 11(11): 1468–1474. doi:10.1074/mcp.O112.017699
19. Guan S, Price JC, Prusiner SB et al (2011) A data processing pipeline for mammalian proteome dynamics studies using stable isotope metabolic labeling. Mol Cell Proteomics 10(12):M111.010728. doi:10.1074/mcp.M111.010728
20. Guan S, Price JC, Ghaemmaghami S et al (2012) Compartment modeling for mammalian protein turnover studies by stable isotope metabolic labeling. Anal Chem 84(9):4014–4021. doi:10.1021/ac203330z
21. Eng JK, Searle BC, Clauser KR et al (2011) A face in the crowd: recognizing peptides through database search. Mol Cell Proteomics 10(11):R111.009522. doi:10.1074/mcp.R111.009522
22. Chalkley RJ, Baker PR, Huang L et al (2005) Comprehensive analysis of a multidimensional liquid chromatography mass spectrometry dataset acquired on a quadrupole selecting, quadrupole collision cell, time-of-flight mass spectrometer: II. New developments in Protein Prospector allow for reliable and comprehensive automatic analysis of large datasets. Mol Cell Proteomics 4(8):1194–1204. doi:10.1074/mcp.D500002-MCP200
23. Kessner D, Chambers M, Burke R et al (2008) ProteoWizard: open source software for rapid proteomics tools development. Bioinformatics 24(21):2534–2536. doi:10.1093/bioinformatics/btn323
24. Sykes MT, Williamson JR (2008) Envelope: interactive software for modeling and fitting complex isotope distributions. BMC Bioinformatics 9:446. doi:10.1186/1471-2105-9-446
25. Zhang Y, Reckow S, Webhofer C et al (2011) Proteome scale turnover analysis in live animals using stable isotope metabolic labeling. Anal Chem 83(5):1665–1672. doi:10.1021/ac102755n
26. Molina E, Fernandez J, Acien FG et al (2001) Tubular photobioreactor design for algal cultures. J Biotechnol 92(2):113–131
27. Price JC, Khambatta CF, Li KW et al (2012) The effect of long term calorie restriction on in vivo hepatic proteostasis: a novel combination of dynamic and quantitative proteomics. Mol Cell Proteomics 11(12):1801–1814. doi:10.1074/mcp.M112.021204

Chapter 8

Stable Isotope Labeling in Mammals (SILAM)

Daniel B. McClatchy and John R. Yates III

Abstract

Analysis of animal models of disease is essential to the understanding of human disease and the identification of potential targets for clinical drugs. Global analysis of proteins by mass spectrometry is an important tool for these studies. Stable isotope labeling in mammals (SILAM) was developed to quantitate the proteomes of rodents using mass spectrometry. The crux of SILAM analysis is the complete labeling of all proteins in a rodent with heavy nitrogen (^{15}N). These ^{15}N tissues are then employed as an internal standard for quantitative proteomics analysis using a high-resolution and mass-accuracy mass spectrometer.

Key words Proteomics, Mass spectrometry, Quantitation, Stable isotopes, SILAM

1 Introduction

Quantitative proteomics has become an essential tool for biologists. There is a large variety of methods now available, and they can be divided into three categories: label free, in vitro labeling, and metabolic labeling [1]. The advantage of metabolic labeling is that two samples are mixed prior to any sample preparation, such as organelle isolation or immunoprecipitation, thus reducing any systematic errors that could accumulate during sample preparation and distort the quantitative analysis. Metabolic labeling is performed by incorporating heavy amino acids or heavy stable isotopes such as ^{13}C, ^{18}O, or ^{15}N into all the proteins in the proteome. The heavy label does not alter the biochemical characteristics (except the mass), so that after protein digestion heavy and light peptide pairs co-elute from the liquid chromatography (LC) capillary into the mass spectrometer. Co-elution coupled with a predictable mass difference allows algorithms to easily extract the light and heavy ion chromatogram pairs and calculate a light/heavy ratio for relative quantification. Metabolic labeling is simple when employing rapidly dividing cells with an accessible nutritional source (e.g., yeast, bacteria, or cultured cells) but more challenging when slow-growing organisms have a complex diet. *S*table *i*sotope *la*beling in

Daniel Martins-de-Souza (ed.), *Shotgun Proteomics: Methods and Protocols*, Methods in Molecular Biology, vol. 1156, DOI 10.1007/978-1-4939-0685-7_8, © Springer Science+Business Media New York 2014

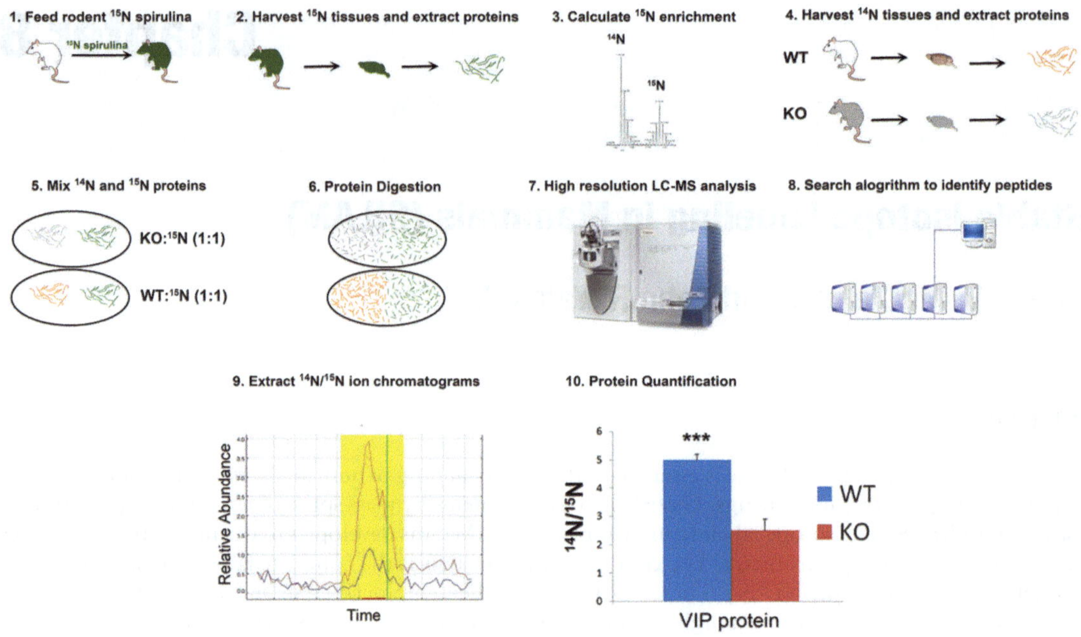

Fig. 1 Schematic of SILAM workflow

*m*ammals (SILAM) was developed to quantitate proteomic differences in animal models of disease. In SILAM, rodents are fed a diet consisting of ^{15}N-labeled spirulina (i.e., blue-green algae) as the only protein source mixed with nonprotein material for a balanced diet. Complete labeling is required for efficient quantification [2]. Incomplete labeling generates overlapping heavy and light isotopic envelopes which jeopardizes the accuracy of the ion chromatogram extraction and the calculated light/heavy ratios. The time required to label specific tissues depends upon tissue-specific protein turnover rates [2]. For example, brain and muscle take longer to label because they have the slowest turnover rates among mammalian tissues. SILAM has been reported for quantitative proteomic analysis of organelles [3, 4], phosphorylation [5, 6], protein turnover [7], and primary cells [8].

Labeling cells or tissues with heavy stable isotopes may result in changes to the proteome, known as an isotopic effect [9]. To eliminate any deleterious effects of heavy isotope labeling, the ^{15}N tissues are not used as a control, but as an internal standard for comparing two or more conditions in unlabeled or ^{14}N rodents (Fig. 1). For example, to determine the effects of a gene knockout (KO) on the brain proteome, ^{15}N brain homogenate is mixed with wild-type (WT) brain homogenate. In parallel, ^{15}N brain homogenate is mixed with KO brain homogenate. By comparing the two ratios (^{14}N KO brain/^{15}N brain and ^{14}N WT brain/^{15}N brain) for a given protein, the proteomic effect of the gene KO can be determined and any isotopic effect (i.e., the denominator) is removed [10].

A previous SILAM protocol has been published on preparing the ^{15}N diet in the laboratory [11]. We have determined that it is more cost effective to have the food prepared by professional rodent nutritionists. This chapter covers (1) generational and non-generational labeling with ^{15}N, (2) calculation of ^{15}N tissue enrichment, and (3) data analysis. SILAM analysis is applicable to any proteomic pipeline, but this chapter describes SILAM analysis of total brain homogenate of KO mice using MudPIT analysis [12] on an LTQ-Orbitrap Velos mass spectrometer.

2 Materials

1. ^{15}N spirulina food (*see* **Note 1**).
2. Mass spectrometer with high resolution and high mass accuracy (*see* **Note 2**).
3. BCA protein assay.
4. Rodents (number and species are dependent upon biological hypothesis that is being tested and the statistical analysis that will be performed on the data) (*see* **Note 3**).
5. Tryptic digestion.
   (a) ProteaseMAX (Promega) (*see* **Note 4**).
   (b) Sequencing-grade modified trypsin.
   (c) 50 mM ammonium bicarbonate: This needs to be prepared on the day of the experiment.
   (d) 8 M urea: This needs to be prepared on the day of the experiment.
   (e) Tris (2-carboxyethyl)-phosphine hydrochloride (TCEP).
   (f) Iodoacetamide (IAA): This needs to be prepared on the day of the experiment.
6. Software tools (*see* **Note 5**).
   (a) RawExtract (data extraction).
   (b) Prolucid [13] (search algorithm).
   (c) DTASelect2 to filter and organize Prolucid results (detailed description [14]).
   (d) Census for ^{15}N data analysis (manual [15]) (*see* **Note 6**).

3 Methods

3.1 Generational Labeling (See Note 7)

1. Obtain two newly weaned female rodents (*see* **Note 8**).
2. Feed two female ^{15}N pellets with ^{15}N spirulina at a 1:3 ratio with nonprotein material. House those rodents in individual cages (*see* **Note 9**).

3. Optional: Feed a female ^{14}N pellets with ^{14}N spirulina at a 1:3 ratio with nonprotein material (*see* **Note 10**).
4. After 1 month when the rodent reaches sexual maturity, introduce a "proven" male breeder rodent to each female rodent until the female is pregnant (*see* **Note 11**).
5. Keep the pregnant female on the ^{15}N through pregnancy and the weaning of the pups, which is about postnatal day 25 (p25) (*see* **Note 12**).
6. After weaning, house the rodents in 3–4 per cage separately by sex with the ^{15}N diet.
7. The female mother rodent can be sacrificed or used for the generation of more ^{15}N pups.
8. Keep the rodents on the ^{15}N diet at least until p45.

3.2 Non-generational Labeling

1. Obtain the desired number of newly weaned rodents needed for the SILAM analysis (*see* **Note 3**).
2. Feed the rodents with ^{15}N diet with ^{15}N spirulina at a 1:2 ratio with nonprotein material (*see* **Note 9**).
3. Keep the rodents on the diet for 12 weeks.

3.3 Tissue Homogenization for Enrichment Analysis

1. Thaw one flash-frozen unlabeled mouse brain and one flash-frozen ^{15}N-labeled mouse brain on ice.
2. Perform the following protocol separately for each brain (*see* **Note 13**).
3. Place brain in a 2 ml Teflon dounce grinder (*see* **Note 14**).
4. Add 1 ml of cold 4 mM HEPES, pH 7.5, 0.32 M sucrose, with protease and phosphatase inhibitors (homogenization buffer).
5. Homogenize the brain on ice with 20 strokes or until no chunks are observed.
6. Remove homogenate and place in an Eppendorf tube.
7. Add 0.5 ml of homogenization buffer to dounce grinder to remove residual homogenate.
8. After ten strokes, combine residual homogenate with first homogenate.
9. Perform protein assay on the ^{15}N and ^{14}N tissue homogenates.
10. Add the two homogenates at a 1:1 ratio based on protein concentration for a total of 100 μg.

3.4 Protein Digestion

1. Add 100 μl of fresh 50 mM ammonium bicarbonate to a vial of ProteaseMAX surfactant to produce a 1 % solution. Dilute 1:5 with 50 mM ammonium bicarbonate for a 0.2 % solution.
2. Add 25 μl of 0.2 % ProteaseMAX to 100 μg of $^{14}N/^{15}N$ brain homogenate.

3. Vortex for 30 s.
4. Add 25 μl of 8 M urea to the $^{14}N/^{15}N$ brain homogenate.
5. Vortex for 30 s.
6. Add 500 mM TCEP for a final concentration of 5 mM.
7. Incubate for 20 min on a 37 °C shaker.
8. Add 500 mM IAA for a final concentration of 10 mM.
9. Incubate for 20 min at room temperature protected from light.
10. Add 75 μl of 50 mM ammonium bicarbonate.
11. Add 1.25 μl of 1 % of ProteaseMAX.
12. Add 4 μg trypsin (1:25 with regard to the amount of protein).
13. Incubate at 37 °C for 3 h while shaking.
14. Freeze at −80 °C if not using immediately.
15. Add 5 % formic acid.
16. Centrifuge at room temperate at 18,000 × *g* for 30 min.
17. Transfer supernatant to new tube (most likely there will be a pellet).

3.5 MudPIT Column (See Note 15)

1. Cut 10–12 cm of 250 μm i.d. fused-silica capillary (PolymicroTechnologies, Phoenix, AZ) column, and prepare one fritted end with immobilized Kasil 1624 (PQ Corporation, Valley Forge, PA).
2. Pack 2.5 cm length of the column, at the fritted end, with a 5 μm Luna strong cation exchange resin (Phenomenex, Torrance, CA) and a 2.5 cm length with 10 μm Jupiter C18 resin (Phenomenex, Torrance, CA) using a stainless steel pneumatic pressure bomb.
3. Pre-equilibrate the biphasic nanocolumn with buffer A (5 % acetonitrile, 0.1 % formic acid, and 94.9 % H_2O) for 5–10 min. Pressure-load the tryptic digest (i.e., the supernatant from **step 17** in Subheading 3.4) onto the nanocolumn and desalt for 25–30 min with buffer A.
4. For preparation of analytical column, pull a 100 μm i.d. × 365 μm O.D. fused-silica capillary with a CO2 P-2000 laser puller (Sutter Instrument, Novato, CA) to get a 3–5 μm i.d. tip.
5. Pack 15 cm of the tip length with a reversed-phase 4 μm Jupiter C18 resin in a stainless steel pneumatic pressure bomb.
6. Attach the analytical column to the SCX end of the biphasic column with a zero-dead-volume (ZDV) union, and then place the entire column setup inline with Agilent1100 quaternary HPLC.

3.6 LC/LC-MS/MS Analysis

1. MudPIT analysis (*see* **Note 15**): The peptides are fractionated using a modified 12-step separation and are eluted directly into a hybrid dual-cell quadrupole linear ion trap-Orbitrap mass spectrometer (LTQ Orbitrap Velos, Thermo-Fisher) with the application of a distal 2.4 kV spray voltage. The buffer solutions are 5 % acetonitrile/0.1 % formic acid (buffer A), 80 % acetonitrile/0.1 % formic acid (buffer B), and 500 mM ammonium acetate/5 % acetonitrile/0.1 % formic acid (buffer C). *Step 1* consists of a 100-min gradient from 0 %–100 % buffer B. *Steps 2–11* have the following profile: 3 min of 100 % buffer A, 5 min of *X*% buffer C, 10 min of 100 % Buffer A, a 10-min gradient from 0 to 15 % buffer B, a 97-min gradient from 15 to 45 % buffer B, and 10 min of 100 % buffer A. The 2-min buffer C percentages (*X*) were 10, 15, 20, 25, 30, 35, 40, 45, 50, and 60 % for the 12-step analysis. In the final step, the gradient contained 3 min of 100 % buffer A, 20 min of 100 % buffer C, a 10-min gradient from 0 to 15 % buffer B, and a 107-min gradient from 15 to 70 % buffer B.
2. High-resolution mass spectrometer parameters: Peptides are analyzed using a data-dependent "Top-20" method in which fragmentation spectra are acquired for the 20 most abundant peptide ions. For each cycle, one full MS scan of m/z 300–1,600 is acquired in the Orbitrap at a resolution of 60,000 at m/z 400 with an AGC target of 10^6 and the maximal injection time of 250 ms. Each full scan is followed by the selection of up to 20 of the most intense ions for CID MS/MS analysis in the LTQ. The target value is 10,000 ions with injection time of 25 ms. The selected peptide ions are dynamically excluded from further analysis for 120 s to allow for the selection of lower abundance ions for subsequent fragmentation and detection. Ions with single or unassigned charge states are rejected. Ion selection threshold is 500 counts for MS/MS; an activation $q = 0.25$ and activation time of 10 ms are used. The m/z isolation width for MS/MS fragmentation is set to 2 Th. Normalized collision energy of 35 % is used while acquiring the MS/MS spectra.

3.7 Data Extraction

1. Open RawExtract software.
2. Select the .raw files generated from the Xcalibur (Thermo-Fisher software) data system that need to be extracted using the button "Files to convert."
3. Select the folder where the extracted MS1 and MS2 files will be located using the "Destination Directory" button.
4. Choose "Data Dependent" under "Experiment Type."
5. Select "MS1, MS2, MS3 etc." in the "Output File Format" drop-down menu.

6. Click on the buttons for both "MS1" and "MS2."
7. Click on the "Convert Files" button and the MS1 and MS2 files will generated in the folder selected as the "Destination Directory."

3.8 Database Search

The MS/MS spectra are searched with the Prolucid algorithm against the UniProt mouse database that is concatenated to a decoy database in which the sequence for each entry in the original database is reversed. The search parameters include 50 ppm peptide precursor (MS1) mass tolerance, 0.6 Da fragment (MS2) mass tolerance, and static cysteine modification of 57.02146 amu. Full trypsin specificity and length of at least seven amino acids are required for all peptides. The ^{15}N peptides are not searched for (*see* **Note 16**).

3.9 Data Filtering

The database search produces a large quantity of data. These data are assembled and filtered using the DTASelect2.0 program that groups related spectra by protein and removes those that do not pass basic data quality criteria. DTASelect2.0 combines XCorr and DeltaCN measurements using a quadratic discriminant function to compute a confidence score to achieve a user-specified false discovery rate (FDR). Identified proteins are required to have at least one peptide with a mass deviation of no more than 5 ppm and 0.01 % false-positive rate at the spectra level. Under such filtering conditions, the estimated FDR is below 1 % at the peptide and protein level. The FDR is estimated by the program from the number and quality of spectral matches to the decoy database.

3.10 Determining ^{15}N Percent Enrichment with Census (See Note 17)

High-resolution MS1 scans are employed to determine the ^{15}N peptide isotopic distributions, which are characteristic of the ^{15}N atomic percent enrichment of the peptide. Census finds the ^{15}N peptide which corresponds to the ^{14}N identification in the MS1 scan. Census predicts isotope distributions over a range of enrichments and compares the predicted distributions to experimental peptide isotope distributions. In addition, it generates a correlation factor between the matching ^{14}N and ^{15}N peptide, with zero being no correlation and 1.0 being a perfect correlation. ^{15}N peptides with a correlation factor greater than 0.7 are used to determine the average ^{15}N enrichment.

3.11 Preparing SILAM Samples (See Note 18)

Preparation of SILAM samples is identical to the protocol for ^{15}N enrichment sample described in the previous steps. To quantitate the effects of the gene KO on the brain proteome, 3 WT and 3 KO brain homogenate will be mixed separately 1:1 with the ^{15}N internal standard, resulting in 6 MS analyses.

3.12 MS Analysis of SILAM Samples

The MS analysis of the SILAM samples is identical to the ^{15}N enrichment sample described in the previous steps.

3.13 Database Searching

The database search is identical as described in Subheadings 3.8 and 3.9, except both ^{14}N (unmodified) and ^{15}N (modified) amino acids are searched for. The Prolucid search algorithm can automatically search for ^{15}N peptides, but the average ^{15}N enrichment is required (*see* **Notes 19** and **20**).

3.14 Data Analysis: $^{14}N/^{15}N$ Ratios

1. Ion chromatogram extraction: Census extracts $^{14}N/^{15}N$ ratios using the MS1 files generated from RawExtract and protein/peptide identifications from DTASelect2.0. The identified peptides are used to calculate the elemental composition and predict the corresponding isotopic distribution for the unlabeled or the labeled peptide. This information is then used to determine the appropriate m/z range from which to extract ion intensities, and all isotopic peaks with greater than 5 % of the calculated isotope cluster base peak abundance are included. The final intensities are the summation of all the isotopic peaks. Ion intensities are extracted from spectral files using a 10 ppm mass accuracy tolerance to generate chromatograms from the m/z range surrounding both the unlabeled and labeled precursor peptides in the MS1 files.
2. Filtering and outlier analysis: Census uses a linear least-square correlation to calculate the $^{14}N/^{15}N$ peptide ratios [16]. It allows users to remove poor-quality peptide ratio measurements with a correlation factor (R^2, values between 0 and 1, where 1 represents perfect correlation). This defines how well light and heavy peptides correlate in the chromatogram and is set to 0.5 as default (*see* **Note 21**). Census assigns different weights to the peptides based on their correlation factor, and then the average of these weighted peptide ratios is the protein $^{14}N/^{15}N$ ratio. In theory, every labeled peptide pair from the same protein should generate the same ratio. However, sometimes outlier peptides are generated due to incorrect sequence identification, splice isoforms not in the database, posttranslational modification, or poor S/N chromatogram. Census applies the Grubbs test to filter these peptides prior to the calculation of the protein ratio and is applicable only when there are more than three peptides detected per protein. The user can define the p-value for the Grubbs test with a higher p-value removing more "outlier" peptides.
3. Normalization: The mixing of light and heavy labeled samples should be in a 1:1 ratio based on protein content. When samples are mixed incorrectly often due to human error, quantitation will be affected. Based on the assumption that the majority of the proteome should not be affected by the biological paradigm being tested, the median of all the peptide ratios for each biological sample or MS analysis should be one [17]. If this is not true, Census allows the user to input the deviation from this median and shifts all the peptide ratios so that it has the expected median. Before any statistical analysis, all datasets are normalized so that the medians of all $^{14}N/^{15}N$ peptide ratios are one.

4. Statistical analysis: The type of statistical analysis used depends upon the experimental design and should be chosen prior to collecting the data. The census output file can be opened in Microsoft Excel and be easily manipulated for statistical analysis in Excel or other statistical programs. For quantitating the differences between WT and KO brain homogenate, a student's *t*-test is performed on proteins that were quantitated in at least two biological KO and WT replicates. Since this analysis is usually performed on hundreds to thousands of proteins, a pos hoc test is performed to control for the increase in false positives due to the large number of comparisons being performed (*see* **Note 22**).

3.15 Data Analysis: Singletons

Census quantifies all peptides (i.e., calculates a $^{14}N/^{15}N$ ratio) and generates a correlation, ranging from 0 to 1, to reflect the correlation of the ^{14}N and ^{15}N peptides. A peptide with a score greater than 0.5 is deemed a confidently quantified peptide. Peptide ratios below 0.5 may be due to noisy uninterruptible data or detection of only a light peptide and not the heavy (or vice versa), which is described as a singleton peptide [5, 6]. Census can separate these singleton peptides from noise using two parameters. First, a large fold difference between the ^{14}N and ^{15}N peptides is required and this "threshold difference" is defined by the user. A minimum of a fivefold difference is required for quality data. Second, there is a composite score ranging from 0 to 1 with 1 representing a highly confident singleton peptide. A minimum composite score of 0.5 is required. One caveat to singleton analysis is that there is a possibility that singleton peptides are actually misidentified peptides and thus there is no corresponding peptide to be found. To avoid this possibility, a singleton protein should be identified in multiple biological replicates to be considered valid (*see* **Note 23**). For example, singleton peptides from the same protein would need to be identified in all three KO biological replicates and not in any WT samples to be confident.

3.16 Data Analysis: Spectral Count

The number of spectral counts for a given protein corresponds to protein abundance [18]. Normalized spectral count comparison is a common method for many label-free strategies, such as NSAF [19]. While spectral count (SC) analysis is not as accurate as SILAM analysis [20], it can be very useful to correlate with SILAM analysis or to identify large changes (>5-fold) in the study. To perform SC analysis, compare normalized SC of only the ^{14}N identifications of the WT and KO samples (*see* **Note 24**). The same statistical methods can be employed as in the $^{14}N/^{15}N$ analysis. In addition, SC analysis can also be used to confirm the singleton analysis. For example, some proteins may only be identified in the WT brain homogenate and not in the KO brain homogenate. As in singleton analysis using Census, confident singleton protein identification using SC requires identification of proteins in all three KO biological replicates and not in any WT samples.

4 Notes

1. ^{15}N spirulina can be purchased from Cambridge Isotopes Laboratories or Sigma-Aldrich. For a minimal fee, Cambridge Isotopes Laboratories will send ^{15}N spirulina to Harlan Laboratories where nutritionists will mix the ^{15}N spirulina with the optimal nonprotein rodent feed after analyzing your specific batch of ^{15}N spirulina. Since there is some variation of ^{15}N spirulina production, this ensures the healthiest ^{15}N diet. The ^{15}N diet prepared in our laboratory resulted in less final product compared to Harlan Laboratories using the same amount of starting material. Thus, having the ^{15}N diet prepared saves time and money. Alternatively, Silantes sells ^{15}N rodent feed with the chemolithoautotrophic bacteria, *Ralstonia eutropha*, as ^{15}N-protein source. To determine the amount of ^{15}N spirulina needed, consult with veterinarians at your animal facilities or Harlan Laboratories to estimate the amount of feed needed for the specific species that will be labeled. In our hands, 2 kg of ^{15}N diet (^{15}N spirulina and nonprotein combined purchased from Cambridge Isotopes) can label ten C57BL/6 male mice for 12 weeks. Finally, both Cambridge Isotopes and Silantes sell frozen rodent tissues fully enriched with ^{15}N.
2. For efficient quantification, a high-resolution and high-mass-accuracy mass spectrometer is required, such as Orbitrap, Velos, or TOF. However, a method was developed to perform SILAM analysis on a low-resolution instrument (e.g., LTQ) using data-independent acquisition [21].
3. The goal of ^{15}N labeling is to generate an internal standard for the entire study. When calculating the number of rodents to be labeled, it is always better to have too much ^{15}N internal standard than too little. Systematic errors in quantitation can arise when there are two different pools of internal standards in one study.
4. ProteaseMAX is an MS-compatible detergent and when combined with urea generates better MS results than other MS-compatible detergents, which we previously published [22]. We routinely use this protocol for digestion of tissues. Notably, using ProteaseMAX reduces the digestion time to 3 h.
5. All software tools (RawXtract1.9.9.2, Prolucid, DTASelect2.0, and Census) are freely available for individual use and evaluation through an Institutional Software Transfer Agreement (for details, *see* http://fields.scripps.edu/downloads.php).
6. We only know of two algorithms that can analyze SILAM data: Census or QuantiSpec. QuantiSpec is restricted to MALDI TOF data [23]. Census is a modification of an earlier program from RelEx [16].

7. Generational labeling was initially performed on rats. Since mice are more prone to eat their young, non-generational labeling is less risky with mice. Also, non-generational labeling allows the researcher to control the exact number of labeled rodents. However, mice have been successfully labeled with the multiple generations using the same female [4].
8. The size of a rodent litter can range from 2 to 14 pups. There is also the possibility that the female rodent will not get pregnant. In order to ensure an adequate number of ^{15}N rodents, always perform generational labeling with a minimum of two female rodents.
9. To conserve on the expensive ^{15}N pellets, only put a few pellets on top of the cage every day, so that only one or two pellets are remaining from the previous day. Placing large amounts of ^{15}N pellets on top of the cage can lead to contamination or ruined food that needs to be thrown away. The number of pellets consumed will increase during labeling as the rodent gets older or becomes pregnant.
10. This control experiment was performed initially to assess any adverse effects of the ^{15}N labeling. We have never observed any adverse effects with the ^{15}N labeling, and the ^{14}N and ^{15}N spirulina rodents were phenotypically identical [10]. We no longer perform a control with ^{14}N spirulina, although some adverse effects have been reported [24]. Furthermore, since the ^{15}N tissues are used as an internal control, ^{14}N spirulina tissues are not used in the SILAM analysis.
11. Rodents reach sexual maturity around 6 weeks, but this varies depending on species. The male can be given regular rodent chow and be placed in the female cage in the evening and removed in the morning. The female will often get pregnant the first week. Once the female is pregnant, males can remain in their own cage or be sacrificed. Observation of a vaginal plug (a waxy congealed plug formed in the vagina of the female from the secretions of the accessory sex glands in the male) is a sign of pregnancy. The vaginal plug is transient and can easily be overlooked by someone without experience. In most cases, the increased weight (visually or calculated) of the female rodent is used to determine pregnancy.
12. The weaning of the pups and the exact time are left up to the staff in the vivarium in our studies.
13. It is recommended to sacrifice only one ^{15}N rodent for the enrichment analysis. If enrichment does not meet expected results, then the other ^{15}N rodents can remain in the ^{15}N diet longer. The unlabeled tissue does not need to be the same unlabeled tissue that will be quantitated in the subsequent SILAM analysis.

14. When working with tissues, the grinder, all tubes, and buffers need to be pre-chilled.
15. Delahunty et al. [25] is a detailed protocol on preparing MudPIT biphasic column, interfacing to the mass spectrometer and running the MudPIT MS method.
16. The ^{15}N peptides are not searched to ensure a completely unbiased enrichment calculation. Search algorithms tend to perform poorly with ^{15}N peptides with low enrichment. Thus, if the enrichment calculation is only performed on ^{15}N peptides that are identified, then it may be falsely elevated. Instead, the corresponding ^{15}N peptide to the ^{14}N peptide identification in the MS1 is employed.
17. At least 500 ^{14}N peptide identifications are required for accurate ^{15}N enrichment. Following the above labeling protocols should generate a completely ^{15}N-enriched rodent (>95 %). Fast turnover tissues, such as liver, should have an enrichment close to 98 %, while slow protein turnover tissues such as brain or muscle will have an enrichment around 95 % [2]. The enrichment analysis will confirm that the labeling was successful and provide a percent enrichment which is required for data analysis using Census and some search algorithms.
18. The details of these procedures are dependent on the tissue of interest and the biological experiment. The researcher should follow the exact procedures that will be performed on the unlabeled tissues. Some protocols will allow the ^{15}N to be flash-frozen, while others may require coordinating the ^{15}N tissue harvest with the unlabeled tissue harvest. At the earliest possible point, a protein assay should be performed on the ^{15}N and unlabeled tissues and then mixed 1:1 before any extensive sample preparation (e.g., isolation of an organelle or a protein complex). The ^{15}N tissues should be processed together to generate one ^{15}N internal standard for the entire study. To ensure proper 1:1 mixing, one protein assay with multiple replicates should be performed with all the ^{14}N samples and the one pooled ^{15}N sample. The number of $^{14}N/^{15}N$ samples depends upon the experimental design and the statistical analysis that will be performed in the data analysis. Routinely, to compare brains of WT and KO mice, a minimum of three WT and three KO brain homogenates will be mixed separately 1:1 with the ^{15}N internal standard. Therefore, this study will result in six MS analyses. Increasing the number of WT/KO mice will increase the statistical power of the study. If enough biological material is available, increasing the number of MS technical replicates will increase the number of protein identified

and quantified. For example, each $^{14}N/^{15}N$ mixture could be analyzed by two or three MS analyses and then combined into one biological replicate.

19. The ^{15}N rodents do not always have to be identical to the ^{14}N rodents. The crucial point is that the majority of the ^{14}N peptides need to have corresponding ^{15}N peptides, and this is the advantage of employing the ^{15}N tissues as an internal standard. How much the ^{15}N internal standard can deviate from the ^{14}N samples depends upon the specific biology being examined. For example, in many studies the ^{15}N rodents are a different strain than the ^{14}N rodents. This allows one ^{15}N internal standard pool to be used for multiple studies. We have successfully quantitated primary neuronal culture with ^{15}N rodent brain [8]. Furthermore, we have quantitated differences between different ^{14}N brain regions using the entire ^{15}N whole brain as the internal standard [3]. To the extreme, we have quantitated the differences between the liver and brain proteomes using ^{15}N liver as an internal standard [5]. When the ^{15}N internal standard does deviate from the ^{14}N samples, it is important to perform singleton (Subheading 3.15) and spectral count analysis (Subheading 3.16).
20. Most search algorithms should allow the user to manually input the ^{15}N modification for each amino acid based on the number of nitrogen for each amino acid. To our knowledge, these algorithms will assume 100 % ^{15}N enrichment.
21. Although ^{14}N and ^{15}N samples are mixed 1:1, there are usually less ^{15}N identifications than ^{14}N identifications.
22. Census generates a $^{14}N/^{15}N$ ratio for every peptide. After filtering, the quantitation efficiency (i.e., the percentage of identified peptides with a confident $^{14}N/^{15}N$ ratio) usually ranges between 70 and 95 %. The more complex sample (e.g., total homogenate) will have lower quantitation efficiency compared to a less complex sample (e.g., isolated mitochondria).
23. There are many different post hoc tests such as Tukey, Bonferroni, Dunnett, and Benjamini–Hochberg (BH). The downside of these tests is that they are very conservative and reduce the number of significant comparisons considerably. For this reason, many published quantitative proteomic datasets omit these tests. Bonferroni is the most conservative; BH is less so and appears to be better suited for proteomic datasets.
24. Due to the very large difference in the ^{14}N and $1^{15}N$ chromatograms, it is difficult to generate an accurate $^{14}N/^{15}N$ ratio for a singleton peptide. Thus, there is no value assigned to the difference and is categorized as a singleton.

Acknowledgements

We acknowledge funding from NIH grants R01 MH067880 and P41 RR011823. We also like to thank Dr. Claire Delahunty for her comments on the manuscript.

References

1. Liao L, McClatchy DB, Yates JR (2009) Shotgun proteomics in neuroscience. Neuron 63(1):12–26
2. McClatchy DB et al (2007) 15 N metabolic labeling of mammalian tissue with slow protein turnover. J Proteome Res 6(5): 2005–2010
3. McClatchy DB et al (2012) Dynamics of subcellular proteomes during brain development. J Proteome Res 11(4):2467–2479
4. Butko MT et al (2013) In vivo quantitative proteomics of somatosensory cortical synapses shows which protein levels are modulated by sensory deprivation. Proc Natl Acad Sci U S A 110(8):E726–E735
5. McClatchy DB et al (2011) Differential proteomic analysis of mammalian tissues using SILAM. PLoS One 6(1):e16039
6. Liao L et al (2008) Quantitative analysis of brain nuclear phosphoproteins identifies developmentally regulated phosphorylation events. J Proteome Res 7(11):4743–4755
7. Price JC et al (2010) Analysis of proteome dynamics in the mouse brain. Proc Natl Acad Sci U S A 107(32):14508–14513
8. Liao L et al (2012) 15 N-labeled brain enables quantification of proteome and phosphoproteome in cultured primary neurons. J Proteome Res 11(2):1341–1353
9. Filiou MD et al (2012) The 15 N isotope effect in Escherichia coli: a neutron can make the difference. Proteomics 12(21):3121–3128
10. Wu CC et al (2004) Metabolic labeling of mammalian organisms with stable isotopes for quantitative proteomic analysis. Anal Chem 76(17):4951–4959
11. McClatchy DB, Yates JR III (2008) Stable isotope labeling of mammals (SILAM). CSH Protoc 2008
12. Washburn MP, Wolters D, Yates JR III (2001) Large-scale analysis of the yeast proteome by multidimensional protein identification technology. Nat Biotechnol 19(3):242–247
13. Xu T et al (2006) ProLuCID, a fast and sensitive tandem mass spectra-based protein identification program. Mol Cell Proteomics 5(10):1(S174)
14. Cociorva D, L Tabb D, Yates JR (2007) Validation of tandem mass spectrometry database search results using DTASelect. Curr Protoc Bioinformatics Chapter 13:Unit 13 4
15. Park SK, Yates JR III (2010) Census for proteome quantification. Curr Protoc Bioinformatics Chapter 13:Unit 13 12 1-11
16. MacCoss MJ et al (2003) A correlation algorithm for the automated quantitative analysis of shotgun proteomics data. Anal Chem 75(24): 6912–6921
17. Ting L et al (2009) Normalization and statistical analysis of quantitative proteomics data generated by metabolic labeling. Mol Cell Proteomics 8(10):2227–2242
18. Liu H, Sadygov RG, Yates JR III (2004) A model for random sampling and estimation of relative protein abundance in shotgun proteomics. Anal Chem 76(14): 4193–4201
19. Zhang Y et al (2010) Refinements to label free proteome quantitation: how to deal with peptides shared by multiple proteins. Anal Chem 82(6):2272–2281
20. Li Z et al (2012) Systematic comparison of label-free, metabolic labeling, and isobaric chemical labeling for quantitative proteomics on LTQ Orbitrap Velos. J Proteome Res 11(3):1582–1590
21. Venable JD et al (2004) Automated approach for quantitative analysis of complex peptide mixtures from tandem mass spectra. Nat Methods 1(1):39–45
22. Chen EI et al (2008) Comparisons of mass spectrometry compatible surfactants for global analysis of the mammalian brain proteome. Anal Chem 80(22):8694–8701
23. Haegler K et al (2009) QuantiSpec–Quantitative mass spectrometry data analysis of (15) N-metabolically labeled proteins. J Proteomics 71(6):601–608
24. Frank E et al (2009) Stable isotope metabolic labeling with a novel N-enriched bacteria diet for improved proteomic analyses of mouse models for psychopathologies. PLoS One 4(11):e7821
25. Delahunty C, Yates JR III (2005) Protein identification using 2D-LC-MS/MS. Methods 35(3):248–255

Chapter 9

Analysis of Individual Protein Turnover in Live Animals on a Proteome-Wide Scale

Stefan Reckow and Christian Webhofer

Abstract

Classical quantitative proteomics studies focus on the relative or absolute concentration of proteins at a given time. In contrast, the investigation of protein turnover reveals the dynamics leading to these states. Analyzing the balance between synthesis and degradation of individual proteins provides insights into the regulation of protein concentration and helps understanding underlying biological processes. Comparing the half-lives of proteins allows detecting functional relationships and common regulation mechanisms. Moreover, comparing turnover of individual brain and plasma proteins between control- and treatment-groups indicates turnover changes induced by the treatment.

Here, we describe a procedure for determining turnover information of individual proteins in mice on a proteome-wide scale based on partial ^{15}N metabolic labeling. We will outline the complete experimental workflow starting from ^{15}N labeling the animals over sample preparation and mass spectrometric measurement up to the analysis of the data.

Key words Protein turnover, Synthesis, Degradation, Partial ^{15}N metabolic labeling, Mass spectrometry

1 Introduction

Protein turnover describes the balance between protein synthesis and degradation. The investigation of protein turnover has been recognized as crucial addition to the classical quantitative proteomics studies, where relative or absolute concentrations of proteins at a given time point are determined. Elucidating the dynamics, which lead to potential concentration differences, is indispensable for a deeper understanding of the biology [1].

Typically, the analysis of proteome dynamics is based on the incorporation or loss of a tracer [2]. The rate of the gain or loss of the tracer provides a measure of the protein turnover. The advent

Stefan Reckow and Christian Webhofer contributed equally.

Daniel Martins-de-Souza (ed.), *Shotgun Proteomics: Methods and Protocols*, Methods in Molecular Biology, vol. 1156, DOI 10.1007/978-1-4939-0685-7_9, © Springer Science+Business Media New York 2014

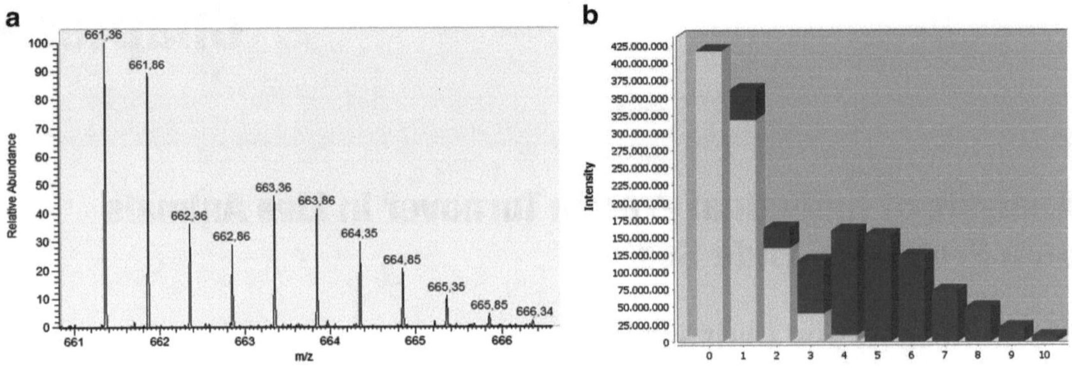

Fig. 1 Deconvolution of signals of the peptide VLGTAFDPFLGGK after 5 days of labeling. (**a**) A typical mass spectrum from a partial labeling experiment, where peaks of the unlabeled peptide significantly overlap with signals from labeled molecules. (**b**) Assuming the monoisotopic peak (all ^{14}N) is unaffected by the overlap, the intensity distribution of the unlabeled peptides can be estimated by the peptide's natural frequency distribution scaled to the monoisotopic intensity (*light peaks*). Simple subtraction from the whole observed distribution then yields the intensity of the labeled fraction (*dark peaks*)

of mass spectrometry (MS) based proteomics made stable isotope labeled tracers, mostly heavy labeled amino acids, the preferred way of labeling. The extent of labeling with a stable isotope is expressed as the relative isotope abundance (RIA). In most variants of the label incorporation approach, it is essential to measure the RIA of the target products (i.e., proteins) as well as the RIA of the precursor pool (i.e., aminoacyl-tRNA) [3]. This is straightforward in cell culture experiments, where the RIA of the precursor pool rapidly equilibrates with the RIA of the media. In contrast, metabolic labeling of multicellular species, in particular higher animals, is complicated by the significantly more complex metabolism. In general, a fast and complete labeling cannot be assumed and an accurate estimate of the precursor RIA is difficult to obtain. Throughout the labeling period, different molecules of a protein display different RIAs, depending on the time point of their synthesis. During mass spectrometry, signals of those proteins overlap significantly making the analysis difficult (Fig. 1a). Alternative procedures and analysis tools are required to address this challenge.

Here, we present a workflow to measure turnover in mice utilizing partial stable isotope ^{15}N metabolic labeling combined with high resolution mass spectrometry. The animal diet is fully heavy nitrogen (^{15}N) labeled, leading to efficient labeling of newly synthesized proteins. After sample collection and preparation peptides of digested proteins are analyzed using mass spectrometry. Protein turnover information is extracted from the MS data by quantifying the relative abundance of unlabeled and labeled peptide populations. Due to the high label content in the diet, most peptides of newly synthesized proteins will have at least one heavy nitrogen incorporated soon after the start of the labeling period. It follows that for each peptide, its monoisotopic signal in the mass spectrum largely represents the unlabeled (i.e., previously existing) population of that peptide, while remaining peaks of its isotopologue distribution may be heavily overlapping and convoluted with signals from labeled (i.e., newly synthesized) populations. For an identified peptide with known amino acid sequence, the total intensity of the unlabeled fraction T_u can be estimated by

$$T_u = \frac{i_0}{f_0}$$

where i_0 is the observed monoisotopic intensity and f_0 is the expected fractional monoisotopic frequency calculated from the peptide's elemental composition (Fig. 1b). A measure of turnover is derived by means of the *labeled peptide fraction*

$$\mathrm{LPF} = 1 - \frac{T_u}{T},$$

where T is the peptide's total observed intensity.

The observed LPF increases with time as the fraction of unlabeled protein is continuously decreased by degradation and the fraction of labeled protein is continuously increased by protein synthesis. Under steady state conditions, where the total concen tration of the protein does not change, the degradation rate constant λ of a protein can be estimated by

$$\lambda = \frac{-\ln(1 - \mathrm{LPF}_t)}{t}$$

where LPF_t is the observed LPF at time point t.

2 Materials

2.1 Partial ^{15}N Metabolic Labeling

1. ^{15}N-labeled bacteria-based diet (U-^{15}N-SILAM-Mouse, Silantes Gmbh) and ^{14}N (unlabeled) bacteria-based diet (U-1^{4}N-SILAM-Mouse, Silantes). Keep bacteria based diet at 4 °C.
2. Standard mouse diet.
3. Isoflurane.
4. 0.9 % NaCl perfusion solution.

2.2 Proteomics Sample Preparation

1. Electrophoretic equipment.
2. SDS sample buffer: 1.25 ml 1 M Tris–HCl pH 6.8, 4.6 ml 20 % SDS, 0.6 g dithiothreitol, 4 ml glycerol, 0.01 % Bromophenol Blue. Add dH_2O up to 10 ml.
3. Coomassie stain solution: 1 g Coomassie Brilliant Blue R-250, 500 ml methanol, 100 ml acetic acid, 400 ml dH_2O.

4. Gel destain solution: 100 ml isopropanol, 100 ml acetic acid, 800 ml dH_2O.
5. 25 mM ammonium bicarbonate in 50 % acetonitrile (ACN).
6. 10 mM dithiothreitol in 25 mM ammonium bicarbonate.
7. 50 mM iodoacetamide in 25 mM ammonium bicarbonate.
8. 5 ng/μl trypsin in 25 mM ammonium bicarbonate.
9. 2 % formic acid in 50 % ACN.

2.3 LC-MS/MS

1. Eluent A: 0.1 % formic acid.
2. Eluent B: 0.1 % formic acid in 95 % ACN.
3. Nano HPLC-2D system (e.g., Eksigent).
4. Zorbax-C18 (5 μm) guard column, 300 μm × 5 mm (Agilent Technologies).
5. In-house packed fused silica 3 μm RP-C18 (Dr. Maisch GmbH), 75 μm × 15 cm analytical column.
6. LTQ-Orbitrap mass spectrometer (Thermo Fisher Scientific).

2.4 Software

A large variety of different software packages are available for analyzing mass spectrometry data, both commercial and free/open source. Our descriptions are based on the following selection of programs to streamline our data analysis. However, the technique of inferring turnover information can be used in combination with other tools as well. In the descriptions and Notes, we will point out important aspects of the individual steps, which are independent of specific software (*see* **Note 1**).

1. *Sequest* for identification of peptide signals from .raw MS data files [4]; implemented in BioWorks v3.3 (Thermo Fisher Scientific).
2. *PeptideProphet* for propabilistic analysis and validation of peptide identifications [5]; implemented in the Trans Proteomic Pipeline v4.5 [6].
3. *ProteinProphet* for probabilistic inference of identified proteins and protein groups [7]; implemented in the Trans Proteomic Pipeline v4.5.
4. *ProTurnyzer* (available from the authors upon request) to extract ion chromatograms and infer turnover information for individual proteins in a high-throughput manner [8].

3 Methods

3.1 Mouse Labeling Protocol and Sample Collection

1. Feed mice with unlabeled ^{14}N bacteria-based diet for up to 14 days in order to habituate the mice to the novel diet.
2. Switch the mouse diet to labeled ^{15}N bacteria-based diet.

3. Depending on the research question, label mice for different time periods. Typically 1–3 days of labeling for plasma protein turnover analysis and 7–14 days for brain protein turnover analysis may be a good starting point (*see* **Note 2**).
4. Monitor diet uptake and mouse weights in order to assure for comparable ^{15}N uptake (*see* **Note 3**).
5. Anesthetize animals with isoflurane and draw blood by heart puncture. Perfuse organs using 0.9 % NaCl to remove residual blood. Sample the brain and dissect brain regions of interest. Centrifuge blood samples at 1,300 × *g* for 10 min at 4 °C to separate plasma.
6. Immediately deep-freeze brain regions and plasma in liquid nitrogen and store tissues at −60 °C until further analysis.

3.2 Sample Preparation

1. Enrich subproteomes of interest using the corresponding homogenization buffers and standard procedures (*see* **Note 4**).
2. Mix 100 μg of the protein fraction with SDS sample buffer, resolve proteins by 12 % SDS-PAGE, briefly stain the gel with Coomassie Brilliant Blue R-250 and destain the gel thoroughly (*see* **Notes 5** and **6**).
3. Slice every gel lane into 10–20 fractions that are cut into 1 mm^3 pieces. Wash gel pieces twice with 25 mM ammonium bicarbonate/50 % ACN, reduce proteins with 10 mM dithiothreitol for 30 min at 56 °C and derivatize reduced cysteines with 50 mM iodoacetamide for 30 min at RT in the dark. Wash again twice, dry and rehydrate gel pieces in trypsin solution (5 ng/μl) for 15 min on ice.
4. Remove the remaining trypsin solution, cover gel pieces with 25 mM ammonium bicarbonate, and allow digestion to proceed overnight at 37 °C.
5. Extract peptides twice with 2 % formic acid/50 % acetonitrile and lyophilize peptide samples. Store lyophilized samples at −20 °C until further analysis.

3.3 LC-MS/MS Analysis

1. Dissolve lyophilized peptides in 10 μl 0.1 % formic acid prior to LC-MS/MS analysis.
2. Load 5 μl of each fraction onto a C18 guard column and online desalt samples for 10 min at 3 μl/min 98 % Eluent A using a nanoflow HPLC-2D system coupled online to an LTQ-Orbitrap mass spectrometer via a nano-electrospray ion source (*see* **Note 7**).
3. Separate peptides on an in-house packed C18 analytical column using a linear gradient from 2 % B to 45 % B within 90 min at a flow rate of 200 nl/min.

4. The mass spectrometer is operated in positive mode (spray voltage 1.7–2.1 kV) applying a data-dependent automatic scan switch between MS and MS/MS acquisition. Full scans are recorded in the Orbitrap mass analyzer (profile mode, m/z 380–1,600, resolution $R = 30{,}000$ at m/z 400, AGC target value 1e6). The CID MS/MS analyses of the five most intense peptide ions for each scan are recorded in the LTQ mass analyzer (centroid mode, AGC target value 3e4, dynamic exclusion 120 s) (*see* **Note 8**).

3.4 Peptide Identification and Turnover Calculation

1. Subject mass spectrometry result files (.raw) to the Sequest software to identify peptide signals using a concatenated forward/reverse (decoy) database (Uniprot) (*see* **Note 9**). Set mass accuracy to 10 ppm and 1 Da for MS and MS/MS, respectively. Allow two missed cleavages and set carboxyamidomethylation of cysteine as fixed and oxidation of methionine as variable modifications.
2. Run *PeptideProphet* and *ProteinProphet* on the search results. Here, use the decoy peptide hits to probabilistically score each peptide and protein/protein group identification (*see* **Note 10**).
3. Use ProTurnyzer to extract ion chromatograms for all peaks of the identified peptides and calculate corresponding labeled peptide fractions. For peak extraction, set the maximum RIA to the expected maximum label enrichment in your experiment (*see* **Note 11**). Also set the mass accuracy according to your instrument settings (*see* **Note 12**).

4 Notes

1. The *ProTurnyzer* software is optimized for Thermo's .RAW files combined with TPP identification data. The binary .RAW files can be efficiently processed and contain noise information, which can be used to filter spectra. Alternatively, the mzXML format can be used with limited support. TPP's prot.xml files naturally includes protein grouping, which is explicitly supported in *ProTurnyzer*. Using default .OUT files from *Sequest* as input for the TPP is straightforward, but for large datasets it is more convenient to bundle Sequest's output in binary .SRF files. However, a conversion of .SRF to the required pep.xml format needs to be done programmatically, e.g., with help of the MS library *mspire* [9]. Alternatively, *Mascot* [10] search results can be used directly in *ProTurnyzer* (without using the TPP).
2. Different mouse strains may exhibit distinct feeding behavior and metabolism and labeling time may need to be adjusted accordingly.

3. Comparable uptake of ^{15}N label is important for the generation of consistent data facilitating data analysis, data interpretation and manual data validation. The described principle of protein turnover determination, however, does not require identical ^{15}N uptake for accurate protein turnover calculation.
4. Protein turnover analyses have revealed that proteins derived from distinct subcellular compartments turn over differently and therefore careful choice of subproteome of interest is advisable.
5. Performing *SDS gel electrophoresis* followed by in-gel digestion in contrast to for example in-solution digestion followed by *Isoelectric Focusing* retains protein molecular weight information which may be indispensable for turnover analysis of protein isoforms.
6. For additional determination of relative changes in protein concentration a fully ^{15}N labeled standard could be mixed into partially labeled samples. However, the labeling time must be carefully adjusted to avoid overlaps between signals from the fully labeled standard and from partially labeled peptide.
7. Instead of an Orbitrap, other high-resolution mass analyzers may be used for protein turnover studies. For instance, Quadrupole Time-of-Flight mass spectrometers may be favorable as they have been shown to provide improved measurements of isotopologue ratios even for low abundant peptide signals.
8. There is a trade-off between mass resolution and signal quality. Higher mass resolution will lead to sharper peaks, but also reduces sensitivity, signal-to-noise ratio and scan-to-scan isotopologue signal stability. Spectral quality is especially important for accurate protein turnover calculation, so care must be taken to limit mass resolution to reasonable values. For further increasing spectra quality, mass acquisition may be run in Selected Ion Monitoring (SIM) mode.
9. It is recommended to perform searches against a protein database which contains decoy (either reversed or random) protein sequences along with the actual protein sequences to be able to estimate the false discovery rate (FDR) (*see* **Note 10**). For PeptideProphet, all protein names of the decoy entries in the concatenated .fasta database must start with a common prefix such as "DECOY". Tools within the TPP assist in generating an appropriate database file.
10. To avoid large numbers of false positive protein identifications, the results of database searches need to be thoroughly scored and filtered. A score cutoff can be determined for a given FDR (typically 1 %) and only peptides with higher scores should be considered in the further analysis. For search results from the TPP, ProTurnyzer accepts the unfiltered results and automatically filters the list of proteins to process for an FDR of 1 %.

11. The parameter *maxRIA* [0…1] controls the number of peaks which are extracted from the spectra for each identified peptide. When set to 0.5 (50 % ^{15}N), all potential peaks from the monoisotopic peak up to the highest peak of a 50 % labeled peptide are considered. Limiting the number of extracted peaks can remove interfering noise signals. However, if the actual labeling in the data is higher than the configured value, signals might be missed, leading to erroneous results.
12. The parameter *detectPPM* determines the maximum allowed mass difference between two signals in consecutive scans, so that both signals are considered the same peak during the creation of extraction ion chromatograms. This value should match the instrument's mass accuracy. A value of 7 ppm is reasonable when using an LTQ Orbitrap.

References

1. Pratt JM, Petty J, Riba-Garcia I et al (2002) Dynamics of protein turnover, a missing dimension in proteomics. Mol Cell Proteomics 1:579–591
2. Schoenheimer R, Rittenberg D, Foster GL et al (1938) The application of the nitrogen isotope 15N for the study of protein metabolism. Science 88:599–600
3. Doherty MK, Beynon RJ (2006) Protein turnover on the scale of the proteome. Expert Rev Proteomics 3:97–110
4. Eng J, McCormack A, Yates J (1994) An approach to correlate tandem mass spectral data of peptides with amino acid sequences in a protein database. J Am Soc Mass Spectrom 5:976–989
5. Keller A, Nesvizhskii AI, Kolker E, Aebersold R (2002) Empirical statistical model to estimate the accuracy of peptide identifications made by MS/MS and database search. Anal Chem 74: 5383–5392
6. Deutsch EW, Mendoza L, Shteynberg D et al (2010) A guided tour of the trans-proteomic pipeline. Proteomics 10:1150–1159
7. Nesvizhskii AI, Keller A, Kolker E, Aebersold R (2003) A statistical model for identifying proteins by tandem mass spectrometry. Anal Chem 75:4646–4658
8. Zhang Y, Reckow S, Webhofer C et al (2011) Proteome scale turnover analysis in live animals using stable isotope metabolic labeling. Anal Chem 83:1665–1672
9. Prince JT, Marcotte EM (2008) mspire: mass spectrometry proteomics in Ruby. Bioinformatics 24(23):2796–2797
10. http://www.matrixscience.com

Part IV

In Vitro Stable Isotope Labeling

Chapter 10

Determining Protein Subcellular Localization in Mammalian Cell Culture with Biochemical Fractionation and iTRAQ 8-Plex Quantification

Andy Christoforou, Alfonso Martinez Arias, and Kathryn S. Lilley

Abstract

Protein subcellular localization is a fundamental feature of posttranslational functional regulation. Traditional microscopy based approaches to study protein localization are typically of limited throughput, and dependent on the availability of antibodies with high specificity and sensitivity, or fluorescent fusion proteins. In this chapter we describe how Localization of Organelle Proteins by Isotope Tagging (LOPIT), a mass spectrometry based workflow coupling biochemical fractionation and iTRAQ™ 8-plex quantification, can be applied for the high-throughput characterization of protein localization in a mammalian cell culture line.

Key words Isobaric tagging, iTRAQ, LOPIT, Organelle proteomics, Protein localization, Spatial proteomics, Subcellular localization

1 Introduction

Subcellular compartmentalization of proteins and other biomolecules is a fundamental feature of eukaryotic cells. By localizing and sequestering proteins within subcellular niches, cells establish a diverse range of microenvironments, enabling the myriad of biological processes required for cellular function to proceed. Re-localization of proteins between these compartments in response to a stimulus or perturbation also provides a time and energy efficient mechanism for signal transduction events. Consequently, protein subcellular localization is a fundamental feature of posttranslational regulation of cell behavior.

Subcellular localization of proteins is typically determined by microscopy, utilizing either fluorophore coupled antibodies or fluorescent fusion proteins. Despite the ubiquity of these techniques, they are not without limitations. The former approach is heavily dependent on the availability of high specificity and high sensitivity antibodies, while the latter is vulnerable to aberrant protein

Daniel Martins-de-Souza (ed.), *Shotgun Proteomics: Methods and Protocols*, Methods in Molecular Biology, vol. 1156, DOI 10.1007/978-1-4939-0685-7_10, © Springer Science+Business Media New York 2014

trafficking due to the effects of the bulky fusion moiety on protein topology, or of nonnative rates of protein synthesis and degradation. Both approaches are also limited in throughput. The versatility and throughput of shotgun proteomics therefore make it highly complementary to these approaches.

Localization of Organelle Proteins by Isotope Tagging (LOPIT) is a quantitative mass spectrometry (MS) workflow for determining protein subcellular localization in high throughput, which combines biochemical fractionation techniques and shotgun proteomics with isobaric tag quantification [1, 2]. This approach was first developed to identify resident proteins of the Golgi apparatus in *Arabidopsis thaliana* callus, but has also been applied to numerous other model organisms, including DT40 lymphocytes [3] and whole *Drosophila melanogaster* embryos [4].

LOPIT is based on the principles of analytical fractionation pioneered by Albert Claude and Christian de Duve. While it is near impossible to achieve complete purification of organelles by biochemical means, it is possible to generate quantitative distributions of organelles across a fractionation scheme that are characteristic and distinctive. Quantifying these distributions therefore provides a more robust and nuanced classification of protein localization than can be achieved by methods which are dependent on organelle "purification", thus making this approach less susceptible to technical artifact [5]. In the analytical fractionation experiments of Claude and de Duve, enzymatic assays were used as a means to quantify the distributions of organelles following fractionation. With LOPIT, these enzymatic assays are substituted with shotgun LC-MS/MS analysis and relative quantification by isobaric tagging, allowing the subcellular distribution of hundreds or thousands of proteins to be characterized in a single experiment.

In this chapter we describe a generic LOPIT workflow for the analysis of protein localization in HEK293T cells, utilizing iTRAQ 8-plex quantification, and analyzed on an LTQ Orbitrap Velos. Many of the techniques and procedures described are transferable to other model organisms.

2 Materials

2.1 Cell Culture

1. Eight culture flasks (175 cm^2 growth area) of confluent HEK293T cells.
2. Versene™ (EDTA-saline).
3. Phosphate buffered saline (PBS), without calcium or magnesium, pH 7.2.

2.2 Stock Solutions

1. 0.1 M HEPES–HCl (pH 7.4), store at 4 °C.
2. 0.1 M EDTA (pH 8), store at 4 °C.
3. 0.1 M magnesium acetate tetrahydrate, store at 4 °C.

2.3 Cell Lysis

1. Detergent-free lysis buffer (50 mL per experiment): 0.25 M sucrose, 10 mM HEPES–HCl (pH 7.4), 2 mM EDTA, 2 mM magnesium acetate tetrahydrate, protease inhibitor cocktail. Store at 4 °C (*see* **Note 1**).
2. 6× concentrated lysis buffer (10 mL per experiment): 60 mM HEPES–HCl (pH 7.4), 12 mM EDTA, 12 mM magnesium acetate tetrahydrate, protease inhibitor cocktail. Store at 4 °C.
3. Ball bearing homogenizer (Isobiotec, Heidelberg, Germany).
4. Benzonase® endonuclease, >90 % purity.
5. 1 and 2 mL syringes, without needles.

2.4 Subcellular Fractionation

1. Iodixanol working solution (15 mL per experiment): 12.5 mL iodixanol (60 % w/v), 2.5 mL 6× concentrated lysis buffer (*see* **Note 2**).
2. Optima™ XL-80 ultracentrifuge (Beckman) or equivalent model.
3. SW55Ti swinging bucket rotor (Beckman).
4. VTi65.1 vertical angle rotor (Beckman).
5. Polyallomer or Ultra-Clear™ ultracentrifuge tubes, 5.2 mL capacity (Beckman).
6. Polyallomer Quick-Seal® tubes, 11.2 mL capacity (Beckman).
7. Optima™ MAX-XP benchtop ultracentrifuge (Beckman) or equivalent model.
8. TLA-55 fixed angle rotor (Beckman).
9. Polyallomer ultracentrifuge tubes, 1.5 mL capacity (Beckman).
10. Auto Densi-Flow® peristaltic pump (Labconco, Kansas City, MO, USA).
11. Hand held refractometer.
12. Wide bore needle with length greater than 100 mm.
13. Analytical grade acetone.

2.5 Protein Digestion and iTRAQ 8-Plex Labeling

1. iTRAQ dissolution buffer: 8 M urea, 0.1 % SDS, 25 mM triethylammonium bicarbonate (TEAB, pH 8.5) (*see* **Note 3**).
2. Bicinchoninic acid (BCA) assay reagents.
3. iTRAQ 8-plex labeling kit (AB Sciex).
4. Sequencing grade porcine trypsin (40 μg).
5. 1 M TEAB, pH 8.5.
6. Vacuum centrifuge with cold trap.
7. SepPak C18 cartridges, 100 mg bed volume (Waters).
8. SepPak loading buffer: 0.1 % (v/v) trifluoroacetic acid (TFA) in HPLC grade water.
9. SepPak desalting buffer: 0.05 % (v/v) acetic acid in HPLC grade water.

10. SepPak elution buffer: 70 % (v/v) acetonitrile + 0.05 % (v/v) acetic acid.
11. HPLC grade acetonitrile.

2.6 LC-MS/MS Analysis

1. LTQ Orbitrap Velos (Thermo Fisher Scientific) with PicoView electrospray source (New Objective) (*see* **Note 4**).
2. nanoAcquity UPLC (Waters).
3. Acquity UPLC with diode array detector (Waters).
4. Acquity UPLC BEH C18 column (2.1 mm i.d. × 150 mm, 1.7 μm particle size) (Waters).
5. nanoAcquity UPLC Symmetry C18 trap column (180 μm i.d. × 20 mm, 5 μm particle size) (Waters).
6. nanoAcquity UPLC BEH C18 column (75 μm i.d. × 150 mm, 1.7 μm particle size) (Waters).
7. 20 mM ammonium formate (pH 10.0) in HPLC grade water.
8. 20 mM ammonium formate (pH 10.0) + 80 % acetonitrile.
9. HPLC grade water + 0.1 % formic acid.
10. HPLC grade acetonitrile + 0.1 % formic acid.

2.7 Data Analysis Software

1. Mascot (Matrix Science, London, UK).
2. msconvert (ProteoWizard Project) [6].
3. MSFileReader 2.2 library (Thermo).
4. Proteome Discoverer (Thermo), optional.
5. R with data analysis packages installed (*see* **step 4** in Subheading 3.8), optional.

3 Methods

An overview of the LOPIT workflow is outlined in Fig. 1.

3.1 Cell Harvesting and Lysis

In order to minimize organelle degradation, all steps from cell lysis up to the end of Subheading 3.3 are ideally performed without pauses. In practice, this may require a continuous effort of 20 h or more, including a centrifugation step of 8 h. Steps at which the protocol may be paused are however noted.

1. Decant culture medium from cell culture flasks and wash cells with PBS. Decant PBS and add 4 mL Versene to each 175 cm^2 culture flask. Incubate at 37 °C for 5 min to allow Versene to disrupt cell adhesion (*see* **Note 5**). Detach cells from the flask by gentle tapping. Transfer detached cells into two 50 mL tubes, dilute to tube capacity with PBS, and place on ice.

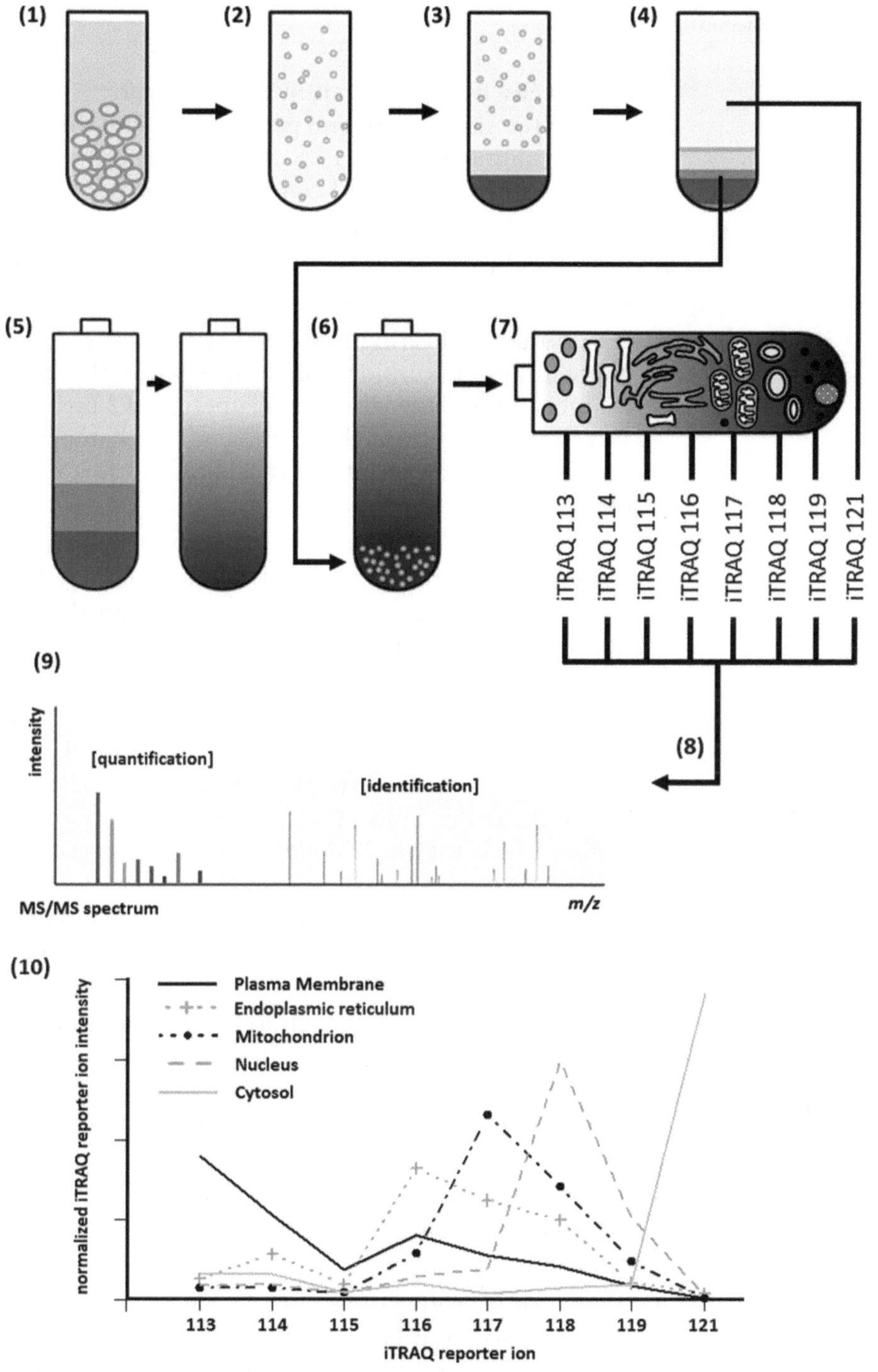

Fig. 1 Schematic of the LOPIT workflow. (*1*) Harvest adherent cells from culture flasks or plates. (*2*) Lyse cells with minimal disruption to organelle integrity. (*3*) Layer lysate over a two layers of iodixanol and ultracentrifuge to separate membranes from cytosol. (*4*) Collect membrane fractions from the iodixanol interface, and an aliquot of the cytosol enriched supernatant. (*5*) Pre-form a resolving density gradient by layering solutions of iodixanol and allowing to diffuse to linearity. (*6*) Lay the membranes beneath the pre-formed gradient and ultracentrifuge to separate organelles. (*7*) Select subcellular fractions displaying differential organelle enrichment, extract and digest proteins, and derivatize peptides with iTRAQ 8-plex reagents. (*8*) Pool iTRAQ labeled peptides and analyze by two dimensional LC-MS/MS. (*9*) MS/MS spectra contain both sequence ions used for identification, and low *m/z* iTRAQ reporter ions which recapitulate the quantitative distribution of subcellular compartments across the fractionation scheme. (*10*) Different organelles display overlapping but distinctive quantitative profiles. Protein subcellular localization can be classified by comparison of these distributions to those of known organelle residents

2. Pellet cells by centrifugation at 200×g for 5 min at 4 °C. Decant the supernatant and gently resuspend the cell pellets in a total volume of 50 mL PBS. Transfer the cell suspension into a single 50 mL tube, and centrifuge as before to re-pellet the cells. Repeat this wash procedure once more.
3. Resuspend the cell pellet in 15 mL detergent-free lysis buffer and place on ice for 5 min.
4. Set up the ball bearing homogenizer on ice; place the appropriate tungsten carbide ball bearing into the homogenizer chamber so that the clearance diameter is 12 μm (*see* **Note 6**), and displace air from the homogenizer chamber by passing 1 mL detergent-free lysis buffer back and forth between the two 1 mL syringes.
5. Gently invert the cell suspension several times to resuspend any settled material. Draw 1 mL aliquots of the solution into a 1 mL syringe and pass the solution through the ball bearing homogenizer and between two syringes 15 times (*see* **Note 7**).
6. To reduce sample viscosity, add 500 U Benzonase to the cell lysate and incubate at room temperature for 20 min, with occasional mixing by inversion. Place the lysate back on ice for an additional 10 min to cool. The lysate may be stored at 4 °C overnight at this point if necessary. The lysate cannot be frozen, as freeze-thawing disrupts organelle integrity.

3.2 Pre-forming an Iodixanol Density Gradient

1. Mix iodixanol working solution and detergent-free lysis buffer to prepare 5 mL stocks with final concentrations of 8, 12, 16, 20, and 25 % iodixanol (w/v).
2. Take a 20 μL aliquot of each iodixanol solution and measure the refractive index with a handheld refractometer. If the value deviates from the expected by more than 0.25°Bx, then adjust with iodixanol working solution or lysis buffer and re-measure (*see* **Note 8**).
3. Using a 2 mL syringe and wide bore needle, add 2.2 mL of 8 % iodixanol solution to two 11 mL polyallomer Quick-Seal ultracentrifuge tubes.
4. Again using a wide bore needle, carefully layer 2.2 mL of 12 % iodixanol solution beneath the 8 % layer. Repeat this process for the 16 % and 20 % solutions, so that each tube contains approximately 8.8 mL in four visibly distinct iodixanol layers of increasing density towards the bottom of the tube (*see* **Note 9**).
5. Place the tubes with layered iodixanol at 4 °C for at least 6 h, or at room temperature for at least 1 h, to allow the iodixanol layers to diffuse into a continuous gradient (*see* **Note 10**).

3.3 Membrane Fractionation

All processes in this section, including centrifugation steps, should be performed on ice or at 4 °C.

1. Centrifuge the cell lysate at 200 × *g* for 5 min to pellet unlysed cells and debris (*see* **Note 11**). Transfer the supernatant to a new tube and repeat this centrifugation step twice more.
2. Divide the supernatant equally into four 5.2 mL polyallomer ultracentrifuge tubes (Beckman). Using a wide bore needle, carefully underlay the supernatant in each tube with 0.8 mL 6 % iodixanol solution. Underlay once again with 0.8 mL 25 % iodixanol solution. The result should be three distinct layers in each tube, with the lysate on top. Balance pairs of ultracentrifuge tubes to within ±10 mg with lysis buffer.
3. Centrifuge at 100,000 × g_{max} for 90 min in a SW55Ti swinging bucket rotor. This centrifugation step pellets membranes out of the lysate and concentrates them at the interface of the two iodixanol layers.
4. Carefully pipette 0.5 mL from the middle of the supernatant layer in each tube, taking care not to disturb the gradient, and collect into a single 15 mL tube. Add 4 volumes of chilled acetone and place at −20 °C for several hours to allow proteins to precipitate. This fraction is strongly enriched with soluble cytosolic proteins.
5. Pipette membranes from the interface of the 6 and 25 % iodixanol layers and collect into a single 15 mL tube. If a membrane band has also collected at the interface of the supernatant and 6 % iodixanol layers, or pelleted to the bottom of the tube, then these may also be collected and pooled with the other membranes (*see* **Note 12**).
6. Dilute any residual cytosolic material contaminating the membrane collection by addition of 5 volumes of lysis buffer, and mix with gentle pipetting or by inversion (*see* **Note 13**). Transfer the membrane suspension into two or four 5.2 mL polyallomer ultracentrifuge tubes and balance to within ±10 mg.
7. Centrifuge at 200,000 × g_{max} for 60 min in a SW55Ti swinging bucket rotor. This centrifugation step pellets membranes to the bottom of the tube, and away from the residual cytosolic contamination in the initial membrane collection.
8. Decant the supernatant and resuspend the membrane pellets in a total volume of 2.0–2.5 mL 25 % iodixanol solution, as prepared in Subheading 3.2.
9. Using a wide bore needle and 2 mL syringe, carefully lay the membrane suspension underneath one of the two pre-formed iodixanol gradients (*see* Subheading 3.2). Underlay the second pre-formed gradient with an approximately equivalent volume of 25 % iodixanol solution to act as a balancer during centrifugation. Balance the two tubes to within ±10 mg.

10. Centrifuge at 100,000 × g_{max} for 8 h in a VTi65.1 vertical rotor. During this ultracentrifugation step, membranes float to and equilibrate at their respective buoyant densities.
11. Using an Auto Densi-Flow peristaltic pump with a meniscus tracking probe, collect 0.5 mL fractions of the density gradient into 1.5 mL polyallomer ultracentrifuge tubes (Beckman), from the top of the gradient downwards (*see* **Note 14**).
12. Take a 20 μL aliquot of each fraction and measure the refractive index using a hand held refractometer to ascertain the final shape of the density gradient. The gradient should be almost linear, except for the densest 2 or 3 fractions, which may have higher iodixanol concentrations.
13. Add 0.8 mL lysis buffer to each gradient fraction, and mix gently by inversion. Balance pairs of tubes to within ±10 mg. Samples can be kept at 4 °C overnight at this stage if necessary.
14. Centrifuge at 186,000 × g_{max} for 20 min with a TLA-55 fixed angle rotor in a benchtop ultracentrifuge. During this centrifugation step membranes are pelleted from the iodixanol solution.
15. Examine each tube carefully to identify the membrane pellet. Pellets in the lower density fractions are typically smaller and more translucent, whereas those in the higher density fractions are larger and golden brown in color due to the abundance of cytochromes in the endoplasmic reticulum and mitochondria. Pipette 1.0 mL of the supernatant from the top of each tube and discard, taking care not to collect any membranes (*see* **Note 15**). Add 1.0 mL lysis buffer to each tube to dilute the remaining iodixanol, and resuspend membrane pellets with gentle pipetting. Balance pairs of tubes to within ±10 mg and repeat centrifugation **step 14**.
16. Remove all of the supernatant from each tube and discard. Freeze membrane pellets to −80 °C for storage, or proceed with sample preparation as described in Subheading 3.4.

3.4 Sample Preparation for iTRAQ Labeling

1. Resuspend membrane pellets in 20–100 μL iTRAQ dissolution buffer.
2. Estimate protein concentrations by BCA assay, as per the manufacturer's instructions.
3. Select eight subcellular fractions for iTRAQ 8-plex labeling (*see* **Note 16**). Aliquot 100 μg protein from each selected fraction into separate 1.5 mL tubes. Reduce and alkylate cysteine residues with TCEP and MMTS respectively, as supplied with the iTRAQ 8-plex kit and per the manufacturer's instructions.
4. If a protein precipitation step is to be performed, then add 4 volumes of chilled acetone to each sample and place at −20 °C

for several hours to allow proteins to precipitate (*see* **Note 17**). Precipitated protein pellets can be resuspended in 0.2 mL 50 mM TEAB (pH 8.5) + 0.02 % (w/v) SDS. If a protein precipitation step is not performed, then reduce the concentrations of urea and SDS prior to protein digestion by adding 6–9 volumes of 50 mM TEAB (pH 8.5) to each fraction.

5. Resuspend 100 μg porcine trypsin in 0.1 mL 50 mM TEAB (pH 8.5). Add 2.5 μg trypsin to each of the eight samples and incubate at 37 °C for 1 h. Prepare a second aliquot of trypsin and add a further 2.5 μg to each sample. Incubate at 37 °C for a further 15 h.
6. Transfer the protein digests to a microfuge, and centrifuge at 16,000 × *g* for 5 min to pellet any insoluble material. Transfer the supernatants into new 1.5 mL tubes, and reduce to dryness by vacuum centrifugation at 10 °C.

3.5 iTRAQ 8-Plex Labeling

1. Allow one set of iTRAQ 8-plex labeling reagents (1 U/vial) to equilibrate to room temperature before opening.
2. Resuspend peptide samples with 30 μL 1 M TEAB (pH 8.5). Add 70 μL isopropanol to each tube, then vortex for 1 min (*see* **Note 18**). Collect the solution to the bottom of the tube with a microfuge pulse.
3. Transfer each of the peptide solutions into a different iTRAQ 8-plex labeling vial. Close and vortex the labeling vials for 1 min each. Place on a shaker at room temperature for 2 h to allow iTRAQ derivatization of peptides to occur.
4. Quench the labeling reaction by addition of 0.2 mL HPLC grade water to each tube, and mix by vortexing. Place back on a shaker at room temperature for a further 2 h (*see* **Note 19**).
5. Pool the eight differentially labeled samples into two 1.5 mL tubes, and reduce to dryness by vacuum centrifugation at 10 °C.
6. Resuspend the peptides in each tube in 0.8 mL 0.1 % TFA. Sample a small aliquot of the solution to ensure that the pH of the solution is below 3 (*see* **Note 20**).
7. Equilibrate a SepPak C18 cartridge with the following solutions, using a vacuum manifold or by manually applying gentle backpressure with a syringe to drive the mobile phase through the column: 2 mL acetonitrile, 2 mL SepPak elution buffer, 2 mL SepPak desalting buffer, 4 mL SepPak loading buffer.
8. Load peptides onto the SepPak C18 cartridge at an approximate flow-rate of 0.1–0.2 mL/min, collecting column eluant into 1.5 mL tubes. Wash the column through with three 0.8 mL aliquots of SepPak loading buffer, followed by two 0.8 mL aliquots of SepPak desalting buffer, at a flow-rate of 0.2–0.4 mL/min. Collect eluants from each wash into separate 1.5 mL tubes.

9. Elute peptides from the cartridge with two 0.8 mL aliquots of SepPak elution buffer. Almost all of the peptides should be eluted in the first 0.8 mL aliquot, so this fraction should be used for downstream analysis. Perform a final elution with 0.8 mL acetonitrile.
10. Reduce collected fractions to dryness by vacuum centrifugation at 10 °C.

3.6 Peptide Separation by High pH Reversed-Phase UPLC

1. Resuspend peptides in 0.1 mL 4 % acetonitrile + 20 mM ammonium formate (pH 10.0). Sample a small aliquot of the solution to ensure that pH is above 9 (*see* **Note 20**).
2. Set up an Acquity UPLC method as follows: Solvent A is HPLC grade water + 20 mM ammonium formate (pH 10.0), and Solvent B is 80 % acetonitrile + 20 mM ammonium formate (pH 10.0). Load 0.05–0.1 mL of peptide solution onto a C18 column (Acquity UPLC BEH; 2.1 mm i.d. × 150 mm, 1.7 μm particle size). Resolve peptides with a linear gradient of 5–60 % B over 60 min at a flow-rate of 0.25 mL/min. Set the photodiode array to scan wavelengths between 200 and 400 nm.
3. Collect eluate fractions into 1.5 mL tubes at 1 min intervals. Peptides should be expected to elute across a 20–30 min window that is readily identifiable in the photodiode array UV chromatogram.
4. Reduce collected fractions to dryness by vacuum centrifugation at 10 °C.

3.7 LC-MS/MS on an LTQ Orbitrap Velos

1. Resuspend peptide fractions in 0.1 mL 4 % acetonitrile + 0.1 % formic acid. Load 2–5 μL of each fraction for LC-MS/MS analysis.
2. Set up a nanoAcquity UPLC method as follows: Solvent A is HPLC grade water + 0.1 % formic acid, and Solvent B is acetonitrile + 0.1 % formic acid. Load peptides onto a C18 trap column (nanoAcquity UPLC Symmetry; 180 μm i.d. × 20 mm, 5 μm particle size) with 5 % buffer B at a flow-rate of 5 μL/min, with the post-column flow directed to waste. After 5 min loading time, switch the solvent flow onto a C18 column (nanoAcquity UPLC BEH; 75 μm i.d. × 150 mm, 1.7 μm particle size) at 0.3 μL/min. Resolve peptides with a linear gradient of 5–45 % buffer B over 70 min, followed by 20 min of high organic wash (80 % buffer B), and 20 min re-equilibration at 5 % B. Eluate from the analytical column should be introduced to the LTQ Orbitrap Velos.
3. Set up an LTQ Orbitrap Velos method as follows: *N*th order double play data-dependent MS/MS experiment setup, operated in positive ion mode. Full scans ranging from 380 to 1,600 m/z is performed in the Orbitrap with nominal resolution of 3×10^4. Up to 10 of the most intense multiply charged

monoisotopic ion peaks from each full scan, above a threshold of 750 counts, are selected for MS/MS analysis, using a precursor ion selection window of 0.5 m/z (*see* **Note 21**). HCD fragmentation is performed with two stepped normalized collision energies of 40 % and 50 %, and a 0.1 ms activation time. MS/MS fragment ions are analyzed in the Orbitrap with a nominal resolution of 7.5×10^3. The automatic gain control (AGC) is set to 1×10^6 and 5×10^4 ions, and maximum ion accumulation times to 500 ms and 250 ms for full scan and MS/MS scans respectively. Two microscans are performed for each full scan, and one microscan is performed for each MS/MS scan. Precursor ions that are singly charged, or for which the charge state cannot be determined, are excluded from MS/MS analysis. A dynamic exclusion window of ±10 ppm m/z for 300 s is applied to any precursor ion sampled three times within a 60 s period.

3.8 Data Processing and Analysis

The primary requirement of data processing is to convert raw MS data into a list of identified proteins, coupled with iTRAQ quantification. Many software packages have been developed to handle isobaric tag data, and perhaps the most straightforward of these are those that seamlessly encompass identification and quantification, such as Proteome Discoverer (Thermo) and MaxQuant [7]. Intuitive software packages such as these do however offer limited control over how the data is processed, and decisions on data handling are often made on behalf of the analyst. Alternatively, exquisite control can be attained if many data processing steps are performed using customized programs and scripts, though such an approach is less accessible to analysts that are not proficient coders. There are therefore multiple pipelines through which data may be passed, which can have a notable impact on the final output [8]. In this section we therefore provide a brief overview of generic data processing steps for the analysis of LOPIT data, and suggest software packages that may be used to achieve these steps.

1. Generate Mascot generic file (MGF) peak lists from raw data using msconvert (ProteoWizard Project) or Proteome Discoverer (Thermo). If MS/MS spectra were acquired in profile, then utilize the default vendor centroiding option within msconvert. Proteome Discoverer will generate centroided peak lists by default.
2. Using perl or python scripting, create a filtered version of the MGF peak lists in which the iTRAQ reporter ions have been removed. While this step is optional, omitting the reporter ions improves the identification performance of the Mascot search algorithm.
3. Query the filtered peak lists against the SwissProt human FASTA database with the Mascot search algorithm (Matrix Science), using the following parameters: precursor ion tolerance ±20 ppm (*see* **Note 22**), fragment ion tolerance ±0.2 Da,

fixed modifications: iTRAQ 8-plex (N-term, K), methylthio (C), variable modifications: iTRAQ 8-plex (Y), oxidation (M), ion scoring series ESI-Q-TOF with immonium ions, run Mascot Percolator on resulting data.

4. Pair iTRAQ reporter ion quantification to identification for peptide-spectrum matches (PSMs) with a posterior error probability less than or equal to 0.01, and for which the peptide is nonredundant (only found in a single database entry). This pairing can be performed by perl or python scripting, or using packages such as Proteome Discoverer, or with the R-based MSnBase package [9]. Reporter ions can be measured either as peak areas (if the MS/MS spectra were acquired in profile) or summed peak intensities (for profile or centroided spectra). All ions within ±0.015 *m/z* of the expected *m*/z for each reporter ion should be incorporated into the peak measurement. Using broader quantification windows will result in near isobaric contaminant ions, such as the phenylalanine immonium ion $[M+1]^+$ isotope, interfering with quantification.

5. Generate full scan peak lists from raw data using msFileReader (Thermo) or pXtract [10]. Using perl or python scripting, and for each significant PSM, determine the isolation specificity of the precursor selection event (i.e., the proportion of total ion current in the precursor selection window attributable to the target peptide), and the position at which the MS/MS spectrum was acquired relative to the peptide's chromatographic apex. Similar properties can be extracted using the precursor.mz python script distributed as part of the multiplierz package [11].

6. Using the data obtained in the previous step, and perl, python, or R scripting, or in a spreadsheet editor, filter out PSMs for which the isolation specificity was less than 90 % (i.e., 90 % of the ion current in the 0.5 *m/z* selection window was attributable to the target peptide). In cases where multiple spectra have been acquired for the same peptide, filter out all except the PSM closest to the chromatographic apex (i.e., with the highest signal:noise ratio), as this typically yields more reliable quantification than MS/MS spectra acquired at lower signal to noise ratios. Filter out any PSMs with more than two missing reporter ion values. Treat remaining missing values as zero. This filtering step is not possible if using data processing software such as Proteome Discoverer or MaxQuant.

7. Merge spectral level iTRAQ quantification into protein level estimations by taking the median of remaining spectral level measurements for that protein.

8. Compile training sets of well curated markers of each organelle of interest that are observed in the dataset. Using these training data, classify protein localization by partial least squares-discriminant analysis (PLS-DA, [2]), or with a nonlinear

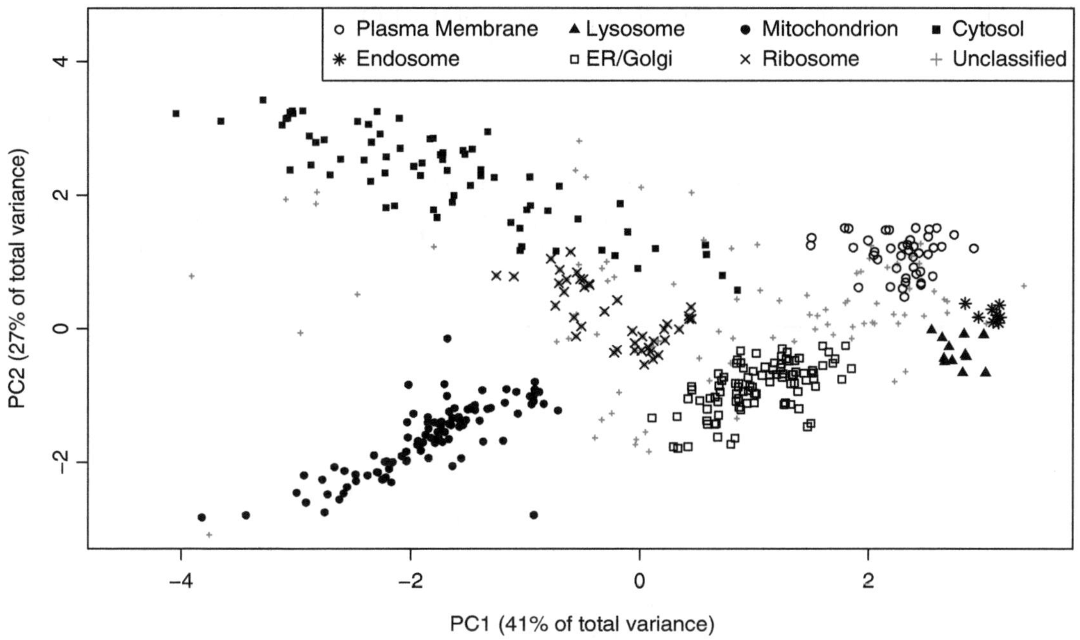

Fig. 2 Principal component analysis of iTRAQ 8-plex data from a post-nuclear fractionation of HEK293T cells, annotated with organelle classification by PLS-DA. *Each point* represents a single protein, and proteins with similar subcellular distributions cluster together in multivariate space. The rate of corroboration between LOPIT classification of protein localization and curated annotation in the UniProt database was greater than 90 % for all classified organelles

support vector machine algorithm (SVM, [12]). Some proteins may fail to be classified. This may be due to a mixed subcellular localization, residence in organelle(s) not present in the training data, or deviation from the expected distribution due to technical imprecision of iTRAQ quantification.

9. For visualization purposes, the multivariate data is most easily presented in two dimensions after principal component analysis, using R or MATLAB (Fig. 2). Care should be taken to also evaluate principal components beyond the first two, as biologically meaningful patterns are often also evident in the lower components for iTRAQ 8-plex data.

10. Several R packages built upon the MSnBase data structure [9] have been developed for the purpose of analyzing subcellular localization experiment data. For example, "pRoloc" [13] enables visualization and a variety of statistical analyses to be performed on LOPIT data. Another tool, "*phenoDisco*", performs semi-supervised phenotype discovery analysis to identify unannotated organelles or sub-organellar groupings of proteins within the multivariate data [14]. For an integrated data analysis pipeline we recommend utilizing the MSnBase and pRoloc packages for LOPIT data analysis.

The final outcome of the LOPIT experiment is a list of proteins and their steady-state subcellular localization, as determined by the supervised classification methods in **step 8**. Localization of proteins may then be validated by traditional techniques such as immunocytochemistry, or in high throughput against protein database annotation. Changes in protein localization may be investigated by performing LOPIT experiments under different culture or treatment conditions.

4 Notes

1. Detergent-free lysis buffer can be made up in advance and stored at 4 °C, without adding protease inhibitors. Protease inhibitors should be added on the day of use. Detergents must not be used in these buffers, as these would disrupt membrane integrity.
2. Iodixanol is a contrast agent with several properties that are highly suited to density gradient centrifugation applications. Solutions of iodixanol will rapidly sediment into "self-generating" gradients under ultracentrifugal forces, offering speed and versatility in gradient design. High concentrations of iodixanol also place membranes under less osmotic stress than sucrose, preserving the integrity of organelles.
3. If proteins from membrane fractions are to be precipitated to remove lipids and other molecules that may interfere with MS analysis, then urea should be omitted from this buffer. SDS may also be substituted with lithium dodecyl sulfate (LDS) to avoid detergent precipitation at −20 °C.
4. While the pipeline described here utilizes the LTQ Orbitrap Velos platform, LOPIT experiments can also be performed on quadrupole-time of flight instruments, quadrupole-Orbitraps, and on linear ion traps with pulsed Q dissociation. The major advantages of the hybrid ion trap-Orbitrap platform are the ability to select ions for fragmentation with high specificity, and to perform MS^3 experiments to improve quantitative performance (*see* **Note 21**).
5. The effectiveness of Versene for detaching cells is dependent on the adhesion proteins expressed by a particular cell line. For example, while Versene is very effective at detaching fibroblasts such as HEK293T, E14TG2a murine embryonic stem cells cultured on gelatinized flasks are not effectively dissociated. Other enzyme and detergent-free dissociation buffers may be suitable alternatives to Versene for such cell lines. Cells may also be detached by rapid trypsin treatment. To minimize plasma membrane protein degradation, care should be taken to quench trypsin with an excess of a serum-containing

medium as soon as cells are detached, after which the cells should be placed on ice immediately. The trypsin is then diluted out in subsequent PBS washes, and the cells are resuspended in a lysis buffer containing protease inhibitors that will nullify residual trypsin activity.

6. The clearance space is the diameter between the tungsten carbide ball bearing and the sides of the stainless steel homogenizer chamber. This can be attenuated by using ball bearings of different sizes, and should be optimized for a particular cell line or application. The aim is to set the clearance space so that it is too narrow for cells to pass through intact, but permissive enough to allow organelles such as nuclei and mitochondria to pass through with minimal disruption. Narrower clearances will increase the back pressure required for homogenization, while wider clearances may reduce lysis efficiency. Ball bearing homogenizers are highly suited to this application because of their consistent performance and high lysis efficiency. If a ball bearing homogenizer is unavailable, then alternative mechanical homogenization methods such as Dounce homogenization may be suitable.
7. The optimal number of passages through the ball bearing homogenizer may require some optimization for each cell line.
8. The expected relationship between iodixanol concentration and refractive index can be established by measuring the refractive index for a series of iodixanol solutions and generating a calibration curve.
9. When layering solutions of iodixanol, care must be taken not to disrupt the gradient shape. Sequential underlaying of solutions is typically more robust than overlaying, as overlaying tends to result in solutions mixing. A wide bore needle and syringe allow easy access to the bottom of the gradient with minimal disruption, and decrease the probability of accidentally introducing air bubbles.
10. The procedures outlined in this section of the protocol describe manually generating a pre-formed iodixanol gradient. If an automated gradient maker is available, then this can be used to generate the density gradient instead.
11. If the nucleus is not an organelle of interest, then this centrifugation step can be performed at $2{,}000 \times g$ instead of $200 \times g$. The post-nuclear supernatant may then be used for subcellular fractionation.
12. Lower density membranes (e.g., plasma membrane fragments, endosomes etc.) may enrich at the interface of the supernatant and 6 % iodixanol layers, so collecting this secondary membrane layer will improve the final yields of these subcellular fractions.

13. Never mix membrane suspensions by vortexing, as this will disrupt membrane integrity.
14. A peristaltic pump with meniscus tracking probe allows for very reproducible gradient collection with minimal mixing. If such a device is unavailable, then fractions may be collected by pipetting from the top of the tube downwards, or by piercing the bottom of the tube and collecting fractions from the bottom of the gradient upwards.
15. For fractions from the bottom of the density gradient, membranes may not completely pellet due to the high iodixanol concentration. In such cases, remove as much of the supernatant as possible without disrupting the partially sedimented membranes.
16. Two criteria should be considered for optimal fraction selection: sufficient protein (50–100 μg) for iTRAQ labeling, and differential enrichment of subcellular niches of interest. The distribution of organelles can be determined by western blotting with antibodies raised against organelle marker proteins, organelle-specific enzymatic assays, or by cataloguing proteins in each fraction by LC-MS/MS and performing GO term enrichment analysis on identified proteins. In practice, the labeling capacity of iTRAQ 8-plex is sufficiently high that simply selecting fractions from across the fractionation scheme (including the cytosolic fraction prepared in **step 4** in Subheading 3.3), pooling adjacent fractions to collect sufficient protein if necessary, will provide reasonable resolution of most organelles for a global analysis. If two or more organelles are the specific focus of the analysis then more experimentation may be required to optimize the density gradient and fraction selection.
17. Protein precipitation is generally recommended, as lipids and other molecules in the membrane preparations may reduce the robustness and sensitivity of iTRAQ labeling and downstream LC-MS/MS analysis. Protein yields from subcellular fractions are however often limited, and so to avoid sample losses and additional technical variability this preparative step can be skipped if necessary.
18. The volumes used should be sufficient to completely resuspend peptides and other non-volatile components of the iTRAQ dissolution buffer such as urea. If the sample cannot be fully resuspended then the volume can be increased from 0.1 to 0.2 mL, provided that the final concentration of isopropanol remains 70 % (v/v).
19. Dilution with water increases the rate of hydrolysis of unreacted iTRAQ reagents, and after 2 h the remaining proportion of unquenched tag is negligible. This quenching step can also be performed more rapidly by the addition of 5 % hydroxylamine

(v/v), followed by a 30 min incubation on a shaker at room temperature.

20. If the pH is not as expected, then the likely cause is that the volatile buffers used in previous steps (e.g., TEAB or acetic acid) were not completely evaporated during the vacuum centrifugation step. In such cases the samples should be dried once more by vacuum centrifugation so that the appropriate pH can be reached.

21. An established limitation of isobaric tagging strategies is that precursor ions selected for fragmentation in complex peptide mixtures are typically co-selected with a population of contaminants, which then contribute to, and distort, the isobaric tag reporter ion ratios [15–19]. Since the LOPIT methodology is dependent on consistently high quantitative performance, this effect reduces the reliability with which subcellular localization can be deduced. Setting the precursor ion selection window to 0.5 *m/z* notably improves the accuracy and precision of quantification, but at the expense of spectral quality and therefore sensitivity [18]. Conversely, applying a precursor ion selection window of 1.2 *m/z* will allow a notably higher rate of identification, but the robustness of quantification is typically poor. Similar approaches are also applicable to other MS platforms typically used for isobaric tag experiments, such as hybrid quadrupole-time of flight and quadrupole-Orbitrap mass spectrometers, by attenuating the quadrupole *m/z* transmission window.

    Alternative approaches utilizing MS^3 on Orbitrap Elite and Orbitrap Fusion platforms (both Thermo) have also been proposed, which promise high performance quantification without compromising the peptide identification rate [19–21].

22. While the nominal masses of the iTRAQ 8-plex reagents are identical, there are fractional mass differences (approximately 6 mDa) between some of the isotopologues, owing to differences in the number of heavy isotopes of carbon, nitrogen, and oxygen residues. Since the iTRAQ 8-plex modification listed by Mascot is based on the monoisotopic mass of one of these isotopologue variants, the observed *m/z* may deviate slightly from that expected by the search algorithm. The precursor ion tolerance used for peptide-spectrum matching should therefore be broadened to accommodate this.

Acknowledgements

The authors would like to thank Daniel Nightingale and Julie Howard for their helpful comments to improve the clarity of the protocol. A.C. was funded by BBSRC grant BB/D526088/1.

References

1. Dunkley TP, Hester S, Shadforth IP, Runions J, Weimar T, Hanton SL et al (2006) Mapping the Arabidopsis organelle proteome. Proc Natl Acad Sci U S A 103(17):6518–6523
2. Sadowski PG, Dunkley TP, Shadforth IP, Dupree P, Bessant C, Griffin JL et al (2006) Quantitative proteomics approach to study subcellular localization of membrane proteins. Nat Protoc 1(4):1778–1789
3. Hall SL, Hester S, Griffin JL, Lilley KS, Jackson AP (2009) The organelle proteome of the DT40 lymphocyte cell line. Mol Cell Proteomics 8(6):1295–1305
4. Tan DJ, Dvinge H, Christoforou A, Bertone P, Martinez Arias A, Lilley KS (2009) Mapping organelle proteins and protein complexes in Drosophila melanogaster. J Proteome Res 8(6):2667–2678
5. de Duve C (1971) Tissue fraction-past and present. J Cell Biol 50(1):20
6. Kessner D, Chambers M, Burke R, Agus D, Mallick P (2008) ProteoWizard: open source software for rapid proteomics tools development. Bioinformatics 24(21):2534–2536
7. Cox J, Mann M (2008) MaxQuant enables high peptide identification rates, individualized p.p.b.-range mass accuracies and proteome-wide protein quantification. Nat Biotechnol 26(12):1367–1372
8. Lacerda CM, Xin L, Rogers I, Reardon KF (2008) Analysis of iTRAQ data using Mascot and Peaks quantification algorithms. Brief Funct Genomic Proteomic 7(2):119–126
9. Gatto L, Lilley KS (2012) MSnbase-an R/Bioconductor package for isobaric tagged mass spectrometry data visualization, processing and quantitation. Bioinformatics 28(2):288–289
10. Wang LH, Li DQ, Fu Y, Wang HP, Zhang JF, Yuan ZF et al (2007) pFind 2.0: a software package for peptide and protein identification via tandem mass spectrometry. Rapid Commun Mass Spectrom 21(18):2985–2991
11. Savitski MM, Mathieson T, Zinn N, Sweetman G, Doce C, Becher I et al (2013) Measuring and managing ratio compression for accurate iTRAQ/TMT quantification. J Proteome Res 12(8):3586–3598
12. Trotter MW, Sadowski PG, Dunkley TP, Groen AJ, Lilley KS (2010) Improved sub-cellular resolution via simultaneous analysis of organelle proteomics data across varied experimental conditions. Proteomics 10(23):4213–4219
13. Gatto L, Breckels LM, Wieczorek S, Burger T, Lilley KS (2014) Mass-spectrometry-based spatial proteomics data analysis using pRoloc and pRolocdata. Bioinformatics (E-pub ahead of print)
14. Breckels LM, Gatto L, Christoforou A, Groen AJ, Lilley KS, Trotter MW (2013) The effect of organelle discovery upon sub-cellular protein localisation. J Proteomics 88:129–140
15. Ow SY, Salim M, Noirel J, Evans C, Rehman I, Wright PC (2009) iTRAQ underestimation in simple and complex mixtures: "the good, the bad and the ugly". J Proteome Res 8(11):5347–5355
16. Griffin TJ, Xie H, Bandhakavi S, Popko J, Mohan A, Carlis JV et al (2007) iTRAQ reagent-based quantitative proteomic analysis on a linear ion trap mass spectrometer. J Proteome Res 6(11):4200–4209
17. Evans C, Noirel J, Ow SY, Salim M, Pereira-Medrano AG, Couto N et al (2012) An insight into iTRAQ: where do we stand now? Anal Bioanal Chem 404(4):1011–1027
18. Christoforou AL, Lilley KS (2012) Isobaric tagging approaches in quantitative proteomics: the ups and downs. Anal Bioanal Chem 404(4):1029–1037
19. Ting L, Rad R, Gygi SP, Haas W (2011) MS3 eliminates ratio distortion in isobaric multiplexed quantitative proteomics. Nat Methods 8(11):937–940
20. Wenger CD, Lee MV, Hebert AS, McAlister GC, Phanstiel DH, Westphall MS et al (2011) Gas-phase purification enables accurate, multiplexed proteome quantification with isobaric tagging. Nat Methods 8(11):933–935
21. Dayon L, Sonderegger B, Kussmann M (2012) Combination of gas-phase fractionation and MS^3 acquisition modes for relative protein quantification with isobaric tagging. J Proteome Res 11(10):5081–5089

Chapter 11

Brain Quantitative Proteomics Combining GeLC-MS and Isotope-Coded Protein Labeling (ICPL)

Giuseppina Maccarrone, Maria Lebar, and Daniel Martins-de-Souza

Abstract

Proteomics has been revolutionized by the rapid advance of mass spectrometric instrumentations and techniques. Parallel methodologies for the quantification of proteomes also evolved, including in vitro stable isotope labeling. Here, we present a protocol for employing isotope-coded protein labeling (ICPL) as part of a shotgun proteomics workflow denoting its advantages and disadvantages. This protocol is suitable to studying any proteome of interest, only requiring a specific sample preparation and protein identification. Given our expertise, descriptions here are centered on the study of brain disorders.

Key words Brain, ICPL, Proteomics, GeLC-MS, Mass spectrometry

1 Introduction

Proteomics has rapidly grown as a science in the past 15 years, promoted by the quick methodological and instrumentation development. While the first 5 years of this science has been centered in the combination of two-dimensional gel electrophoresis (2DE) for protein separation and quantification followed by mass spectrometry (MS) for protein identification by peptide mass fingerprinting [1], from the 2000s on, mass spectrometry-based shotgun proteomics [2] started to take the lead progressively as the central methodology for qualitative and quantitative proteome analyses.

One of the main advantages of employing shotgun proteomics relies in the simultaneous identification and quantitation of thousands of proteins. There are several quantitative approaches for shotgun proteomics, mainly divided in two categories: label-free and stable isotope labeling [3]. The second category can be subdivided in two additional categories: in vivo and in vitro labeling. While the first relies in incorporating stable isotopes in situ via cell culture media (SILAC—Chapter 5) or via feeding (SILAM—Chapter 7), the second relies in binding covalently to the sample of

Daniel Martins-de-Souza (ed.), *Shotgun Proteomics: Methods and Protocols*, Methods in Molecular Biology, vol. 1156, DOI 10.1007/978-1-4939-0685-7_11, © Springer Science+Business Media New York 2014

interest compounds (tags) with the same molecular structure but different masses due to the presence of isotopes (^{15}N, ^{13}C). Stable isotope labeling, which has been used since decades in analytical chemistry [4], provide the opportunity of tracking and quantifying peptides and proteins when combined to liquid chromatography-mass spectrometry (LC-MS).

The first application of stable isotope labeling to quantitative proteomics was the isotope-coded affinity tags (ICAT) launched in 1999 [5]. Basically, ICAT relies in heavy and light mass tags containing either eight or no deuterium atoms, respectively, allowing the comparison of two samples in one experiment. While this methodology requires multi-chromatographic steps to purify labeled proteins or peptides and only quantifies two samples, other methods have been developed requiring less experimental steps and allowing the quantification of more samples at once—i.e., iTRAQ [6] (Chapter 8), which allows the quantification of up to 8 samples and TMT (Chapters 10 and 11) which enables the quantification of six samples at MS/MS level [7]. Alternatively, there is the isotope-coded protein labeling (ICPL) which may allow the quantification of up to 4 samples at MS level [8].

The principle of ICPL is based in the isotope labeling of free amino groups of intact proteins or peptides with four different forms of the molecule N-nicotinoyloxy-succinimide (commonly referred as "tag") according to its isotope composition (Fig. 1) [9]. Each sample of interest is labeled with a different tag which presents the same molecular structure with different masses. Due to the presence of the tag with different masses, labeled peptides will present expected shifts in mass (Fig. 2). Peptide quantitation is simply a ratio among the monoisotopic intensities of each labeled species, performed by a computational algorithm, which will also calculate a weighted average of the intensities of all identified peptides of a given protein, in order to generate a total intensity per protein per sample. These can be therefore compared across samples. ICPL labeling is suitable for any given proteome and any given protein/peptide separation method. Our group have used in the past the

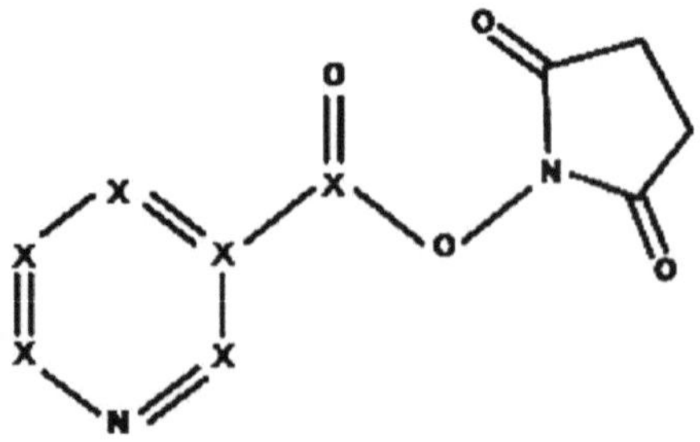

Fig. 1 *N*-Nicotinoyloxy-succinimide, the ICPL tag. This molecule can present four different configurations based on their isotopic constitution. ICPL_0 is the natural form; ICPL_4 has four deuteriums; ICPL_6 has six ^{13}C; and ICPL_10 has six ^{13}C and four deuteriums

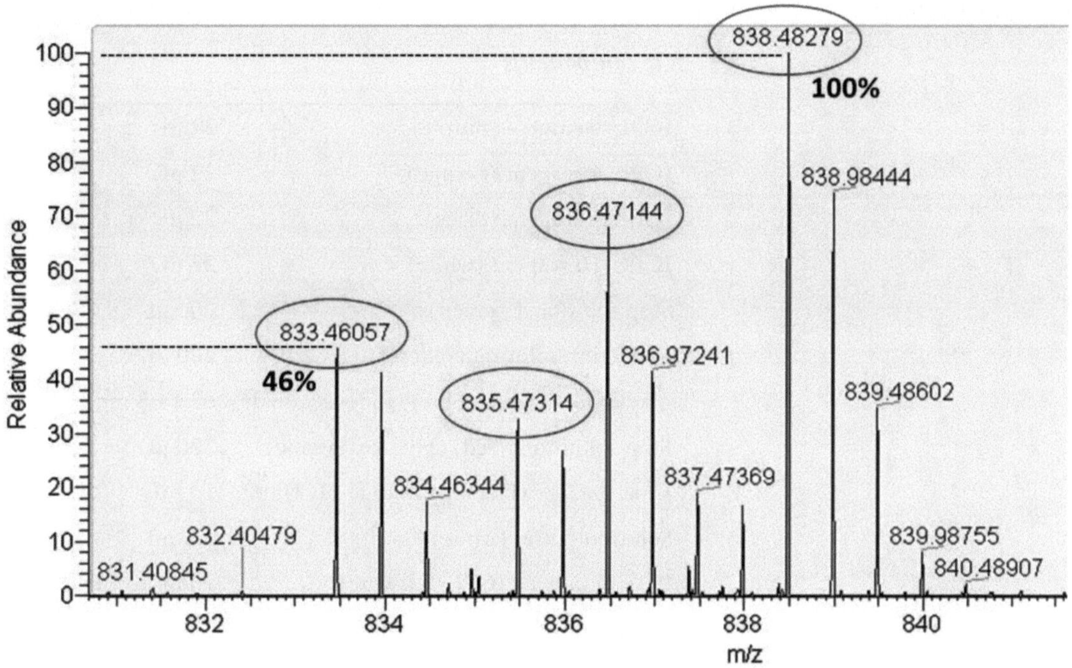

Fig. 2 Example demonstrating a given peptide doubly charged from four different samples labeled with different ICPL tags. Its quantitation is calculated by comparing the monoisotopic intensities. In this case, samples labeled with ICPL_10 is almost twice more abundant than the samples labeled with ICPL_0

ICPL 2-plex, which enables to label 2 samples, in studies of brain tissue from schizophrenia patients [10, 11]. More recently, we have been employing the 4-plex version, which allows for the simultaneous quantitative analysis of four independent proteomes.

Here, we present a protocol for ICPL-labeling for quantifying brain proteomes analyzed in a shotgun platform denominated "GeLC-MS," which combines SDS-PAGE for protein fractionation, followed by nano-liquid chromatography coupled online to tandem mass spectrometry (nLC-MS/MS). This protocol can be employed for any type of proteome analyses.

2 Materials

2.1 Brain Tissue Preparation

1. Sample Grinding Kit (GE Healthcare).
2. Protein extraction buffer (7 M Urea, 2 M Thiourea, 4 % (v/w) 3-[(3-Cholamidopropyl)dimethylammonio]-1-propanesulfonate (CHAPS), 2 % (v/w) Tetradecanoylamidopropyl-dimethylammonio-butanesulfonate (ASB-14), 70 mM DTT) [12].
3. Cool centrifuge.
4. Bradford dye-binding assay.

2.2 ICPL Labeling

1. SERVA ICPL™ Quadruplex Kit (SERVA Electrophoresis)
   - *Kit components*

ICPL_0 reagent (purple)	22 μL
ICPL_4 reagent (orange)	22 μL
ICPL_6 reagent (black)	22 μL
ICPL_10 reagent (white)	22 μL
Stop solution 1 (green)	200 μL
Reduction solution (yellow) in TCEP	200 μL
Alkylation reagent (blue) (Iodoacetamide)	6 × 14.8 mg
Stop solution 2 (red) (Hydroxylamine)	200 μL
Lysis buffer (white) (guanidine-HCl)	10 mL
Solution buffer (white)	1.5 mL
Protein mix for ICPL_0 (white)	20 μL lyophilized
Protein mix for ICPL_4 (white)	20 μL lyophilized
Protein mix for ICPL_6 (white)	20 μL lyophilized
Protein mix for ICPL_10 (white)	20 μL lyophilized
ICPL standard PLUS (white)	12 μg lyophilized
Trypsin NB sequencing grade, mod. (white)	1 × 25 μg lyophilized
Endoproteinase Glu-C (red)	3 × 25 μg lyophilized

2. Nitrogen or Argon.
3. 2 N NaOH.
4. 2 N HCl.
5. 100 % acetone (ice cold).
6. pH paper (pH 0–14).
7. Cool microcentrifuge (4 °C).
8. Thermo mixer (up to 60 °C).

2.3 Proteomics Sample Preparation

1. Electrophoresis equipment.
2. Laemmli sample buffer 2× (0.125 M Tris–HCl pH 6.8, 4 % SDS, 20 % glycerol, 10 % 2-mercaptoethanol, 0.004 % bromophenol blue).
3. Staining solution: 0.1 % of Coomassie brilliant blue R 250, 5 % acetic acid, 10 % methanol.
4. Destaining solution: 80 % distilled water, 10 % acetic acid, 10 % isopropanol.
5. 10 mM dithiothreithol.
6. 50 mM iodoacetamide.

7. 25 mM ammonium bicarbonate in 50 % acetonitrile (ACN).
8. 25 mM ammonium bicarbonate.
9. Trypsin Sequencing Grade (5 ng/μL) in 50 mM ammonium bicarbonate.
10. 50 % ACN, 2 % formic acid.

2.4 nLC ESI-MS/MS

1. Nano LC system (Eksigent).
2. Trap column RP-C18.
3. In-house packed nano-column (75 μm i.d. × 15 cm).
4. Column packing material (RP-C18, 3 μm).
5. Picofrit self-packed column (No coated, 75 μm i.d, 10 μm tip).
6. 0.1 % Formic acid in LiChrosolv distilled water.
7. 95 % Chromatographic grade ACN in 0.1 % formic acid.
8. Centrifuge filters (0.22 μm).
9. LTQ Orbitrap XL mass spectrometer (Thermo Fisher, Germany).
10. Nano electrospray ionization (ESI) source (Thermo Fisher).

2.5 Mass Spectrometry Software

1. LC-ESI-MS/MS data acquisition software: Xcalibur v. 2.07 (Thermo Fisher).
2. Nano-LC controlling software: LC-Eksigent v .09 (Eksigent).
3. Raw MS and MS/MS spectra processing software: Mascot Distiller 2.3.2 (Matrix Science).
4. Protein database search engine: Mascot Daemon 2.2.2 (Matrix Science).

3 Methods

3.1 Protein Extraction from Brain Tissue

This protocol is an alternative for total proteome extraction which has been proved to be effective for human and mammalian brain tissue [12, 13]. Other proteome extraction protocols and sub-proteomes are also suitable to ICPL labeling and quantitation.

1. Separate 20 mg of brain tissue (human or mammalian).
2. Homogenize it in 100 μL of the protein extraction buffer using the Sample Grinding Kit for 2 min.
3. Centrifuged for 10 min at 20,000 × *g* at 4 °C.
4. Collect the supernatant and discard the pellet.
5. Determine protein concentrations using the Bradford dye-binding assay.

3.2 Sample Preparation for Protein Labeling

(*See* **Notes 1** and **2**.)

Firstly, 100 μg of each protein sample to be analyzed must be dissolved in lysis buffer (white).

If protein samples are dissolved in another buffer, take 100 μg of total protein sample and submit to cold acetone precipitation as it follows:

1. Add 4× volumes of ice cold acetone to the sample.
2. Incubate for at least 1 h (but we recommend overnight) at −20 °C.
3. Centrifuge the sample at 20,000 × *g*, 10 min at RT.
4. Discard the supernatant and leave the tube open for 1–2 min so that the remaining acetone can evaporate (*see* **Note 3**).
5. Resuspend the sample in 20 μL of ICPL lysis buffer (*see* **Note 4**).

3.3 ICPL Protein Labeling

The carbamidomethylation protocol is identical for all four samples named here as A, B, C, and D.

3.3.1 Carbamidomethylation

1. Check the pH of the sample buffer and if necessary, adjust to 8.5 ± 0.1 by addition of HCl or NaOH.
2. Add 0.5 μL Reduction solution (yellow) to 20 μL sample solution (equivalent to 100 μg protein), overlay sample with argon (or N_2) to avoid oxidation. Vortex and spin down. Incubate proteins for 30 min at 60 °C and keep shaking.
3. Cool sample to 25 °C and spin down to remove condensed solution from the lid.
4. Dissolve one vial of Alkylation reagent (blue) in 200 μL Solution buffer (white) and wrap in aluminum foil (light protection).
5. Add 0.5 μL of the freshly prepared alkylation solution to each sample, vortex and spin down. Wrap samples quickly in aluminum foil for light protection and leave samples for 30 min at 25 °C and keep shaking.
6. Quench the reaction by adding 0.5 μL Stop solution 1 (green) to each sample, vortex and spin down. Incubate for 15 min at 25 °C. *Optional break point*: At this point the samples can be stored under argon (or N_2) at −20 °C for a later processing.
7. Check pH and adjust to 8.3 before starting the label procedure.
8. Add:
   3 μL of ICPL_0 Reagent (purple) to sample A.
   3 μL of ICPL_4 Reagent (orange) to sample B.
   3 μL of ICPL_6 Reagent (black) to sample C.
   3 μL of ICPL_10 Reagent (white) to sample D.
9. Overlay the four samples with argon (or N_2) to avoid oxidation, vortex (10 s) and sonicate for 1 min in ultrasound bath. Spin down samples.

10. Incubate samples for 2 h at 25 °C and keep shaking.
11. Add 2 μL of Stop solution 2 (red) to each sample, vortex and spin down.
12. Incubate for 20 min at 25 °C in a shaking mode to neutralize the excess reagent.
13. Combine all four ICPL labeled samples, vortex and spin down.
14. Adjust the pH of the mixture to 11.9±0.1 by adding 2 N NaOH (about 4 μL for 4×20 μL sample volume) to destroy possible esterification products.
15. After 20 min at 25 °C in a shaking mode add the same amount of 2 N HCl to neutralize sample (usually it is not necessary to check the pH).

3.4 Acetone Precipitation

This is an essential step to reduce salt concentration and to prevent SDS precipitation in further protein separation steps such as 1D SDS-PAGE gel run.

1. Add the same volume of distilled water to the pooled ICPL-labeled samples, [i.e., 114 μL=4×(20 μL sample+1.5 μL reduction+3 μL reagent+2 μL stop solution)+8 μL for pH adjustment]. Vortex and spin down.
2. Add five fold excess (related to total volume of sample and water) of ice-cold acetone to the ICPL-labeled sample, i.e., 1,140 μL acetone (=228 μL×5), vortex. Leave the sample at −20 °C for 1 h to overnight.
3. Spin down precipitated proteins at 20,000×*g* for 30 min at 4 °C.
4. Remove the acetone and let the remaining solution evaporate at room temperature (let the tube stand open for 1–2 min).
5. Dissolve the salt-depleted samples in 30 μL of Laemmli sample buffer 2× (*see* **Note 5**). At this point, your samples must contain 400 μg protein.

3.5 Proteomics Sample Preparation

1. Before running gel electrophoresis, the sample is cooked for 5 min at 95 °C.
2. Loaded 60 μg of labeled samples and run in a 12 % SDS-PAGE. The gel is stained with Coomassie Brilliant Blue R-250 solution for at least 2 h at room temperature with gentle agitation (*see* **Note 6**).
3. The stained gel lane is sliced into ≈20 bands and submitted to a trypsin in-gel digestion protocol **Note 7** [14].
4. Each fraction is diced into smaller pieces, and washed with 25 mM ammonium bicarbonate in 50 % ACN for 10 min. Repeat the washes until the gel is completely distained.

5. Each diced gel fraction is incubated with 10 mM dithiothreithol for 30 min at 56 °C followed by carboxyamidomethylation with 50 mM iodoacetamide for 30 min at room temperature.
6. Samples are washed twice, dried, and then rehydrated in trypsin solution (5 ng/μL) for 15 min in ice.
7. The remaining trypsin solution is removed, the gel pieces are covered with 25 mM ammonium bicarbonate, and digestion proceeds for 4–6 h at 37 °C (*see* **Note 8**).
8. Peptides are extracted twice with 50 % ACN, 2 % formic acid and then lyophilized.

3.6 nLC ESI-MS/MS

1. Dried peptides are dissolved in 10 μL 0.1 % formic acid, filtered with the filter devices before being analyzed by nLC ESI-MS/MS (*see* **Note 9**).
2. The peptides extracted from each one of the 20 fractions are separated by using a nano HPLC system equipped with a trap column and a picofrit nano-column coupled online to an LTQ-Orbitrap mass spectrometer via a nano electrospray ion source (ESI) (*see* **Note 10**).
3. 5 μL of each sample is loaded onto the RP-C18 trap column and washed for 10 min with H_2O/ACN (95/5) 0.1 % formic acid (eluent A) at a flow-rate of 3 μL/min. Peptides are separated on the RP-C18 nano-column by applying a linear gradient of 95 % ACN in 0.1 % formic acid (eluent B) from 2 to 10 % in 5 min and 10 to 40 % in 98 min at flow rate of 200 nL/min (*see* **Note 11**).
4. The mass spectrometer is operated in the positive mode. A method is set that includes a data-dependent scan mode acquisition of full scan (FS) and MS/MS scan. The FS scan is recorded at the Orbitrap analyzer in the mass range from m/z 380 to 1,600 at resolution 60,000 (FMHW, m/z 400). The FS data are acquired in profile mode.
5. The fragment spectra (MS/MS) are acquired at the ion trap (LTQ) in centroid mode (*see* **Note 12**).
6. The fragments are generated in collision-induced dissociation (CID) mode; no fragmentation at the source is considered. The five most intensive ions per FS are selected for the fragmentation. Each mass ion with intensity over 500 counts is fragmented once and inserted into a dynamic exclusion list for 60 s. Mono-charged ions are rejected for MS/MS. Other parameters set for the fragmentation are as follows: (a) 30 s repeat duration time; (b) 2 uma isolation width (m/z); (c) 30 ms activation time; (d) 35 V normalized collision energy; and (e) Q = 0.250 activation.

3.7 Protein Identification: Mascot Distiller

1. The MS RAW data can be directly loaded within Mascot Distiller. A new "multi file experiment" is created and all obtained *.RAW files are selected to be part of this.
2. Go to Processing/Processing options and configure the settings of your preference for your experiment in this window. As a standard procedure, go to "Load" and choose the default settings according to your instrument. Hit "OK."
3. Go to Processing/Process all scans.
4. Next, go to Analysis/Mascot Search/All peak lists. The suggested searching parameters are: trypsin as enzyme allowing one missed cleavage; cysteine carbamidomethylation (fixed), methionine oxidation and protein N-terminal 4-ICPL tags (variable) are chosen as modifications. Precursor and fragment ion mass accuracies are set to 10 ppm and 0.5 Da, respectively. Quantitation method is "ICPL 4-plex" (*see* **Note 13**). Check the box *Decoy* to run the search using the Decoy database. Choose the appropriated species of the organism used in the experiment.
5. Peptides are only accepted if the confidence interval (CI) is 95 % and present scores above the threshold set by Mascot. Protein identifications are considered to be reliable if at least two unique and no redundant peptides match the protein.

3.8 Protein Quantitation

See **Note 14**.

1. After having a protein list generated in Mascot Distiller, go to tools/quantitation. Select all families.
2. Mascot Distiller will automatically generate a list of proteins and the paired-ratios among the different labels: light/medium, light/medium 2, light/heavy, medium/medium2, medium/heavy and medium2/heavy (*see* **Note 15**).
3. In the default mode, Mascot Distiller will only calculate proteins ratios for those proteins that are present in all four samples analyzed and when a given protein is at least quantified by two peptides.

4 Notes

1. If lysis buffers different from the provided lysis buffer (white) are used, they should be tested with the protein standard mixture before usage with the samples. Make sure that the pH is 8.5. Avoid the presence of primary amino group containing agents in buffers. Urea, thiourea, CHAPS can be used.
2. The recommended protein concentration for the labeling reaction is 5 mg/mL. 100 μg total protein is needed.

3. Don´t let the sample dry too much—this will result in difficulties in resuspending the sample.
4. If there are problems in dissolving the pellet, put the sample in an ultrasonic bath for ~5 min to disrupt the protein pellet.
5. The samples can be stored at −20 °C for 1 month.
6. The Coomassie staining and destaining solutions are prepared in house.
7. Since lysine residues are labeled, the user might want to use Glu-C to digest the proteins for having one more cleavage site. Since we did not observe any increment in the protein sequence coverage when using Glu-C, we recommend only the use of trypsin, which makes the protocol faster and simpler.
8. Overnight trypsin digestion does not yield more peptides but may result in increased autolysis trypsin peptides.
9. Before filtering the samples, wet the filter membrane device by adding 0.1 % formic acid, then centrifuge and discard it. This additional step increases sample recovery.
10. The picofrit columns are not coated. In our system, the ionization occurs via the liquid junction and is highly stable across the LC-MS/MS analysis.
11. The use of HPLC-grade bottled water reduces the likelihood of clogging the nano-column by particulate contamination. Particulate contamination can be carbon particles released from the in-house meg-ohm water system.
12. We employed centroid spectra for reducing the raw file size since the MS/MS information in centroid spectra is sufficient to identify the fragment sequence.
13. Set a ICPL quantification method in Mascot—(a) Go to *Configuration Editor* accessible from the Mascot home page (home-server); (b) Choose *Quantitation methods*; (c) select *New quantitation method*; (d) Define *Method*: protein identification (minimum two peptides) and protein ratio (median); (e) *Protocol*: use intensities of precursors within a single dataset. (f) *Component*: use for the ratio calculation, insert the name, select the modification group; (g) *Report ratio*: name the Report ratios (e.g., Light/heavy), select the numerator (light) and denominator (heavy); (h) *Integration*. Select the parameters according to the setup of the LC-MS run.
14. Alternatively, ICPL labeled samples can be quantified by using the ICPL-Quant [15], which can be freely downloaded from http://sourceforge.net/projects/icplquant/.
15. Ratios can be changed according to the settings established in the quantification method set in **Note 13**.

References

1. Mann M, Hojrup P, Roepstorff P (1993) Use of mass spectrometric molecular weight information to identify proteins in sequence databases. Biol Mass Spectrom 22:338–345
2. Link AJ, Eng J, Schieltz DM, Carmack E, Mize GJ, Morris DR, Garvik BM, Yates JR 3rd (1999) Direct analysis of protein complexes using mass spectrometry. Nat Biotechnol 17:676–682
3. Martins-de-Souza D, Guest PC, Vanattou-Saifoudine N, Harris LW, Bahn S (2011) Proteomic technologies for biomarker studies in psychiatry: advances and needs. Int Rev Neurobiol 101:65–94
4. Crowther J, Adusumalli V, Mukherjee T, Jordan K, Abuaf P, Corkum N, Goldstein G, Tolan J (1994) Determination of nanogram levels of peptide drug in rabbit and human plasma using high-performance liquid chromatography coupled with electrospray ionization mass spectrometry. Anal Chem 66:2356–2361
5. Gygi SP, Rist B, Gerber SA, Turecek F, Gelb MH, Aebersold R (1999) Quantitative analysis of complex protein mixtures using isotope-coded affinity tags. Nat Biotechnol 17:994–999
6. Ross PL, Huang YN, Marchese JN, Williamson B, Parker K, Hattan S, Khainovski N, Pillai S, Dey S, Daniels S, Purkayastha S, Juhasz P, Martin S, Bartlet-Jones M, He F, Jacobson A, Pappin DJ (2004) Multiplexed protein quantitation in Saccharomyces cerevisiae using amine-reactive isobaric tagging reagents. Mol Cell Proteomics 3:1154–1169
7. Thompson A, Schafer J, Kuhn K, Kienle S, Schwarz J, Schmidt G, Neumann T, Johnstone R, Mohammed AK, Hamon C (2003) Tandem mass tags: a novel quantification strategy for comparative analysis of complex protein mixtures by MS/MS. Anal Chem 75:1895–1904
8. Lottspeich F, Kellermann J (2011) ICPL labeling strategies for proteome research. Methods Mol Biol 753:55–64
9. Schmidt A, Kellermann J, Lottspeich F (2005) A novel strategy for quantitative proteomics using isotope-coded protein labels. Proteomics 5:4–15
10. Martins-de-Souza D, Gattaz WF, Schmitt A, Rewerts C, Maccarrone G, Dias-Neto E, Turck CW (2009) Prefrontal cortex shotgun proteome analysis reveals altered calcium homeostasis and immune system imbalance in schizophrenia. Eur Arch Psychiatry Clin Neurosci 259:151–163
11. Martins-de-Souza D, Gattaz WF, Schmitt A, Rewerts C, Marangoni S, Novello JC, Maccarrone G, Turck CW, Dias-Neto E (2009) Alterations in oligodendrocyte proteins, calcium homeostasis and new potential markers in schizophrenia anterior temporal lobe are revealed by shotgun proteome analysis. J Neural Transm 116:275–289
12. Martins-de-Souza D, Guest PC, Guest FL, Bauder C, Rahmoune H, Pietsch S, Roeber S, Kretzschmar H, Mann D, Baborie A, Bahn S (2012) Characterization of the human primary visual cortex and cerebellum proteomes using shotgun mass spectrometry-data-independent analyses. Proteomics 12:500–504
13. Martins-de-Souza D, Alsaif M, Ernst A, Harris LW, Aerts N, Lenaerts I, Peeters PJ, Amess B, Rahmoune H, Bahn S, Guest PC (2012) The application of selective reaction monitoring confirms dysregulation of glycolysis in a preclinical model of schizophrenia. BMC Res Notes 5:146
14. Shevchenko A, Wilm M, Vorm O, Mann M (1996) Mass spectrometric sequencing of proteins silver-stained polyacrylamide gels. Anal Chem 68:850–858
15. Brunner A, Kellermann J, Lottspeich F (2013) A practical guide to the ICPL_ESIQuant software for ICPL-based quantitative proteomics. Biochim Biophys Acta

Chapter 12

Employing TMT Quantification in a Shotgun-MS Platform

Darragh P. O'Brien and John F. Timms

Abstract

The blood serum proteome may be an ideal source of disease biomarkers, although its complexity necessitates novel strategies to enrich and quantify lower-abundance proteins with biomarker utility. Herein, serum samples from pre-diagnosis pancreatic cancer cases and controls were compared using a workflow of immunodepletion, multi-lectin fractionation, and peptide tandem mass tag (TMT) labeling. Samples were then subjected to SCX and high pH reversed-phase fractionation and analyzed by liquid chromatography tandem mass spectrometry (LC-MS/MS). The aim was the discovery of candidate serum biomarkers of pancreatic cancer, although the method is applicable to any comparative proteomic analysis of serum samples.

Key words Proteomic profiling, Serum, Protein quantification, LC-MS/MS, Tandem mass tags (TMT), Mass spectrometry, Pancreatic cancer

1 Introduction

There is an urgent need for accurate disease biomarkers which may provide novel approaches to screening and diagnosis. The blood serum proteome may be an ideal source of such biomarkers, although its complexity necessitates novel strategies to enrich and quantify lower-abundance proteins which may have the biomarker utility.

Shotgun proteomics is the technique of identifying proteins in complex biological mixtures using high performance liquid chromatography (HPLC) in combination with tandem mass spectrometry (MS/MS) [1]. LC-MS/MS has become the method of choice for high-coverage protein expression profiling, and in combination with differential labeling strategies, it provides proteomic quantification at either the peptide or protein level. Such strategies permit the mixing of samples prior to LC-MS/MS; in some cases upstream of any fractionation. Thus, multiple specimens can be run simultaneously with the same peptides (or proteins) being identically separated and co-eluted into the mass spectrometer, with ion intensities being directly compared in the same MS or MS/MS scans.

Daniel Martins-de-Souza (ed.), *Shotgun Proteomics: Methods and Protocols*, Methods in Molecular Biology, vol. 1156, DOI 10.1007/978-1-4939-0685-7_12, © Springer Science+Business Media New York 2014

Differential labeling falls into two categories; those which use chemical derivatization or enzymatic modification of proteins (or peptides) following sample collection and those which use incorporation of isotope-labeled amino acids in vivo. Isobaric chemical tags are chemical molecules with identical structures which label the ε-amino group of lysine residues and N-termini of peptides and proteins [2, 3]. Labeling is performed post-digestion; therefore, theoretically all the peptides in a given sample have the potential to be labeled. This results in multiple peptide measurements per protein, which increases the confidence of the protein identification and quantitative measurement as compared to isotopic labeling approaches. Another benefit of isobaric labeling is the ability to multiplex, allowing for several comparisons within a single experiment, i.e., up to eight for isobaric tags for relative and absolute quantitation (iTRAQ) [4–6], and up to six for tandem mass tags (TMT) [7–9]. Each isobaric tag consists of an N-hydroxysuccinimide moiety or ''peptide reactive group'' for peptide and protein labeling; a mass normalization group and a reporter group. Introduction of different numbers of the ''heavy'' isotopes ^{13}C or ^{15}N into the reporter and mass normalization groups produce the different versions of the tag, with each generating a specific fragment "reporter" ion in MS/MS.

For comparative analyses in biomarker discovery, peptide populations are differentially labeled (usually at the peptide level, following digestion) and mixed. During MS analysis, samples are indistinguishable from each other due to the isobaric nature of the tag. Upon peptide fragmentation by collision-induced dissociation (CID) both identification and relative quantitation of a peptide is achieved with release of the reporter group, giving rise to a unique singly charged ion signature in an ion-sparse region of the MS/MS spectra of the corresponding labeled peptides. Peptide quantitation is performed by comparing the reporter ion signal intensities between different tags and is made easier due to the perfect co-elution of the differentially labeled peptides. Ratios of reporter ions for assigned peptides are computed and integrated into protein ratios which are then evaluated statistically. The major advantage of these MS/MS-dependent quantification strategies is that the multiplex labeling does not increase the mass complexity of the sample and only peptides subjected to CID fragmentation are quantified. In addition, higher signal-to-noise ratios can be achieved with MS/MS-based detection versus MS-mode measurements.

High proteomic coverage is essential in biomarker discovery studies. The highly complex nature of the serum proteome necessitates the use of multiple, orthogonal fractionation steps based on different physicochemical properties of the proteins/peptides. The workflow described herein incorporates removal of highly abundant serum proteins by immunodepletion, fractionation of the glyco-proteome using lectin affinity chromatography, protein digestion (usually tryptic) and several strategies of peptide LC, e.g., strong cation exchange (SCX) followed by reversed-phase LC. Whilst isobaric protein labeling workflows have been described [9], TMT has been optimized for peptide-based labeling, and the labeling should be applied at the earliest possible point in the workflow to minimize differences due to sample processing. There are several drawbacks to isobaric labeling that are worth mentioning. The TMT reagents are expensive and only give signals when peptides are subjected to fragmentation. Thus, the strategy misses peptides not selected for MS/MS, lowering proteomic coverage. Dedicated software must be used for data analysis, although commercially available software packages (Mascot, Proteome Discoverer) now integrate reporter ion quantification and MS/MS data search output. It is also evident that higher energy CID methods and careful tuning are required for optimal "reporter" fragmentation to provide highest quantitative accuracy [4, 5]. The method described herein, integrates CID with higher energy collisional dissociation (HCD) for the same precursor ions on an LTQ Orbitrap XL instrument, although other instrumentation can be used (see below).

Herein, sera from a unique cohort of pre-diagnosis pancreatic cancer subjects and noncancer controls from the UK Collaborative Trial of Ovarian Cancer Screening (UKCTOCS) [10], were pooled into different pre-diagnosis time groups (up to 3+ years prior to the diagnosis of the cancer) and compared with a matched noncancer control group using a workflow of immunodepletion, multilectin fractionation, digestion, and peptide TMT labeling. Mixed labeled peptide samples were then subjected to SCX and high pH reversed-phase fractionation prior to LC-MS/MS analysis. The aim was the discovery of candidate serum biomarkers for the early detection of pancreatic cancer (Fig. 1), although the method is applicable to any comparative analysis of multiple serum samples.

2 Materials

2.1 Immunodepletion of High Abundance Serum Proteins

1. Bradford microtiter plate assay (Pierce) and plate reader.
2. Bovine serum albumin (BSA) standards.
3. Proteome Purify™12 Immunodepletion Resin (R&D Systems). A 50 % slurry of a proprietary absorption gel in Phosphate Buffered Saline (PBS) with 0.02 % sodium azide (pH 7.4).
4. Spin-X®Filter Units (Corning Costar). Centrifuge tube filters with a 0.22 μm pore cellulose acetate membrane.
5. 5,000 Da MWCO filters (Sigma).
6. Rotary shaker or mixer.
7. Vortex.
8. Centrifuge.

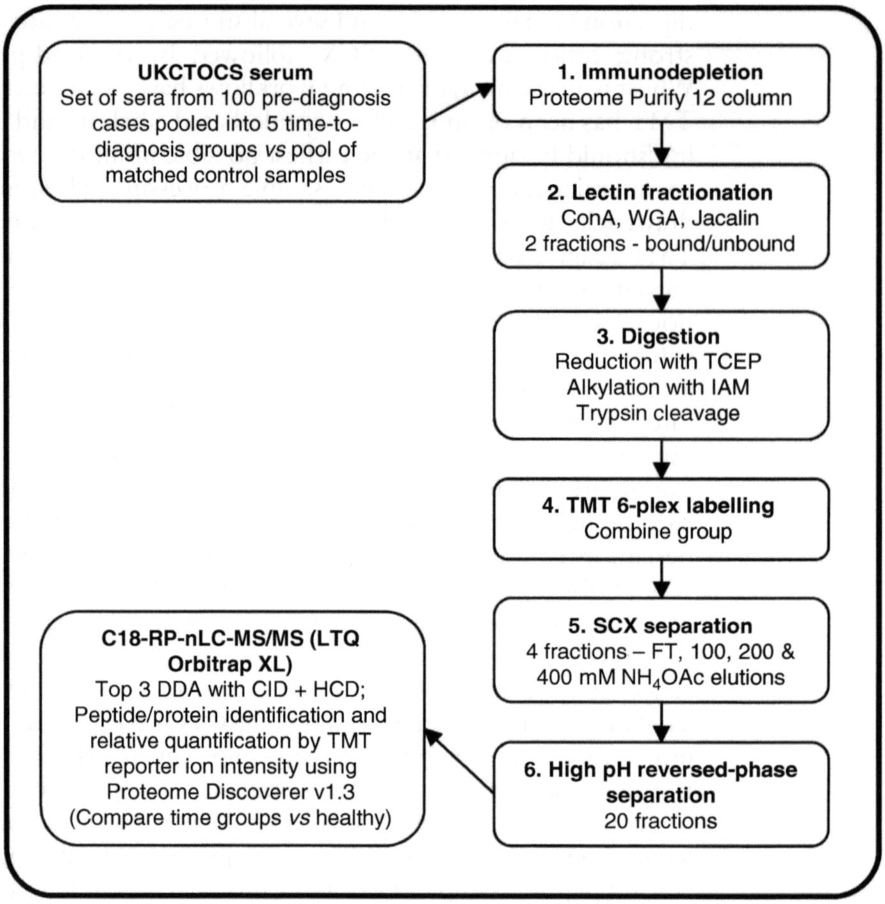

Fig. 1 Proteomic workflow for the discovery of serum biomarkers using multidimensional fractionation and TMT relative quantification

2.2 Lectin Affinity Chromatography

1. Concanavalin A (ConA) lectin resin (Thermo Fisher Scientific); 1.1 mL of settled resin supplied as a 50 % slurry.
2. Wheat Germ Agglutinin (WGA) lectin resin (Thermo Fisher Scientific); 1.1 mL of settled resin supplied as a 50 % slurry.
3. Jacalin lectin resin (Vector Laboratories); 2.0 mL of settled resin supplied as a 50 % slurry.
4. Glycoprotein Isolation Kit Binding/Wash Buffer (Thermo Fisher Scientific). A total volume of 6.5 mL of a 5× stock solution.
5. Elution Buffer; 2 % formic acid in HPLC grade water.
6. Column Accessory Pack (Thermo Fisher Scientific). Spin columns (10×) with bottom caps and collection tubes (20×).
7. Rotary shaker or mixer.
8. Vortex.
9. Centrifuge.

2.3 Trypsin Digestion and TMT Labeling

1. Siliconized eppendorf tubes.
2. Dissolution buffer: 100 mM triethyl ammonium bicarbonate (TEAB) (pH 8.0), 0.5 % (w/v) SDS. Aliquot and store at –20 °C.
3. Reducing solution: 200 mM tris(2-carboxyethyl) phosphine in 200 mM triethyl ammonium bicarbonate (pH 8.0). Prepare fresh.
4. Alkylation solution: 375 mM iodoacetamide in 100 mM TEAB (pH 8.0). Prepare fresh.
5. Trypsin solution: 500 ng/μL sequencing-grade modified trypsin (Promega) in 5 mM acetic acid. Store at –20 °C.
6. SpeedVac.
7. Acetonitrile (ACN): 100 % HPLC grade.
8. TMTsixplex Isobaric Mass Tagging Kit (Thermo Fisher Scientific).
9. Quenching solution: 5 % hydroxylamine in 200 mM TEAB (pH 8.0).

2.4 Strong Cation Exchange Chromatography

1. Siliconized eppendorf tubes.
2. Acetonitrile (ACN): 100 % HPLC grade.
3. Trifluoroacetic acid (TFA): 100 % HPLC grade.
4. Buffer A: 50 % ACN, 0.1 % TFA made with HPLC grade water.
5. Buffer B: 25 % ACN, 0.1 % TFA made with HPLC grade water.
6. Buffer C: Individual solutions of 100, 200, and 400 mM ammonium acetate, each in 25 % ACN, made with HPLC grade water.
7. Disposable plastic columns (Thermo Scientific).
8. SP Sepharose Fast Flow slurry (Sigma-Aldrich).

2.5 Basic Reversed-Phase Chromatography

1. 1100 Series HPLC (Agilent).
2. ZORBAX Poroshell 300Extend—C18 Analytical column, 2.1 × 75 mm (Agilent).
3. Solvent A: 20 mM ammonium formate in HPLC grade water.
4. Solvent B: 20 mM ammonium formate in 80 % ACN made with HPLC grade water.
5. Screw-thread amber autosampler vials (Agilent).
6. 96 well plates, 0.5 mL, PP (Agilent).

2.6 LC-MS/MS

1. Ultimate 3000 nano LC system (Dionex) linked to a LTQ Orbitrap XL mass spectrometer (Thermo Scientific) (*see* **Note 1**).
2. C18 PepMap guard column: 300 μm i.d. × 5 mm, 5 μm bead size, 100 Å pore size (LC Packings).
3. C18 PepMap nano LC column: 75 μm i.d. × 150 mm, 3 μm bead size, 100 Å pore size (LC Packings).

4. Solvent A: 0.1 % formic acid in HPLC grade water.
5. Solvent B: 0.1 % formic acid, 99.9 % ACN.
6. Xcalibur 2 (Thermo Scientific), Proteome Discoverer 1.3 (Thermo Scientific) and Mascot Server 2.2 (Matrix Science) software.

3 Methods

3.1 Preparation of Serum Pools

1. PC sera: combine 10 μL of sera from 100 pre-diagnosis cases into five time-to-diagnosis pools: 0–0.5, 0.5–1, 1–2, 2–3, and 3+ years.
2. Noncancer control sera: combine 10 μL of sera from 100 controls into one pool.

3.2 Immunodepletion of High Abundance Serum Proteins

1. Determine protein concentration using the Bradford microtiter plate assay. Make a 5 mg/mL stock of BSA in HPLC grade water and prepare serial dilutions of 0, 0.25, 0.5, 1.0, 2.5, and 5.0 mg/mL to make a standard curve. Use a 96-well flat-bottomed assay plate and make triplicate measurements for the BSA standards and four replicates for the experimental samples. For this, add 2 μL of sample per well and 200 μL of assay reagent and mix without introducing bubbles. Use a plate reader at a wavelength of 595 nm and calculate protein concentrations using the standard curve (*see* **Note 2**).
2. For each PC time group, add 30 μL of serum to a test tube.
3. Add 3.0 mL of the suspended Immunodepletion Resin to each serum sample (*see* **Note 3**).
4. Incubate on a rotary shaker with end-over-end mixing for 30 min. The mixing speed should be adequate to keep the Immunodepletion Resin in suspension.
5. After the incubation period, pipette equal volumes of the Immunodepletion Resin slurry into the upper chamber of six Spin-X Filter Units.
6. Centrifuge for 2 min at 2,000 × *g* and collect immunodepleted serum in a microcentrifuge tube. The combined volume of the depleted serum filtrates will be approximately 1,000–1,500 μL.
7. Dispose of the used Immunodepletion Resin (*see* **Note 4**).
8. Aliquot equal volumes of each immunodepleted serum sample into 3 × 5 kDa MWCO filtration devices. Concentrate each by centrifugation at 13,000 × *g* for 20 min.
9. Combine concentrated triplicate samples of the same PC time group. The volume of the combined filtrates should be approximately 150 μL.

10. Determine protein concentration of each immunodepleted serum sample using the Bradford microtiter plate assay, as described in **step 1** above (*see* **Notes 5** and **6**).

3.3 Lectin Affinity Chromatography

1. Equilibrate all buffers to room temperature.
2. Dilute immunodepleted serum samples 4:1 with 5× Binding/Wash Buffer stock solution. The total volume after dilution must not exceed 800 μL.
3. From the 5× stock solution, prepare 1× Binding/Wash Buffer with HPLC grade water.
4. Gently swirl the bottle of ConA lectin resin to obtain a homogeneous suspension. Use a wide-bore or cut pipette tip to transfer 100 μL of 50 % resin slurry to a 1 mL eppendorf tube.
5. Repeat **step 4** for WGA and Jacalin lectin resins, adding all three resins to the same eppendorf tube. Gently vortex the multi-lectin resin mixture to obtain a homogenous suspension.
6. Insert a column into a collection tube. Pipette 200 μL of the multi-lectin resin solution into this.
7. Centrifuge for 1 min at 1,000 ×*g* and discard the storage buffer. Reuse the collection tube through to **step 18**.
8. Place column in collection tube. Wash the multi-lectin resin solution with 200 μL of the 1× Binding/Wash Buffer by centrifugation for 1 min at 1,000 ×*g*. Discard the flow-through. Repeat this step two times.
9. Place bottom cap on column and add immunodepleted serum sample to the resin. Close the top cap.
10. Incubate for 10 min at room temperature with end-over-end mixing.
11. Remove top cap and then bottom cap from column. Place column in the collection tube and replace top (*see* **Note 7**).
12. Centrifuge for 1 min at 1,000 ×*g*. Collect the flow-through containing unbound/non-glycosylated proteins. Store on ice.
13. Reinsert column and add 400 μL 1× Binding/Wash Buffer to the resin. Cap column and centrifuge for 1 min at 1,000 ×*g*. Collect the flow-through and combine with that of **step 12**. Repeat this step.
14. Place bottom cap on column and add 400 μL 1× Binding/Wash Buffer to the resin. Cap column and incubate for 5 min at room temperature with end-over-end mixing.
15. Remove top cap and then bottom cap from column. Place column in the collection tube, and replace top cap. Centrifuge column for 1 min at 1,000 ×*g*. Discard the flow-through.

16. Repeat **steps 14** and **15**.
17. Replace bottom cap on column. Add 200 μL Elution Buffer to resin and cap column. Incubate column for 5 min at room temperature with end-over-end mixing.
18. Remove top cap and then bottom cap from column. Place column in a new collection tube. Replace top cap and centrifuge column for 1 min at 1,000 × *g*.
19. Carefully set aside the collection tube and remove top cap.
20. Repeat **steps 17–19**. Collect eluate in the same collection tube containing eluate from the first elution. Store eluted glycoprotein fraction on ice for immediate use or freeze for later analysis.
21. Determine protein concentration of both glycoprotein and non-glycoprotein fractions using the Bradford microtiter plate assay, as described in Subheading 3.2, **step 1** above.

3.4 Trypsin Digestion and TMT Labeling (See Note 8)

1. Dry down 100 μg of protein from each of the fractions in a SpeedVac and resuspend the protein pellets in 100 μL of dissolution buffer (100 mM TEAB) (pH 8.0), 0.5 % (w/v) SDS.
2. Add 5 μL of reducing solution (9.5 mM final concentration of tris(2-carboxyethyl) phosphine) and incubate for 1 h at 55 °C.
3. Add 5 μL of freshly prepared alkylation solution (17 mM final concentration of iodoacetamide) and incubate for 30 min at room temperature in the dark.
4. Add 5 μL of trypsin solution (2.5 μg/100 μg protein) and incubate overnight at 37 °C.
5. Reconstitute a tube of each TMT label in 41 μL of 100 % ACN and carefully add to the appropriate sample tube.
6. Allow labeling reactions to proceed for 1 h at room temperature.
7. Quench reactions by adding 8 μL of quenching solution and incubate for 15 min at room temperature.
8. Combine samples at equal amounts and dry down in a SpeedVac.

3.5 Strong Cation Exchange Chromatography (See Note 9)

1. Prepare cartridges using disposable plastic columns filled with 650 μL of SP Sepharose Fast Flow slurry and a frit placed on top of the resin bed.
2. Prior to peptide loading, wash cartridges with Buffer B (3 × 1 mL).
3. Load samples in 3 mL Buffer A. Collect unbound material for later analysis (*see* **Note 10**).

4. Wash bound peptides with 4 mL of Buffer B (4 × 1 mL).
5. Elute peptides from the cartridge with 2 mL of Buffer C; ammonium acetate at 100, 200, and 400 mM.
6. Lyophilize the samples to dryness in a vacuum centrifuge.
7. To ensure the removal of the volatile salt used for SCX, resuspend the samples in Buffer B and re-lyophilize to dryness.
8. Store samples at −80 °C until further analysis.

3.6 Basic Reversed-Phase Chromatography

1. Resuspend sample in 42 μL of 3 % buffer B and transfer to a screw-thread amber autosampler vials.
2. Inject 40 μL of the sample onto a Poroshell 300Extend C-18 analytical column (pre-equilibrated for 20 min in 3 % buffer B) using an Agilent 1100 Series HPLC system loading pump.
3. Apply a 3–60 % gradient of buffer B at 200 μL/min over 50 min using the loading pump, and collect 300 μL fractions in a 96 well plate every 2.6 min to generate 20 fractions.
4. Dry fractions down in a SpeedVac.

3.7 LC-MS/MS

1. Resuspend samples in 12 μL of 0.1 % formic acid.
2. Inject 5 μL of sample from the Dionex Ultimate 3000 autosampler onto the C18 PepMap guard column and wash for 3 min with 100 % solvent A at a flow rate of 25 μL/min.
3. Switch to the analytical C18 PepMap nano LC column with 10 % solvent B and apply a linear gradient of 10–50 % B over 90 min, then to 100 % B over 3 min. Continue with 100 % B for 20 min and then reduce to 10 % B over 0.5 min and continue for a further 20 min to re-equilibrate the column for the next injection.
4. Operate the mass spectrometer (LTQ Orbitrap XL) in the data-dependent mode for automated switching between MS and MS/MS acquisition. Acquire survey full scan MS spectra (from *m/z* 400 to 2,000) in the Orbitrap (FT) with a resolution of 60,000 at *m/z* 400. Select the "top 3" most intense ions for both CID and HCD. Select a target ion value of 1×10^6 and maximum scan time of 500 ms for the survey full scan in the FT. Select target ion values of 1×10^5 and 1×10^4 and scan time settings of 300 and 150 ms for FT-MSn (HCD) and IT-MSn (CID), respectively (*see* **Note 11**). Set resolution of fragment ion detection (in the FT) at 7,500. Acquire centroid data in both detectors.
5. Dynamically exclude ions selected for MS/MS for 60 s.
6. Enable the lock mass option for accurate mass measurement, using the polydimethylcyclosiloxane ion (*m/z* 455.120025) as an internal calibrant.

3.8 Data Analysis

1. Process the raw data files generated in Xcalibur using Proteome Discoverer 1.3. Set up the workflow shown in Fig. 1 using the following instructions and parameters.
2. Input RAW files in the Spectrum Files node.
3. In the Spectrum Selector, use MS1 precursor, minimum precursor mass 350 Da, maximum precursor mass 6,500 Da, select MS order as MS2 and a S/N threshold as 1.5. All other settings as default or "0."
4. In the Scan Event Filter for HCD, set analyzer to FTMS, activation type to HCD and all other parameters to "Any."
5. In the Scan Event Filter for CID, set analyzer to ITMS, activation type to CID and all other parameters to "Any."
6. In the Mascot node for HCD, select the most recent version of the SwissProt database (updated in Mascot Server), select trypsin as the enzyme, number of missed cleavages as 1, instrument type as ESI-FTICR, taxonomy as human, precursor mass tolerance as 10 ppm, fragment mass tolerance as 20 mmu and do not average precursor masses. Set methionine oxidation and N-terminal acetylation as dynamic modifications and carbamidomethylation of cysteines and TMT modification of peptide N-termini and lysine residues as static modifications.
7. In the Mascot node for CID, select the parameters as described above, but using ESI-TRAP as instrument type and a fragment mass tolerance of 0.6 Da.
8. In the Percolator used for decoy database searching and estimation of false discovery rate (FDR), apply a maximum delta Cn of 0.05, a target FDR (strict) of 0.01, a target FDR (relaxed) of 0.05 and validation based on *q*-value.
9. In the Reporter Ions Quantifier, choose an integration window tolerance of 10 ppm and "most confident centroid" for integration method. Set the scan event filters to mass analyzer FTMS, MS order to MS2 and activation type HCD. Set up the quantification method using TMT6plex/+229.163 Da on lysine (K) for residue modification and TMT6plex/+229.163 Da for N-terminal modification. Input masses of the tags and reporter ion isotopic distributions (this information should be provided with each kit). Set the desired ratio reporting: here ratios 127/126, 128/126, 129/126, 130/126, and 131/126 were used to compare case sample pools from the different time-to-diagnosis groups with the noncancer control pool labeled with the 126 tag. For reporter ion ratio calculation, apply quantity value correction and set the fold-change threshold and maximum allowed fold-change (*see* **Note 12**). For protein quantification, consider protein groups for peptide uniqueness and use only unique peptides.

10. Run the analysis (*see* **Note 13**).
11. Apply search result filters, only allowing peptides with a score >20 and equal to or above the Mascot significance threshold of $P < 0.05$.
12. Accept only quantitative information for proteins with at least 2 peptide matches, using the median value of the peptide ratios and a % variability of less than 20 %.
13. Validate changes of interest by immunoblotting with specific antibodies or ELISA.

4 Notes

1. Other instrumentation and modes of operation can be used for peptide identification and MS/MS-based quantification using TMT isobaric tags. Fragmentation and quantification of reporter ions can be achieved using HCD/CID (as described here), MS3 [11], or proton transfer reaction (PTR)-MS [12] on the Orbitrap Velos and by pulsed Q collision induced dissociation/CID (PQD/CID) [4] on the LTQ-Orbitrap Discovery or LTQ ion trap instruments. The TMT approach has also been successfully applied using Q-TOF [13], TOF/TOF [7], Triple Quadrupole [14, 15] instruments and those equipped with electron transfer dissociation (ETD) capability [16].
2. It is recommended that at least three replicate assays are performed for each sample for accurate protein determination.
3. It is essential that the Immunodepletion Resin be a homogeneous suspension prior to pipetting.
4. The Spin-X Filter Units may be cleaned with deionized water followed by 20 % ethanol and saved for reuse.
5. Depletion of the serum or plasma will result in the removal of greater than 90 % of the twelve most abundant proteins. If higher levels of depletion are required, the sample should be concentrated (5,000 Da MWCO spin concentrator) and depleted a second time.
6. The depleted sample will be in a buffer of PBS with 0.02 % sodium azide. Spin concentrators are recommended if desalting is required.
7. Remove top cap before bottom cap to prevent sample leaking from the bottom of the column.
8. Reagents and protocol are provided in the TMTsixplex Isobaric Mass Tagging Kit. Carry out all steps from this point forward using siliconized tubes.
9. SCX is used here to remove excess TMT reagents and SDS and to fractionate the TMT-labeled peptides.

10. During loading and elution of the peptides, maintain the flow rate of the system at approximately 1 drop/s
11. Use Tune Plus V2.4 SP2 in instrument software for adjusting these parameters. For example, if the sample amount is low, then increasing the MSn scan times is advisable. Tune Plus V2.4 SP2 automatically adjusts the HCD normalized collision energy depending upon the charge state of TMT-tagged precursor ions.
12. A fold-change threshold of 2.0 and maximum allowed fold-change of 100 were used in this study.
13. The experiment should be run in triplicate and the data from each processed separately in Proteome Discoverer. The results files can then be uploaded together in Proteome Discoverer to compare the experiments and identify reproducible changes.

Acknowledgement

This work was funded by CRUK Project Grant C12077/A12790 and was supported by the National Institute for Health Research University College London Hospitals Biomedical Research Centre.

References

1. Washburn MP, Wolters D, Yates JR 3rd (2001) Large-scale analysis of the yeast proteome by multidimensional protein identification technology. Nat Biotechnol 19:242–247
2. Kuhn K, Thompson A, Prinz T, Muller J, Baumann C, Schmidt G, Neumann T, Hamon C (2003) Isolation of N-terminal protein sequence tags from cyanogen bromide cleaved proteins as a novel approach to investigate hydrophobic proteins. J Proteome Res 2:598–609
3. Ross PL, Huang YN, Marchese JN, Williamson B, Parker K, Hattan S, Khainovski N, Pillai S, Dey S, Daniels S, Purkayastha S, Juhasz P, Martin S, Bartlet-Jones M, He F, Jacobson A, Pappin DJ (2004) Multiplexed protein quantitation in Saccharomyces cerevisiae using amine-reactive isobaric tagging reagents. Mol Cell Proteomics 3:1154–1169
4. Bantscheff M, Boesche M, Eberhard D, Matthieson T, Sweetman G, Kuster B (2008) Robust and sensitive iTRAQ quantification on an LTQ Orbitrap mass spectrometer. Mol Cell Proteomics 7:1702–1713
5. Griffin TJ, Xie H, Bandhakavi S, Popko J, Mohan A, Carlis JV, Higgins L (2007) iTRAQ reagent-based quantitative proteomic analysis on a linear ion trap mass spectrometer. J Proteome Res 6:4200–4209
6. Guo T, Gan CS, Zhang H, Zhu Y, Kon OL, Sze SK (2008) Hybridization of pulsed-Q dissociation and collision-activated dissociation in linear ion trap mass spectrometer for iTRAQ quantitation. J Proteome Res 7:4831–4840
7. Dayon L, Hainard A, Licker V, Turck N, Kuhn K, Hochstrasser DF, Burkhard PR, Sanchez JC (2008) Relative quantification of proteins in human cerebrospinal fluids by MS/MS using 6-plex isobaric tags. Anal Chem 80:2921–2931
8. Sinclair J, Metodieva G, Dafou D, Gayther SA, Timms JF (2011) Profiling signatures of ovarian cancer tumour suppression using 2D-DIGE and 2D-LC-MS/MS with tandem mass tagging. J Proteomics 74:451–465
9. Sinclair J, Timms JF (2011) Quantitative profiling of serum samples using TMT protein labelling, fractionation and LC-MS/MS. Methods 54:361–369
10. Menon U, Kalsi J, Jacobs I (2012) The UKCTOCS experience-reasons for hope? Int J Gynecol Cancer 22(Suppl 1):S18–S20
11. Ting L, Rad R, Gygi SP, Haas W (2011) MS3 eliminates ratio distortion in isobaric multiplexed quantitative proteomics. Nat Methods 8:937–940

12. Wenger CD, Lee MV, Hebert AS, McAlister GC, Phanstiel DH, Westphall MS, Coon JJ (2011) Gas-phase purification enables accurate, multiplexed proteome quantification with isobaric tagging. Nat Methods 8:933–935
13. van Ulsen P, Kuhn K, Prinz T, Legner H, Schmid P, Baumann C, Tommassen J (2009) Identification of proteins of Neisseria meningitidis induced under iron-limiting conditions using the isobaric tandem mass tag (TMT) labeling approach. Proteomics 9:1771–1781
14. Stella R, Cifani P, Peggion C, Hansson K, Lazzari C, Bendz M, Levander F, Sorgato MC, Bertoli A, James P (2011) Relative quantification of membrane proteins in wild-type and prion protein (PrP)-knockout cerebellar granule neurons. J Proteome Res 11:523–536
15. Byers HL, Campbell J, van Ulsen P, Tommassen J, Ward MA, Schulz-Knappe P, Prinz T, Kuhn K (2009) Candidate verification of iron-regulated Neisseria meningitidis proteins using isotopic versions of tandem mass tags (TMT) and single reaction monitoring. J Proteomics 73:231–239
16. Viner RI, Zhang T, Second T, Zabrouskov V (2009) Quantification of post-translationally modified peptides of bovine alpha-crystallin using tandem mass tags and electron transfer dissociation. J Proteomics 72:874–885

Chapter 13

Employing TMT Quantification in Shotgun-MS Proteomic Analysis: A Focus on Skeletal Muscle

Bruno Menezes de Oliveira

Abstract

Quantify proteins present in a sample is the ultimate goal of a proteomic experiment. Although it might reveal countless physiological and /or pathological process on a studied model, it is still a daunting task to perform. Of the protein present in a sample, only a small number can be identified and even a lesser number quantified, each with its own weaknesses and strengths. Presented here are the "tandem mass tags" isobaric labels (TMT) and a protocol for its utilization in the quantification of skeletal muscle, which might be adapted also to other animal tissues.

Key words Shotgun proteomics, Tandem mass tags, TMT, Isobaric tags, Mass spectrometry, Relative quantification

1 Introduction

Shotgun proteomics per se came to revolutionize the proteomics field. The promise of thousands of proteins capable of being identified in a single experiment was something that amazed the researchers from this specific field in late 1990s [1]. On the quest for providing answers for studies about pathological or physiological processes, at the molecular level, high-throughput technologies are fundamental [2, 3].

Identifying proteins present in a given sample is usually not enough; it is also necessary to quantify them, especially when considering the highly dynamic nature of organisms' proteome [4].

Since LC-MS is not inherently quantitative, due to the wide range of physiochemical properties of proteolytic peptides, which leads to differences in mass spectrometry responses, it must be associated with a quantitative method. Basically, quantification techniques can be classified as label based and label free depending on the use of tags to achieve this aim. Label-based techniques have as "Rosetta's Stone" the stable isotope dilution theory, which postulates that a stable isotope-labeled peptide is chemically similar to

Daniel Martins-de-Souza (ed.), *Shotgun Proteomics: Methods and Protocols*, Methods in Molecular Biology, vol. 1156, DOI 10.1007/978-1-4939-0685-7_13, © Springer Science+Business Media New York 2014

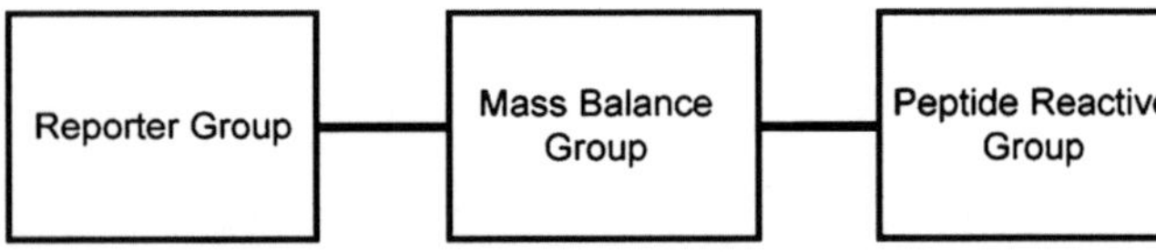

Fig. 1 Typical structure of an isobaric label

its native counterpart and therefore have the same behavior in a chromatographic run and/or MS analysis [5].

Thompson and coworkers introduced the isobaric tags, named Tandem Mass Tag™ reagents (TMT™), slightly different from the isotopic labeling, though similar in concept. In this case, each reagent consists of a mass reporter tag (126–131 Da for TMT6plex™) with a unique number of ^{13}C mass substitutions, a mass normalizer group, with a unique mass that balances the reagent mass, and last but not least, a reactive site that allows the incorporation of the reagent to the peptides in the sample (Fig. 1). Tagged peptides are chemically identical and, after individually labeling samples and pooling them, tandem MS (MS/MS) is performed. Upon peptide fragmentation, differences can be detected and quantification can be done. Inn MS/MS not only identification of peptides is achieved but the reporter ion from each tag is also released, providing an accurate quantification of the selected peptide/protein in that sample. An advantage of the method is that the MS signal is not split, by the contrary the signal from same peptide specie signal is reinforced due to its presence across different samples, increasing sensitivity [6, 7].

Initially conceived as two channel or six channel, TMT2plex™ and TMT6plex™ respectively, recently it has been demonstrated the possibility of expanding the multiplexity capabilities of 6plex reagents to 8plex and at least to 10plex, and up to 18plex, significantly improving the parallelization of the method [8].

Here, we describe a detailed protocol for sample preparation and labeling when using TMT™ reagents for the accurate quantification of peptides/proteins from mouse muscle samples labeled with TMT6plex™ reagents, which can be extended to most animal tissue with few modifications. Besides this, further steps used by our group for data analysis are presented.

2 Materials

2.1 Sample Preparation

1. Dissected mice muscles rinsed well in 0.9 % NaCl solution (*see* **Note 1**).
2. Mortar, pestle, and liquid nitrogen.
3. Lysis Buffer: 100 mM $NaHCO_3$, 20 % SDS, protease and phosphatase inhibitors (*see* **Note 2**).

4. Ultrasonic Homogenizer.
5. Benchtop Centrifuge.
6. Protein quantification assay (e.g., Bradford, BCA).
7. Microplate reader (562 nm).

2.2 TMT Labeling

1. HPLC grade water.
2. Dissolution buffer: 0.2 M Triethylammonium bicarbonate (TEAB).
3. Denaturing reagent: 2 % Sodium dodecyl sulfate SDS (w/v).
4. Reducing reagent: 0.2 M TCEP.
5. Cysteine blocking reagent: 375 mM iodoacetamide.
6. Quenching reagent: 5 % Hydroxylamine.
7. Pre chilled (−20 °C) acetone.
8. Sequencing Grade Modified Trypsin.
9. TMT™ label reagents: 126, 127, 128, 129, 130, 131 (Pierce Biotechnology, Rockford, IL, USA).
10. SpeedVac Concentrator.

2.3 Sample Analysis by 2D Liquid Chromatography and Tandem Mass Spectrometry

2.3.1 SCX Fractionation

1. Strong cation exchange chromatography buffer A (SCX-A): 5 mM KH_2PO_4, 25 % acetonitrile, pH 2.9 (adjust using H_3PO_4), made with HPLC grade water. Lifetime (2–3 days).
2. Strong cation exchange chromatography buffer B (SCX-B): 5 mM KH_2PO_4, 25 % acetonitrile, 600 mM KCl, pH 2.9, made with HPLC grade water. Lifetime (2–3 days).
3. Strong cation exchange chromatography buffer B2 (SCX-B2): 5 mM KH_2PO_4, 25 % acetonitrile, 1 M KCl, pH 2.9, made with HPLC grade water. Lifetime (2–3 days).
4. 1 M H_3PO_4.
5. Harvard Syringe Pump.
6. SCX Cartridge (Applied Biosystems).
7. Cartridge-holder.
8. Glass Syringe.

2.3.2 SCX Peptide Desalting

1. C18 spin columns (Pierce Biotechnology, Rockford, IL, USA).
2. Activation Buffer: 50 % acetonitrile (ACN).
3. Loading Buffer: 5 % ACN, 0.5 % trifluoroacetic acid (TFA).
4. Eluting Buffer: 70 % ACN.
5. 0.1 % formic acid.

2.3.3 LC-MS/MS Analysis on LTQ-OrbitrapXL

1. LTQ-OrbitrapXL (Thermo Fisher Scientific) interfaced with a commercially available nano-LC column.
2. HTC-PAL autosampler (CTC Analytics AG).

3. Agilent 1200 binary pump (Agilent Technologies).
4. In-house packed trapping column (40 × 0.075 mm i.d.) with 3 μm Reprosil-Pur C18-AQ particles.
5. In-house packed reversed phase column (200 × 0.050 mm i.d.) with 3 μm Reprosil-Pur C18-AQ particles.
6. Reverse phase chromatography buffer A (RP-A): 0.1 % (v/v) formic acid, made with HPLC-grade water. Lifetime (2–3 days).
7. Reverse phase chromatography buffer B (RP-B): 80 % (v/v) acetonitrile, 0.1 % (v/v) formic acid, made with HPLC-grade water. Lifetime (2–3 days).

3 Methods

3.1 Sample Preparation

For dissection, animals must be sacrificed using appropriated methods (*see* **Note 3**).

1. Muscles must be carefully pulverized in liquid Nitrogen using a mortar and a pestle.
2. Extract protein in lysis buffer (100 mM $NaHCO_3$, 20 % SDS) (*see* **Note 4**), a ratio ¼ (w/v) must be reasonable by ultrasonication (3 × 5″), at 4 °C, protease inhibitors must be freshly added to the buffer.
3. Centrifuge samples for 5 min at 15,000 × *g*, and collect the supernatant and determine protein concentration using a suitable assay (*see* **Note 5**).

3.2 TMT Labeling

1. Pipette 100 μg of protein from each different sample to be analyzed and equalize sample volumes with used lysis buffer.
2. Add 45 μL of TEAB 100 mM to the samples and adjust the final volume to 100 μL with ultra-pure water.
3. To each sample add 5 μL of TCEP 200 mM and incubate them at 55 °C for 1 h.
4. To complete an efficient step of reduction of dissulphide bonds add 5 μL of 375 mM iodoacetamide, freshly made. Vortex, spin and incubate for 30 min protected from light.
5. Precipitate protein overnight by adding six volumes of pre chilled acetone at −20 °C.
6. Centrifuge samples at 8,000 × *g* for 10 min at 4 °C. Carefully discharge supernatant and dissolve pellets in 100 μL of TEAB 100 mM.
7. Compose a standard sample by adding equal amounts of each sample. This sample will be the reference to define relative protein amounts for each analyzed sample

8. Reconstitute just before use a vial of trypsin with 20 μL Trypsin Storage Solution. Apply 2.5 μg of trypsin per 100 μg of protein allowing sample to digest overnight at 37 °C (*see* **Note 6**).
9. Allow TMT reagents to come at room temperature. Add 41 μL of anhydrous acetonitrile, vortex and spin.
10. Add exactly 41 μL of TMT Label Reagent to each sample according to experiment design (*see* **Note** 7).
11. Incubate for 1 h at room temperature.
12. Stop reactions with 8 μL of 5 % hydroxylamine and add 100 μL of water.
13. Leave the samples on the bench for 30 min and pool them together or store isolated at −80 °C (*see* **Note 8**).

3.3 Sample Analysis by 2D Liquid Chromatography and Tandem Mass Spectrometry

Pooled peptides must be first fractionated by strong cation exchange (SCX). This procedure will allow the cleaning up of excess of any TMT reagents.

3.3.1 SCX Fractionation

1. Dry the combined sample on a SpeedVac, reducing its volume to around 100 μL.
2. Add 500 μL (SCX-A), 5 mM KH_2PO_4, 25 % acetonitrile, pH 2.9.
3. Confirm if sample's pH is acidic (below pH 3), if not adjust it with 1 M phosphoric acid.
4. Allow the SCX cartridge (Applied Biosystems) to reach room temperature.
5. Set syringe diameter (individual for each syringe) and flow rate to 50 μL/min on the Syringe Pump.
6. Wash the needle three times with methanol and three times with water.
7. Mixing buffers SCX-A and SCX-B prepare solutions with final KCL concentrations of 30, 60, 90, 120, 240, 300, 420, and 500 mM.
8. Activate and wash the resin with 500 μL of SCX-B2.
9. Wash the resin with at least 1.5 mL buffer SCX-A.
10. Centrifuge sample at 16,100 × *g* for 5 min.
11. Load all sample in the syringe and inject it in the column (*see* **Note 9**).
12. Collect the flow-through (FT1).
13. Inject 500 μL of SCX-A and collect flow-through (FT2).
14. Inject one by one the all the eluting buffers prepared at different concentrations and collect fractions 1–8.
15. 15 Inject 500 μL of SCX-B2 and collect the flow-through (FT-after) (*see* **Note 10**).

16. 15 Inject another 500 μL of SCX-B2 followed by 1 mL of SCX-A to clean and equilibrate resin.
17. Reduce the volumes of fractions 1–5 to ca. 100 to 200 μL, getting rid of all ACN.

3.3.2 SCX Peptide Desalting

1. Activate resin wetting it with 200 μL 50 % acetonitrile (ACN). Centrifuge for 5 min and repeat procedure.
2. Equilibrate column with 200 μL 5 % ACN, 0.5 % TFA. Centrifuge for 5 min and repeat procedure.
3. Load sample, centrifuge for 5 min and repeat procedure.
4. Wash resin with 200 μL 5 % ACN, 0.5 % TFA. Centrifuge for 5 min and repeat procedure two times.
5. Elute peptides with 200 μL 70 % ACN.
6. Lyophilize or dry the fractions on a SpeedVac. Dissolve them 30 μL of 0.1 % formic acid.

3.3.3 LC-MS/MS Analysis on LTQ-OrbitrapXL

1. Inject 2 μL per with an HTC-PAL autosampler (CTC Analytics AG) connected to an Agilent 1200 binary pump (Agilent Technologies).
2. Set flow through the analytical column to 10 nL/min.
3. Gradient used by us follow: 0–6 min 0.1 % formic acid, 6–76 min 7–35 % ACN, 0.1 % formic acid, 76–79 min 40–80 % ACN 0.1 % formic acid.
4. LTQ-OrbitrapXL settings are: spray voltage 1.4 kV, 1 microscan for MS1 scans at 60,000 resolutions (*m/z* 400), full MS mass range *m/z* 400–2,000. LTQ-Orbitrap XL is operated in a data-dependent mode with one MS1 FTMS scan of precursor ions followed by CID (collision induced dissociation) and HCD (high energy collision dissociation), MS2 scans of the three most abundant doubly, triply, and quadruply protonated ions in each FTMS scan. Settings for the MS2 were as follows: 1 microscans for HCD-MS2 at 7,500 resolution (at *m/z* 400), mass range *m/z* 100–2,000 with a collision energy of 50 %, 1 microscans for CID-MS2 with a collision energy of 30 %. Dynamic exclusion of a precursor selected for MS2 is used for 120 s after one repeat, enabling most of the co-eluting precursors to be selected for MS2.
5. All samples are analyzed a second time as described above and also a third time using an exclusion list of all *m/z* within a 3 min retention window, already passing the identification criteria within the TMT set (8 fractions) in the database search.

3.4 Data Analysis

1. MS raw data files from all SCX fractions per one TMT set and all technical or biological replicates, if present, are merged for relative quantification and identification using Proteome Discoverer version 1.3 (Thermo Fisher Scientific).

2. Database search is performed by Mascot search engine using the following criteria: *Mus musculus* (our animal model) in Swissprot protein database, MS peptide tolerance as 10 ppm, MS/MS tolerance as 0.5 Da, trypsin digestion allowing one missed cleavages with variable modifications; methionine oxidation, cysteine methylthiol, and fixed modifications; N-terminal TMT6-plex label, lysine TMT6-plex label.
3. The detected protein threshold in the software is set to a confidence using the FDR 1 % method and identified proteins are grouped by sharing the same sequences to minimize redundancy.
4. For quantification, the ratios of TMT reporter ion intensities in MS/MS spectra (*m/z* 126.12, 127.13, 128.13, 129.14, 130.14) from raw data sets are used to calculate fold changes between samples via the relative ratio to the reference pool. Only peptides unique for a given protein are considered for relative quantitation, excluding those common to other isoforms or proteins of the same family.
5. Normalization on protein median is used. The median of peptides was used for protein ratio and the resulting ratios were then exported into Excel for manual data interpretation (*see* **Note 11**).

4 Notes

1. Muscles must be dissected and cleaned on ice. All fur must be removed to avoid contamination on subsequent steps. If needed, perfusion can be made to remove blood before extraction.
2. Lysis buffer can be modified and optimized according to the proteins that one desires to contemplate in the sample (e.g., mitochondrial, cytosolic or extracellular proteins) [9].
3. Sacrifice methods should be adequately chosen to the analysis. Certain methods (e.g., use of drugs) can induce the insertion of differences not related to studied condition, but instead be a side effect of the method.
4. Sample preparation must be performed without presence of compounds containing free amine groups (e.g., Tris, ammonium salts) which can promote interference on labeling reaction.
5. Protein concentration determination is a critical step. Technical errors here will compromise further quantification and all the experiment. It is highly recommended to monitor these results using an SDS-PAGE.
6. Acetone precipitated might not completely dissolve immediately, but after 12 h peptides will be completely soluble.

Table 1
Experimental setup for two muscles under two conditions (control and disease) and with three biological replicates (Pool 1, 2, and 3)

	Groups	TMT label
Pool 1 ($n=5$)	Muscle1 control	127
	Muscle2 control	128
	Muscle1 disease	129
	Muscle2 disease	130
	Internal standard	126
Pool 2 ($n=5$)	Muscle1 control	130
	Muscle2 control	129
	Muscle1 disease	128
	Muscle2 disease	127
	Internal standard	126
Pool 3 ($n=5$)	Muscle1 control	129
	Muscle2 control	130
	Muscle1 disease	127
	Muscle2 disease	128
	Internal standard	126

The internal standard is a mixture of all samples

7. In order to avoid possible bias, leading to technical errors, it is recommended to perform technical replicates applying a label swapf (Table 1).
8. This step as all others that requires equal amounts of the samples is critical, since technical errors from the user can compromise the entire quantification.
9. When setting the volume of the syringe at the pump give some margin to the syringe does not break. After injection stops, dismount syringe from pump and manually inject what is left.
10. Usually no peptides should be seen on fraction FT1, FT2, or FT-after, but it is recommended to freeze and store them as a control.
11. Additionally we usually check differences among samples using Student's t-test, considering only proteins with a variation of less than 30 %, a fold change higher or lesser than 1.25 and a statistical power above 0.8 [10].

References

1. Link AJ, Eng J, Schieltz DM et al (1999) Direct analysis of protein complexes using mass spectrometry. Nat Biotechnol 17:676–682
2. Wilkins MR (2009) Hares and tortoises: the high-versus low-throughput proteomic race. Electrophoresis 30(Suppl 1):S150–S155
3. Oliveira BM, Schmitt A, Falkai P et al (2013) Is clinical proteomics heading towards to "bench to bedside"?, Translational Proteomics 1:53–56.
4. Wilkins MR, Pasquali C, Appel RD et al (1996) From proteins to proteomes: large

scale protein identification by two-dimensional electrophoresis and amino acid analysis. Biotechnology (Nature Publishing Company) 14:61–65

5. Bantscheff M, Schirle M, Sweetman G et al (2007) Quantitative mass spectrometry in proteomics: a critical review. Anal Bioanal Chem 389:1017–1031
6. Martins de Souza D, Oliveira BM, Castro-Dias E et al (2008) The untiring search for the most complete proteome representation: reviewing the methods. Brief Funct Genomic Proteomic 7:312–321
7. Thompson A, Schäfer J, Kuhn K et al (2003) Tandem mass tags: a novel quantification strategy for comparative analysis of complex protein mixtures by MS/MS. Anal Chem 75:1895–1904
8. McAlister GC, Huttlin EL, Haas W et al (2012) Increasing the multiplexing capacity of TMTs using reporter ion isotopologues with isobaric masses. Anal Chem 84:7469–7478
9. Martins-de-Souza D, Martins D, Menezes de Oliveira B et al (2007) The use of ASB-14 in combination with CHAPS is the best for solubilization of human brain proteins for two-dimensional gel electrophoresis. Brief Funct Genomic Proteomic 6:70–75
10. Matsumura CY, Menezes de Oliveira B, Durbeej M et al (2013) Isobaric tagging-based quantification for proteomic analysis: a comparative study of spared and affected muscles from mdx mice at the early phase of dystrophy. PloS One 8:e65831

Part V

Label-Free Proteomics

Chapter 14

Spectral Counting Label-Free Proteomics

Liisa Arike and Lauri Peil

Abstract

Label-free proteome quantification methods used in bottom-up mass-spectrometry based proteomics are gaining more popularity as they are easy to apply and can be integrated into different workflows without any extra effort or cost. In the label-free proteome quantification approach, samples of interest are prepared and analyzed separately. Mass-spectrometry is generally not recognized as a quantitative method as the ionization efficiency of peptides is dependent on composition of peptides. Label-free quantification methods have to overcome this limitation by additional computational calculations. There are several algorithms available that take into account the sequence and length of the peptides and compute the predicted abundance of proteins in the sample. Label-free methods can be divided into two categories: peptide peak intensity based quantification and spectral counting quantification that relies on the number of peptides identified from a given protein.

This protocol will concentrate on spectral counting quantification—exponentially modified protein abundance index (emPAI). Normalized emPAI, most commonly derived from Mascot search results, can be used for broad comparison of entire proteomes. Absolute quantification of proteins based on emPAI values with or without added standards will be demonstrated. Guidelines will be given on how to easily integrate emPAI into existing data; for example, calculating emPAI based absolute protein abundances from iTRAQ data without added standards.

Key words Label-free quantification, Relative protein quantification, Absolute protein quantification, Spectral counting, emPAI

1 Introduction

Mass spectrometry-based proteomics turned quantitative shortly after its birth [1] as it was realized that protein identification only provides very limited information [2, 3]. Two main approaches are used in quantitative mass spectrometry based proteomics: stable isotope labelling and label-free quantification. Quantification with stable isotopes is based on the mass difference between labelled and unlabelled ions in MS analysis. After mixing samples the intensity ratio between the isotope variants reflects the fold-difference

Daniel Martins-de-Souza (ed.), *Shotgun Proteomics: Methods and Protocols*, Methods in Molecular Biology, vol. 1156, DOI 10.1007/978-1-4939-0685-7_14, © Springer Science+Business Media New York 2014

between their abundances. Label-free quantification compares separately prepared samples and separately analyzed LC-MS/MS runs. Differences in sample preparation may cause inaccurate quantification; therefore, shot-gun proteomics with a minimum sample preparation steps is ideal for label-free quantification.

The most simple label-free quantification technique is spectral counting. This is based on the observation that the more abundant the protein is the more peptides can be identified from it [4]. Because larger proteins produce more peptides and therefore also more MS/MS events that identify peptides, spectral counting methods must take into account the size of proteins. This has been implemented in several spectral counting methods: protein abundance index (PAI) [5], exponentially modified protein abundance index (emPAI) [6], normalized spectral abundance factor (NSAF) [7], absolute protein expression (APEX) [8], normalized spectral index (SIN) [9], etc.

When it comes to choosing a spectral counting method, there are conflicting reports as to which method preforms the best. A recent study reported that APEX generally provided better quantification accuracy than emPAI [10]. However, we have found that emPAI outperforms APEX in terms of quantification reproducibility [11]. Another study suggests that NSAF yields more reproducible counts across technical and biological replicates than emPAI [12]. SINQ and emPAI were reported to be easily implemented and superior for identifying significant differences between samples compared to chromatographic peak intensity based methods [13].

Detailed protocols for implementing APEX [14] and NSAF [15] have been published in Methods in Molecular Biology before. Current protocol will focus on relative and absolute quantification by emPAI.

The emPAI method is based on comparing the number of experimentally observed peptides and calculated number of observable peptides [6]. It is an improvement of PAI [5], which is defined as follows:

$$\mathrm{PAI} = \frac{N_{\mathrm{obsd}}}{N} \tag{1}$$

where N_{obsd} is the number of experimentally observed peptides per protein and N is the number of theoretically observable peptides per protein [5]. The emPAI is defined as follows [6]:

$$\mathrm{emPAI} = 10^{\mathrm{PAI}} - 1 \tag{2}$$

Absolute protein amounts in complex mixtures can be estimated with emPAI by calculating the protein fraction from all identified proteins (normalized emPAI):

$$\mathrm{Normalized\,emPAI} = \frac{\mathrm{emPAI}}{\sum \mathrm{emPAI}} \tag{3}$$

where $\sum$emPAI is the sum of emPAI values for all identified proteins. When the total amount of protein (C) analyzed in MS is known, protein absolute amount can be calculated as follows:

$$\text{Protein amount} = \text{Normalized emPAI} \times \text{C} \tag{4}$$

This method of concentration determination may not be very precise compared when using isotopically labelled standards; however, it is useful for large-scale screening of protein concentrations. The accuracy of abundance measurements using emPAI values has been demonstrated to lie within the same error range or better than protein concentration measurements based on staining methods such as the Bradford assay [6]. Protein abundance for more than 1,000 *E. coli* proteins has previously been measured with emPAI [16]. This research showed a strong correlation between emPAI-derived data and the known abundance of 40 *E. coli* enzymes [16]. We have demonstrated that emPAI with other label-free techniques accurately estimates standard protein amounts in the range of 10 fmol to 10 pmol [11]. We have used emPAI to compare protein abundances for *E. coli* proteins measured with three different label-free approaches (emPAI, APEX, and peak intensity) and found a good correlation between those methods ($R^2 = 0.8$) [11].

Because emPAI is implemented into the Mascot database search platform (Matrix Science, UK) it is easy to apply this approach to existing data in order to add quantitative information without any additional experimental steps. In this protocol we will concentrate on the spectral counting approach using emPAI from Mascot search and shotgun samples. However, it is also possible to use emPAI Calc (http://empai.iab.keio.ac.jp/), an open-source Web application, to calculate emPAI values taking into account also LC conditions and peptide retention times [17].

It is claimed that label-free quantification has a deeper proteome coverage for protein identifications but is less reproducible than label based approaches [18]. Label-free quantification combined with data from labelling experiments can provide valuable information on protein abundances without the need for any internal standards. For example, isobaric tags for relative and absolute quantification (iTRAQ) has been combined with APEX [19] and emPAI [20] to produce absolute protein quantification for complex samples without using internal standards. Here, we also cover the process of calculating protein concentrations from labelling experiment (iTRAQ) without the use of standard proteins.

Although methods for spectral counting-based absolute protein abundance estimation are easy to implement, they have been reported not to be particularly sensitive to small changes in protein abundance [10, 13, 21]. Furthermore, at higher protein concentrations (>100 fmol on column) and in complex samples spectral counting methods suffer from saturation effects [10, 16, 22]. The poor performance of spectral counting methods for extremely low or high protein concentrations is most probably caused by the dynamic exclusion settings during data dependent LC-MS acquisition analysis. Low abundance proteins are under-sampled and for high abundance proteins the degree of sampling reaches maximum sequence coverage. Optimization of dynamic exclusion settings in data dependent analysis mode can increase the reproducibility of spectral counting and also the quantification of low or very high abundant proteins [23]. Despite of the mentioned drawbacks the spectral counting method is still a valuable tool for the rapid screening of differences between samples and the broad estimation of protein concentrations within the proteome (see also review [24]).

The protocol described below is applicable to all label-free quantitative proteomics methods in order to interpret absolute protein amounts. Here, we present calculations based solely on the spectral counting method emPAI.

2 Materials

2.1 Data

1. Mass spectrometry data from a shotgun proteomics experiment (*see* also **Note 1**), processed with Mascot search engine. The information required for emPAI is always present in the Mascot search result (without any additional parameter settings), as long as the MS/MS search contains at least 100 spectra. In case of missing emPAI values *see* **Note 2**. Example data (published in [11]) is provided, consisting of two Mascot search results and RAW files available to practice. These are biological replicates of *E. coli* cell lysate supplemented with UPS2 standard (Sigma) in w/w ratio of 3:1. The mass spectrometry proteomics data have been deposited to the ProteomeXchange Consortium (http://proteomecentral.proteomexchange.org) via the PRIDE partner repository [25] with the dataset identifier PXD000283 and DOI 10.6019/PXD000283. All calculations mentioned in this protocol are available in Supplementary Table 1 (see on Springer Extras at extras.springer.com).
2. The information of the protein concentration in the sample in order to correct the relative measurements to absolute amounts (*see* **Note 3**). Another option is to include protein standards into the analysis, for example Sigma UPS2 standard (*see* **Note 4**).

2.2 Equipment

1. Mac, PC, or Linux/Unix workstation.
2. Excel or similar software.

3 Methods

3.1 Normalized emPAI

1. Export data from Mascot Web browser to Excel or other preferred calculation software. Note that the usual transfer to CSV format can take a lot of time if large data sets are analyzed, therefore, simplest way is to copy everything (Ctrl-A and Ctrl-C) from the Protein Family Summary view Quantitation tab (Mascot v2.3) and paste as a text to Excel workbook. In Mascot v2.4 Quantitation tab has been replaced by Report Builder that allows exporting data to CSV form with one click. Example data exported from Mascot v2.3 result to Excel is provided as Supplementary Table 1 (extras.springer.com).
2. Delete contaminants and reverse hits.
3. In order to compare between different samples normalize emPAI values from Mascot search to the sum of all emPAI values Eq. 3. Normalized emPAI values are suitable for comparing datasets that have similar sample preparation and composition, e.g., shotgun sample preparation of cell lysates (*see* **Note 5**).

3.2 Calculation of Absolute Protein Amounts Without Standards

1. Start with normalized emPAI values.
2. Multiply normalized emPAI values with the estimated amount of protein in the sample Eq. 4. This value can be taken from previous studies (Lu et al. [8]) or calculated for current study conditions (*see* **Note 3**). Although protein concentrations calculated by this way may seem an over-simplification, we have seen very good correlation with true protein concentrations [11]. *See* also **Notes 6** and **7**.

3.3 Calculation of Absolute Protein Amounts by Using Standard Proteins with Known Concentrations

1. If standard proteins are added to the mass spectrometry analysis, linear regression can be used to determine the absolute protein concentrations for other proteins.
2. Filter out standard proteins to another Excel sheet.
3. Transform the emPAI values and spike-in standard protein concentrations to logarithmic scale. The log transformation is applied to stabilize the variance caused by very high or low MS signals [10].
4. Calculate intercept and slope based on linear regression of log-transformed standard protein concentrations versus log-transformed emPAI values.
5. Extrapolate protein concentrations based on the following equation:

$$\log_{10} C = \text{SLOPE} \times \log_{10}(\text{emPAI}) + \text{INTERCEPT} \quad (5)$$

6. Transform protein concentrations from log scale to normal scale and calculate the protein concentration in the sample, taking into account all the sample preparation steps. In current data example (Supplementary Table 1, extras.springer.com) 0.75 % of the total sample was injected onto the column.

3.4 Calculate Protein Abundances from iTRAQ Experiment

1. If iTRAQ quantification is specified in Mascot search then software reports iTRAQ quantification for each protein as ratios of 115/114, 116/114 and 117/114, together with the emPAI score of the protein. This report should be exported to Excel or other software for further calculations like in previous sections.
2. In iTRAQ quantification emPAI for each protein is a sum of pooled differentially labelled protein emPAI values; therefore, protein emPAI will be multiplied with each label contribution to protein abundance. Mascot expresses the four iTRAQ channels relative to channel 114, this means that 114/114 = 1 and each sample would have following contributions:

$$\mathrm{emPAI}_{114} = \frac{1}{1+\frac{115}{114}+\frac{116}{114}+\frac{117}{114}} \times \mathrm{emPAI}$$

$$\mathrm{emPAI}_{115} = \frac{\frac{115}{114}}{1+\frac{115}{114}+\frac{116}{114}+\frac{117}{114}} \times \mathrm{emPAI}$$

$$\mathrm{emPAI}_{116} = \frac{\frac{116}{114}}{1+\frac{115}{114}+\frac{116}{114}+\frac{117}{114}} \times \mathrm{emPAI}$$

$$\mathrm{emPAI}_{117} = \frac{\frac{117}{114}}{1+\frac{115}{114}+\frac{116}{114}+\frac{117}{114}} \times \mathrm{emPAI}$$

3. Normalize each channel emPAI values to the sum of that label emPAI values, e.g.,:

$$\mathrm{Normalized\,emPAI}_{114} = \frac{\mathrm{emPAI}_{114}}{\sum \mathrm{emPAI}_{114}}$$

4. Multiply normalized emPAI values with the estimated amount of protein in the sample Eq. 4 (*see* **Note 3**).

4 Notes

1. It has been noticed by several studies that spectral counting methods APEX and emPAI values saturate at higher protein concentrations [10, 11, 16, 22]. Optimization of dynamic exclusion (DE) settings improves the quality of quantification of low abundance proteins [23]. Optimal DE depends on the chromatographic peak width and mass spectrometry parameters [23, 26]. General recommendation would be to try different DE settings on a reference sample if planning spectral counting experiment. However, if applying spectral counting quantification to an already acquired data the effect of DE should be kept in mind when interpreting results.
2. There are some cases when Mascot search result is missing emPAI values. Those instances are described in Mascot homepage and referred below:
   (a) No emPAI values if there is less than 100 queries in the search.
   (b) Semi-specific enzyme or no enzyme searches result in meaningless emPAI values.
   (c) No emPAI values if manual error tolerant search is performed.
   (d) For an integrated decoy search, no emPAI values in the report for hits in the decoy database.
   (e) emPAI values are only reported for the primary protein in the hit, not for same set or sub-set proteins.
   (f) An emPAI value may be missing if the protein was a very weak hit and the protein mass was not saved to the result file and the sequence database is no longer on-line or the protein has disappeared from the sequence database.
3. In order to calculate the total concentration of individual protein molecules in the cell or sample, following measures should be determined:
   (a) Amount of cells in sample. This can be determined by plate counting of cells or microscopy or flow cytometry.
   (b) Amount of protein in the sample. This can be determined by colorimetric assays (Bradford, BCA, Lowry etc.) or amino acid analysis.

   If measurements above are not possible in some reason one could also use reference values from other publications (Lu et al. [8]), good source is BioNumbers database (http://bionumbers.hms.harvard.edu/) [27].

4. If standard proteins are going to be used it is important to balance between the number of standard proteins and experimental cost. Recent study found that at least 20 reference protein measurements are required for a stable quantification model [10]. For all nonhuman samples, we recommend using Sigma UPS2 standard, a mixture of 48 precisely quantified human proteins with a dynamic concentrations range spanning five orders of magnitude. If human proteome is studied UPS2 standard should be substituted with standard proteins from other organism in order not to interfere with the quantification.
5. Label-free quantification is suitable for samples that have not been prepared with fractionation or enrichment; as sample losses in additional steps are hard to predict. Therefore, shotgun sample preparation is ideal for label-free quantification methods.
6. Absolute amount determination without added standards is based on the assumption that proteins detected make up most of the protein content in the sample and unidentified proteins are at low abundance and have only minor effect on the total protein amount. This simplification has been used successfully for APEX [8], Total Protein Approach (TPA) [28] and emPAI. Such simplification has resulted in good agreement with spiked in standard protein concentrations [8, 11, 28]. Recently published work comparing different label-free absolute quantification strategies is an excellent reference where large scale proteome quantification is made with or without standard proteins, referred as calibration curve model and the direct proportionality model, respectively [10].
7. Spiked in standard proteins enables to evaluate the magnitude of absolute protein abundances: in dataset provided (Supplementary Table 1, extras.springer.com) the sum of all proteins in a cell according to emPAI was 8–9 % less than the value derived from the Lowry total protein analysis. This very small difference between the total protein amount measured by the colorimetric assay and the label-free quantitative proteomics method emPAI indicates that with MS based proteomics most of the abundant proteins are covered, Therefore, simplified assumption made when calculating proteins concentrations without standards does not affect the large scale quantitative proteomics results [11]. Adding standard proteins will improve the accuracy of the method [10]; however, if there is no possibility to use calibration curve, for example if interpolating absolute concentrations from previously acquired datasets, estimation of protein abundances could also be made using by normalizing identified protein abundances to total protein concentration [28]. A comparison of approaches is presented on Fig. 1.

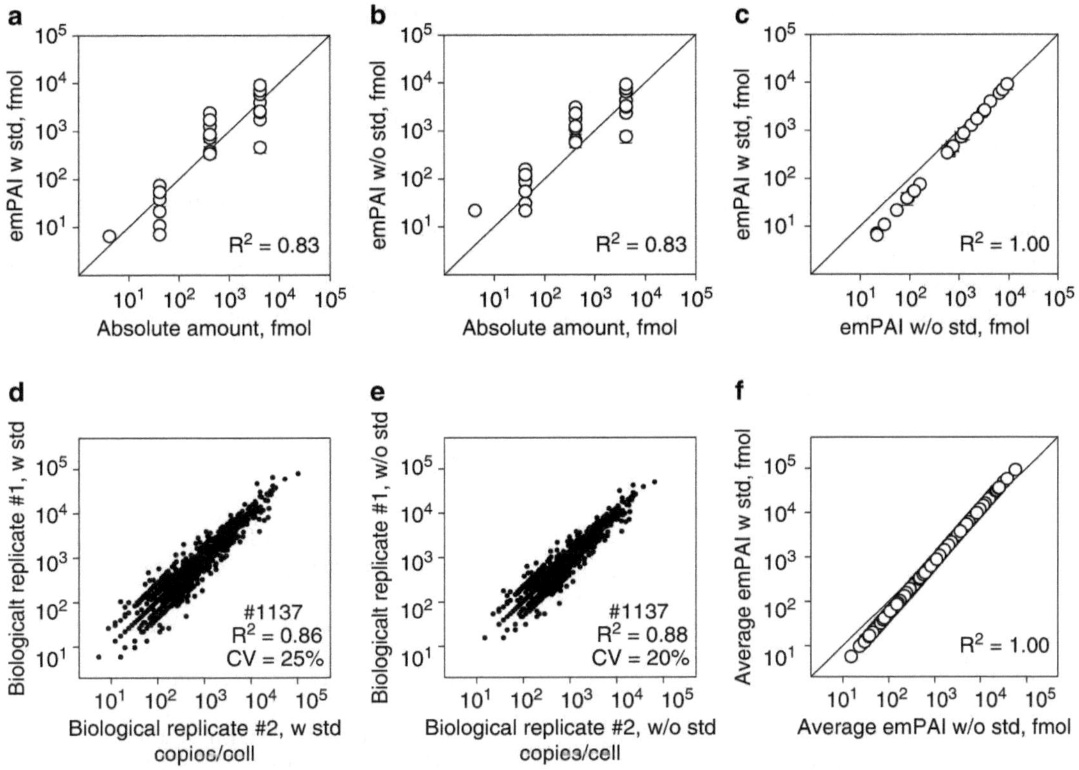

Fig. 1 Absolute quantification of proteins with or without spiked in standards. (**a**, **b**) emPAI showed a high correlation ($R^2 = 0.83$) with the actual protein amount in UPS2 standard with a wide dynamic range from 10 to 10^4 fmol calculated either with (w) or without (w/o) standard proteins. Including calibration based standard proteins improved quantification for the low abundant proteins. (**c**, **f**) Comparison of protein concentrations calculated with (w) or without (w/o) added standards shows good agreement, slight disagreement from perfect correlation for low and high abundant proteins. (**d**, **e**) When biological replicates were correlated the values calculated without standards showed slightly better results for emPAI in terms of Pearson squared correlation (R^2) and coefficient on variation (CV)

References

1. Wilm M, Shevchenko A, Houthaeve T et al (1996) Femtomole sequencing of proteins from polyacrylamide gels by nano-electrospray mass spectrometry. Nature 379:466–469
2. Mann M (1999) Quantitative proteomics? Nat Biotechnol 17:954–955
3. Ong S-E, Mann M (2005) Mass spectrometry-based proteomics turns quantitative. Nat Chem Biol 1:252–262
4. Washburn MP, Wolters D, Yates JR et al (2001) Large-scale analysis of the yeast proteome by multidimensional protein identification technology. Nat Biotechnol 19:242–247
5. Rappsilber J, Ryder U, Lamond AI et al (2002) Large-scale proteomic analysis of the human spliceosome. Genome Res 12:1231–1245
6. Ishihama Y, Oda Y, Tabata T et al (2005) Exponentially modified protein abundance index (emPAI) for estimation of absolute protein amount in proteomics by the number of sequenced peptides per protein. Mol Cell Proteomics 4:1265–1272
7. Zybailov B, Mosley AL, Sardiu ME et al (2006) Statistical analysis of membrane proteome expression changes in Saccharomyces cerevisiae. J Proteome Res 5:2339–2347
8. Lu P, Vogel C, Wang R et al (2007) Absolute protein expression profiling estimates the relative contributions of transcriptional and translational regulation. Nat Biotechnol 25:117–124
9. Griffin NM, Yu J, Long F et al (2010) Label-free, normalized quantification of complex

mass spectrometry data for proteomic analysis. Nat Biotechnol 28:83–89
10. Ahrné E, Molzahn L, Glatter T et al (2013) Critical assessment of proteome-wide label-free absolute abundance estimation strategies. Proteomics 13(17):2567–2578
11. Arike L, Valgepea K, Peil L et al (2012) Comparison and applications of label-free absolute proteome quantification methods on Escherichia coli. J Proteomics 75: 5437–5448
12. McIlwain S, Mathews M, Bereman MS et al (2012) Estimating relative abundances of proteins from shotgun proteomics data. BMC Bioinformatics 13:308
13. Trudgian DC, Ridlova G, Fischer R et al (2011) Comparative evaluation of label-free SINQ normalized spectral index quantitation in the central proteomics facilities pipeline. Proteomics 11:2790–2797
14. Vogel C, Marcotte EM (2012) Label-free protein quantitation using weighted spectral counting. Methods Mol Biol 893:321–341
15. Neilson KA, Keighley T, Pascovici D et al (2013) Label-free quantitative shotgun proteomics using normalized spectral abundance factors. Methods Mol Biol 1002:205–222
16. Ishihama Y, Schmidt T, Rappsilber J et al (2008) Protein abundance profiling of the Escherichia coli cytosol. BMC Genomics 9:102
17. Shinoda K, Tomita M, Ishihama Y (2010) emPAI Calc-for the estimation of protein abundance from large-scale identification data by liquid chromatography-tandem mass spectrometry. Bioinformatics 26:576–577
18. Li Z, Adams RM, Chourey K et al (2012) Systematic comparison of label-free, metabolic labeling, and isobaric chemical labeling for quantitative proteomics on LTQ Orbitrap Velos. J Proteome Res 11:1582–1590
19. Adamberg K, Seiman A, Vilu R (2012) Increased biomass yield of Lactococcus lactis by reduced overconsumption of amino acids and increased catalytic activities of enzymes. PloS One 7:e48223
20. Kim PD, Patel BB, Yeung AT (2012) Isobaric labeling and data normalization without requiring protein quantitation. J Biomol Tech 23:11–23
21. Hendrickson EL, Xia Q, Wang T et al (2006) Comparison of spectral counting and metabolic stable isotope labeling for use with quantitative microbial proteomics. Analyst 131:1335–1341
22. Grossmann J, Roschitzki B, Panse C et al (2010) Implementation and evaluation of relative and absolute quantification in shotgun proteomics with label-free methods. J Proteomics 73:1740–1746
23. Hoehenwarter W, Wienkoop S (2010) Spectral counting robust on high mass accuracy mass spectrometers. Rapid Commun Mass Spectrom 24:3609–3614
24. Lundgren DH, Hwang S-I, Wu L et al (2010) Role of spectral counting in quantitative proteomics. Expert Rev Proteomics 7:39–53
25. Vizcaíno JA, Côté RG, Csordas A et al (2013) The PRoteomics IDEntifications (PRIDE) database and associated tools: status in 2013. Nucleic Acids Res 41:D1063–D1069
26. Zhang Y, Wen Z, Washburn MP et al (2009) Effect of dynamic exclusion duration on spectral count based quantitative proteomics. Anal Chem 81:6317–6326
27. Milo R, Jorgensen P, Moran U et al (2010) BioNumbers – the database of key numbers in molecular and cell biology. Nucleic Acids Res 38:D750–D753
28. Wiśniewski JR, Ostasiewicz P, Duś K et al (2012) Extensive quantitative remodeling of the proteome between normal colon tissue and adenocarcinoma. Mol Syst Biol 8:611

Chapter 15

Quantification of Proteins by Label-Free LC-MS^E

Alon Savidor and Yishai Levin

Abstract

Quantitative proteomics by LC-MS/MS is a widely used approach for quantifying a significant portion of any complex proteome. Among the different techniques used for this purpose, one is by use of Data Independent Acquisition (DIA). We present a descriptive protocol for label-free quantitation of proteins by one DIA method termed LC-MS^E, which facilitates large-scale quantification of proteins without the need for isotopic labelling and with no theoretical limit to the number of samples included in an experiment.

Key words Label-free, Proteomics, LC-MS, MS^E, Time-alignment, nanoLC-MS/MS, Quantitation, Relative quantitation, DIA

1 Introduction

The field of proteomics is constantly evolving as new technologies emerge and advance at a rapid rate. These new technologies enable researchers to break new frontiers and reach uncharted territories in the investigation of biological systems. Proteomic investigation has been applied essentially to the entire range of biological systems: from humans [1–5] to bacteria [6–8], from plants [9–13] to fungi [14–16], from insects [17–20] to viruses [21–23]. The sample to be analyzed can be as simple as a few purified proteins [24, 25] or as complex as an infected tissue containing thousands of proteins from both host and pathogen [26–28]. The information that can be obtained from proteomic experiments can go beyond the qualitative analysis of protein identification, and provide quantitative information regarding protein expression, as well as information about post-translational protein modifications and protein localization. Recent advances in analytical instrumentation and bioinformatics now enable the relative quantitation of thousands of proteins across dozens of samples in a single experiment [29–35]. Experiment design can be tailored to answer specific biological questions, taking advantage of the most recent advances in the field. In clinical proteomics, for example, the improved accuracy

Daniel Martins-de-Souza (ed.), *Shotgun Proteomics: Methods and Protocols*, Methods in Molecular Biology, vol. 1156, DOI 10.1007/978-1-4939-0685-7_15, © Springer Science+Business Media New York 2014

and sensitivity of the new instruments are increasingly exploited for the purpose of disease biomarker discovery [36–39]. Newly developed bioinformatics tools along with the improvement in instrumentation, now allow for the establishment of metaproteomics, where entire populations of organisms from environmental samples are analyzed simultaneously [40–42]. Newly developed or improved methods in click-chemistry and protein labelling improve the ability to identify post-translational modification as well as identify protein quantification, dynamics and localization [43–45]. Obtaining both qualitative and quantitative information also provides deeper and more comprehensive insights into the origin and structure of biological systems by allowing global proteomic profiling. Regardless of the method of choice, quantitative analysis requires an added degree of stringency, especially when analyzing complex samples. In order for proteomic profiling experiments to be statistically powered, it is necessary to investigate large sample cohorts.

There are several approaches used for global proteomic quantitative profiling. Among the most popular are two-dimensional gel-based and isotopic-labelling methods. However, these come with difficulties in meeting some of the requirements for comprehensive and reproducible analysis [32]. These requirements are [46, 47]: (1) the ability to detect as many proteins as possible, (2) achieving a dynamic range that is wide enough to detect low abundance proteins, (3) high reproducibility and consistency of the platform performance so that biological differences can be sufficiently distinguished from instrumental ones, along with successful validation of quantitative significance (4) the ability to profile and compare a large number of non-pooled samples. The necessity to analyze as many discrete (non-pooled) samples as possible cannot be overestimated. Greater *n*-numbers enable researchers to apply a variety of statistical methods and increase the statistical power of the analysis that are not possible when analyzing only a handful of samples or pooled samples.

To avoid some of the downfalls of protein labelling, several label-free methodologies have been developed for the relative quantitation of proteins. The basic strategy is similar for all of these methods, where the samples are analyzed sequentially and discretely, and neither proteins nor peptides are labelled.

There are two basic approaches in label-free quantitation: one is by counting the number of spectra acquired for a given protein in a give sample. This method is termed Spectral Counting (http://www.expert-reviews.com/doi/abs/10.1586/epr.09.69). The second methodology is by use of the precursor ion intensity, measured by MSI. The latter has been shown to produce a more accurate quantification and can be used when acquiring data in either data directed acquisition (DDA) or data independent acquisition (DIA).

One of the earliest DIA approaches, termed MS^E [48], is a unique acquisition mode where the collision energy is alternated such that two channels are collected. The first channel includes

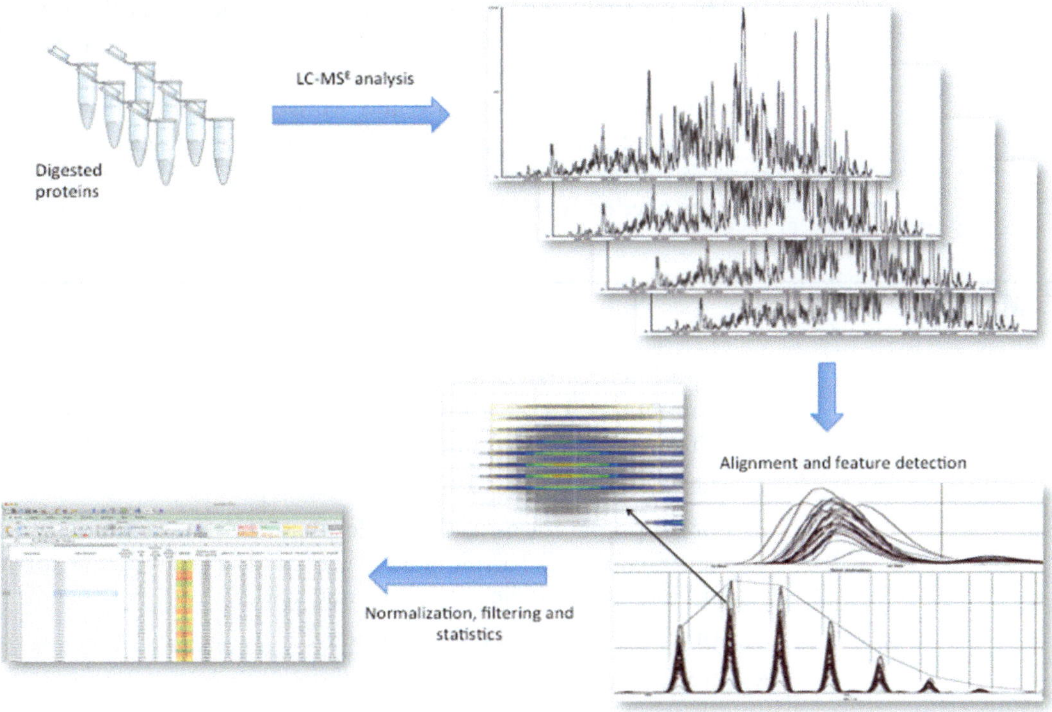

Fig. 1 A typical workflow for label-free quantitation. Samples are analyzed sequentially and randomly. Each raw data file is processed and searched then all data files are combined by time alignment of all detected features

the abundance measurements of the intact peptides and a second channel for the fragmented peptides. Both are acquired at a high sampling rate. In this type of acquisition the chromatographic profile is reproducibly maintained throughout the run time and thus across the sample set, allowing for the precursor ions to be aligned in time. Once aligned the intensities of precursor ions can be normalized and directly compared across all injections of all samples. This method, therefore, enables relative quantitation and identification of peptides (and thus proteins) in the same experiment without the need to reanalyze samples. Furthermore, data analysis can be performed on unidentified peptides as well as those that were identified.

Whichever method is used, time alignment of data must be performed to summarize the information from all samples (*see* Fig. 1).

In this chapter we present the methodology for MS^E based label-free proteomics. It enables the reliable identification and quantification of thousands of proteins with a high dynamic range across more than three orders of magnitude [29, 49]. Furthermore, the reproducibility of the platform is high enough to perform accurate relative quantitation of proteins across dozens of samples at a time, which makes this approach ideal for large scale biomarker discovery as well as systems biology [30].

The protocol for label-free LC-MS/MS includes three steps: (1) nanoLC-MS^E conditions, (2) data processing/database search, and (3) data analysis. The first part includes information of optimum nano Liquid Chromatography and Mass Spectrometry conditions for analysis of complex samples. The second part describes the data processing and database searching parameters. The last part describes the steps needed to analyze a typical data set.

2 Materials

1. Liquid chromatography: nanoAcquity; trapping column: Symmetry C18 180 μm × 20 mm 5 μm particles; Analytical column: T3 HSS 75 μm × 250 mm 1.8 μm particles (Waters).
2. Mass spectrometer: Q-IMS-ToF (Synapt G2, Waters).
3. For all preparations and mobile phase ULC-MS grade water, acetonitrile and formic acid were used: H_2O + 0.1 % formic acid and acetonitrile + 0.1 % formic acid.
4. 500 fmol/μL Glu-Fibrinopeptide B.
5. 25 fmol/μL digested Yeast Enolase.
6. PicoTip emitter: 360 μm O.D., 20 μm i.d., 10 μm tip (New Objective).
7. ProteinLynx Global Server software package version 2.5.2 (Waters).
8. Elucidator© (Rosetta Biosoftware).
9. Progenesis LC-MS© (Nonlinear Dynamics).
10. Expressionist (Genedata).
11. TransOmics© (Waters).

3 Methods

3.1 Liquid Chromatography–Mass Spectrometry

The protocol for label-free quantitation of proteins by LC-MS is a "bottom-up" approach, which means proteins must be digested into peptides. The analytical process is then performed at the peptide level.

Each biological sample should be injected and analyzed in triplicate followed by a blank injection (to ensure there is no carryover of peptides from one sample to the other in this sequential process). Samples should be run randomly and blinded. For each sample 1.5 μg of total protein digest is loaded using split-less nano Ultra Performance Liquid Chromatography (10kpsi nanoAcquity). The autosampler is maintained at 8 °C to prevent degradation and evaporation and yet not to cause condensation in the HPLC vial.

Buffers used for the mobile phase A: H_2O + 0.1 % formic acid; B: Acetonitrile + 0.1 % formic acid.

1. Prior to analysis of biological samples a system verification protocol is run using a digested protein standard mix which includes Bovine Serum Albumin, Alcohol Dehydrogenase, Glycogen Phosphorylase from rabbit muscle, and Yeast Enolase. This ensures the nanoLC-MS system is in peak performance. Initially, Five consecutive injections of 25 fmol/µl yeast enolase digest are injected and processed, (*see* **Note 1**). Minimum system requirements include retention time RSD, intensity RSD and mass accuracy all based on measurement of the five most intense enolase peptides (*see* a typical table filled in Table 2 and Subheading 4 for further details). Once these requirements are fulfilled two protein mixtures containing four standard proteins are analyzed in triplicate injections to assess the accuracy of relative quantitation. *See* Table 1.
2. Desalting of the samples is performed online with 99.5 %A, flow rate of 15 µL/min for 2 min, using an online Reverse-Phase C18 trapping column (180 µm i.d., 20 mm length, and 5 µm particle size).
3. The peptides are then separated using a T3 HSS nanoColumn (75 µm i.d., 250 mm length, 1.8 µm particle size), at 300 nl/min with column temperature maintained at 50 °C. The gradient used for complex samples is shown in Table 2 (*see* **Note 3**).
4. The nanoUPLC is coupled online through a pre-cut nanoESI emitter of 7 cm length and 10 µm tip to a Quadrupole Ion Mobility Time-of-Flight Mass Spectrometer (Synapt G2 HDMS). Data is acquired in positive, Resolution, HDMS^E mode. In this mode, the quadrupole is set to transfer all ions while the collision cell switches from low to high collision energy intermittently throughout the acquisition time. In the low energy scans, "transfer" collision energy is set to 4 eV while in the high energy scans it is ramped from 25 eV to 50 eV. Mass accuracy is maintained throughout the analysis by the use of a LockSpray apparatus. A reference compound (Glu-Fibrinopeptide B) is continuously infused using the LockSpray and scanned intermittently every 30 s. During data processing, the analyte spectra are corrected automatically based on the difference between the detected *m/z* peak and the theoretical *m/z* peak (785.8426 [m + 2H]+) of Glu-Fibrinopeptide B.

3.2 Data Processing and Protein Identification

Raw data, acquired in continuum format, is processed using the ProteinLynx Global Server software. Both quantitative and qualitative information are produced automatically by the software, using the default parameters.

Table 1

A typical "system verification" table is shown. The verification protocol includes calculation of retention time variation, intensity variation, and mass accuracy

Enolase peptide	*1,755.9487*	*1,578.801*	*1,416.722*	*1,412.822*	*1,288.7107*	*1,286.7103*	*Spec*
Injection1 (Intensity)	105,069	81,385	52,072	19,656	135,892	89,877	
Injection2 (Intensity)	92,140	69,280	63,556	16,436	127,266	96,019	
Injection3 (Intensity)	85,609	81,181	55,907	17,746	115,688	95,311	
Injection4 (Intensity)	80,341	69,139	54,935	17,040	110,997	92,092	
Injection5 (Intensity)	73,926	69,166	64,162	13,053	107,105	94,735	
Average intensity	11,929.43	6,621.47	5,423.92	2,412.51	11,931.22	2,559.39	
SD intensity	87,417.00	74,030.20	58,126.40	16,786.20	119,389.60	93,606.80	
%RSD intensity	*13.65*	*8.94*	*9.33*	*14.37*	*9.99*	*2.73*	*<15 %*
Injection1 (Ret. time)	27.113	29.4882	21.8523	27.9082	22.4245	25.8871	
Injection2 (Ret. time)	27.0873	29.465	21.827	27.8833	22.4033	25.9071	
Injection3 (Ret. time)	27.0153	29.3821	21.8025	27.8161	22.3583	25.8134	
Injection4 (Ret. time)	27.033	29.4104	21.8319	27.8302	22.3952	25.8255	
Injection5 (Ret. time)	27.0033	29.3652	21.7711	27.7893	22.3495	25.7636	
SD Ret. time	27.05	29.42	21.82	27.85	22.39	25.84	
Average Ret. time	0.05	0.05	0.03	0.05	0.03	0.06	
%RSD Ret. time	*0.18*	*0.18*	*0.14*	*0.18*	*0.14*	*0.22*	*<2 %*

Mass accuracy (PPM)	**Precursor RMS mass error**	**Spec**	**Mass accuracy (PPM)**	**Fragment RMS mass error**	**Spec**
Injection1	4.9	<10	Injection1	10.3	<20
Injection2	4.1		Injection2	10.1	
Injection3	4.7		Injection3	10.7	
Injection4	4.6		Injection4	10.8	
Injection5	4.7		Injection5	10.7	

Table 2
(a) The LC gradient for tissue extracts and cell lines. (b) The LC gradient for depleted serum or plasma

Time (min)	Flow (μl/min)	%A	%B	Curve
Initial	0.3	94	6	Initial
1	0.3	94	6	Linear
60	0.3	65	35	Linear
65	0.3	10	90	Linear
70	0.3	10	90	Linear
71	0.3	94	6	Linear

3.2.1 Quantitative Information

Intensity measurements are obtained by integration of the total ion volume of each extracted, charge-state-reduced, deisotoped, and mass-corrected ions across the mass spectrometric and chromatographic volume. The algorithm calculates the observed mass and intensity measurement deviation for every detected component. The chromatographic area associated with each component is calculated using an integration algorithm similar to the ApexTrack peak integration algorithm provided in the MassLynx software. If a particular component exists in more than one charge-state, the corresponding area for any given monoisotopic ion is reported as the summed area from all contributing charge states. The retention time is determined for each reported monoisotopic ion at the moment it reaches its maximum intensity (apex). This process is performed for both the low collision energy and the high energy scans (saved as separate channels). The ion detection thresholds should be optimized for each instrument. The values we typically use are as follows: Low energy ion detection threshold of 200 counts; high energy ion detection threshold was set to 50, bin count of 500 (*see* **Note 2**).

3.2.2 Protein Identification

Proteinlynx Global Server version 2.5.2 (Identity^E) is also used for database searches. The database should be chosen based on the biological samples being analyzed. The database should contain species specific nonredundant sequence entries.

The database search algorithm of the software was described by Li et al. [48]. Briefly, the software detects the 250 most abundant peptides and performs an initial pass through the database in order to identify those peptides (with mass tolerance of 10 ppm of precursor ions and 20 ppm for fragment ions). It then calculates the precursor-ion mass tolerance, fragment-ion mass tolerance and chromatographic peak widths for these 250 peptides. These peptides are then depleted from the database and the remaining peptides are searched based on these criteria. The cycle continues to the next abundant peptides, which are identified and then temporarily deleted from the database. These tentative peptide identifications are ranked and scored by how well they conform to 14 predetermined models of specific, physicochemical attributes (such as retention time and fragmentation pattern). All tentative peptides are collapsed into their parent proteins utilizing only the highest scoring peptides that contribute to the total protein score. Once a protein has been identified, all top ranked precursor ions and their corresponding product ions are removed from all other tentatively identified proteins. The remaining unidentified peptides—and tentatively identified proteins—are then re-ranked and re-scored, and the process is repeated until a 1 % false positive rate is reached. The false discovery rate is determined by the number of random or reverse identifications identified (false positive rate—FPR) divided by the number of correct identifications, (true positive rate—TPR) expressed as a percentage. Therefore, the false discovery rate (FDR) equals the $FDR = FPR/TPR \times 100$.

All protein identifications should be based on at least two peptides.

3.2.3 Data Analysis

A typical quantitative analysis includes dozens of samples and thousands of peptides. Thus the data analysis scheme should be automated and unbiased. This can be done using statistical software packages such as the free software R (http://www.r-project.org). The output of the analysis will highlight the most significantly changing proteins and peptides depending on significance thresholds. There are five major steps to perform: alignment, normalization, filtering, annotation and combining. The very last step is statistical analysis. Since the choice of statistical methods can vary based on the experimental design and purpose, it is excluded from this protocol. It is, however, recommended that for a comparison of two treatment groups, univariate statistics such as two-tailed, unpaired Student's *t*-Test is used after a logarithmic transformation to approximate a normal distribution.

1. *Time alignment*: A key step in label-free LC-MS based quantitation is the time alignment and annotation of data. Since all samples are analyzed sequentially and separately, the data must be combined and summarized. There are several commercially available software packages that can perform this alignment. These include Elucidator©, Progenesis LC-MS©, Expressionist and TransOmics©. The result of the time alignment is a two dimensional table that includes all detected peaks, their average mass, average retention time and detected intensity in all replicates of all samples. This table is the basis for all consequent data analysis steps that follow (*see* Fig. 2). This table can then

			Intensity											
Index	Mass	Retention Time	sample 1_01	sample 1_02	sample 1_03	sample 2_01	sample 2_02	sample 2_03	sample 3_01	sample 3_02	Sample 3_03	sample 4_01	sample 4_02	sample 4_03
1	1821.9407	81.3128	30026	28035	27818	38122	36191	26999	22124	23621	24762	31701	30763	25922
2	1030.5615	41.6935	9386	8613	8205	11105	11430	8162	6743	7290	7488	10520	10017	7273
3	2256.114	93.3242	5793	5428	5154	7762	7693	5689	4335	4549	5388	7089	6857	4824
4	1755.9407	73.388	58465	61019	81471	19216	18325	14476	15625	18004	17724	30839	29024	23864
5	1578.7996	81.4449	67010	86289	36055	33020	43292	67721	91690	76578	111797	87561	83053	99582
6	1286.7109	67.144	10930	11917	24733	24095	23915	10846	11556	15825	12732	12318	11948	11519

Sum	181609	201302	183437	133318	140845	133893	152073	145867	179891	180028	171662	172983
Normalization factor	1.11	1.00	1.10	1.51	1.43	1.50	1.32	1.38	1.12	1.12	1.17	1.16

normalized data

Index	Mass	Retention Time	sample 1_01	sample 1_02	sample 1_03	sample 2_01	sample 2_02	sample 2_03	sample 3_01	sample 3_02	sample 3_03	sample 4_01	sample 4_02	sample 4_03
1	1821.9407	81.3128	33281	28035	30527	57561	51725	40592	29286	32597	27709	35447	36074	30165
2	1030.5615	41.6935	10403	8613	9004	16767	16336	12272	8925	10060	8379	11764	11747	8464
3	2256.114	93.3242	6421	5428	5656	11721	10995	8553	5739	6277	6030	7926	8041	5613
4	1755.9407	73.388	64805	61019	89406	29014	26191	21764	20683	24847	19833	34483	34036	27770
5	1578.7996	81.4449	74276	86289	39566	49857	61874	101815	121372	105681	125103	97908	97393	115885
6	1286.7109	67.144	12115	11917	27142	36381	34180	16306	15297	21840	14247	13774	14011	13405

Fig. 2 The table shows partial and hypothetical data matrices before and after normalization. The rows are the detected features (peptides) and the columns include the mass, retention time and intensity in all injections of all samples. In this example, sample 1_02 had the highest sum of intensities; therefore, its normalization factor was set to 1. All the other samples are referenced to it. In actual practice, the sum includes all detected features

be annotated using all search results of all analyzed samples. The peak annotation is performed using accurate mass and retention time measurements.

2. *Normalization (scaling)*: Normalization should be applied to each experiment independently. The normalization should account for experimental variation leaving biological variation unchanged. The choice of normalization method depends on the experimental design. There are two basic options. The first includes normalization based on an exogenous internal standard which is spiked into each sample either before LC-MS analysis, in which case it should be digested first, or in the initial sample preparation stage, in which case it should be spiked as an intact protein. During the normalization step in the data analysis only the signal from the internal standard peptides are used to calculate the normalization factor.

   The second method is total-ion-current normalization. In this method the intensity measurements of all detected peptides in a given sample is used to calculate the normalization factor.

   Whichever method is used, the sum of peptide intensities in each sample is used to calculate the normalization factor.

			Intensity											
			Group 1						Group 2					
Index	Mass	Retention Time	sample 1_01	sample 1_02	sample 1_03	sample 2_01	sample 2_02	sample 2_03	sample 3_01	sample 3_02	sample 3_03	sample 4_01	sample 4_02	sample 4_03
5879	1159.5985	46.7278	2584	0	3138	5158	4733	4587	3084	3159	0	4635	4202	4936
5880	1835.8864	81.7817	0	5597	0	0	0	0	4303	0	4002	0	0	9349
5882	1755.9407	73.388	21410	0	19075	17314	14532	11603	58658	53074	54405	18957	19810	25479
5883	1288.7094	56.1242	8213	0	1676	0	1652	7144	5404	4786	0	0	0	28978
5881	807.4214	47.7552	23759	22035	21031	22388	0	35004	0	0	0	0	0	0
5884	1286.7109	67.144	9794	13750	0	16004	17793	14722	19131	19894	16199	13605	9968	11231

Filtered Data

			Group 1						Group 2					
Index	Mass	Retention Time	sample 1_01	sample 1_02	sample 1_03	sample 2_01	sample 2_02	sample 2_03	sample 3_01	sample 3_02	sample 3_03	sample 4_01	sample 4_02	sample 4_03
5879	1159.5985	46.7278	2584	0	3138	5158	4733	4587	3084	3159	0	4635	4202	4936
5881	807.4214	47.7552	23759	22035	21031	22388	0	35004	0	0	0	0	0	0
5882	1755.9407	73.388	21410	0	19075	17314	14532	11603	58658	53074	54405	18957	19810	25479
5883	1288.7094	56.1242	8213	0	1676	0	1652	7144	5404	4786	0	0	0	28978
5884	1286.7109	67.144	9794	13750	0	16004	17793	14722	19131	19894	16199	13605	9968	11231

Fig. 3 The table shows partial and hypothetical data matrices before and after filtering. In this example samples 1 and 2 belong to one treatment group and the other two samples belong to a different treatment group. The filtering criteria are detection in at least 2 of 3 technical replicates and at least 1 sample in any one of the treatment groups. Feature no. 5880 is the only one that does not satisfy these criteria. Feature 5881 does fulfil the criteria since it is detected in samples 1 and 2 which are part of the same treatment group

The sample with the largest sum is set to 1 and all other sums are divided by the intensity of the highest sample (*see* Fig. 2). Each intensity value is then multiplied by the appropriate normalization factor.

3. *Filtering*: The summarized data matrix includes all detected peptides and their corresponding intensity in all replicates of all samples, as mentioned previously. However, for various reasons not all peptides are detected in all replicates or samples. The number of times a given peptide was detected reflects the reliability of the measurement. For example, a peptide that was detected in only one replicate of three can be regarded not reliable and should be excluded. The same is true for peptides that are not detected in several samples. Thus a data set can be filtered to exclude those peptides that are not detected in at least two of three replicates for each sample. The second filtering is on the sample level such that a peptide will be included in the down-stream analysis only if it replicates in the majority of samples in each treatment group (i.e., diseased and control). This method of filtering ensures that a qualitative difference, e.g., detection only in one treatment group, will be noted for further analysis. This is shown for peptide 5881 in Fig. 3.

Index	Mass	Retention Time	sample 1_01	sample 1_02	sample 1_03	sample 2_01	sampl e 2_02	sampl e 2_03	sampl e 3_01	sampl e 3_02	sampl e 3_03	sampl e 4_01	sampl e 4_02	sampl e 4_03
5879	1159.599	46.7278	2584	0	3138	5158	4733	4587	3084	3159	0	4635	4202	4936
5881	807.4214	47.7552	23759	22035	21031	22388	0	35004	0	0	0	0	0	0
5882	1755.941	73.388	21410	0	19075	17314	14532	11603	58658	53074	54405	18957	19810	25479
5883	1288.709	56.1242	8213	0	1676	0	1652	7144	5404	4786	0	0	0	28978
5884	1286.711	67.144	9794	13750	0	16004	17793	14722	19131	19894	16199	13605	9968	11231

Mass	1286.711
RT	67.144
Score	176.38
Sequence	NVNDVIAPAFVK
Protein	P00924|ENO1_YEAST Enolase 1

Mass	1288.709
RT	56.1242
Score	192.87
Sequence	VNQIGTLSESIK
Protein	P00924|ENO1_YEAST Enolase 1

Mass	1755.9407
RT	73.388
Score	250.53
Sequence	TAGIQIVADDLTVTNPK
Protein	P00924|ENO1_YEAST Enolase 1

Mass	807.4214
RT	47.7552
Score	77.81
Sequence	TFAEALR
Protein	P00924|ENO1_YEAST Enolase 1

Mass	1159.599
RT	46.7278
Score	33.51
Sequence	IGSEVYHNLK
Protein	P00924|ENO1_YEAST Enolase 1

Index	Mass	Retention Time	Sequence	Score	Protein	sample 1_01	sample 1_02	sample 1_03	sample 2_01	sample 2_02	sample 2_03	sample 3_01	sample 3_02	sample 3_03	sample 4_01	sample 4_02	sample 4_03
5879	1159.599	46.7278	IGSEVYHNLK	33.51	P00924|ENO1_YEAST Enolase 1	2584	0	3138	5158	4733	4587	3084	3159	0	4635	4202	4936
5881	807.4214	47.7552	TFAEALR	77.81	P00924|ENO1_YEAST Enolase 1	23759	22035	21031	22388	0	35004	0	0	0	0	0	0
5882	1755.941	73.388	TAGIQIVADDLTVTNPK	250.53	P00924|ENO1_YEAST Enolase 1	21410	0	19075	17314	14532	11603	58658	53074	54405	18957	19810	25479
5883	1288.709	56.1242	VNQIGTLSESIK	192.87	P00924|ENO1_YEAST Enolase 1	8213	0	1676	0	1652	7144	5404	4786	0	0	0	28978
5884	1286.711	67.144	NVNDVIAPAFVK	176.38	P00924|ENO1_YEAST Enolase 1	9794	13750	0	16004	17793	14722	19131	19894	16199	13605	9968	11231

Fig. 4 The figure exemplifies the annotation procedure. The individual peptide identifications are matched to the data matrix based on the accurate mass and retention time measurements

Index	Mass	Retention Time	Sequence	Score	Protein	sample 1_01	sample 1_02	sample 1_03	sample 2_01	sample 2_02	sample 2_03	sample 3_01	sample 3_02	sample 3_03	sample 4_01	sample 4_02	sample 4_03
5879	1159.599	46.7278	IGSEVYHNLK	33.51	P00924|ENO1_YEAST Enolase 1	2584	0	3138	5158	4733	4587	3084	3159	0	4635	4202	4936
5881	807.4214	47.7552	TFAEALR	77.81	P00924|ENO1_YEAST Enolase 1	23759	22035	21031	22388	0	35004	0	0	0	0	0	0
5882	1755.941	73.388	TAGIQIVADDLTVTNPK	250.53	P00924|ENO1_YEAST Enolase 1	21410	0	19075	17314	14532	11603	58658	53074	54405	18957	19810	25479
5883	1288.709	56.1242	VNQIGTLSESIK	192.87	P00924|ENO1_YEAST Enolase 1	8213	0	1676	0	1652	7144	5404	4786	0	0	0	28978
5884	1286.711	67.144	NVNDVIAPAFVK	176.38	P00924|ENO1_YEAST Enolase 1	9794	13750	0	16004	17793	14722	19131	19894	16199	13605	9968	11231

Index	Mass	Retention Time	Sequence	Score	Protein	sample 1	sample 2	sample 3	sample 4
5879	1159.599	46.7278	IGSEVYHNLK	33.51	P00924|ENO1_YEAST Enolase 1	1908	4826	2081	4591
5881	807.4214	47.7552	TFAEALR	77.81	P00924|ENO1_YEAST Enolase 1	22275	19131	0	0
5882	1755.941	73.388	TAGIQIVADDLTVTNPK	250.53	P00924|ENO1_YEAST Enolase 1	13495	14483	55379	21415
5883	1288.709	56.1242	VNQIGTLSESIK	192.87	P00924|ENO1_YEAST Enolase 1	3296	2932	3397	9659
5884	1286.711	67.144	NVNDVIAPAFVK	176.38	P00924|ENO1_YEAST Enolase 1	7848	16173	18408	11601

Fig. 5 Intensity measurements of technical replicates are averaged

4. *Annotation*: The database search results for all injections are used to annotate the data set. This is an automated process in which the identity of the peptides is matched to the entries in the data matrix based on the accurate molecular weight and retention time measurements. The peptide with the highest identification score of all search results is used for annotation (*see* Fig. 4). Since not all peptides are fragmented efficiently, identification will be available for only a subset of the detected peptides. The proportion of identified peptides versus detected peptides can vary according to the type of sample and preparation.

5. *Combining*: Since the analysis includes technical replicates of each sample these replicates must be combined. This is done by calculating the mean intensity value including all three replicates such that for a given peptide there is only one intensity value for each sample (*see* Fig. 5).
6. *Peptides to protein*: The last step in the analysis includes moving from the peptide level to the proteins level. While there are different approaches to do this, the most accurate method is by summing the intensity of the three most abundant peptides per protein (termed the Hi3 approach). This is done by summing the intensity of a given peptide across all samples. The next step is to calculate which are the three most abundant peptides for every given protein and summing the intensity of those peptides within each sample. This produces a quantitative index for every protein, which is the basis of all downstream univariate and multivariate statistics.

4 Notes

1. System verification: This protocol should be run prior to every quantitative study as it ensures the optimal operational performance of the nanoLC-MS/MS system.
   (a) Inject five consecutive runs of 50 fmol digested Yeast Enolase using the gradient in Table 2.
   (b) Process the five raw data files using the data processing and searching software.
   (c) Create a table as shown in Table 1 and fill it out based on the results from the processing software.
2. Thresholds: Intensity thresholds are set in PLGS processing template to indicate which signal is to be used for quantification and identification. These values are low energy threshold, elevated energy threshold, and bin count. These values should be tested and can differ from instrument to instrument. One should adjust the values to obtain the highest number of peptide identifications at a maximum 1 % FDR. Typical values are 150–300 for low energy values. For elevated energy the threshold will be 40–80. Bin count will be optimized between 500 and 2,000.
3. Gradients: The optimal gradient is a function of column loading, throughput and required coverage. The longer the gradient, the higher the proteomic coverage up to a point where chromatographic peaks are too wide and thus too low in intensity. Optimal column loading should be established and should be used for every analysis to obtain peak performance.

References

1. Alcazar O, Hawkridge AM, Collier TS et al (2009) Proteomics characterization of cell membrane blebs in human retinal pigment epithelium cells. Mol Cell Proteomics 8:2201–2211
2. Corrales FJ (2009) Proteomics and human proteome. From bench to bedside. J Proteomics 73:178–179
3. Koutroukides TA, Guest PC, Leweke FM et al (2011) Characterization of the human serum depletome by label-free shotgun proteomics. J Sep Sci 34:1621–1626
4. Oppermann FS, Gnad F, Olsen JV et al (2009) Large-scale proteomics analysis of the human kinome. Mol Cell Proteomics 8:1751–1764
5. Schrimpe-Rutledge AC, Fontes G, Gritsenko MA et al (2012) Discovery of novel glucose-regulated proteins in isolated human pancreatic islets using LC-MS/MS-based proteomics. J Proteome Res 11:3520–3532
6. Burton E, Martin VJ (2012) Proteomic analysis of Clostridium thermocellum ATCC 27405 reveals the upregulation of an alternative transhydrogenase-malate pathway and nitrogen assimilation in cells grown on cellulose. Can J Microbiol 58:1378–1388
7. Fabietti A, Gaspari M, Krishnan S et al (2013) Shotgun proteomic analysis of two Bartonella quintana strains. Proteomics 13:1375–1378
8. Ibrahim M, Shi Y, Qiu H et al (2012) Differential expression of in vivo and in vitro protein profile of outer membrane of Acidovorax avenae subsp. avenae. PLoS One 7:e49657
9. Carter C, Pan S, Zouhar J et al (2004) The vegetative vacuole proteome of Arabidopsis thaliana reveals predicted and unexpected proteins. Plant Cell 16:3285–3303
10. Lee J, Garrett WM, Cooper B (2007) Shotgun proteomic analysis of Arabidopsis thaliana leaves. J Sep Sci 30:2225–2230
11. Chalivendra SC, Lopez-Casado G, Kumar A et al (2013) Developmental onset of reproductive barriers and associated proteome changes in stigma/styles of Solanum pennellii. J Exp Bot 64:265–279
12. Lopez L, Carbone F, Bianco L et al (2012) Tomato plants overexpressing cryptochrome 2 reveal altered expression of energy and stress-related gene products in response to diurnal cues. Plant Cell Environ 35: 994–1012
13. Manaa A, Ahmed HB, Smiti S et al (2011) Salt-stress induced physiological and proteomic changes in tomato (Solanum lycopersicum) seedlings. OMICS 15:801–809
14. Adav SS, Ravindran A, Chao LT et al (2011) Proteomic analysis of pH and strains dependent protein secretion of Trichoderma reesei. J Proteome Res 10:4579–4596
15. Espino JJ, Gutierrez-Sanchez G, Brito N et al (2010) The Botrytis cinerea early secretome. Proteomics 10:3020–3034
16. Kim JM, Park JA, Kim DH (2012) Comparative proteomic analysis of chestnut blight fungus, Cryphonectria parasitica, under tannic-acid-inducing and hypovirus-regulating conditions. Can J Microbiol 58:863–871
17. Cammarato A, Ahrens CH, Alayari NN et al (2011) A mighty small heart: the cardiac proteome of adult Drosophila melanogaster. PLoS One 6:e18497
18. Frohlich T, Arnold GJ, Fritsch R et al (2009) LC-MS/MS-based proteome profiling in Daphnia pulex and Daphnia longicephala: the Daphnia pulex genome database as a key for high throughput proteomics in Daphnia. BMC Genomics 10:171
19. Peiren N, de Graaf DC, Vanrobaeys F et al (2008) Proteomic analysis of the honey bee worker venom gland focusing on the mechanisms of protection against tissue damage. Toxicon 52:72–83
20. Wolschin F, Amdam GV (2007) Comparative proteomics reveal characteristics of life-history transitions in a social insect. Proteome Sci 5:10
21. Manes NP, Estep RD, Mottaz HM et al (2008) Comparative proteomics of human monkeypox and vaccinia intracellular mature and extracellular enveloped virions. J Proteome Res 7:960–968
22. Radhakrishnan A, Yeo D, Brown G et al (2010) Protein analysis of purified respiratory syncytial virus particles reveals an important role for heat shock protein 90 in virus particle assembly. Mol Cell Proteomics 9:1829–1848
23. Saphire AC, Gallay PA, Bark SJ (2006) Proteomic analysis of human immunodeficiency virus using liquid chromatography/tandem mass spectrometry effectively distinguishes specific incorporated host proteins. J Proteome Res 5:530–538
24. Foster DB, Rucker JJ, Marban E (2008) Is Kir6.1 a subunit of mitoK(ATP)? Biochem Biophys Res Commun 366:649–656
25. Kishore D, Kayastha AM (2012) A beta-galactosidase from chick pea (Cicer arietinum) seeds: its purification, biochemical properties and industrial applications. Food Chem 134: 1113–1122
26. Xia Q, Wang T, Taub F et al (2007) Quantitative proteomics of intracellular Porphyromonas gingivalis. Proteomics 7:4323–4337

27. Savidor A, Teper D, Gartemann KH et al (2012) The Clavibacter michiganensis subsp. michiganensis-tomato interactome reveals the perception of pathogen by the host and suggests mechanisms of infection. J Proteome Res 11:736–750
28. Yang F, Melo-Braga MN, Larsen MR et al (2013) Battle through signaling between wheat and the fungal pathogen Septoria tritici revealed by proteomics and phosphoproteomics. Mol Cell Proteomics 12:2497–508
29. Levin Y, Schwarz E, Wang L et al (2007) Label-free LC-MS/MS quantitative proteomics for large-scale biomarker discovery in complex samples. J Sep Sci 30:2198–2203
30. Levin Y, Wang L, Ingudomnukul E et al (2009) Real-time evaluation of experimental variation in large-scale LC-MS/MS-based quantitative proteomics of complex samples. J Chromatogr B Analyt Technol Biomed Life Sci 877: 1299–1305
31. van der Greef J, Martin S, Juhasz P et al (2007) The art and practice of systems biology in medicine: mapping patterns of relationships. J Proteome Res 6:1540–1559
32. Wang G, Wu WW, Zeng W et al (2006) Label-free protein quantification using LC-coupled ion trap or FT mass spectrometry: reproducibility, linearity, and application with complex proteomes. J Proteome Res 5:1214–1223
33. Sitek B, Waldera-Lupa DM, Poschmann G et al (2012) Application of label-free proteomics for differential analysis of lung carcinoma cell line A549. Methods Mol Biol 893:241–248
34. Tang Y, Mackey J, Lai R et al (2013) Quantitative proteomic analysis of HER2 normal and overexpressing MCF-7 breast cancer cells revealed proteomic changes accompanied with HER2 gene amplification. J Proteomics 91:200–9
35. Wisniewski JR, Ostasiewicz P, Mann M (2011) High recovery FASP applied to the proteomic analysis of microdissected formalin fixed paraffin embedded cancer tissues retrieves known colon cancer markers. J Proteome Res 10: 3040–3049
36. Chung K, Nishiyama N, Wanibuchi H et al (2012) AGR2 as a potential biomarker of human lung adenocarcinoma. Osaka City Med J 58:13–24
37. Han NY, Kim EH, Choi J et al (2012) Quantitative proteomic approaches in biomarker discovery of inflammatory bowel disease. J Dig Dis 13:497–503
38. Indovina P, Marcelli E, Pentimalli F et al (2013) Mass spectrometry-based proteomics: the road to lung cancer biomarker discovery. Mass Spectrom Rev 32:129–142
39. Kaur P, Schlatzer D, Cooke K et al (2012) Pairwise protein expression classifier for candidate biomarker discovery for early detection of human disease prognosis. BMC Bioinformatics 13:191
40. Becher D, Bernhardt J, Fuchs S et al (2013) Metaproteomics to unravel major microbial players in leaf litter & soil environments: challenges & perspectives. Proteomics 13: 2895–909
41. Chourey K, Nissen S, Vishnivetskaya T et al (2013) Environmental proteomics reveals early microbial community responses to biostimulation at a uranium- and nitrate-contaminated site. Proteomics 13:2921–30
42. Kolmeder CA, de Vos WM (2013) Metaproteomics of our microbiome—developing insight in function and activity in man and model systems. J Proteomics 97:3–16
43. Dieterich DC, Lee JJ, Link AJ et al (2007) Labeling, detection and identification of newly synthesized proteomes with bioorthogonal non-canonical amino-acid tagging. Nat Protoc 2:532–540
44. Ikeda-Boku A, Kondo K, Ohno S et al (2013) Protein fishing using magnetic nanobeads containing calmodulin site-specifically immobilized via an azido group. J Biochem 154:159–165
45. Ngai MH, Yang PY, Liu K et al (2010) Click-based synthesis and proteomic profiling of lipstatin analogues. Chem Commun (Camb) 46:8335–8337
46. Mischak H, Apweiler R, Banks RE et al (2007) Clinical proteomics: a need to define the field and to begin to set adequate standards. Proteomics Clin Appl 1:148–156
47. Qian WJ, Jacobs JM, Liu T et al (2006) Advances and challenges in liquid chromatography-mass spectrometry-based proteomics profiling for clinical applications. Mol Cell Proteomics 5:1727–1744
48. Li GZ, Vissers JP, Silva JC et al (2009) Database searching and accounting of multiplexed precursor and product ion spectra from the data independent analysis of simple and complex peptide mixtures. Proteomics 9:1696–1719
49. Schwarz E, Levin Y, Wang L et al (2007) Peptide correlation: a means to identify high quality quantitative information in large-scale proteomic studies. J Sep Sci 30:2190–2197

Part VI

Proteomics Informatics and Quantitation

Chapter 16

Bioinformatics for Proteomics: Opportunities at the Interface Between the Scientists, Their Experiments, and the Community

Marc Vaudel, Harald Barsnes, Lennart Martens, and Frode S. Berven

Abstract

Within the last decade, bioinformatics has moved from command line scripts dedicated to single experiments towards production grade software integrated in experimental workflows providing a rich environment for biological investigation. Located at the interface between the scientists, their experiments, and the community, bioinformatics acts as a gateway to a wide source of information. This chapter does not list tools and methods, but rather hints at how bioinformatics can help in improving biological projects, all the way from their initial design to the dissemination of the results.

Key words Bioinformatics, Experimental design

1 Introduction

What Ronald Fischer so famously stated for the statistician is equally true for the bioinformatician: *To consult the bioinformatician after an experiment is finished is often merely to ask him to conduct a post mortem examination. He can perhaps say what the experiment died of.* A famous example from the field of proteomics is the finding of viral and fungi contamination in bee colonies affected by Colony Collapse Disorder [1]. The conclusions of this experiment were strongly questioned in the scientific community due to suspected erroneous identification of the peptides derived from the obtained spectra [2]. Here, a typical problem in bioinformatics applied to proteomics is observed: the user blindly trusted the validation of the peptide identifications without accounting for the inaccuracies (and intricacies) of the process—inaccuracies which were subsequently and independently characterized in more detail in the literature [3–5].

Daniel Martins-de-Souza (ed.), *Shotgun Proteomics: Methods and Protocols*, Methods in Molecular Biology, vol. 1156, DOI 10.1007/978-1-4939-0685-7_16, © Springer Science+Business Media New York 2014

It is clear that the bioinformatics tools used should have flagged the result of the peptide identifications as uncertain, thus suggesting reconsideration of the identification procedure and notably the database usage [6]. This example also shows a vital strength of bioinformatics: the possibility to reprocess data efficiently, and compare it to existing knowledge in the field, and through this enable scientific discussion of the results—perhaps leading to new discoveries as well.

Indeed, bioinformatics is no longer just a command line script run at the end of an experiment to identify peptides and estimate *p*-values. In this chapter, we will show how bioinformatics can help scientists design better experiments, gain insight in the results and easily access vast amounts of existing biological knowledge. We will also list typical pitfalls and introduce good practices for the integration of bioinformatics in proteomics projects.

2 Use of Existing Knowledge

Proteomics knowledge is available in two forms: raw experimental data and processed published data. The latter is integrated into databases with a specific focus, including interactions [7, 8], functions [9], pathways [10, 11], modifications [12], specific diseases [13, 14], or general protein information [15–17]. For a guide to these (and many other) resources, *see* Vizcaíno et al. [18].

Additionally, the raw data—the unprocessed readout of an experiment, typically a collection of spectra—is also available and can be reprocessed to validate a preliminary assumption or refine a biological question [19, 20]. As stated in the introduction, this data is crucial for the afterlife of a project [21]. But even more than the critical discussion of the original conclusions, reprocessing of raw data makes it possible to answer new questions which might not have been the original focus of the experiment [22, 23]. In this way the researcher can save time by reusing already available data.

It is now mandatory in most proteomics journals to make experimental data available in one of the many repositories [24–26] or via the unified ProteomeXchange consortium (http://www.proteomexchange.org). For PRIDE (the Proteomics Identifications database), a user-friendly data browsing tool has been developed called PRIDE Inspector [27], allowing detailed inspection and even download of data in publicly available experiments. Such a downloaded dataset can subsequently be extracted and processed like one's own data.

Similarly to literature benchmarking, browsing and reprocessing freely available data brings essential preliminary knowledge and allows better experimental planning of the biological investigation [28].

3 Experimental Design

As demonstrated in the introduction, even a well-planned and well-executed experiment can easily lead to wrong conclusions if combined with inaccurate or inappropriate bioinformatics. It is therefore crucial to consider the bioinformatics approach already during the experimental planning. Notably, the number of replicates is a central issue in proteomics: the more replicates, the lower the chances of drawing conclusions based on random experimental or computational artifacts [29]. However, a high number of replicates also means more intensive computational work—it is thus crucial to ensure a priori that the complexity of the task can be handled using the available bioinformatics tools and/or expertise. Deciding on the number of biological replicates requires a fine balance between statistical requirements and experimental and computational constraints [30]. To circumvent this, pooling strategies [31], or a pooled reference, are widely used [32]. Pooling should however be used with care as the approach is very sensitive toward outliers if few patients is included in each pool, and in addition information about individual variation is lost.

The accuracy of the bioinformatics workflow can also influence the choice of the experimental approach. For example, in absolute protein quantification, one can chose between the use of intensity based quantification aided by spiked-in labeled peptides or proteins, or estimate the protein abundance directly using spectrum counting or intensities as reviewed elsewhere [33, 34]. The different approaches come at different costs and accuracies which are crucial to consider when designing the experiment [35].

For relative quantification, considering the biological variability and the technical variability is central for the interpretation of regulations [30]. In some cases one can even adapt the experimental design to the complexity of the sample [36]. Similarly, the process of protein identification is not perfectly accurate. Notably, there is a strong dependence on the size and content of the database used [37], which has a strong impact in fields like metaproteomics [38] and plant proteomics [39]. The low accuracy of the identification process may then dictate tuning the mass spectrometer towards a higher sequencing rate rather than maximal mass accuracy—as the benefits of the high mass accuracy will be lost during data interpretation.

Bioinformatics can also help in optimizing the experimental protocol, for example the chromatographic gradient [40], and plays a key role in the design of targeted quantitative experiments based on Selected Reaction Monitoring (SRM) [41, 42] or Data Independent Acquisition (DIA) [43]. Here, the quantified compounds are taken from a list of targeted ions. These can be adapted from previous experiments [44], notably using fragmentation pattern analysis [45, 46] or fragment ion intensity prediction [47], and retention time prediction [48]. Such lists can also be further optimized using other bioinformatics tools [49].

4 Data Workflow Design

When designing an experiment, special care must be given to the data processing workflow. It is crucial to ensure that the software required to answer the desired biological question actually exists, or can be developed in a reasonable timeframe using the available resources to the lab. One should also check that the tools to be used are compatible with the instrumentation, as some software can only handle data from specific kinds of mass spectrometers, e.g., MaxQuant [50] is dedicated to Thermo instruments. It is also important to verify that the software is properly documented and maintained, allowing reliable use. Finally, it should abide by the field's expectations in terms of data processing, file format standards, and data sharing [51]—as detailed in the instructions for authors in most proteomics journals [52–55].

When designing an experiment, one should furthermore always consider the generated data load and verify that the capability to handle it is available. Large experiments can be especially demanding in terms of hardware and storage requirements. Some software also requires advanced informatics skills to install and operate. Because of all these issues, choices with respect to the software pipeline should be settled and tested before the experiment is started. In order to simplify the data flow and management, laboratory information management systems (LIMS) were developed for proteomics [56, 57], with several such systems reviewed by Piggee [58] and by Stephan et al. [59].

Given that they are fully transparent and sometimes performing better than commercial tools [60, 61], it is preferable to choose open source software [62]. Notably, the Trans Proteomic Pipeline (TPP) [63] and the OpenMS [64] packages allow the design of complex and flexible workflows. Their use has been greatly simplified with the development of converters [65] and graphical user interfaces [66]. Moreover, the groups developing these tools also provide teaching materials, courses, and active support.

Finally, advanced workflows require connecting different software. It is vital to verify upfront that the tools are compatible. Also, it is important to verify that the chaining of the methods does not compromise the quality of the results, e.g., the use of multiple pass searches for peptide identification which was found to be incompatible with the widely employed target decoy method [67] to control the quality of the identification results [68].

In summary, advanced workflows often require advanced setups. For more complex designs it is therefore highly recommended to test the workflow before conducting the real experiments. Various standard datasets are available [69] with different levels of complexity and realism, or specific benchmark datasets can be designed. It is important to keep in mind that time and hardware requirements rarely increase linearly with the amount of input but often exponential increases is observed.

5 Quality Control

After the measurements, bioinformatics strongly empowers the quality control of the experiments [70–72]. Virtually all parts of an experiment can be inspected—on a single experiment or on a day to day basis [73]. Elements that can be monitored are as follows: (1) deviation in protein concentration estimation and pipetting typically introduce distribution deviations in relative quantification and small deviation can be corrected by normalization; (2) labeling efficiency can be verified by searching the data with the label as variable; (3) digestion efficiency can be verified by inspecting the Total Ion Chromatogram (TIC) and screening for missed cleavages; (4) the elution and the spray can also be inspected in the TIC; and (5) the mass calibration is straightforward to inspect after the identification of the spectra [74].

However, these control steps require good skills at manipulating large mass spectrometry datasets, but the task is simplified by commercial and open source viewers [75, 76]. Data can also be imported into Perseus (http://www.perseus-framework.org) for advanced visualization and processing. In addition, the SIMPATIQCO software has been specifically designed to provide bioinformatics tools to the service of experiment quality control [77]. Finally, data can also be quality controlled on a large scale as a preprocessing step for downstream use [78, 79].

6 Results Interpretation

The final goal of a mass spectrometry based proteomics experiment is typically the identification and quantification of large protein sets, complete with their post-translational modifications (PTMs) [80]. Making the large number of results in such datasets intuitively navigable is a big challenge for bioinformaticians, and most tools do not support easy browsing of proteomics datasets. Visual navigation of proteomics datasets can however provide the researcher with unique insights. Crucially, the display of error rates and sample distributions is vital for reliable downstream biological interpretation. Most interfaces allow an export to text files which can subsequently be interpreted in tools like Perseus (http://www.perseus-framework.org), where error rates and associated statistics can easily be computed. It is also possible to export results to graph viewers [81, 82] that provide a graph-based approach to complex problems like protein inference [83].

Furthermore, many bioinformatics resources allow the mapping of proteins to a variety of knowledge bases [84]. Proteins are typically linked to genes and their ontology, e.g., as facilitated by the Ensembl database [85]. Proteomics results can also be combined with transcriptomics findings in order to provide an across-omics dimension to the results [86]. Finally, it is possible to annotate the results with existing knowledge about the identified proteins and their modifications. Doing this requires excellent connectivity between the bioinformatics tools and the different sources of information; a task that was greatly simplified by the creation of Biomarts supporting the querying of different databases simultaneously [87–89], notably via Web interfaces such as Dasty [90].

7 Results Dissemination

Results dissemination is an often neglected yet vital part of any proteomics investigation. As detailed in the previous sections, browsing and reprocessing acquired knowledge has become a key aspect of current biological experiments. It is possible to upload experimental results into public repositories, which in turn can be queried through online resources making the hard-won knowledge immediately available to the community at large. In many cases, such as for the ProteomeXchange consortium (http://www.proteomexchange.org), this process has been facilitated by the development of dedicated converters [91, 92]. However, the communication between the different elements of this system is only possible if they speak the same language. Hence, controlled vocabularies (CVs) have been developed [93] as well as convenient interfaces [94, 95] to these, allowing uniform annotation of the data. It is clear that with better annotation, the value for future investigations has improved [96].

8 Tools Development

Despite the best efforts of the bioinformatics community to cover the computational needs of life scientists, sometimes a required bioinformatics tool does not exist, is not maintained anymore, or is no longer compatible with current hardware, software or data formats. It is then very tempting to simply get someone to write a small piece of software (a script) that will do the job. As demonstrated in the previous sections, however, bioinformatics has matured to a level that such small-scale, short-term projects simply no longer match the stringent requirements of the field, neither in terms of user-friendliness nor in terms of compatibility towards standards and integration into complex workflows.

Also, a major yet underestimated task in bioinformatics tool development is the maintenance and support of an application. Indeed, user-friendly graphical user interfaces, maintenance and support are major weaknesses of most academic software [97]. In a workflow, when a building block is missing, it is therefore advised to preferentially create a solid partnership with an experienced bioinformatics group to integrate the new piece of software in a larger environment to ensure long term usability of the tool. This also makes optimal use of the entire proteomics community's efforts at solving the many bioinformatics related challenges in proteomics. Challenges that can only be properly solved if both users and developers work together as a team [28].

Acknowledgements

H.B. is supported by the Research Council of Norway. L.M. acknowledges the support of Ghent University (Multidisciplinary Research Partnership "Bioinformatics: from nucleotides to networks"), the PRIME-XS project, grant agreement number 262067, and the "ProteomeXchange" project, grant agreement number 260558, both funded by the European Union 7th Framework Program. The authors have no competing financial or commercial interests.

References

1. Bromenshenk JJ, Henderson CB, Wick CH et al (2010) Iridovirus and microsporidian linked to honey bee colony decline. PLoS One 5:e13181
2. Foster LJ (2011) Interpretation of data underlying the link between colony collapse disorder (CCD) and an invertebrate iridescent virus. Mol Cell Proteomics 10:M110.006387
3. Ma K, Vitek O, Nesvizhskii AI (2012) A statistical model-building perspective to identification of MS/MS spectra with PeptideProphet. BMC Bioinformatics 13 Suppl 16:S1
4. Vaudel M, Burkhart JM, Sickmann A et al (2011) Peptide identification quality control. Proteomics 11:2105–2114
5. Colaert N, Degroeve S, Helsens K et al (2011) Analysis of the resolution limitations of peptide identification algorithms. J Proteome Res 10:5555–5561
6. Knudsen GM, Chalkley RJ (2011) The effect of using an inappropriate protein database for proteomic data analysis. PLoS One 6:e20873
7. Szklarczyk D, Franceschini A, Kuhn M et al (2011) The STRING database in 2011: functional interaction networks of proteins, globally integrated and scored. Nucleic Acids Res 39:D561–D568
8. Kerrien S, Aranda B, Breuza L et al (2012) The IntAct molecular interaction database in 2012. Nucleic Acids Res 40:D841–D846
9. Sherman BT, da Huang W, Tan Q et al (2007) DAVID Knowledgebase: a gene-centered database integrating heterogeneous gene annotation resources to facilitate high-throughput gene functional analysis. BMC Bioinformatics 8:426
10. Haw R, Hermjakob H, D'Eustachio P et al (2011) Reactome pathway analysis to enrich biological discovery in proteomics data sets. Proteomics 11:3598–3613
11. Kanehisa M, Goto S (2000) KEGG: Kyoto encyclopedia of genes and genomes. Nucleic Acids Res 28:27–30
12. Hornbeck PV, Kornhauser JM, Tkachev S et al (2012) PhosphoSitePlus: a comprehensive

resource for investigating the structure and function of experimentally determined post-translational modifications in man and mouse. Nucleic Acids Res 40:D261–D270

13. Reddy TB, Riley R, Wymore F et al (2009) TB database: an integrated platform for tuberculosis research. Nucleic Acids Res 37:D499–D508
14. Forbes SA, Bindal N, Bamford S et al (2011) COSMIC: mining complete cancer genomes in the Catalogue of Somatic Mutations in Cancer. Nucleic Acids Res 39:D945–D950
15. Apweiler R, Bairoch A, Wu CH et al (2004) UniProt: the Universal Protein knowledgebase. Nucleic Acids Res 32:D115–D119
16. Craig R, Cortens JP, Beavis RC (2004) Open source system for analyzing, validating, and storing protein identification data. J Proteome Res 3:1234–1242
17. Lane L, Argoud-Puy G, Britan A et al (2012) neXtProt: a knowledge platform for human proteins. Nucleic Acids Res 40:D76–D83
18. Vizcaino JA, Mueller M, Hermjakob H et al (2009) Charting online OMICS resources: a navigational chart for clinical researchers. Proteomics Clin Appl 3:18–29
19. Hahne H, Moghaddas Gholami A, Kuster B (2012) Discovery of O-GlcNAc-modified proteins in published large-scale proteome data. Mol Cell Proteomics 11:843–850
20. Matic I, Ahel I, Hay RT (2012) Reanalysis of phosphoproteomics data uncovers ADP-ribosylation sites. Nat Methods 9:771–772
21. Martens L, Nesvizhskii AI, Hermjakob H et al (2005) Do we want our data raw? Including binary mass spectrometry data in public proteomics data repositories. Proteomics 5: 3501–3505
22. Fannes T, Vandermarliere E, Schietgat L et al (2013) Predicting tryptic cleavage from proteomics data using decision tree ensembles. J Proteome Res 12:2253–2259
23. Vandermarliere E, Martens L (2013) Protein structure as a means to triage proposed PTM sites. Proteomics 13:1028–1035
24. Desiere F, Deutsch EW, King NL et al (2006) The PeptideAtlas project. Nucleic Acids Res 34:D655–D658
25. Martens L, Hermjakob H, Jones P et al (2005) PRIDE: the proteomics identifications database. Proteomics 5:3537–3545
26. Vizcaino JA, Foster JM, Martens L (2010) Proteomics data repositories: providing a safe haven for your data and acting as a springboard for further research. J Proteomics 73:2136–2146
27. Wang R, Fabregat A, Rios D et al (2012) PRIDE Inspector: a tool to visualize and validate MS proteomics data. Nat Biotechnol 30:135–137
28. Barsnes H, Martens L (2013) Crowdsourcing in proteomics: public resources lead to better experiments. Amino Acids 44:1129–1137
29. Levin Y (2011) The role of statistical power analysis in quantitative proteomics. Proteomics 11:2565–2567
30. Oberg AL, Vitek O (2009) Statistical design of quantitative mass spectrometry-based proteomic experiments. J Proteome Res 8: 2144–2156
31. Karp NA, Lilley KS (2009) Investigating sample pooling strategies for DIGE experiments to address biological variability. Proteomics 9:388–397
32. Geiger T, Cox J, Ostasiewicz P et al (2010) Super-SILAC mix for quantitative proteomics of human tumor tissue. Nat Methods 7:383–385
33. Bantscheff M, Schirle M, Sweetman G et al (2007) Quantitative mass spectrometry in proteomics: a critical review. Anal Bioanal Chem 389:1017–1031
34. Vaudel M, Sickmann A, Martens L (2010) Peptide and protein quantification: a map of the minefield. Proteomics 10:650–670
35. Domon B, Aebersold R (2010) Options and considerations when selecting a quantitative proteomics strategy. Nat Biotechnol 28:710–721
36. Vaudel M, Burkhart JM, Radau S et al (2012) Integral quantification accuracy estimation for reporter ion-based quantitative proteomics (iQuARI). J Proteome Res 11:5072–5080
37. Vaudel M, Burkhart JM, Breiter D et al (2012) A complex standard for protein identification, designed by evolution. J Proteome Res 11:5065–5071
38. Muth T, Benndorf D, Reichl U et al (2013) Searching for a needle in a stack of needles: challenges in metaproteomics data analysis. Mol Biosyst 9:578–585
39. Castellana NE, Payne SH, Shen Z et al (2008) Discovery and revision of Arabidopsis genes by proteogenomics. Proc Natl Acad Sci U S A 105:21034–21038
40. Moruz L, Pichler P, Stranzl T et al (2013) Optimized nonlinear gradients for reversed-phase liquid chromatography in shotgun proteomics. Anal Chem 85:7777–7785
41. Jenden DJ, Cho AK (1979) Selected ion monitoring in pharmacology. Biochem Pharmacol 28:705–713
42. Yost RA, Enke CG (1979) Triple quadrupole mass spectrometry for direct mixture analysis and structure elucidation. Anal Chem 51: 1251–1264
43. Purvine S, Eppel JT, Yi EC et al (2003) Shotgun collision-induced dissociation of peptides using a time of flight mass analyzer. Proteomics 3:847–850

44. Craig R, Cortens JP, Beavis RC (2005) The use of proteotypic peptide libraries for protein identification. Rapid Commun Mass Spectrom 19:1844–1850
45. Barsnes H, Eidhammer I, Martens L (2011) A global analysis of peptide fragmentation variability. Proteomics 11:1181–1188
46. Mallick P, Schirle M, Chen SS et al (2007) Computational prediction of proteotypic peptides for quantitative proteomics. Nat Biotechnol 25:125–131
47. Degroeve S, Martens L (2013) MS2PIP: a tool for MS/MS peak intensity prediction. Bioinformatics 29(24):3199–3203
48. Moruz L, Staes A, Foster JM et al (2012) Chromatographic retention time prediction for posttranslationally modified peptides. Proteomics 12:1151–1159
49. Nahnsen S, Kohlbacher O (2012) In silico design of targeted SRM-based experiments. BMC Bioinformatics 13 Suppl 16:S8
50. Cox J, Mann M (2008) MaxQuant enables high peptide identification rates, individualized p.p.b.-range mass accuracies and proteome-wide protein quantification. Nat Biotechnol 26:1367–1372
51. Orchard S, Jones P, Taylor C et al (2007) Proteomic data exchange and storage: the need for common standards and public repositories. Methods Mol Biol 367:261–270
52. Kinsinger CR, Apffel J, Baker M et al (2012) Recommendations for mass spectrometry data quality metrics for open access data (corollary to the Amsterdam Principles). J Proteome Res 11:1412–1419
53. Kinsinger CR, Apffel J, Baker M et al (2012) Recommendations for mass spectrometry data quality metrics for open access data (corollary to the Amsterdam principles). Proteomics 12:11–20
54. Kinsinger CR, Apffel J, Baker M et al (2011) Recommendations for mass spectrometry data quality metrics for open access data (corollary to the Amsterdam principles). Proteomics Clin Appl 5:580–589
55. Kinsinger CR, Apffel J, Baker M et al (2011) Recommendations for mass spectrometry data quality metrics for open access data (corollary to the Amsterdam Principles). Mol Cell Proteomics 10:O111.015446
56. Martens L (2011) Data management in mass spectrometry-based proteomics. Methods Mol Biol 728:321–332
57. Hakkinen J, Vincic G, Mansson O et al (2009) The proteios software environment: an extensible multiuser platform for management and analysis of proteomics data. J Proteome Res 8:3037–3043
58. Piggee C (2008) LIMS and the art of MS proteomics. Anal Chem 80:4801–4806
59. Stephan C, Kohl M, Turewicz M et al (2010) Using laboratory information management systems as central part of a proteomics data workflow. Proteomics 10:1230–1249
60. Weisser H, Nahnsen S, Grossmann J et al (2013) An automated pipeline for high-throughput label-free quantitative proteomics. J Proteome Res 12(4):1628–1644
61. Lange E, Gropl C, Reinert K et al (2006) High-accuracy peak picking of proteomics data using wavelet techniques. Pac Symp Biocomput 243–254
62. Martin SF, Falkenberg H, Dyrlund TF et al (2013) PROTEINCHALLENGE: crowd sourcing in proteomics analysis and software development. J Proteomics 88:41–46
63. Keller A, Eng J, Zhang N et al (2005) A uniform proteomics MS/MS analysis platform utilizing open XML file formats. Mol Syst Biol 1:2005.0017
64. Sturm M, Bertsch A, Gropl C et al (2008) OpenMS—an open-source software framework for mass spectrometry. BMC Bioinformatics 9:163
65. Kessner D, Chambers M, Burke R et al (2008) ProteoWizard: open source software for rapid proteomics tools development. Bioinformatics 24:2534–2536
66. Junker J, Bielow C, Bertsch A et al (2012) TOPPAS: a graphical workflow editor for the analysis of high-throughput proteomics data. J Proteome Res 11:3914–3920
67. Elias JE, Gygi SP (2010) Target-decoy search strategy for mass spectrometry-based proteomics. Methods Mol Biol 604:55–71
68. Everett LJ, Bierl C, Master SR (2010) Unbiased statistical analysis for multi-stage proteomic search strategies. J Proteome Res 9:700–707
69. Ivanov AR, Colangelo CM, Dufresne CP et al (2013) Interlaboratory studies and initiatives developing standards for proteomics. Proteomics 13:904–909
70. Martens L, Vizcaino JA, Banks R (2011) Quality control in proteomics. Proteomics 11:1015–1016
71. Tabb DL (2013) Quality assessment for clinical proteomics. Clin Biochem 46:411–420
72. Martens L (2013) Bringing proteomics into the clinic: the need for the field to finally take itself seriously. Proteomics Clin Appl 7: 388–391
73. Burkhart JM, Premsler T, Sickmann A (2011) Quality control of nano-LC-MS systems using stable isotope-coded peptides. Proteomics 11: 1049–1057

74. Staes A, Vandenbussche J, Demol H et al (2013) Asn3, a reliable, robust and universal lock mass for improved accuracy in LC-MS and LC-MS/MS. Anal Chem 85(22):11054–11060
75. Cote RG, Reisinger F, Martens L (2010) jmzML, an open-source Java API for mzML, the PSI standard for MS data. Proteomics 10:1332–1335
76. Sturm M, Kohlbacher O (2009) TOPPView: an open-source viewer for mass spectrometry data. J Proteome Res 8:3760–3763
77. Pichler P, Mazanek M, Dusberger F et al (2012) SIMPATIQCO: a server-based software suite which facilitates monitoring the time course of LC-MS performance metrics on Orbitrap instruments. J Proteome Res 11:5540–5547
78. Gonnelli G, Hulstaert N, Degroeve S et al (2012) Towards a human proteomics atlas. Anal Bioanal Chem 404:1069–1077
79. Foster JM, Degroeve S, Gatto L et al (2011) A posteriori quality control for the curation and reuse of public proteomics data. Proteomics 11:2182–2194
80. Domon B, Aebersold R (2006) Mass spectrometry and protein analysis. Science 312:212–217
81. Shannon P, Markiel A, Ozier O et al (2003) Cytoscape: a software environment for integrated models of biomolecular interaction networks. Genome Res 13:2498–2504
82. Bastian M, Heymann S, Jacomy M (2009) Gephi: an open source software for exploring and manipulating networks. International AAAI Conference on Weblogs and Social Media. https://gephi.org/
83. Nesvizhskii AI, Aebersold R (2005) Interpretation of shotgun proteomic data: the protein inference problem. Mol Cell Proteomics 4:1419–1440
84. Vaudel M, Sickmann A, Martens L (2013) Introduction to opportunities and pitfalls in functional mass spectrometry based proteomics. Biochim Biophys Acta 1844(1 Pt A):12–20
85. Flicek P, Amode MR, Barrell D et al (2011) Ensembl 2011. Nucleic Acids Res 39: D800–D806
86. Cox J, Mann M (2012) 1D and 2D annotation enrichment: a statistical method integrating quantitative proteomics with complementary high-throughput data. BMC Bioinformatics 13 Suppl 16:S12
87. Kasprzyk A, Keefe D, Smedley D et al (2004) EnsMart: a generic system for fast and flexible access to biological data. Genome Res 14: 160–169
88. Kasprzyk A (2011) BioMart: driving a paradigm change in biological data management. Database (Oxford) 2011:bar049
89. Smedley D, Haider S, Ballester B et al (2009) BioMart—biological queries made easy. BMC Genomics 10:22
90. Villaveces JM, Jimenez RC, Garcia LJ et al (2011) Dasty3, a WEB framework for DAS. Bioinformatics 27:2616–2617
91. Barsnes H, Vizcaino JA, Eidhammer I et al (2009) PRIDE Converter: making proteomics data-sharing easy. Nat Biotechnol 27: 598–599
92. Cote RG, Griss J, Dianes JA et al (2012) The PRoteomics IDEntification (PRIDE) Converter 2 framework: an improved suite of tools to facilitate data submission to the PRIDE database and the ProteomeXchange consortium. Mol Cell Proteomics 11: 1682–1689
93. Martens L, Palazzi LM, Hermjakob H (2008) Data standards and controlled vocabularies for proteomics. Methods Mol Biol 484:279–286
94. Cote R, Reisinger F, Martens L et al (2010) The Ontology Lookup Service: bigger and better. Nucleic Acids Res 38:W155–W160
95. Barsnes H, Cote RG, Eidhammer I et al (2010) OLS dialog: an open-source front end to the ontology lookup service. BMC Bioinformatics 11:34
96. Klie S, Martens L, Vizcaino JA et al (2008) Analyzing large-scale proteomics projects with latent semantic indexing. J Proteome Res 7:182–191
97. (2013) In need of an upgrade. Nat Biotechnol 31:857. doi: 10.1038/nbt.2717

Part VII

Protein–Protein Interactions

Chapter 17

Identification of DNA Damage Checkpoint-Dependent Protein Interactions in *Saccharomyces cerevisiae* Using Quantitative Mass Spectrometry

Francisco M. Bastos de Oliveira and Marcus B. Smolka

Abstract

The DNA damage checkpoint (DDC) is an evolutionarily conserved signaling pathway that is crucial to maintain genomic integrity. In response to DNA damage, DDC kinases are rapidly activated and phosphorylate an elaborate network of substrates involved in multiple cellular processes. An important role of the DDC response is to assemble protein complexes. However, for most of the DDC substrates, how the DDC-dependent phosphorylation modulates their network of interactions remains to be established. Here, we present a protocol for the identification of DDC-dependent protein–protein interactions based on *S*table *I*sotope *L*abeling of *A*mino acids in *C*ell culture (SILAC) followed by affinity-tagged protein purification and quantitative mass spectrometry analysis. Based on a model study using *Saccharomyces cerevisiae*, we provide a method that can be generally applied to study the role of kinases in mediating protein–protein interactions.

Key words Protein–protein interaction, DNA damage checkpoint, *Saccharomyces cerevisiae*, SILAC, Affinity-tagged protein purification, Quantitative mass spectrometry

1 Introduction

In response to DNA damage, eukaryotic cells trigger a highly conserved signaling pathway known as the DNA damage checkpoint (DDC). The DDC is critical to prevent the accumulation of genomic instability and cancer [1–3]. DDC signaling is initiated by the upstream PI3K-like sensor kinases Mec1 and Tel1 (orthologs of human ATR and ATM, respectively) [4]. Once activated, these kinases phosphorylate numerous proteins to coordinate a wide range of cellular processes including DNA repair, cell cycle progression, transcription regulation, and DNA replication among others [5–8].

Daniel Martins-de-Souza (ed.), *Shotgun Proteomics: Methods and Protocols*, Methods in Molecular Biology, vol. 1156, DOI 10.1007/978-1-4939-0685-7_17, © Springer Science+Business Media New York 2014

A prevalent role for phosphorylation in the DDC response is believed to be the assembly of protein complexes [9–11]. Consistent with this notion, many proteins that participate in the DDC response contain phosphopeptide binding domains, such as BRCT (*BR*CA1 *C*-*T*erminus) or FHA (*F*ork*h*ead-*A*ssociated), that recognize sites phosphorylated by DDC kinases [12, 13]. Recently, ATR/ATM or Mec1/Tel1-dependent phosphorylation sites have been identified in hundreds of proteins [14–16], but in the vast majority of these substrates the role of phosphorylation is not known. In most cases, establishing whether these phosphorylation events are modulating the network of interactions of the substrates will be important to understand how the DDC coordinates the DNA damage response.

Mass spectrometry analysis of protein pull-down assays has been widely used for the identification of protein–protein interactions [17]. However, due to the high sensitivity of mass spectrometry, any pull-down assay results in identification of a large number of nonspecific background proteins. Recently, our laboratory has developed a two-step protocol based on quantitative mass spectrometry analysis to identify protein–protein interactions that are mediated by the DDC (*see* Fig. 1). In the first step we combine SILAC [18] and protein affinity tag purification to compare the relative abundance of co-purified proteins between a tagged or untagged version of any protein of interest. This step allows the discrimination between background and specific interactions with the protein of interest (*see* Fig. 1a). In the second step, we compare the relative abundance of co-purified proteins between a wild type and a checkpoint kinase null mutant strain to reveal kinase-dependent interactions (*see* Fig. 1b).

For the general purpose of this protocol, here we use the Mec1-dependent interactions mediated by the BRCT domain-containing protein Rtt107 as an example [10]. However, with minor modifications, this protocol can be adapted to provide a reliable method for the identification of protein–protein interactions in budding yeast.

2 Materials

2.1 Stable Isotope Labeling of Amino Acids in Cell Culture (SILAC)

1. Strains MBS164 (*MATa, ura3-52, leu2Δ1, trp1Δ63, his3Δ200, lys2ΔBgl, hom3-10, ade2Δ1, ade8, arg4Δ, sml1::TRIP1, bar1::HIS3*); MBS266 (*MATa, ura3-52, leu2Δ1, trp1Δ63, his3Δ200, lys2ΔBgl, hom3-10, ade2Δ1, ade8, arg4Δ, sml1::TRIP1, bar1::HIS3, Rtt107-6xhis-3xHA::KanMX6*) and MBS402 (*MATa, ura3-52, leu2Δ1, trp1Δ63, his3Δ200, lys2ΔBgl, hom3-10, ade2Δ1, ade8, arg4Δ, sml1::TRIP1, bar1::HIS3, Rtt107-6xhis-3xHA::KanMX6, mec1::URA3*) (*see* **Note 1**).

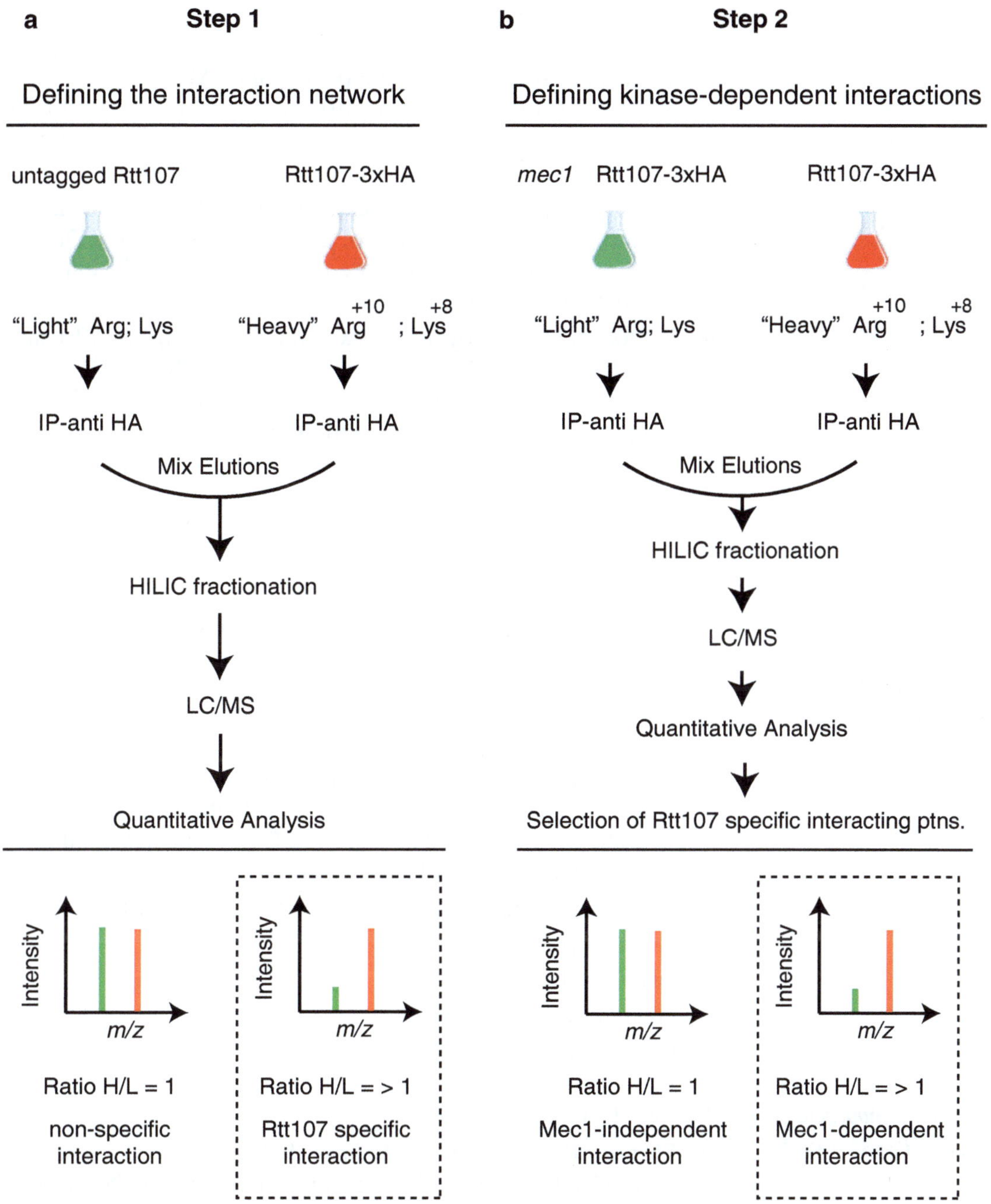

Fig. 1 Overview of the methodology for the identification of checkpoint-dependent interactions of Rtt107-3xHA during replication stress. (**a**) Untagged and 3xHA tagged Rtt107 strains were grown in "Light" and "Heavy" media, respectively. After cultures were treated with 0.04 % methyl-methanesulfonate (MMS), proteins were extracted and subjected to anti-HA immunoprecipitation (IP). Immunoprecipitation elutions were mixed, fractionated by HILIC, and subjected to LC–quantitative mass spectrometry analysis. Peptides ratio significantly higher than 1 indicates Rtt107 specific interactions while peptides ratio around 1 indicates nonspecific interactions. (**b**) *mec1*Δ and wild type cells, both containing Rtt107-3xHA were grown in "light" and "heavy" media, respectively. After LC–quantitative mass spectrometry analysis, only the proteins previously identified as Rtt107 specific interactors were considered. Heavy-to-light peptide ratio significantly higher than 1 indicates checkpoint-dependent Rtt107 interactions while heavy-to-light peptide ratio around 1 indicates checkpoint-independent Rtt107 interactions. Proteins 1–11 were identified with a heavy-to-light ratio greater than 10 and considered to be Rtt107 specific interacting protein

2. SD-arg-lys media for SILAC: 0.67 g of CSM-arg-his-lys, 6.7 g of YNB, 80 mg of L-proline, 40 mg of L-histidine, 960 ml of distilled water. After autoclave, add 40 ml of a 50 % dextrose solution to a final concentration of 2 % (*see* **Note 2**).
3. 1,000× "Light" arg/lys stock solution: 30 mg/ml of L-arginine (0.2 M), 20 mg/ml of L-lysine (0.115 M). Dissolve amino acids in distilled water and filtrate using a 0.22 μm disposable syringe filter.
4. 1,000× "Heavy" arg/lys stock solution: 38 mg/ml of L-arginine $^{13}C_6$, $^{15}N_4$ (0.2 M), 25 mg/ml of L-lysine $^{13}C_6$, $^{15}N_2$ (0.115 M). Dissolve amino acids in distilled water and filtrate using a 0.22 μm disposable syringe filter.
5. To prepare 1 l of SD "Light" or SD "Heavy" add 1 ml of 1,000× "Light" arg/lys or 1,000× "Heavy" arg/lys stock solution to 1 l of SD-arg-lys.

2.2 Methyl Methanesulfonate (MMS)-Induced Replication Stress and Cells Harvesting

1. Methyl methanesulfonate.
2. TE buffer: 10 mM Tris–HCl pH 8.0, 5 mM EDTA.

2.3 Cell Lysis

1. Lysis buffer: 50 mM Tris–HCl pH 7.5, 0.2 % Tergitol, 150 mM NaCl, 5 mM EDTA, 1 mM phenylmethylsulfonyl fluoride (PMSF), 1× complete EDTA-free protease inhibitor cocktail, and 1× phosphatase inhibitor (*see* **Note 3**).
2. 100× phosphatase inhibitor: 100 mM Na_3VO_4, 500 mM NaF, and 1 M β-glycerophosphate.
3. Glass beads, 0.05 mm dia.
4. Protein assay dye reagent.
5. BeadBeater type cell disrupter.

2.4 Immunoprecipitation of 3xHA-Tagged Rtt107

1. Anti-HA affinity resin.
2. Elution Buffer: 100 mM Tris–HCl pH 8.0, 1 % sodium dodecyl sulfate, 10 mM dithiothreitol (DTT) (*see* **Note 4**).
3. Micro spin type chromatography columns.

2.5 Protein Precipitation and Tryptic Digestion

1. Alkylation solution: 1 M Tris–HCl pH 8.0, 0.5 M iodoacetamide (*see* **Note 5**).
2. Precipitation solution: 50 % acetone, 49.9 % ethanol, 0.1 % biochemistry grade acetic acid.
3. Water bath sonicator.
4. Urea/Tris solution: 1 M Tris–HCl pH 8.0, 8 M urea (*see* **Note 5**).
5. $NaCl_2$/Tris solution: 50 mM Tris–HCl pH 8.0, 150 mM $NaCl_2$.
6. Mass spec grade trypsin 1 μg/μl in 0.3 % biochemistry grade acetic acid (*see* **Note 6**).

2.6 Protein Sample Cleanup

1. 10 % LC-MS grade formic acid (*see* **Note** 7).
2. 10 % Trifluoroacetic acid (*see* **Note** 7).
3. C_{18} cartridges Vac 1 cc, 50 mg.
4. C_{18} buffer A: 0.1 % trifluoroacetic acid (*see* **Note** 7).
5. C_{18} buffer B: 0.1 % biochemistry grade acetic acid (*see* **Note** 7).
6. C_{18} buffer C: 80 % HPLC grade acetonitrile, 0.1 % biochemistry grade acetic acid (*see* **Note** 7).
7. Silanized conical base insert vial, 300 μl.

2.7 Hydrophilic Interaction Liquid Chromatography (HILIC)

1. This protocol assumes access to high-performance liquid chromatography (HPLC) equipment including an elution gradient programmer with data processing software, UV absorbance detector, and automated fraction collector.
2. 99.9 % HPLC grade acetonitrile (*see* **Note** 7).
3. Column: 2.0 × 150 mm TSK gel Amide-80 5 μm particle.
4. Buffer D: 90 % HPLC grade acetonitrile (*see* **Note** 7).
5. Buffer E: 80 % HPLC grade acetonitrile with 0.005 % trifluoroacetic acid (*see* **Note** 7).
6. Buffer F: 0.025 % trifluoroacetic acid (*see* **Note** 7).

2.8 Reverse Phase Liquid Chromatography–Tandem Mass Spectrometry

1. The protocol assumes access to an online nano LC system interfaced to a mass spectrometer capable of performing tandem MS/MS. The present study was performed by using an online Nano LC-Ultra® system (Eksigent) coupled with a Q Exactive mass spectrometer (Thermo Fisher Scientific).
2. Reverse phase analytical column: 125 μm ID × 20 cm in-house packed with 3 μm C_{18} resin (*see* **Note 8**).
3. Buffer I: 0.1 % LC-MS grade formic acid (*see* **Note** 7).
4. Buffer II: 0.1 % LC-MS grade formic acid with 80 % HPLC grade acetonitrile (*see* **Note** 7).

2.9 Database Searching and Protein Quantification

1. For database search: SEQUEST software on a SORCERER system (Sage-N Research).
2. For quantification of relative protein abundances: XPRESS software, part of the Trans-Proteomic Pipeline (Seattle Proteome Center-SPC).

3 Methods

3.1 Stable Isotope Labeling of Amino Acids in Cell Culture (SILAC)

The basic principle of SILAC for quantitative proteomics analysis consists on growing two cultures of cells, one in a medium complemented with normal ("light") amino acids and the other in a medium complemented with stable-isotope labeled ("heavy") amino acids. The incorporation of "heavy" amino acids allows each peptide to be detected as a pair in the mass spectra, with a known mass shift compared with the peptide labeled with the "light" version. In this case, because the "light" and "heavy" amino acids are chemically identical, except for their mass difference, the ratio of peak intensities in the mass spectrometer directly yields the ratio of protein abundance between the two cultures.

1. For step 1 of the analysis, inoculate 100 ml of SD "light" and SD "heavy" media with fresh colonies from either strains MBS164 and MBS266, respectively, and grow cultures overnight for at least 12 h, at 30 °C with constant shaking. For the step 2 of the analysis, inoculate 100 ml of SD "light" and SD "heavy" media with fresh colonies from either MBS402 or MBS266, respectively (*see* **Note 9**).
2. On the next day the cultures OD_{600} should be around 0.4–0.5. Dilute cultures in their respective SD media to an $OD_{600}=0.1$ in a final volume of 250 ml (*see* **Note 10**).

3.2 Methyl Methanesulfonate (MMS)-Induced Replication Stress and Cells Harvesting

Previous work has shown that methyl methanesulfonate (MMS) concentrations up to 0.04 % are enough to efficiently activate the DNA damage checkpoint response during S-phase but not during G1 [19]. For the purpose of this protocol we are using 0.04 % of MMS to address checkpoint-dependent Rtt107 protein–protein interactions during S-phase (*see* **Note 11**).

1. When cells reach an $OD_{600}=0.4$, add 100 μl of MMS to final concentration of 0.04 % and keep cultures growing for 2.5 h.
2. After 2.5 h treatment with MMS transfer the 2 × 250 ml cultures to two separated 500 ml conical centrifugation bottles.
3. Centrifuge the cultures for 2,500 × *g* for 5 min at 4 °C. Discard the supernatant.
4. Add 10 ml of ice-cold TE buffer to each centrifugation bottle, resuspend cell pellets and transfer each pellet to a separate 15 ml Falcon tubes
5. Repeat **step 3** from Subheading 3.2.
6. Add 2 ml of ice-cold TE buffer to each 15 ml Falcon tube, resuspend each cell pellet separately and transfer each pellet to 2 × 2 ml conical screw cap tube. Spin down cells for 15 s in a bench top microcentrifuge for 5,000 × *g* at room temperature (RT) and remove supernatant. Each cell pellet should be around 150–200 mg. Cell pellets can be stored at −80 °C for up to 2 weeks.

3.3 Cell Lysis

1. Add 600 μl of ice-cold glass beads to each tube.
2. Add 1 ml of lysis buffer to each tube and break cells for 15 min at 4 °C using the cell disrupter.
3. Puncture the bottom of the 2 ml tube using a hot 21-gauge needle. Place tube on top of a 5 ml syringe and place the

syringe inside a 15 ml Falcon tube. Centrifuge at 250×*g* at 4 °C for 3 min to recover extract.

4. Transfer lysate to a 1.7 ml microcentrifuge and centrifuge for 10 min at 13,000×*g* at 4 °C.
5. Collect the cleared supernatant and determine the protein concentration of both samples by using protein assay dye reagent.
6. Normalize protein concentration between the two samples (MBS164 "Light" and MBS266 "Heavy" or MBS402 "light" and MBS 266 "heavy"). Final concentration should range between 7.5–10 mg/ml and total amount of protein should be around 15–20 mg. After normalization, dilute samples to maximum concentration of 5 mg/ml (final volume of 3–4 ml) and transfer each sample to a different 15 ml Falcon tube. Keep samples on ice.

3.4 Immunoprecipitation of 3xHA-Tagged Rtt107

1. Condition anti-HA resin by washing 3× in at least 500 μl of lysis buffer. Use a minimum of 50 μl of anti-HA resin per 15–20 mg of protein. Add the anti-HA resin to the samples and incubate for 2–4 h at 4 °C under constant agitation (*see* **Note 12**).
2. Collect resin by quick spin at 4 °C for 1 min at 2,500×*g* and wash resin 3× with at least 500 μl of ice-cold lysis buffer (*see* **Note 13**).
3. Add 3 volumes of elution buffer to the resin and heat it at 95 °C for 5 min (*see* **Note 14**).
4. Transfer the resin and elution buffer to a Micro spin type column pre-equilibrated with elution buffer. Collect elution in a 1.7 ml microcentrifuge tube by spinning the column at 650×*g* for 1 min at RT.

3.5 Protein Precipitation and Tryptic Digestion

1. At this point combine both elutions from immunoprecipitations in a 1.7 ml microcentrifuge tube.
2. For protein alkylation, add 15 μl of a 0.5 M iodoacetamide stock solution (final concentration of 25 mM) and incubate for 15 min at RT (*see* **Note 5**).
3. Use a SpeedVac to concentrate sample at 45 °C until it reaches 100 μl (one third of the initial volume) (*see* **Note 15**).
4. Add 300 μl of protein precipitation solution and incubate sample on ice for 30 min.
5. Centrifuge sample at 13,000×*g* for 10 min at RT.
6. Discard supernatant carefully without dislodging the tiny protein pellet.
7. Add 200 μl of protein precipitation solution and sonicate sample in water bath sonicator for 10 s (*see* **Note 16**).
8. Repeat **steps 5** and **6** from Subheading 3.5.

9. Keep microcentrifuge tube upside-down for 5–10 min until acetone smell is gone.
10. Resuspend protein pellet in 50 μl of urea/Tris solution by carefully pipeting up and down (*see* **Note 5**).
11. Add 150 μl $NaCl_2$/Tris solution.
12. Add 1 μg of mass spec grade trypsin, seal microcentrifuge tube with Parafilm and incubate sample ON at 37 °C.

3.6 Protein Sample Cleanup

1. Using a glass syringe, acidify sample by adding 5 μl of 10 % trifluoroacetic acid and 5 μl of 10 % formic acid (0.25 % final concentration each).
2. Spin sample at 13,000×*g* for 1 min at room temperature to remove particulates prior sample cleanup. Transfer supernatant to a new microcentrifuge tube.
3. By applying air pressure with a pipette bulb, condition a 50 mg C_{18} column by adding 1 ml of C_{18} buffer D (*see* **Note 17**).
4. Add 1 ml of C_{18} buffer A to equilibrate column.
5. Apply the sample through the column and let it flow by gravity.
6. Add 2×1 ml of C_{18} buffer B to wash the column. Let it flow by gravity.
7. Wipe residual volume of C_{18} buffer C off column tip using a paper wipe.
8. By applying air pressure with a bulb, elute sample in 200 μl of C_{18} buffer C in a silanized conical base insert vial.
9. Using a gel loader tip, mix sample by pipeting up and down.
10. Dry sample completely at 45 °C using a SpeedVac concentrator.
11. To resuspend sample, add 15 μl of distilled water.
12. By using a glass syringe, add 15 μl of 10 % formic acid and 60 μl of 99.9 % acetonitrile to sample. Using a gel loader tip, mix sample by pipeting up and down (*see* **Note 18**).
13. Transfer the vial to 1.7 ml microcentrifuge tube and spin it at 5,000×*g* for 1 min at room temperature to remove particulates prior HILIC chromatography.
14. Transfer sample (approximately 85 μl) to another silanized conical base insert vial using a gel loader tip.

3.7 Hydrophilic Interaction Liquid Chromatography (HILIC)

The unique separation ability of Hydrophilic Interaction Liquid Chromatography (HILIC) and is orthogonality towards other reverse phase separation techniques make it an ideal method for multidimensional fractionation of complex protein samples prior to MS analysis [20]. The present study was performed using a Dionex Ultimate 3000 HPLC controlled by Chromeleon 6.8

software. A TSK gel Amide-80 column (2 mm × 150 mm, 5 μm; Tosoh Bioscience) was used for the HILIC fractionation experiment. Three buffers were used for the gradient: buffer D (90 % acetonitrile); buffer E (80 % acetonitrile and 0.005 % trifluoroacetic acid) and buffer F (0.025 % trifluoroacetic acid).

1. Load samples into HILIC TSK gel Amide-80 column via a 100 μl loop and set the fraction collector to time-based mode, collecting 1 min fractions between 10 and 22 min of the gradient in 300 μl silanized conical base insert vials. The gradient used consists on a 100 % buffer D at time = 0 min, 98 % of buffer E and 2 % of buffer F at time = 5 min, 82 % of buffer E and 18 % of buffer F at time = 15 min and 5 % of buffer E and 95 % of buffer F from time = 25–27 min in a flow of 150 μl/min.
2. Dry sample fractions completely at 45 °C using a SpeedVac concentrator.
3. Resuspend each fraction in 7 μl of 0.1 % trifluoroacetic acid.

3.8 Reverse Phase Liquid Chromatography-Tandem Mass Spectrometry

As part of our multidimensional separation platform, we perform a C_{18} reverse phase fractionation prior to MS analysis. C_{18} reverse phase is a high resolution fractionation technique that presents an orthogonal selectivity towards HILIC [20]. The present study was performed by using an online Nano LC Ultra system coupled with a Q Exactive mass spectrometer.

1. Load sample fractions into a 20-cm column with 125 μm inner diameter, packed in-house with 3 μm C_{18} particles. Reverse phase chromatography is performed with a binary buffer system consisting of buffer I (0.1 % formic acid) and buffer II (0.1 % formic acid with 80 % acetonitrile). The peptides are separated by a linear gradient of buffer II up to 95 % for 80 min with a flow rate of 200 nl/min.
2. Xcalibur 2.2 software (Thermo Fischer Scientific) was used for the data acquisition and the Q Exactive [21] was operated in the data-dependent mode. Survey scans were acquired in the Orbitrap mass analyzer over the range of 380–2,000 *m/z* with a mass resolution of 70,000 (at *m/z* 200). The maximum ion injection time for the survey scan was 80 ms with a 3e8 automatic gain-control target ions. MS/MS spectra was performed selecting up to the ten most abundant isotopes with a charge state > or = than 2 within an isolation window of 2.0 *m/z*. Selected isotopes were fragmented by higher energy collisional dissociation with normalized collision energies of 27 and the tandem mass spectra was acquired in the Orbitrap mass analyzer with a mass resolution of 17,500 (at *m/z* 200). The maximum ion injection time for the MS/MS scans were 120 ms with a 1e5 automatic gain-control target ions. Repeat sequencing of peptides was kept to a minimum by dynamic exclusion of the sequenced peptides for 30 s.

3.9 Database Searching and Protein Quantification

Raw MS/MS spectra was searched by using SEQUEST software on a SORCERER system using a composite yeast protein database, consisting of both the normal yeast protein sequences and their reversed protein sequences as a decoy to estimate the false discovery rate in the search results.

1. The following parameters were used in the database search: Semi-tryptic requirement, a mass accuracy of 15 ppm for the precursor ions, differential modification of 8.0142 Da for lysine and 10.00827 Da for arginine. Finally, results were filtered based on Probability Score to achieve a 1 % false positive rate.
2. XPRESS software was used to quantify all the identified peptides.
3. In the first step of the analysis the specific interacting partners of Rtt107-3xHA are identified by abundance ratios significantly higher than 1 while co-purified background proteins exhibit abundance ratios of around 1 (*see* Fig. 2a). In the second step of the analysis the Mec1-dependent interacting partners of Rtt107-3xHA are identified by abundance ratios significantly higher than 1 while Mec1-independent interactions exhibit abundance ratios of around 1 (*see* Fig. 2b).

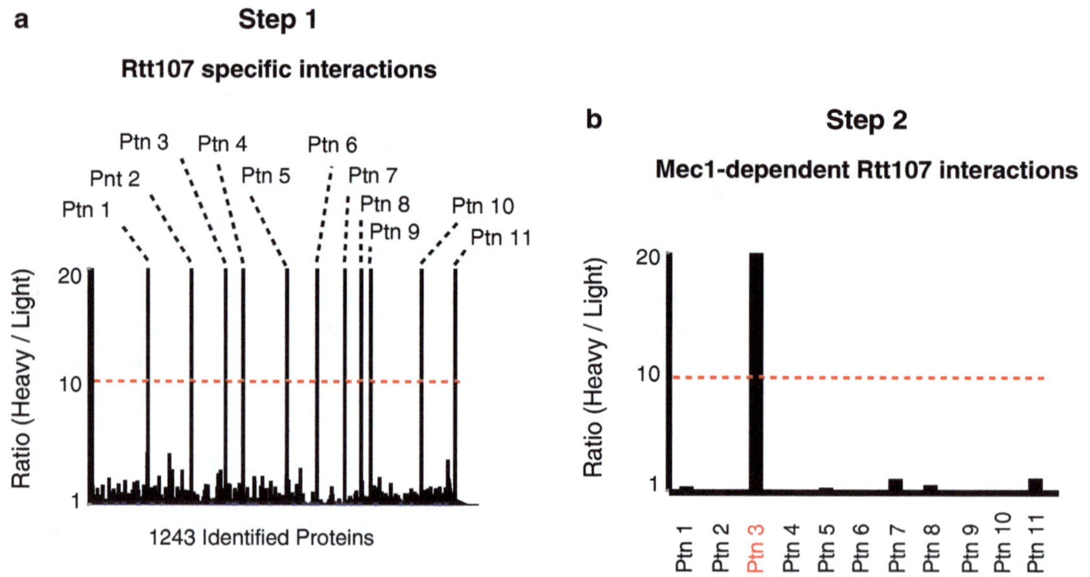

Fig. 2 Results from experiments depicted in Fig. 1. (**a**) Ratios of Rtt107-3xHA IP (*heavy*)/Rtt107 untagged IP (*light*) for the 1,243 proteins identified. Each ratio corresponds to the geometric mean of the ratios from all peptides identified and quantified for each protein. Eleven proteins with ratios greater than 10 were considered to be specific Rtt107-interacting proteins. (**b**) Ratios of *mec1*Δ IP (*light*)/wt IP (*heavy*) for the specific Rtt107-interacting proteins identified. Ratio for each protein was calculated by averaging the ratios of all corresponding peptides identified and quantified in the analysis. The protein Dpb11, assigned here as protein 3 (in *red*), was identified with a heavy-to-light ratio greater than 10 and considered to be a Mec1-dependent Rtt107-interacting protein

4 Notes

1. Strains MBS164, MBS266 and MBS402 are SILAC auxotrophic strains for Lysine and Arginine and are available upon request. These strains carry a deletion of the ribonucleotide reductase inhibitor *SML1* which suppresses the lethality of *mec1Δ* allele [22].
2. Proline is added to prevent the metabolic conversion of heavy arginine to heavy proline.
3. Add PMSF, EDTA-free protease inhibitor cocktail and phosphatase inhibitor to cell lysis buffer right before proceeding to cell lysis.
4. Add DTT to protein elution buffer right before proceeding to protein elution.
5. Prepare a fresh alkylation solution for every experiment.
6. After the trypsin is diluted in 0.3 % acetic acid, keep small aliquots frozen at −80 °C.
7. To avoid sample contamination with polymers always use a glass syringe/glass pipette/glass graduated cylinder to dilute strong acids and high concentrated organic solvents.
8. The analytical column was generated by pulling capillary to 5 μm-ID tip. Reverse-phase particles were packed directly into the pulled column at 1,000 psi until 20 cm long. The column was further packed, washed and equilibrated at 1,000 psi. in buffer II followed by buffer I.
9. Use fresh yeast culture plates.
10. Avoid culture saturation in SD media ($OD_{600} > 0.5$). Once saturated SILAC strains will not grow well. If culture is saturated, redilute and let it grow for at least four generations before proceeding with the experiment.
11. MMS is extremely toxic and should be handle in a chemical fume hood. All glasses and bottles in contact with MMS should be washed with a 10 % sodium thiosulfate solution. Cell culture medium containing MMS should be inactivated by adding an equal volume of 10 % sodium thiosulfate.
12. This is the starting point for depletion of the Rtt107-3xHA. For other proteins the amount of HA resin and time of incubation should be empirically tested.
13. Washing repetitions should be empirically tested.
14. If too much IgG background is detected during the MS analysis, stop adding DTT to the elution buffer. In this case, add DTT to the sample after elution from beads.
15. This step is required to increase the efficiency of protein precipitation.

16. Wash the walls of the microcentrifuge tube carefully to remove all traces of detergent.
17. Avoid letting the C_{18} columns to dry.
18. If sample will not be fractionated immediately, it should be resuspended in 15 μl of water and kept at −80 °C. Add the 10 μl of 10 % formic acid and 60 μl of 99.9 % acetonitrile right before HILIC fractionation.

Acknowledgments

We thank Beatriz S. Almeida for technical support. M.B.S. is supported by grants from the National Institute of Health (R01-GM097272) and American Cancer Society (RSG-11-146-01-DMC) and F.M.B.d.O. is supported by a Cornell Fleming Research Fellowship.

References

1. Hartwell LH, Kastan MB (1994) Cell cycle control and cancer. Science 266(5192): 1821–1828
2. Kolodner RD, Putnam CD, Myung K (2002) Maintenance of genome stability in Saccharomyces cerevisiae. Science 297(5581): 552–557. doi:10.1126/science.1075277
3. Nyberg KA, Michelson RJ, Putnam CW, Weinert TA (2002) Toward maintaining the genome: DNA damage and replication checkpoints. Annu Rev Genet 36:617–656. doi:10.1146/annurev.genet.36.060402.113540
4. Zhou BB, Elledge SJ (2000) The DNA damage response: putting checkpoints in perspective. Nature 408(6811):433–439. doi:10.1038/35044005
5. Foiani M, Pellicioli A, Lopes M, Lucca C, Ferrari M, Liberi G, Muzi Falconi M, Plevani P (2000) DNA damage checkpoints and DNA replication controls in Saccharomyces cerevisiae. Mutat Res 451(1–2):187–196
6. Tercero JA, Diffley JF (2001) Regulation of DNA replication fork progression through damaged DNA by the Mec1/Rad53 checkpoint. Nature 412(6846):553–557. doi:10.1038/35087607
7. Branzei D, Foiani M (2006) The Rad53 signal transduction pathway: replication fork stabilization, DNA repair, and adaptation. Exp Cell Res 312(14):2654–2659. doi:10.1016/j.yexcr.2006.06.012
8. Labib K, De Piccoli G (2011) Surviving chromosome replication: the many roles of the S-phase checkpoint pathway. Philos Trans R Soc Lond B Biol Sci 366(1584):3554–3561. doi:10.1098/rstb.2011.0071
9. Keogh MC, Kim JA, Downey M, Fillingham J, Chowdhury D, Harrison JC, Onishi M, Datta N, Galicia S, Emili A, Lieberman J, Shen X, Buratowski S, Haber JE, Durocher D, Greenblatt JF, Krogan NJ (2006) A phosphatase complex that dephosphorylates gamma-H2AX regulates DNA damage checkpoint recovery. Nature 439(7075):497–501. doi:10.1038/nature04384
10. Ohouo PY, Bastos de Oliveira FM, Almeida BS, Smolka MB (2010) DNA damage signaling recruits the Rtt107-Slx4 scaffolds via Dpb11 to mediate replication stress response. Mol Cell 39(2):300–306. doi:10.1016/j.molcel.2010.06.019
11. Polo SE, Jackson SP (2011) Dynamics of DNA damage response proteins at DNA breaks: a focus on protein modifications. Genes Dev 25(5):409–433. doi:10.1101/gad.2021311
12. Mohammad DH, Yaffe MB (2009) 14-3-3 proteins, FHA domains and BRCT domains in the DNA damage response. DNA Repair (Amst) 8(9):1009–1017. doi:10.1016/j.dnarep.2009.04.004
13. Williams RS, Bernstein N, Lee MS, Rakovszky ML, Cui D, Green R, Weinfeld M, Glover JN (2005) Structural basis for phosphorylation-dependent signaling in the DNA-damage response. Biochem Cell Biol 83(6):721–727. doi:10.1139/o05-153

14. Smolka MB, Albuquerque CP, Chen SH, Zhou H (2007) Proteome-wide identification of in vivo targets of DNA damage checkpoint kinases. Proc Natl Acad Sci U S A 104(25):10364–10369. doi:10.1073/pnas.0701622104
15. Chen SH, Albuquerque CP, Liang J, Suhandynata RT, Zhou H (2010) A proteome-wide analysis of kinase-substrate network in the DNA damage response. J Biol Chem 285(17):12803–12812. doi:10.1074/jbc.M110.106989
16. Matsuoka S, Ballif BA, Smogorzewska A, McDonald ER 3rd, Hurov KE, Luo J, Bakalarski CE, Zhao Z, Solimini N, Lerenthal Y, Shiloh Y, Gygi SP, Elledge SJ (2007) ATM and ATR substrate analysis reveals extensive protein networks responsive to DNA damage. Science 316(5828):1160–1166. doi:10.1126/science.1140321
17. Dunham WH, Mullin M, Gingras AC (2012) Affinity-purification coupled to mass spectrometry: basic principles and strategies. Proteomics 12(10):1576–1590. doi:10.1002/pmic.201100523
18. Ong SE, Blagoev B, Kratchmarova I, Kristensen DB, Steen H, Pandey A, Mann M (2002) Stable isotope labeling by amino acids in cell culture, SILAC, as a simple and accurate approach to expression proteomics. Mol Cell Proteomics 1(5):376–386
19. Tercero JA, Longhese MP, Diffley JF (2003) A central role for DNA replication forks in checkpoint activation and response. Mol Cell 11(5):1323–1336
20. Horvatovich P, Hoekman B, Govorukhina N, Bischoff R (2010) Multidimensional chromatography coupled to mass spectrometry in analysing complex proteomics samples. J Sep Sci 33(10):1421–1437. doi:10.1002/jssc.201000050
21. Michalski A, Damoc E, Hauschild JP, Lange O, Wieghaus A, Makarov A, Nagaraj N, Cox J, Mann M, Horning S (2011) Mass spectrometry-based proteomics using Q Exactive, a high-performance benchtop quadrupole Orbitrap mass spectrometer. Mol Cell Proteomics 10(9):M111.011015. doi:10.1074/mcp.M111.011015
22. Zhao X, Muller EG, Rothstein R (1998) A suppressor of two essential checkpoint genes identifies a novel protein that negatively affects dNTP pools. Mol Cell 2(3):329–340

Chapter 18

Application of Shotgun Proteomics for Discovery-Driven Protein–Protein Interaction

Livia Goto-Silva, Zoltan Maliga, Mikolaj Slabicki, Jimmy Rodriguez Murillo, and Magno Junqueira

Abstract

Affinity purification of protein complexes and identification of co-purified proteins by mass spectrometry is a powerful method to discover novel protein–protein interactions. Application of this method to the study of biological systems often requires the ability to process a large number of samples. Hence, there is great need to generate proteomic workflows compatible with large-scale studies. The major goal of this protocol is to present a fast, reliable, and scalable method to characterize protein complexes by mass spectrometry to overcome the limitations of conventional geLC-MS/MS or MudPIT protocols. This method was successfully employed for the discovery and characterization of novel protein complexes in cultured yeast, mammalian cells, and mice.

Key words Shotgun proteomics, In-solution digestion, IP-LC-MS/MS, Affinity purification, High-throughput screens, Bacterial artificial chromosome, TransgeneOmics

1 Introduction

Protein–protein interactions are crucial for virtually every biological process. Our current knowledge of biomolecular interactions in terms of cataloging complexes and understanding their physiological properties is still limited by throughput and reproducibility of current technologies. The combination of affinity purification of native protein complexes and use of mass spectrometry (MS) for protein identification is one of the most effective methods to study protein interaction [1]. Immunopurification (IP) allows for the enrichment of complexes in natura, and recent developments in MS-based proteomics technology [2] can increase the sensitivity and speed of the analysis, allowing for its application in large-scale studies [3].

A large number of immunopurifications and mass spectrometry experiments are required to systematically characterize the multiple protein complexes required for any biological function.

Daniel Martins-de-Souza (ed.), *Shotgun Proteomics: Methods and Protocols*, Methods in Molecular Biology, vol. 1156, DOI 10.1007/978-1-4939-0685-7_18, © Springer Science+Business Media New York 2014

Several factors increase the number of samples to be analyzed by MS for characterization of a single protein complex. (1) First, in order to prove the specificity of interactions in IP experiments, several reverse pull downs from a few independent "bait-entry" points should be performed in order to get a full picture of the interactome (especially if one is dealing with sub-stoichiometric interactors and hub proteins). (2) Second, since most of the post translational modifications (PTMs) of proteins within a complex depends on the biochemical context (i.e., cell cycle stage, genetic background, growth conditions, among others), the IP of a protein complex in several contexts may be needed in order to correlate biological context with PTMs and possible new interaction partners. (3) In addition, affinity purification eluates may come in a wide range of elution volumes, and hence, some may be diluted or in a buffer which introduces sample-to-sample variability. Consequently, the matching analytical approach should offer high throughput and scalability, low femtomolar sensitivity, and high dynamic range.

Here, we describe a workflow for high-throughput identification of protein complexes to complement functional genomics screens [3–9]. This method relies on enzymatic digestion of affinity purified proteins (IPs) directly in large volumes of affinity column eluates, rapid off-line concentration, and cleanup of peptides and accurate protein identification in a single LC-MS/MS run on a LTQ Orbitrap instrument. Our IP-LC-MS/MS strategy relies on the concept that, assuming moderate complexity of affinity purified complexes, they can be accurately dissected in a single, long gradient on a LTQ Orbitrap.

A critical feature of this strategy includes sorting out unspecific background noise and recognizing unique potential true interactors. This step is essential to the success of any IP-MS analysis because it is reported that background proteins of diverse origin represent a large fraction of affinity purified proteins [10, 11]. Therefore, application of bioinformatics filtering, which are described in the literature, are critical to discriminate true interactors from background [11–18].

The method described below has been applied successfully to identify protein complexes in human and mouse cell lines in studies investigating cell division, DNA repair, and intracellular transport.

2 Materials

2.1 Immunoprecipitation Components

Make sure to include negative controls for the IP reaction (*see* **Note 1**), which will be critical for background filtering.

1. Protein A (for mouse antibodies) or Protein G (for goat and rabbit antibodies) Sepharose Fast Flow.
2. PBS-T, 10 mM sodium phosphate, 0.15 M NaCl, 0.05 % Tween-20, pH 7.5.
3. Borate buffer, 200 mM sodium borate, pH 8.0.
4. Anti-GFP antibody (*see* **Note 2** for alternatives).
5. DMP (dimethyl pimelimidate), 220 mM solution in water.
6. Ethanolamine buffer: 200 mM ethanolamine hydrochloride in water.
7. Trypsin, 0.05 % solution, cell culture grade.
8. Protease inhibitor mix. Leupeptin, chymostatin, and pepstatin 10 mg/ml each dissolved in DMSO.
9. PMSF (phenylmethanesulfonylfluoride), 100 mM solution in ethanol (*see* **Note 3**).
10. Benzonase, 25 units/μL solution in 50 % glycerol containing 20 mM Tris–HCl, pH 8.0, 2 mM $MgCl_2$, and 20 mM NaCl.
11. Lysis buffer, 75 mM HEPES, 150 mM KCl, 1.5 mM $MgCl_2$, 15 % glycerol, 1.5 mM EGTA, 0.075 % NP-40, 1 mM DTT, 1 mM PMSF, 1:10,000 dilution protease inhibitor mix, pH 7.4.
12. Wash buffer A, 50 mM HEPES, 150 mM KCl, 1 mM $MgCl_2$, 10 % Glycerol, 1 mM EGTA, 0.05 % NP-40, 1 mM DTT, 1 mM PMSF, 3 μg/ml protease inhibitor mix, pH 7.4.
13. Wash buffer B, 50 mM HEPES, 150 mM KCl, 1 mM $MgCl_2$, 1 mM EGTA, 1 mM DTT, pH 7.4.

2.2 MS Components

1. **Note 4** outlines general precautions and guidelines on handling chemicals, reagent solutions and plasticware.
2. Water and acetonitrile (LC-MS grade).
3. Formic acid (FA) and trifluoroacetic acid (TFA) (HPLC grade).
4. Modified porcine trypsin (*Trypsin Gold* grade).
5. UltraMicroSpin-C18 (The Nest Group).
6. Solvent A: 95:5 H_2O:ACN (v/v) containing 0.1 % formic acid.
7. Solvent B: 20:80 H_2O:ACN (v/v) containing 0.1 % formic acid.

2.3 Equipment and Software

1. Ultimate 3000 nanoLC system from Dionex equipped with two pre-columns (5 mm × 300 μm ID, packed with C18 PepMAP100 5 μm particles) and analytical column (15 cm × 75 μm ID, packed with C18 PepMAP100 3 μm particles) (*see* **Note 5**).

2. Robotic nanoflow ion source TriVersa from Advion BioSciences Ltd. equipped with a LC coupler (*see* **Note 6**).
3. LTQ Orbitrap Velos mass spectrometer from Thermo Fisher Scientific.
4. NanoLC interfaced online to LTQ Orbitrap hybrid mass spectrometer (Thermo Fisher Scientific) via a robotic nanoflow ion source TriVersa (Advion BioSciences Ltd.) equipped with a LC coupler and a nanoESI chip with the diameter of spraying nozzles of 4.1 μm.
5. The TriVersa source operated at the ionization voltage of 1.8 kV and spraying stability monitored by total electrospray current reported by Chipsoft 6.4. software (Advion Biosciences).
6. Vacuum centrifuge.
7. Benchtop centrifuge.
8. MASCOT software from Matrix Science Ltd.

3 Methods

3.1 Protein Complexes Purification: Cross-linking Antibodies to Sepharose Resin

This protocol describes the coupling of rabbit anti-GFP antibody to G Sepharose resin, make sure to match A/G Sepharose resin to the correct antibody species.

1. For one immunoprecipitation experiment use 100 μl of settled G Sepharose 4 Fast Flow resin. Wash resin three times with 1 ml PBS-T. After every wash, pellet resin in a benchtop centrifuge for 1 min, 1,000 × *g*.
2. Resuspend resin in 500 μl PBS-T and incubate with rotation with 100 μg anti-GFP antibody at room temperature. Wash three times with PBS-T to remove unbound antibodies.
3. Wash resin with sodium borate buffer in order to change pH. Cross-link antibody to resin by adding 100 μl freshly prepared 220 mM DMP dissolved in 900 μl of borate buffer.
4. Incubate with rotation for 30 min at room temperature. Stop the reaction by washing twice with 1 ml ethanolamine buffer. Incubate in 1 ml ethanolamine for 1 h with rotation at room temperature.
5. Wash resin five times with 1 ml wash buffer A. Resuspend resin in 100 μl of wash buffer A. Resin coupled to antibody can be stored at +4 °C for up to 3 months.

3.2 Protein Complexes Purification: Cell Extract Preparation

1. Harvest cells (4 × 1E8 HeLa cells for a typical immunoprecipitation experiment). Construction of cell lines to which this protocol has been applied is described in **Note 7**. Adherent cells may be harvested using 0.05 % trypsin or a cell scraper.

The amount of cells required varies depending on the abundance of the target protein.

2. Rinse harvested cells 3 times with 100 ml ice-cold PBS and once with 40 ml wash buffer B. Resuspend cell pellet in 2 ml wash buffer B and shock-freeze in liquid nitrogen. Frozen cell pellet can be stored at −80 °C.
3. Thaw cell pellet in 37 °C water bath and immediately after thawing place it at 4 °C. All subsequent immunopurification steps are performed at 4 °C.
4. Add 1:1 volume of lysis buffer supplemented with 90 U/ml benzonase (3.6 μl/ml) to wet cell pellet (4 × 108 HeLa cells correspond to 2 ml wet cell pellet). Benzonase is used to digest genomic DNA and RNA and avoid sample aggregation.
5. Homogenize cells on ice using a Dounce homogenizer.
6. Remove insoluble fraction by short centrifugation (10 min, 2,000 × *g*). Recover supernatant for immunoprecipitation.

3.3 Protein Complexes Purification: Immunopurification

All steps performed on ice or in equipment set to 4 °C.

1. Wash 100 μl of anti-GFP antibody coupled to G Sepharose matrix 3 times with wash buffer A.
2. Incubate for 1 h with the soluble protein lysate.
3. Remove unbound supernatant from the resin by centrifugation for 5 min, 500 × *g* in a benchtop centrifuge.
4. Wash resin three times with 1 ml (each time) of wash buffer A and two times with wash buffer B. Pellet resin by centrifugation at full speed in a benchtop centrifuge each time.
5. Elute protein complexes in two steps of incubation with 50 μl Glycine (0.1 M, pH 2.0) for 5 min at room temperature. Neutralize supernatant with 10 μl 1.5 M Tris–Cl pH 9.0. Monitor pH using pH test paper strip (should be around pH 8.0).
6. Combine the two eluted fractions containing purified protein complexes, snap-freeze in liquid nitrogen, and store at −20 °C prior to mass spectrometry analysis.

3.4 Mass Spectrometry Identification of Protein Complex Subunits

1. Digest purified protein complexes with modified porcine trypsin (Promega) at final concentration of 16 ng/μl overnight at 39 °C directly in the eluate of last step immunopurification.
2. Stop the reaction by adding 0.1 % formic acid (final concentration).
3. Concentrate and desalt tryptic peptides off-line using UltraMicroSpin-C18 cartridge following instructions of manufacturer except that peptides are eluted by two steps of 75 μl of 80:20 acetonitrile–water with 0.1 % TFA. If the loading

volume of the sample is more than 200 μl several cycles of loading and re-equilibration might be applied using a maximum volume of 200 μl for each cycle.

4. Dry peptides eluted from the cartridge in a SpeedVac. Store dried peptides at −20 °C.
5. Re-dissolve peptides in 10 μl of 0.5 % FA. Load 4 μl of the peptide mixture in Ultimate 3000 nanoLC system (Dionex) equipped with two parallel trap columns (5 mm × 300 mm i.d. trapping micro-column packed with C18 PepMAP100 5 μm particles, Dionex) working at flow rate of 20 μl/min.
6. Wash the trap column for 5 min with solvent A. Elute peptides onto a 15 cm × 75 μm i.d. nano-column, packed with C18 PepMAP100 3 μm particles (Dionex).
7. Use the following elution gradient for direct analysis in-solution: from 5 to 20 % of solvent B in 100 min, 20–50 % B in 70 min, 50–100 % B in 5 min, 100 % B during 10 min, and back to 5 % B in 10 min (*see* **Note 8**).
8. Aquire MS/MS in data-dependent acquisition (DDA) mode under the control of Xcalibur 2.0 software (Thermo Fisher Scientific). The DDA cycle consists of a survey scan acquired in one μscan within the range of m/z 300–1,600 performed under the target mass resolution of 60,000 FWHM (full width at half-maximum) at Orbitrap analyzer under the target value of $1E^{06}$ ions. Apply MS/MS fragmentation to the ten most abundant multiply charged precursors. MS/MS spectra are acquired in one microscan under the normalized collision energy of 35 % and target value of $2E^{04}$ ions in the linear ion trap (ion selection threshold 400 counts; precursor ions isolation with 4 amu). Use activation parameter $q = 0.25$ and activation time of 10 ms. Previously fragmented precursors are dynamically excluded for 90 s.
9. Save LC-MS/MS runs as .raw files and convert to .mgf (MASCOT generic format) using ms2mz free software (http://www.bioproximity.com).
10. Search .mgf files against protein sequences using The Universal Protein Resource (UniProt) [19] with MASCOT software—(Matrix Science Ltd., London, UK) or against nrNCBI in case of other organisms.
11. Use tolerances for precursor and fragment masses of 10 ppm and 0.6 Da, respectively. Allow up to one missed cleavage. Setting for instrument profile: ESI-Trap. Add variable modifications: oxidation of methionine, acetylation of the N-terminal peptide, and phosphorylation of serine, threonine, and tyrosine or others specific to the sample analyzed. We consider confident protein identifications by MASCOT with a minimum ion score

of 20 for each peptide, each peptide has to be the first hit against database, at list two peptides per protein and 1 % FDR calculated by using a decoy database.

12. MASCOT will retrieve a list of protein identifications. Filtering strategies may be applied (*see* **Note 9**).
13. Protein complexes can be validated using reciprocal immunopurification. Selected hits are tagged within the BAC with GFP and immunopurified for back validation. Consistent analysis should demonstrate that the former immunopurification bait can be retrieved from the pull down of the prey (*see* example). Further validation can be performed using orthogonal methods such as fluorescence microscopy for subcellular localization and functional assays.

3.5 Example Analysis

High-throughput screens can provide unbiased discovery of novel protein functions. Here, we present two examples in which IP-LC-MS was applied to high-throughput screenings enabling association of previously uncharacterized genes with novel cellular processes and pathways. Protein fusions to GFP were performed inserting the genes of interest in bacterial artificial chromosomes containing a LAP tag using the TransgeneOmics method [20]. This approach enables expression of the tagged gene nearly to its physiological level from endogenous regulatory elements. Protein complexes were recovered by IP using anti-GFP antibody and analyzed by LC-MS/MS (IP-LC-MS/MS).

3.5.1 Example 1

The first example demonstrates the use of IP-LC-MS as a complementary orthogonal strategy following an esiRNA screening for homologous recombination DNA double-strand break repair. Previously uncharacterized gene KIAA0415 was identified as one of the hits in this high-throughput screening [8]. The aim was to apply the concept of "guilty-by-association" to uncover the function of KIAA0415. KIAA0415 was tagged with the LAP tag within a BAC and protein complexes recovered by IP. In-gel and in-solution digests of KIAA0415 IPs revealed a previously not reported molecular complex of the bait with SPG11, SPG15, C20orf29, and DKFZp761E198. This complex was further confirmed by reciprocal IP mass spectrometry analysis of C-terminally tagged SPG11, SPG15, and DKFZp761E198 (Fig. 1). Direct comparison of mass spectrometry analysis of in-gel and in-solution digest of KIAA0415 and SPG11 confirmed that in-solution digestion brings better coverage of putative interactors and less abundant background [8]. Two newly discovered interactors of KIAA0415: SPG11 and SPG15 were associated with hereditary spastic paraplegia with thin corpus callosum. This led us to discover novel mutations in KIAA0415 in two patients with hereditary spastic paraplegia. Hence, this example demonstrates the application of

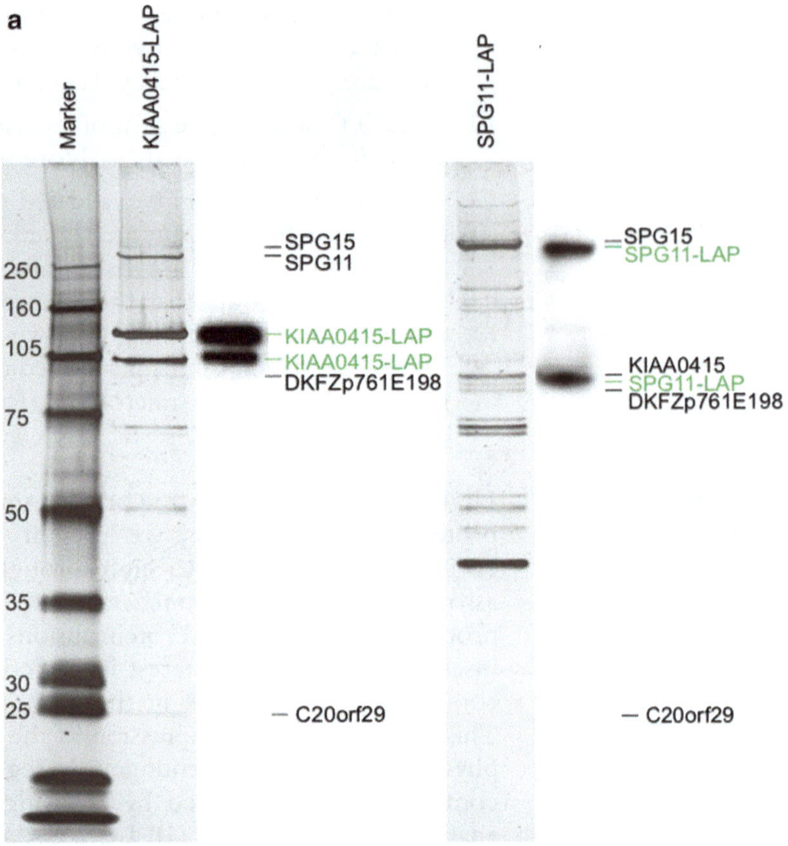

b

		Interaction partner or bait: matched peptides/sequence coverage				
		KIAA0415	SPG11	SPG15	C20orf29	DKFZp 761E198
BAIT	KIAA0415 LAP	**27/43%**	47/20%	43/18%	12/63%	11/20%
	KIAA0415 NFLAP	**18/15%**	10/5%	3/1%	6/39%	6/9%
	SPG11 LAP	20/32%	**65/31%**	40/18%	10/53%	11/18%
	SPG15 LAP	18/23%	45/21%	**59/30%**	9/52%	10/18%
	DKFZp761E198	14/22%	32/15%	36/16%	6/38%	**14/22%**

Fig. 1 Previously uncharacterized protein: KIAA0415 interacts with SPG11, SPG15, DKFZp761E198, and C20orf29. (**a**) SDS-PAGE gels and western blot analysis obtained from the immunoprecipitation of KIAA0415-LAP and SPG11-LAP. Baits (confirmed by western blot analysis, are marked in *green*) and prey (marked in *black*) were identified both by in-gel digestion and nanoLC-MS/MS analysis. (**b**) The composition of KIAA0415 protein complex analyzed as established by shotgun-LC-MS/MS. The number of matched detected peptides and protein sequence coverage are shown. Results for bait proteins are marked in *bold*. Figure modified from [8]. Antibodies: mouse anti-GFP (Roche Diagnostics, 1:4,000 dilution) mouse anti-DM1alpha tubulin (MPI-CBG Antibody Facility, 1:50,000 dilution)

IP-LC-MS/MS in functional assessment of previously uncharacterized protein KIAA0415 to inherited neurological disorders.

3.5.2 Example 2

In the second example, tandem co-IP MS analysis was used as a primary discovery tool to identify protein binding partners for actin and microtubule-based motors uncovering general principles of the regulation of intracellular transport and cell division [3]. Below we describe a background subtraction strategy employed in this screen and biochemical validation of one protein complex by reciprocal co-immunoprecipitation.

Critically, a common recombinant tagging strategy for a structurally related superfamily of proteins permitted background subtraction by cross-comparison of samples, not just control cell lysates, to uncover motor-selective interactors. To this end, we analyzed a collection of 63 HeLa cell lines stably expressing motor proteins tagged at their N or C-termini and recovered these proteins from cell lysates treated to induce distinct cell cycle states using anti-GFP affinity purification. Originally, we included lysates from the parental HeLa line (negative 1) and HeLa expressing GFP (negative 2) as background controls. Analysis of our MASCOT output files revealed that out of 222 affinity purifications, 20 did not detect the recombinant tag or bait protein and another 13 detected GFP-derived peptides but not the expected bait, consistent with recovery of protein fragments. Spectra from these negative AP-MS experiments were included as negatives 1 and 2, respectively, and used to identify "background" proteins. Next, we set a detection threshold of at least two peptides unique for any prey protein and a maximum of five interacting motors. After both steps in this analysis, we defined 154 prey proteins that defined 204 pairwise interactions [3].

Below, we present previously unpublished analysis in which we identified SHCBP1 and MICAL3 as novel components of the centralspindlin (KIF23-RACGAP1) complex and further characterized this interaction across the cell cycle by AP-MS of tagged SHCBP1 and RACGAP1 from stably transfected HeLa cells. First, cells expressing tagged KIF23 or RACGAP1 were released from S-phase arrest and harvested for AP-MS of GFP-associated proteins, revealing these two proteins bind SHCBP1 avidly (Table 1 panel A). Next, HeLa expressing tagged SHCBP1 was treated with protocols to enrich for mitotic cells or to arrest them using either of two independent pharmacologic approaches. In each case, the core complex emerged containing KIF23, RACGAP1, MICAL3, and SHCBP1 (Table 1 panel B).

Table 1

Reciprocal affinity purification across multiple cell cycle states. This data represents the combined MASCOT scores for peptides derived from HeLa expressing the indicated proteins transgene. (A) Reciprocal IP-MS from HeLa cells expressing tagged versions of the two known centralspindlin complex components (KIF23, a kinesin, and RACGAP1, an accessory protein). Cells were previously synchronized in S-phase, released into medium and harvested at 7 h (late G2, early metaphase), 10 h (mixed mitosis from prometaphase to cytokinesis) and 12 h (predominantly post-metaphase cells). The combined MASCOT score for proteins selective for KIF23 are listed. (B) SHCBP1-LAP was affinity purified from pharmacologically mitotically synchronized cells, as follows: untreated, condition 1: 10 h after S-phase block (thymidine) washout, condition 2: 18 h 500 ng/ml nocodazole, condition 3: 18 h 2 μM S-trityl-L-cysteine. Peptides from the affinity tag, EGFP, are indicated. The proposed tetra-protein centralspindlin complex in HeLa cells is *boxed*

(A) Reciprocal immunoprecipitation of KIF23-LAP and RACGAP1-LAP after release from double-thymidine block						
Bait:	**KIF23**	**KIF23**	**KIF23**	**RACGAP1**	**RACGAP1**	**RACGAP1**
Protein	7 h	10 h	12 h	7 h	10 h	12 h
EGFP	216	260	316	394	102	371
KIF23	848	955	952	1,507	805	1,390
RACGAP1	560	734	561	930	697	1,121
SHCBP1	696	623	342	937	900	1,128
MICAL3	134	244	368	385	196	127
CD2AP	123	127	91	0	176	0
ECT2	0	0	0	0	59	0

(B) Immunoprecipitation of SHCBP1-LAP from multiple mitotically synchronized states				
Protein	**Untreated**	**Condition 1**	**Condition 2**	**Condition 3**
EGFP	387	546	370	520
KIF23	*1,096*	*1,571*	*1,664*	*1,146*
RACGAP1	*798*	*1,151*	*953*	*1,083*
SHCBP1	*1,063*	*1,313*	*1,153*	*1,016*
MICAL3	470	548	834	810
CD2AP	0	128	506	389
PLK1	0	0	273	406
ERC1	0	0	0	958

Condition 1: 10 h after thymidine washout, condition 2: 18 h 500 ng/ml nocodazole, condition 3: 18 h 2 μM S-trityl-cysteine

4 Notes

1. Three approaches can be applied to generate the negative control: (1) expressing the tag alone (without the bait) and performing IP in parallel; (2) using another antibody that does not target the tag; (3) using the wild-type cell (without the tag) to perform the negative control IP.
2. Antibodies other than anti-GFP can be coupled to Sepharose resin and used for immunoprecipitation, provided that appropriate controls are performed. Magnetic resin coupled to antibody for retrieval of protein complexes can be used instead of G Sepharose to minimize sample loss throughout multiple wash steps. Magnetic resin coupled to anti-GFP antibody is commercially available (Chromotek-GFP-Trap® magnetic resin).
3. PMSF has a short half-life time in aqueous solutions. Stock solutions in DMSO or anhydrous ethanol are more convenient, and should be stored at –20 °C. PMSF should be diluted into buffer immediately before use. Pefablock is a less toxic and more soluble and stable alternative for PMSF.
4. Most solvents and chemical reagents used in this protocol may produce harm to human health. Make sure that personal safety equipment is used. Wear gloves at all times and rinse them occasionally as they readily accumulate static charge and attract dust and pieces of hair and wool. Perform all operations in a laminar flow hood and use a dedicated set of pipettes, tips, tubes that should be stored in the hood in a dust-free environment. Do not use polymeric detergents (Twin, Triton, etc.) for cleaning flasks and glass plates for electrophoresis. To avoid contamination of samples with human and sheep keratins, all plastic ware (tubes, dishes, etc.) should be washed with methanol, water, and once again with methanol prior to use. Polymeric detergents (Tween, Triton, or NP40) should not be used for cleaning laboratory dishes and tools.
5. To minimize sample carryover within LC system, Ultimate 3000 was operated with two pre-columns in parallel. While one of the two pre-columns was used for desalting and concentration of the loaded sample, another idle pre-column was extensively washed using a separate gradient program at the higher flow rate [21].
6. TriVersa robotic nanoflow ion source used operating electrospray chips with 4.1 μm nozzle diameter and was controlled by Chipsoft 6.4 software (Advion Biosciences). Ionization voltage was set to 1.8 kV and spacing between the chip and ion transfer capillary opening of the LTQ mass spectrometer was maintained within 3–5 mm.

7. This protocol was applied to the immunopurification of proteins with LAP tag from BAC-transgenic HeLa-cells [20]. Bacterial artificial chromosomes (BACs) containing the human target genes were obtained from the BACPAC Resource Center (http://bacpac.chori.org). A modified version of localization and affinity purification (LAP) cassette [22] was inserted as a carboxy-terminal fusion using recombineering [23]. Isolated BAC DNA was transfected and selected for stable integration as described [20]. The large size of the BAC transgenes ensures the presence of most cis regulatory elements and results in expression that closely matches that of the endogenous gene. For control experiments we generated two cell lines (HP1beta-hum_NFLAPneg (ENSG00000108468) and TUBG1-hum_NFLAPneg (ENSG00000131462)) expressing only the cassette LAP by adding three stop codons after the tag so that the only expressed sequence is a result of the tag.
8. Prior to the analysis, check void volume of the nanoLC setup. To this end, we recommend to monitor UV trace of the solvent gradient (typically detected at $\lambda = 214$ nm) and pressure profile at the separation column, as described by Mitulovic et al. [24]. Spray stability can be monitored using total electrospray current reported by TriVersa source.
9. A background library could be produced by combining MS/MS spectra obtained several independent LC-MS/MS runs coming from negative control experiments in order to generate a background library of spectra from common contaminate proteins (for details and software for filtering see ref. [12]). It is recommended that the user maintains several background libraries tuned to specific proteomics experiments, such as immunoprecipitations with different tags. Different libraries can also be pooled together, if necessary.
10. Sometimes is recommended to run MASCOT searches against largest available sequence resource, even if full-length sequences of target proteins are not present in a database. Cross-species protein identification can be achieved by matching MS/MS spectra to several sequences that happen to be identical in the analyzed peptide and reference database entry from other species. The data file in .mgf format can be submitted in parallel to MASCOT search and batch de novo sequencing by PepNovo. Alternatively, MASCOT search could be run first and then .dta files corresponding to positive hits removed from the dataset by a script. Subsequently, the remaining .dta files can be submitted to de novo sequencing and MS BLAST searches for sequence-similarity identifications [25].

Acknowledgements

Z.M. was supported by the Sixth Framework Programme Integrated Project Mitocheck (LSHG-CT-2004-503464) and the Max Planck Society. M.J. was financially supported by the Conselho Nacional de Desenvolvimento Científico e Tecnológico (CNPq), grant no. 483642/2012-6 MCT/CNPq—Universal. The authors acknowledge Frank Buchholz, Andrej Shevchenko, Anthony Hyman, Mirko Theis, and Yusuke Toyoda for previous collaborations and fruitful discussion.

References

1. Kocher T, Superti-Furga G (2007) Mass spectrometry-based functional proteomics: from molecular machines to protein networks. Nat Methods 4(10):807–815
2. Zubarev R, Mann M (2007) On the proper use of mass accuracy in proteomics. Mol Cell Proteomics 6(3):377–381
3. Maliga Z, Junqueira M, Toyoda Y, Ettinger A, Mora-Bermúdez F, Klemm RW, Vasilj A, Guhr E, Ibarlucea-Benitez I, Poser I et al (2013) A genomic toolkit to investigate kinesin and myosin motor function in cells. Nat Cell Biol 15(3):325–334
4. Ding L, Paszkowski-Rogacz M, Nitzsche A, Slabicki MM, Heninger AK, de Vries I, Kittler R, Junqueira M, Shevchenko A, Schulz H et al (2009) A genome-scale RNAi screen for Oct4 modulators defines a role of the Paf1 complex for embryonic stem cell identity. Cell Stem Cell 4(5):403–415
5. Krastev DB, Slabicki M, Paszkowski-Rogacz M, Hubner NC, Junqueira M, Shevchenko A, Mann M, Neugebauer KM, Buchholz F (2011) A systematic RNAi synthetic interaction screen reveals a link between p53 and snoRNP assembly. Nat Cell Biol 13(7):809–818
6. Maffini S, Maia AR, Manning AL, Maliga Z, Pereira AL, Junqueira M, Shevchenko A, Hyman A, Yates JR, Galjart N et al (2009) Motor-independent targeting of CLASPs to kinetochores by CENP-E promotes microtubule turnover and poleward flux. Curr Biol 19(18):1566–1572
7. Matos J, Lipp JJ, Bogdanova A, Guillot S, Okaz E, Junqueira M, Shevchenko A, Zachariae W (2008) Dbf4-dependent CDC7 kinase links DNA replication to the segregation of homologous chromosomes in meiosis I. Cell 135(4): 662–678
8. Słabicki M, Theis M, Krastev DB, Samsonov S, Mundwiller E, Junqueira M, Paszkowski-Rogacz M, Teyra J, Heninger AK, Poser I et al (2010) A genome-scale DNA repair RNAi screen identifies SPG48 as a novel gene associated with hereditary spastic paraplegia. PLoS Biol 8(6):e1000408
9. Theis M, Slabicki M, Junqueira M, Paszkowski-Rogacz M, Sontheimer J, Kittler R, Heninger AK, Glatter T, Kruusmaa K, Poser I et al (2009) Comparative profiling identifies C13orf3 as a component of the Ska complex required for mammalian cell division. EMBO J 28(10):1453–1465
10. Krogan NJ, Cagney G, Yu H, Zhong G, Guo X, Ignatchenko A, Li J, Pu S, Datta N, Tikuisis AP et al (2006) Global landscape of protein complexes in the yeast Saccharomyces cerevisiae. Nature 440(7084):637–643
11. Shevchenko A, Roguev A, Schaft D, Buchanan L, Habermann B, Sakalar C, Thomas H, Krogan NJ, Shevchenko A, Stewart AF (2008) Chromatin central: towards the comparative proteome by accurate mapping of the yeast proteomic environment. Genome Biol 9(11):R167
12. Junqueira M, Spirin V, Santana Balbuena T, Waridel P, Surendranath V, Kryukov G, Adzhubei I, Thomas H, Sunyaev S, Shevchenko A (2008) Separating the wheat from the chaff: unbiased filtering of background tandem mass spectra improves protein identification. J Proteome Res 7(8):3382–3395
13. Gentzel M, Kocher T, Ponnusamy S, Wilm M (2003) Preprocessing of tandem mass spectrometric data to support automatic protein identification. Proteomics 3(8):1597–1610
14. Trinkle-Mulcahy L, Boulon S, Lam YW, Urcia R, Boisvert FM, Vandermoere F, Morrice NA, Swift S, Rothbauer U, Leonhardt H et al (2008) Identifying specific protein interaction partners using quantitative mass spectrometry and bead proteomes. J Cell Biol 183(2): 223–239

15. Schirle M, Heurtier MA, Kuster B (2003) Profiling core proteomes of human cell lines by one-dimensional PAGE and liquid chromatography-tandem mass spectrometry. Mol Cell Proteomics 2(12):1297–1305
16. Andersen JS, Wilkinson CJ, Mayor T, Mortensen P, Nigg EA, Mann M (2003) Proteomic characterization of the human centrosome by protein correlation profiling. Nature 426(6966):570–574
17. Mueller LN, Rinner O, Schmidt A, Letarte S, Bodenmiller B, Brusniak MY, Vitek O, Aebersold R, Muller M (2007) SuperHirn—a novel tool for high resolution LC-MS-based peptide/protein profiling. Proteomics 7(19):3470–3480
18. Rinner O, Mueller LN, Hubalek M, Muller M, Gstaiger M, Aebersold R (2007) An integrated mass spectrometric and computational framework for the analysis of protein interaction networks. Nat Biotechnol 25(3):345–352
19. Apweiler R, Bairoch A, Wu CH, Barker WC, Boeckmann B, Ferro S, Gasteiger E, Huang H, Lopez R, Magrane M et al (2004) UniProt: the Universal Protein knowledgebase. Nucleic Acids Res 32(Database issue):D115–D119
20. Poser I, Sarov M, Hutchins JR, Heriche JK, Toyoda Y, Pozniakovsky A, Weigl D, Nitzsche A, Hegemann B, Bird AW et al (2008) BAC TransgeneOmics: a high-throughput method for exploration of protein function in mammals. Nat Methods 5(5):409–415
21. Schaefer H, Chervet JP, Bunse C, Joppich C, Meyer HE, Marcus K (2004) A peptide preconcentration approach for nano-high-performance liquid chromatography to diminish memory effects. Proteomics 4(9):2541–2544
22. Cheeseman IM, Desai A (2005) A combined approach for the localization and tandem affinity purification of protein complexes from metazoans. Sci STKE 2005(266):l1
23. Zhang Y, Muyrers JP, Testa G, Stewart AF (2000) DNA cloning by homologous recombination in Escherichia coli. Nat Biotechnol 18(12):1314–1317
24. Mitulovic G, Stingl C, Smoluch M, Swart R, Chervet JP, Steinmacher I, Gerner C, Mechtler K (2004) Automated, on-line two-dimensional nano liquid chromatography tandem mass spectrometry for rapid analysis of complex protein digests. Proteomics 4(9):2545–2557
25. Waridel P, Frank A, Thomas H, Surendranath V, Sunyaev S, Pevzner P, Shevchenko A (2007) Sequence similarity-driven proteomics in organisms with unknown genomes by LC-MS/MS and automated de novo sequencing. Proteomics 7(14):2318–2329

Chapter 19

Mapping Protein Complexes Using Covalently Linked Antibodies and Isobaric Mass Tags

Antje Dittmann, Sonja Ghidelli-Disse, Carsten Hopf, and Marcus Bantscheff

Abstract

Affinity enrichment techniques in combination with quantitative proteomics enable the unbiased identification of protein–protein interaction, and thus the delineation of protein complexes and interaction networks. Here, we describe an immunoaffinity enrichment approach that employs covalently immobilized antibodies for the identification of protein–protein interactions of endogenously expressed proteins under near-to-physiological conditions. Specifically enriched proteins are identified using shotgun mass spectrometry and isobaric mass tag-based relative quantification.

Key words IP-MS, Isobaric mass tags, Mass spectrometry, Protein–protein interactions

1 Introduction

In cellular systems, function, activity, and localization of proteins is defined and regulated through interaction with other cellular components. Many proteins currently investigated by the pharmaceutical industry as targets for pharmacological modulation in disease elicit their activity as components of multiprotein complexes. Consequently, the analysis of protein complex components and cellular protein networks provides information on functional implications of the target protein, and has recently emerged as a critical tool in identifying a drug's or probe compound's mode of action [1, 2]. Immunoaffinity purification coupled to mass spectrometry analysis (IP-MS) is commonly the method of choice for the interrogation of physiologically relevant protein–protein interactions. Various immunoaffinity methods have been employed to dissect protein networks and interactions [3], and often include the use of (epitope-)tagged proteins and genetically engineered cell lines [4, 5]. Recent methodological developments have addressed issues of existing tagging strategies such as non-physiological protein

Daniel Martins-de-Souza (ed.), *Shotgun Proteomics: Methods and Protocols*, Methods in Molecular Biology, vol. 1156, DOI 10.1007/978-1-4939-0685-7_19, © Springer Science+Business Media New York 2014

localization and expression levels [6]. However, tagging may still perturb cellular protein interactions and can lead to unspecific protein binding during purification. Disadvantages inherent in using epitope tagging in overexpression systems are offset by using specific antibodies to selectively enrich endogenous proteins and associated entities. Prerequisites are no or little cross-reactivity with other cellular targets, and experimental and statistical methods to differentiate unspecific binding from true interactions especially for sub-stoichiometric and dynamic associations [7].

In typical experimental workflows, antibodies are first immobilized on a solid support followed by incubation with cell extracts, a number of washing steps, elution of bound proteins, proteolytic digestion, and identification of bound proteins using mass spectrometry (MS). Often, antibody immobilization involves the use of Protein A or Protein G beads which allows binding via the antibody Fc domain and thus correct antibody orientation for maximal antigen interaction. However, elution of antibody-bound proteins is usually only achieved under denaturing conditions, which leads to the co-elution of antibody heavy and light chains. Because of their abundance, these peptides can interfere with the identification of lower abundant proteins in the subsequent mass spectrometric analysis. Such effects can be limited by leak-resistant covalent antibody immobilization. Various polymeric bead matrices with different functional groups exist that allow the coupling to antibodies via primary amines (at peptide N-terminus and in lysine side chains), carboxyl-groups (at peptide C-terminus and inside chains of acidic amino acids), sulfhydryl groups of reduced cysteine disulfide bonds, or aldehydes of oxidized glycoconjugates (in antibody Fc domain).

A recurring challenge in one-step immunoaffinity purifications is the differentiation between unspecific and specific protein–protein interactions. Primary sources of false-positive interactions are non-physiological interaction with the target protein due to mixing of proteins from different compartments, unspecific binding of high-abundance proteins to the antibody matrix, antibody cross-reactivity, and unspecific binding of denatured proteins or protein aggregates. Consequently, much effort has been devoted to improve the identification of "true" protein interactions using experimental as well as statistical means. Experimental strategies include increasing buffer stringency (raising salt concentrations, using low concentrations of ionic detergents such as SDS and deoxycholate) and lysate pre-clearing approaches. However, such strategies inevitably impair the antibody-bait binding reaction, and preclude the identification of medium- to low-affinity interactors of the target protein. Other approaches aim at increasing stringency by introducing the requirement of reciprocity and/or the usage of different antibodies generated against the same cellular target [7]. Additionally, recent works have addressed this issue by using large complementary data sets that allow statistical evaluation of co-occurrences [8] or the generation of contaminant repositories [9].

For studies including only a small number of experiments statistical approaches benchmarking the frequency and abundance of individual proteins across a series of immunoaffinity purifications with different antibodies are not applicable or of limited use. In such studies, isotype-matched control antibodies are a powerful tool to differentiate proteins that bind unspecifically to the antibody affinity matrix at comparable levels in all samples from proteins enriched in target-specific IP-MS samples. For differential display and relative quantification of bound proteins a variety of quantitative proteomics approaches are available including label-free and stable isotope labeling-based methods [10]. Methods commonly applied for focused protein–protein interaction studies include stable isotope labeling of amino acids in cell culture (SILAC) [11] and isobaric mass tag-based chemical labeling approaches such as iTRAQ or TMT [12, 13] for the multiplexed analysis of up to eight samples [14].

Here we describe an immunoaffinity purification method that uses reductive amination for antibody immobilization on a sepharose matrix, followed by the binding reaction to endogenous proteins under non-denaturing conditions, and elution in LDS-containing buffer. Eluted proteins are digested with trypsin and peptide mixtures are encoded with isobaric mass tags enabling relative protein quantification by quantitative MS-analysis (Fig. 1). A specificity filter based on relative quantification in target-specific and control purification is used for the identification of specific protein–protein interactions, whereas inferences about overall protein abundance are based on spectral counts. Example data is provided which include the characterization of the two HDAC1 and 2 complexes Sin3 and CoREST (Fig. 2).

2 Materials

2.1 Cell Lysis

1. Dounce Tissue Grinder (Wheaton, Millville, NJ) or mechanized POTTER S (B. Braun, Biotech International).
2. Lysis buffer: 50 mM Tris–HCl, pH 7.5, 5 % (w/v) glycerol, 1.5 mM $MgCl_2$, 150 mM NaCl, 25 mM NaF, 1 mM sodium vanadate detergent (IGEPAL CA-630) at a concentration of 0.8 % (w/v), EDTA-free tablet, protease inhibitor cocktail (Roche Diagnostics) per 25 mL of lysis buffer, filter lysis buffer through a 0.22 μM filter.
3. Conical tubes.
4. UZ-polycarbonate tubes and suitable lids (Beckman Coulter).
5. End-over-end shaker (Roto Shake Genie, Scientific Industries, Inc.).

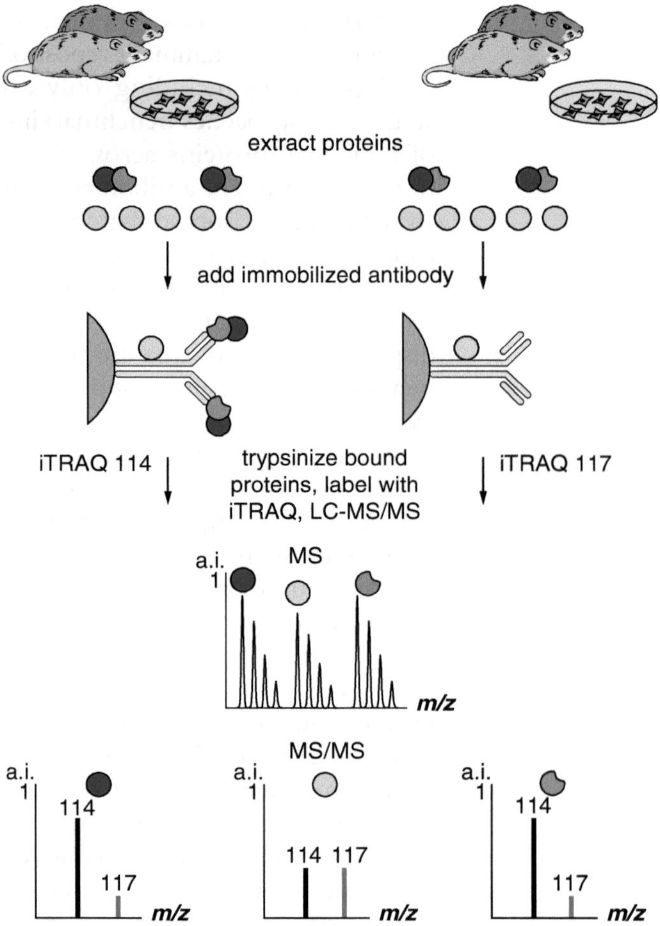

Fig. 1 Schematic representation of the immunoaffinity purification workflow

6. General purpose centrifuge with rotor suitable for 50 and 15 mL conical tubes, e.g., Heraeus Multifuge 3S-R.
7. Ultracentrifuge with matching rotor: e.g., Sorvall 90SE, Sorvall SLA600; rotor, e.g., Ti50.2.
8. Bradford reagent for protein determination.

2.2 Preparation of Antibody Matrix

1. Wash buffers: 1× PBS, 0.01 M Sodium hydroxide, 1 M Tris–HCl pH 7.4, 1 M NaCl.
2. Coupling reagents: Sodium cyanoborohydride, AminoLink Plus coupling resin (Thermo Fisher Scientific), target-specific antibody, same species control antibody of same clonality.
3. Fume hood.
4. Microtube centrifuge, e.g., Eppendorf 5415 D.
5. End-over-end shaker as in Subheading 2.1.

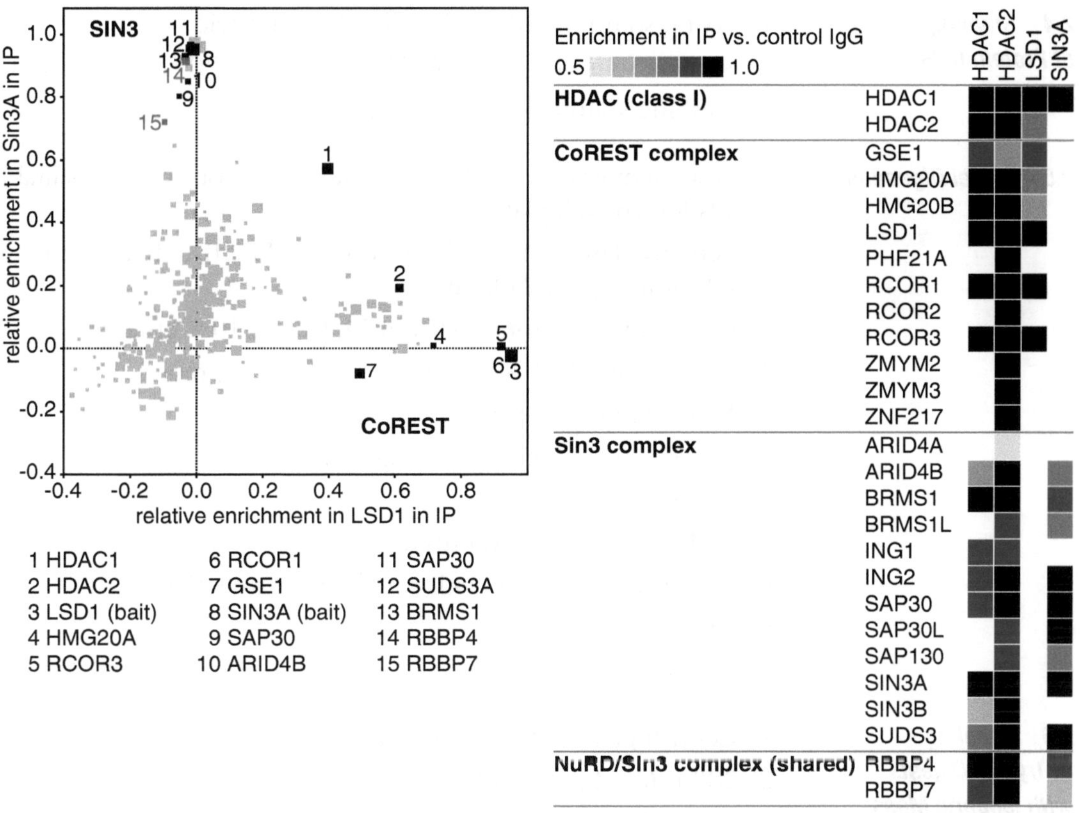

Fig. 2 Identification of specific interactors of histone deacetylases (HDACs) and HDAC complexes. (**a**) Relative protein enrichment in LSD1 and Sin3A immunoaffinity purifications. Symbols represent relative enrichment in the indicated purification based on reporter ion intensities. Symbol sizes correspond to binned spectral counts. Symbols in *dark gray* are known components of the indicated HDAC subcomplex CoREST or Sin3. (**b**) Heatmap representing relative protein enrichment levels in samples in (**a**) and in reciprocal immunoaffinity purifications using HDAC1 and HDAC2 as baits. Shown are proteins annotated as unique CoREST and Sin3 components as well as proteins shared by Sin3 and another HDAC complex NuRD

2.3 Immunoaffinity Purification Reagents

1. IP buffer: lysis buffer as in Subheading 2.1 with EDTA-free tablet, protease inhibitor cocktail.
2. Detergent as in Subheading 2.1.
3. Laboratory centrifuge as in Subheadings 2 and 2.2.
4. Orbital benchtop mixer for microtubes, e.g., Thermomixer (Eppendorf).
5. LDS sample buffer (Life Technologies).
6. 1 M Dithiothreitol (DTT), 1 M Iodoacetamide (IAA).
7. NuPAGE 4–12 % Bis–Tris gel, gel electrophoresis equipment (Life Technologies).

2.4 Staining of Protein Gels

1. Fixing solution: 40 % ethanol, 2 % acetic acid.
2. Staining solution: 16 % colloidal Coomassie, 20 % ethanol/H_2O.
3. Washing solutions: 25 % ethanol, 5 % acetic acid.

2.5 In-Gel Digestion

1. Low-binding v-shaped 96-well plates, polypropylene and suitable lids for 96-well plates.
2. Tetraethylammonium bicarbonate (TEAB), 1 M aqueous solution (Sigma-Aldrich).
3. Destaining solution: 60 % 5 mM TEAB, 40 % ethanol.
4. Acetonitrile, 100 % ethanol, formic acid suprapur.
5. Multichannel pipette.
6. Trypsin, mass spectrometry grade.
7. HPLC grade H_2O.
8. 1 M aqueous HCl solution.
9. Benchtop centrifuge with rotor for 96-well plates.
10. Laminar flow hood.
11. SpeedVac.

2.6 Labeling of Tryptic Digests with Isobaric Mass Tags

1. Centrifuge as in Subheading 2.3.
2. 200 mM TEAB, pH 8.5.
3. Stop solution: 100 mM TEAB/100 mM glycine, pH 8.5.
4. Formic acid solution, 20 %/H_2O (HPLC).
5. iTRAQ reagent multi-plex kit (Applied Biosystems).

2.7 Mass Spectrometry

1. Solution A: 0.1 % formic acid in water.
2. Solution B: 70 % ACN/29.9 % H_2O/0.1 % formic acid.
3. Nano-flow HPLC system, e.g., RLSC nano (Thermofisher Scientific), pre-column and separation column.
4. Ultrasonic bath.
5. Mass spectrometer equipped with online nano-spray ion source, e.g., Q Exactive or Orbitrap Elite (Thermofisher Scientific).

2.8 Data Analysis

1. Mascot™ 2.0 (Matrix Science).

3 Methods

3.1 Whole Cell Lysis

This protocol describes a generic cell lysis protocol for frozen cell pellets using 0.8 % (w/v) IGEPAL CA-630 as detergent (*see* **Note 1**).

1. Mix cell pellets with 3 volumes of lysis buffer/volume cell pellet.
2. Thaw cells in ice-cold lysis buffer at 4 °C on an end-over-end shaker.

3. Transfer suspension to pre-cooled Dounce Tissue Grinder with a tight pestle (alternatively mechanized POTTER S, B. Braun, Biotech International, speed setting: 300 min^{-1}).
4. Dounce 20× and transfer homogenate into pre-cooled 50 mL conical tubes.
5. Incubate for 30 min rotating at 4 °C.
6. Spin cells 10 min at 20,000 × *g* at 4 °C.
7. Transfer supernatant to UZ-polycarbonate tubes.
8. Spin supernatant 1 h at 140,000 × *g*.
9. Transfer supernatant into fresh 50 mL conical tube.
10. Determine protein concentration (Bradford) according to the manufacturer's instruction. Typical protein concentrations range between 5 and 20 mg/mL.
11. Freeze aliquots in liquid nitrogen and store at −80 °C.

3.2 Covalent Antibody Immobilization

This step describes the un-directional immobilization of monoclonal and polyclonal antibodies via primary amines regardless of antibody clonality (*see* **Notes 2** and **3**). Typical antibody amounts for default large scale immunoaffinity purification experiments vary between 50 and 100 μg per 100 μL AminoLink matrix. Caution: $NaCNBH_3$ is toxic and should be handled with care. All steps including this reagent should be performed under the fume hood. For each experiment at least one control and one target-specific antibody is used.

1. Transfer beads into microtube.
2. Wash AminoLink beads three times with 10× bead bed volume of 1× PBS, every time centrifuge for 2 min at 500 × *g* (microtube centrifuge) to collect beads.
3. Prepare 1 M $NaCNBH_3$ in 0.01 M NaOH.
4. Prepare antibody solution by adding 1× PBS to fill up to 200 μL if antibody volume is less than 200 μL or use as is.
5. Add respective amounts of antibody solution to the beads at a ratio of 2:1, i.e., 2 volumes antibody solution/bead bed volume.
6. Add 7.5 μL $NaCNBH_3$ to 300 μL antibody-bead mix (25 mM final concentration, *see* **Note 4**).
7. Incubate overnight at 4 °C, rotating.
8. Centrifuge for 2 min at 500 × *g* in microtube centrifuge, take off and discard supernatant.
9. Wash two times with 10× bead bed volumes of 1 M Tris–HCl pH 7.4, every time centrifuge for 2 min at 500 × g and discard the supernatant.
10. Add 1 M Tris–HCl pH 7.4 at a ratio of 2:1 to the antibody beads.

11. Add 7.5 μL $NaCNBH_3$ to 300 μL of the antibody bead slurry, and incubate for 30 min rotating at room temperature to block remaining reactive groups.
12. Centrifuge for 2 min at 500 × *g* in microtube centrifuge, take off and discard supernatant.
13. Wash three times with 10× bead bed volume of 1 M NaCl, every time centrifuge for 2 min at 500 × *g* and discard the supernatant.
14. Wash three times with 10× bead bed volume of IP buffer, every time centrifuge for 2 min at 500 × *g* and discard the supernatant.

3.3 Antibody Binding Reaction

1. Discard IP buffer after the last washing step.
2. This protocol uses 5 mg of protein for each binding reaction which allows the binding reaction to be carried out in microtubes (*see* **Note 5**).
3. Dilute cell lysate 1:1 with IP buffer without detergent to reduce the detergent concentration to 0.4 %.
4. Dilute lysate further with IP buffer with 0.4 % (w/v) detergent to a final protein concentration of 5 mg/mL.
5. Transfer lysate to UZ-polycarbonate tubes and ultracentrifuge 20 min at 140,000 × *g*.
6. Take off supernatant and discard pellet.
7. Add 1 mL of lysate to antibody beads (*see* **Note 5**).
8. Incubate for 2–3 h rotating at 4 °C (*see* **Notes 6** and 7).
9. Collect beads at 200 × *g*, 2 min and take off small sample of non-bound fraction (=NBF). Discard the remaining supernatant.
10. Wash two times with 10× bead bed volume of IP buffer with 0.4 % detergent, every time centrifuge for 2 min at 200 × *g* and discard supernatant (*see* **Note 8**).
11. Wash one times with 10× bead bed volume of IP buffer without detergent, every time centrifuge for 2 min at 200 × *g* and discard supernatant (*see* **Note 8**).
12. Collect beads at 200 × *g*, 2 min and discard residual wash buffer (*see* **Note 8**).
13. Add one time bead bed volume of LDS sample buffer to beads and incubate 10 min shaking at 95 °C in orbital shaker.
14. Centrifuge for 2 min at 500 × *g* and carefully take off eluate.
15. Transfer eluate into a fresh microtube and centrifuge for 2 min at 500 × *g* to separate carried over beads.
16. Transfer eluate into a fresh microtube, add 2 μL DTT (20 mM) and incubate in an orbital shaker for 5 min at 95 °C.

17. Transfer 45 μL of eluate into a new tube and add 4.5 μL iodoacetamide solution and incubate for 30 min at room temperature.
18. Spin samples for 5 min at maximum speed in microtube centrifuge.
19. Apply 45 μL of each sample onto a 1.5 mm precast gel.
20. Perform SDS gel electrophoresis according to the manufacturer's instructions.

3.4 Processing of Protein Gels

Since proteins are separated along the whole gel range, and digestion is carried out in small gel pieces from that range, an MS-compatible protein stain allows to better align size and positions of gel pieces of different sample lanes (e.g., target-specific and control antibody, *see* **Note 4**).

1. Remove gel from cassette and incubate for 1 h in 30 mL fixing solution.
2. Remove fixing solution and add 30 mL staining solution. Incubate overnight shaking at 4 °C.
3. Remove staining solution and wash several times with 20 % ethanol/2 % acetic acid.
4. Remove washing solution and wash several times with 25 % ethanol.
5. Remove ethanol and wash twice with H_2O.

3.5 In-Gel Digestion

1. Drill holes into the center of 96-well plates using a 0.4 mm drill.
2. Wash plates extensively using ~70 °C 3 % nitric acid (HNO_3) and water.
3. Cut the protein-containing part of each lane obtained in Subheading 3.3 into up to 24 pieces. Ensure that the size and cutting position of corresponding gel pieces are the same across the 2–4 lanes.
4. Place each slice into a separate well of the 96-well plate.
5. Incubate twice with 100 μL destaining solution at 55 °C for 60 min and discard liquid. Ensure that gel pieces are completely destained, repeat destaining step if necessary.
6. Add 100 μL 100 % ethanol to each gel piece and wait for 10 min at room temperature until gel pieces turn white. Centrifuge to remove supernatant.
7. Add 100 μL 5 mM TEAB and incubate for 10 min at room temperature. Gel pieces should turn colorless. Centrifuge to remove supernatant.
8. Repeat **steps 6** and **7** and then again **step 6**.
9. Dissolve Trypsin at 1 mg/mL in 1 mM HCl and dilute 20 μL in 1,980 μL 5 mM TEAB.

10. Add 25 μL of Trypsin working solution to each gel piece.
11. Incubate samples at 4 °C until they are colorless and swollen to their normal size (~10 min). Centrifuge to remove liquid.
12. Put a cleaned collector plate underneath the sample plate and add 20 μL 5 mM TEAB to each gel piece. Incubate in a plastic bag for 4 h at 37 °C.
13. Stop digestion with 5 μL 5 % formic acid.
14. Extract gel pieces twice with 20 μL 1 % formic acid for 30 min, and centrifuge to collect liquid.
15. Extract once with 20 μL 60 % acetonitrile/0.4 % formic acid for 30 min and centrifuge to collect liquid.
16. Extract once with 25 μL pure acetonitrile for 15 min and centrifuge to collect liquid.
17. Dry extracted peptides in collector plate in vacuo using a SpeedVac.

3.6 Labeling of Tryptic Digests with Isobaric Mass Tags

1. Thaw iTRAQ reagent at room temperature, dissolve in 120 μL ethanol, vortex and spin down.
2. Thaw sample plate at room temperature.
3. Resuspend samples in 15 μL of 200 mM TEAB, seal plate with foil and incubate 10 min shaking at 20 °C.
4. Add 15 μL ethanol to each sample and shake for 1–2 min.
5. Add 10 μL iTRAQ reagent of the respective color and incubate for 1 h shaking at 20 °C.
6. Add 5 μL stop solution to each sample and incubate for 15 min shaking at 20 °C.
7. Combine samples of corresponding gel pieces from all gel lanes labeled with different iTRAQ color.
8. Add 5 μL of 20 % formic acid.
9. Dry sample in SpeedVac and store at −20 °C.

3.7 LC-MS/MS Analysis

Appropriate instrumentation for mass spectrometric analysis of the samples generated in Subheading 3.6 is available at many institutions and core facilities. Many different instrument platforms are similarly well suited for this purpose. As an example the following section typical instrument settings for the currently very common Q-Exactive mass spectrometers will be described.

1. Resuspend samples in 10 μL 0.1 % formic acid/H_2O.
2. Inject 50 % of the sample into a nano-LC system coupled online to the mass spectrometer.
3. Peptides are separated on a custom-made 50 cm reversed phase columns (Reprosil) with 100 μm inner diameter using a linear

elution gradient from 2 to 40 % acetonitrile in 0.1 % formic acid within 2–3 h at a flow rate of 250 μL/min.

4. Peptide masses were detected at 50,000 resolution with an ion target value of 1E6 in MS mode and at 17,500 resolution with an ion target of 1E5 in MS/MS mode. Ion accumulation time was kept at 120 ms. Internal mass calibration was performed using the ion signal from polydimethylcyclosiloxane $(Si(CH_3)_2O)_6H+$ at m/z 445.120025 [15].
5. Intact peptide ions were isolated with isolation width set to 1.5 Th and HCD fragmentation was performed using 35 % normalized collision energy.

3.8 Data Analysis

At this stage the objectives are to (a) identify the proteins bound to the affinity matrix, (b) accurately quantify the captured amounts of these proteins in the target IP and the mock IP, and (c) determine those proteins that are specifically enriched in the target IP.

1. For protein identification tandem mass spectra are typically converted to peak lists and submitted to a database search against a nonredundant sequence database of the species of interest (*see* **Note 9**). A variety of free and commercial search engines such as Mascot [16, 17], Sequest [16], and Andromeda [18] are available for this purpose.
2. Enrichment of target proteins is then calculated using relative reporter ion intensities of matching high-confidence spectra and can be expressed as the fraction of target-specific signal in the summed up intensities in target-specific and control purification, and ranges between −1 and 1. Further statistical analysis can then be performed using standard statistics software packages, e.g., R [http://www.r-project.org/].

4 Notes

AU1]

1. The cell lysis protocol should be adapted depending on the subcellular localization (e.g., nuclear, cytoplasmic, membrane) of the protein(s) of interest. Buffer conditions may have to be optimized for individual interactions (type and concentration of detergent, salt concentration).
2. Some antibodies may not be amenable to this procedure since antigen recognition motifs may be modified during the coupling process, or not sufficient lysine residues are readily available.
3. For protein separation along the full gel length the number of samples should not exceed four per gel and MS analysis to minimize inaccurate protein quantification caused by migration differences in gel lanes and variations in excision of protein bands.

4. In cases where antibody solution exceeds 200 μL, $NaCNBH_3$ volume has to be adjusted to 25 mM final concentration.
5. For larger protein amounts or volumes that exceed the maximum microtube capacity, 15 mL conical tubes are used. If lysate volumes exceed the maximum microtube capacity transfer antibody beads to 15 mL conical tubes and add respective lysate volume.
6. Overnight incubation should be avoided. It increases unspecific interactions due to protein aggregation and potentiates complex dissociation during the incubation and increases the likelihood of antibody mediated displacement.
7. High affinity interactions usually lead to high purification efficiency in less than 1 h. Purification efficiency could be determined using a small-scale approach and immunodetection of the candidate protein in the non-bound fraction (=NBF) and the input material.
8. IP binding reaction was carried out in 15 mL conical tubes, wash two times with 30× bead bed volume of IP buffer with 0.4 % detergent, every time centrifuge for 2 min at 200 × *g* and discard supernatant. For the elution, transfer antibody beads to microtubes.
9. Typical settings for database search include a mass tolerance of 10 ppm for peptide precursor ions and 20 mDa for fragment masses. Variable modifications include methionine oxidation, N-terminal acetylation of proteins, and iTRAQ/TMT modification of peptide N termini. Carbamidomethylation of cysteine residues and iTRAQ/TMT modifications of lysine residues are set as fixed modifications.

Acknowledgments

We would like to thank Frank Weisbrodt for help with the figures.

References

1. Bantscheff M, Hopf C, Savitski MM et al (2011) Chemoproteomics profiling of HDAC inhibitors reveals selective targeting of HDAC complexes. Nat Biotechnol 29: 255–265
2. Dawson MA, Prinjha RK, Dittmann A et al (2011) Inhibition of BET recruitment to chromatin as an effective treatment for MLL-fusion leukaemia. Nature 478:529–533
3. Kocher T, Superti-Furga G (2007) Mass spectrometry-based functional proteomics: from molecular machines to protein networks. Nat Methods 4:807–815
4. Gavin A-C, Aloy P, Grandi P et al (2006) Proteome survey reveals modularity of the yeast cell machinery. Nature 440:631–636
5. Krogan NJ, Cagney G, Yu H et al (2006) Global landscape of protein complexes in the yeast Saccharomyces cerevisiae. Nature 440: 637–643
6. Song J, Hao Y, Du Z et al (2012) Identifying novel protein complexes in cancer cells using

epitope-tagging of endogenous human genes and affinity-purification mass spectrometry. J Proteome Res 11:5630–5641

7. Malovannaya A, Lanz RB, Jung SY et al (2011) Analysis of the human endogenous coregulator complexome. Cell 145:787–799
8. Malovannaya A, Li Y, Bulynko Y et al (2010) Streamlined analysis schema for high-throughput identification of endogenous protein complexes. Proc Natl Acad Sci U S A 107: 2431–2436
9. Mellacheruvu D, Wright Z, Couzens AL et al (2013) The CRAPome: a contaminant repository for affinity purification-mass spectrometry data. Nat Methods 10:730–736
10. Bantscheff M, Lemeer S, Savitski M et al (2012) Quantitative mass spectrometry in proteomics: critical review update from 2007 to the present. Anal Bioanal Chem 404:939–965
11. Selbach M, Mann M (2006) Protein interaction screening by quantitative immunoprecipitation combined with knockdown (QUICK). Nat Methods 3:981–983
12. Kruse U, Pallasch CP, Bantscheff M et al (2011) Chemoproteomics-based kinome profiling and target deconvolution of clinical multi-kinase inhibitors in primary chronic lymphocytic leukemia cells. Leukemia 25:89–100
13. Brehme M, Hantschel O, Colinge J et al (2009) Charting the molecular network of the drug target Bcr-Abl. Proc Natl Acad Sci U S A 106:7414–7419
14. Werner T, Becher I, Sweetman G et al (2012) High-resolution enabled TMT 8-plexing. Anal Chem 84:7188–7194
15. Olsen JV, de Godoy LMF, Li G et al (2005) Parts per million mass accuracy on an orbitrap mass spectrometer via lock mass injection into a C-trap. Mol Cell Proteomics 4:2010–2021
16. Eng J, McCormack A, Yates J (1994) An approach to correlate tandem mass spectral data of peptides with amino acid sequences in a protein database. J Am Soc Mass Spectrom 5:976–989
17. Perkins DN, Pappin DJC, Creasy DM et al (1999) Probability-based protein identification by searching sequence databases using mass spectrometry data. Electrophoresis 20:3551–3567
18. Jr C, Neuhauser N, Michalski A et al (2011) Andromeda: a peptide search engine integrated into the MaxQuant environment. J Proteome Res 10:1794–1805

Part VIII

Targeted Proteomics

Chapter 20

Biomarker Verification Using Selected Reaction Monitoring and Shotgun Proteomics

Angel Mauricio Castro-Gamero, Clarice Izumi, and José César Rosa

Abstract

Shotgun proteomics (liquid chromatography-electrospray ionization-mass spectrometry, LC-ESI-MS/MS) has dominated the strategies for global protein expression in subcells, cells, tissues, and whole organisms with several types of approaches, as isobaric tags for relative and absolute quantification (iTRAQ), isotope-coded affinity tags (ICAT), or stable isotope labeling using amino acids in cell culture (SILAC) and non-labeling (label free) methods. Shotgun proteomics practically replaced the classical 2D gel electrophoresis. Selected reaction monitoring (SRM), also denominated multiple reaction monitoring (MRM), is a targeted quantitative technology that uses a complex mixture of tryptic peptides that can be selectively detected by liquid chromatography coupled to electrospray triple-quadrupole mass spectrometer; this system can select precursor ions in combination with their correspondent product ions during collision-induced dissociation to produce specific detection related to a particular protein. Here we describe protocols that are efficient to produce a complete enzymatic trypsin digestion from complex biological matrices and concomitant material to be used for LC-SRM-MS and LC-ESI-MS/MS (labeled or label free).

Key words Shotgun proteomics, Selected reaction monitoring, Multiple reaction monitoring, Electrospray triple-quadrupole, Protein quantitation, Biomarker verification, Plasma, Tissue, Urine and Cerebrospinal Fluid (CSF)

1 Introduction

The development of proteomics in recent years has been boosted by the introduction of new technologies in mass spectrometry. This allowed the identification and quantification of thousands of proteins in a single experiment through different sets of approaches called shotgun proteomics [1]. However, the obstacle that remains is to quantify low-abundance proteins in complex biological matrices [2]. The development of chromatographic separation methods in capillary columns (1.8–3.0 micra particles) gave an increased sensitivity and also three or four orders of magnitude in the detection of a particular peptide ion. On the other hand, high-abundance proteins still remains an issue in shotgun proteomic methods.

Daniel Martins-de-Souza (ed.), *Shotgun Proteomics: Methods and Protocols*, Methods in Molecular Biology, vol. 1156, DOI 10.1007/978-1-4939-0685-7_20, © Springer Science+Business Media New York 2014

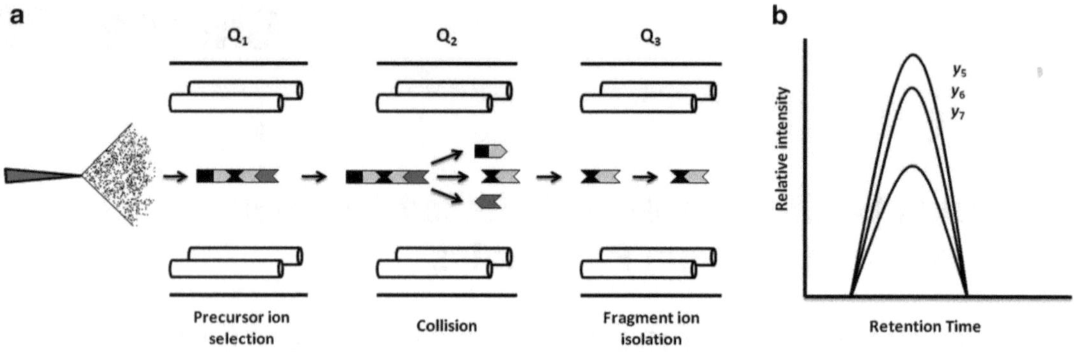

Fig. 1 (**a**) A schematic representation of a triple quadrupole operating in selected reaction monitoring mode. (**b**) A typical spectrum obtained by monitoring a single peptide showing three *y*-transitions

The use of liquid chromatography-tandem selected reaction monitoring-mass spectrometry (LC-SRM-MS), already well established in the analysis of small molecules [3], has been expanded to the area of proteomics. Commonly performed in electrospray triple-quadrupole mass spectrometer, the SRM technique consists in a separation of a precursor ion in the first quadrupole (Q1) which is selectively transmitted to a second quadrupole (Q2) or collision cell, where the precursor ion collides with an inert gas and fragments on various product ions; hereinafter, ions of these products are separated and detected in the third quadrupole (Q3) producing a chromatographic peak corresponding to the pair of precursor–product ions called transition which is directly related to the concentration of a protein of interest [4] (Fig. 1).

Recent publication of the group of MS/Proteomics Resource at Yale University showed the saturation of information obtained from shotgun proteomics (LC-MS/MS) as a consequence of the re-sequencing of same peptides derived from abundant proteins. This explains the tendency of changes in many proteomic laboratories, where many groups are replacing strategies for complete proteome toward targeted proteomics methods as SRM [5].

The goal of an SRM experiment is the selection of peptides and their transitions to effectively serve as a quantitative measure of a specific protein in a complex biological matrix. The importance of this method lies in the possibility of multiplex detection of hundreds of protein components in the same sample by selecting pairs of precursor–product ions that were planned in computer programs (in silico trypsin digestion) or information about peptides extracted from proteomic discovery methods such as LC-ESI-MS/MS [6]. The importance of a complete enzymatic digestion of proteins in complex biological matrix directs the selection of peptides/transitions in SRM experiment due to the large variation in the amount of certain peptides within the same protein, indicating that although the action of trypsin should be directed to peptide bonds with lysine and arginine, protein substrate contains various sub-sites which may interfere with the hydrolysis kinetics of the peptide bond involved. Consequently, a certain protein peptide may differ in some hundred times in concentration to be released during the enzymatic hydrolysis for the same protein and produces effects on sensitivity and dynamic detection capability of the peptide. SRM is a technique that takes time and intensive labor for assay development [6, 7].

Here we describe protocols for sample preparation for LC-SRM-MS and LC-ESI-MS/MS and give some tips for beginners in the field of targeted and shotgun proteomics.

2 Materials

Prepare all solutions using ultrapure water (18 MΩ cm at 25 °C) and analytical and HPLC-grade reagents.

Sample: Biofluids (serum, plasma, liquor, urine), tissue homogenates, cell culture lysates.

Prepare and store all reagents at room temperature (unless indicated otherwise). Diligently follow all waste disposal regulations when disposing waste materials and biological residues.

2.1 Tryptic Digests

1. Solution HCA (8 M urea and 4 M thiourea), a detergent (0.1 % SDS, 0.1 % Triton X-100, or 4 % 3-[(3-cholamidopropyl)dimethylammonio]-1-propanesulfonate, CHAPS), and a cocktail of protease and phosphatase inhibitors.
2. Detergent RapiGest SF (Waters): This reagent is a detergent compatible with electrospray mass spectrometry due to its degradation in acid medium before mass spectrometry analysis, whose composition is sodium-3-[(2-methyl-2-undecyl-1, 3-dioxolan-4-yl)-methoxyl]-1-propanesulfonate.
3. Solution 50 mM ammonium bicarbonate: Dissolve 0.08 g ammonium bicarbonate in 20 mL ultrapure water (make fresh before use every time).
4. Solution 25 mM ammonium bicarbonate: Dissolve 0.04 g ammonium bicarbonate in 20 mL ultrapure water (make fresh before use every time).

2.2 Protocol 1 (General Shotgun and SRM/Label Free)

1. Cold neat acetone.
2. Alternative solution for protein precipitation: 10 % (w/v) trichloroacetic acid (TCA)/acetone: dissolve 10 g of TCA in 100 mL of cold acetone pure.
3. 10 mM dithiothreitol (DTT): 1.54 mg in 1 mL of ultrapure water. Always use fresh solution.

4. 10 mM iodoacetamide (IAA): 1.85 mg in 1 mL of ultrapure water. Always use fresh solution.
5. Sequencing-grade modified trypsin.
6. Trifluoroacetic acid.
7. Stock solution of 1 M ammonium formate, pH 3.2: To prepare a 1 M stock solution of ammonium formate, pH 3.2, in our laboratory we use a mixture of formic acid and ammonium hydroxide, based on a ratio calculated from the Henderson–Hasselbalch equation, $pH = pKa + \log((\text{formate})/(\text{formic acid}))$. For instance, for this solution the pKa of formic acid is 3.75; hence, $\log((\text{formate})/(\text{formic acid})) = 3.2 - 3.75 = -0.55$. The antilog of $-0.55 = 0.28$. Using this ratio, for 100 mL of 1 M ammonium formate stock buffer, transfer 13.6 mL of formic acid (Merck, ACS grade, MW = 46.0, d = 1.22 g/mL, and 98–100 %) to a beaker containing ~60 mL of ultrapure water and titrate to pH 3.2 with ~9 mL ammonium hydroxide (Sigma, ACS grade, MW = 35.04, d = 0.89 g/mL, and 28–30 %). The reaction is exothermic; therefore, cool the mixture to room temperature, finally set the pH, and then complete the volume to 100 mL.
8. Solution 5 mM of ammonium formate pH 3.2 containing 20 % acetonitrile as solvent A for strong cation-exchange (SCX) chromatography. Dilute 200-fold stock solution of 1 M ammonium formate in solution of acetonitrile:water (20:80 v/v).
9. Solution 500 mM ammonium formate pH 3.2 with 20 % acetonitrile as solvent B for SCX chromatography. Dilute twofold stock solution of 1 M ammonium formate in solution of acetonitrile:water (20:80 v/v).

2.3 Protocol 2 (Compatible with iTRAQ Labeling)

1. 2 % sodium dodecyl sulfate (SDS) solution: 1 g SDS in 50 mL ultrapure water.
2. Solution 0.5 M triethylammonium bicarbonate (TEAB) buffer: This TEAB buffer is a volatile buffer commercially available commonly provided at 1 M concentration. Dilute this solution in ultrapure water at a ratio of 1:1 to achieve 0.5 M concentration.
3. Reducing reagent solution, 50 mM tris-2-carboxyethyl-phosphine (TCEP): Dilute 81.5 mg of TCEP in 10 mL ultrapure water.
4. Cysteine blocking reagent, 200 mM methyl methanethiosulfonate (MMTS), in isopropanol: Add 19 μL of MMTS (MW = 126.2 and d = 1.34) to 1.0 mL of isopropanol.
5. Sequencing-grade modified trypsin (Promega): Prepare solution of 0.2 μg/μL in 0.1 M TEAB or MilliQ-Waters or equivalent.

3 Methods

3.1 Selection of Standard Peptides (Proteotypic Peptides) and SRM Transitions

1. The selection of standard peptides should be made through searching in spectral libraries, discovery proteomics, or hypotheses derived from genomic studies (*see* **Notes 1** and **2**) [8]. Some of these search tools are showed in Table 1.
2. If the selected software for starting the modeling of SRM assay and the peptide searching is the Skyline software *see* **Notes 3** and **4**.
3. Create an MS/MS spectral library from part of protein data set downloaded from the Peptide Atlas online data repository, or create new spectral libraries in Skyline using other publicly available data or peptide search results from your laboratory experiments using diverse formats (*see* **Note 5**). It is possible as well to use fasta sequence files to inform Skyline about background matrix in which the experiment will take place (named Background Proteome) (*see* **Note 6**).
4. Digest the background proteome with a given enzyme (commonly trypsin) (*see* **Note 7**).
5. Once the digestion is finished, choose ion types, selecting the precursor m/z and product m/z transitions in Skyline or related software (*see* **Note 8**).

Table 1
List of bioinformatic tools for designing MRM/SRM assays (Boja and Rodriguez [8])

Bioinformatic tools	URL
Global Proteome Machine, GMP	http://www.thegpm.org
PeptideAtlas	http://www.peptideatlas.org
Proteomics IDEntification, PRIDE	http://www.ebi.ac.uk/pride
SRMAtlas	http://www.srmatlas.org
ESP predictor	http://www.broadinstitute.org/cancer/software/genepattern/modules/ESPPpredictor.html
PeptideSieve	http://tools.proteomecenter.org/software.php
MRMer	http://proteomics.fhcrc.org/CPL/MRMer.html
Skyline	https://brendanx-uw1.gs.washington.edu/labkey/project/home/software/Skyline/begin.view
Automated detection of inaccurate and imprecise transitions, AuDIT	http://www.broadinstitute.org/cancer/software/genepattern/modules/AuDIT.html
Automated and targeted analysis with quantitative SRM, ATAQS	http://tools.proteomecenter.org/ATAQS/ATAQS.html

6. Verify the sequence alignment of unique peptide sequences for the protein of interest avoiding isoforms and peptides prone to posttranslational modifications (S/T or sequon N-X-S/T).
7. Open the amino acid sequence, and set the properties of a group of targeted peptides. Using the default setting is a good start, and then refine the SRM protocol for the specific application (*see* **Notes 9** and **10**).

3.2 Preparation of Tryptic Digests

We described two protocols for obtaining tryptic peptides of total soluble protein extract from different types of sample, appropriate for shotgun proteomics and SRM assay.

1. Before cell lysis for protein extraction, culture media should be discarded and cells washed three times with phosphate-buffered saline (PBS) buffer to remove traces of fetal bovine serum, trypsinized (if necessary), and centrifuged at 160 × *g*.
2. Prepare the protein extract from your sample in accordance with a previously established protocol.
3. Starting with mammalian tissues or culture cells, usually a hyposmotic solution containing a chaotropic agent is used (solution HCA, *see* Subheading 2).
4. Sonicate in ice bath for 5 min.
5. Centrifugate for 15 min at 20,000 × *g* at 4 °C to pellet cell debris and insoluble material.
6. For mammalian cell lysis, 0.1 % acid-labile detergent RapiGest SF (*see* Subheading 2) in 25 mM ammonium bicarbonate can be used.
7. Then, the samples should be vortexed and sonicated three times for 30 s in ice bath. Posteriorly, the lysates should be centrifuged for 20 min at 20,000 × *g* at 4 °C as recommended by Drabovich et al. [11] (*see* **Note 11**).
8. In case of plasma or serum, depletion of abundant proteins using a multiple affinity removal system (MARS, an Agilent Technologies spin columns) or Sigma immunodepletion kits should be performed (*see* **Note 12**).
9. After depletion, sample can be exchanged to 50 mM ammonium bicarbonate using a concentrator device of 5,000 molecular weight cutoff [12], and follow with **step 2** of this protocol.
10. Before starting the experiment, estimate the total protein concentration of the protein extracts by Bradford assay [13] or other protein quantification method.

3.3 Protocol 1 (General Shotgun and SRM/Label Free)

1. Pipette the sample (10–100 μg of total protein) into microtubes to perform the trypsin hydrolysis. To eliminate buffer that contains interferences (urea, CHAPS, and others) add cold acetone to the sample in a ratio of 1:60 (sample:acetone,

v/v) to precipitate proteins. Incubate at −20 °C during 4–18 h. Centrifuge the microtubes at low speed to precipitate the proteins. Discard the supernatant, and dry the pellet at room temperature (*see* **Note 13**).

2. Add 200 μL of 0.1 % RapiGest SF in 25 mM ammonium bicarbonate. Cell pellets are vortexed, sonicated three times for 30 s, and centrifuged during 20 min at 15,000 × *g* at 4 °C.
3. To reduce disulfide bonds, add 5 μL of 10 mM DTT, freshly prepared, and incubate at 60 °C for 1 h.
4. To alkylate the proteins, add 5 μL of 20 mM IAA, freshly prepared, and incubate at room temperature for 30 min in a dark place (*see* **Note 14**).
5. Add sequencing-grade modified trypsin (Promega) to the sample at 1:30 ratio, and carry out enzymatic hydrolysis overnight at 37 °C.
6. Add trifluoroacetic acid to 1 % final concentration to cleave RapiGest, centrifuge at 160 × *g* for 10 min, and dilute sample to 500 μL with SCX buffer (5 mM ammonium formate, pH 3.2) (*see* **Notes 15** and **16**).
7. Fractionate sample by SCX, using a polySULFOETHYL A column (4.6 mm I.D. × 100 mm, 5 μm, 200 Å). A suggestion to develop the chromatography is using a flow rate of 1.0 mL/min: start with 5 mM of ammonium formate pH 3.2 with 20 % acetonitrile as solvent A and a linear gradient to 500 mM ammonium formate pH 3.2 with 20 % acetonitrile as solvent B during 70 min (*see* **Note 17**).
8. Collect 1-min fraction and then concentrate to dryness in SpeedVac or a similar vacuum concentrator. Analyze each SCX fraction by reverse-phase chromatography online with ESI-Q-TOF or triple-Q mass spectrometer for shotgun and SRM, respectively.

3.4 Protocol 2 (Compatible with iTRAQ Labeling)

1. Precipitate the proteins as described in **step 1** of the Protocol 1.
2. Add 1 μL of 2 % SDS to the pellet, mix it, and then add 20–30 μL of 0.5 M TEAB buffer. Vortex the mixture during 2–3 min.
3. Add 2 μL of 50 mM TCEP to each sample to reduce disulfide bridges, and incubate at 57 °C for 60 min.
4. Add 1 μL of cysteine blocking reagent (200 mM MMTS) (*see* **Note 18**).
5. Incubate the reaction at room temperature for 30 min in a dark place (*see* **Note 15**).
6. Add sequencing-grade modified trypsin (Promega) to the sample at 1:30 ratio. Carry out the digestion overnight (14–18 h) at 37 °C.

7. Develop the isobaric tags for relative and absolute quantification (iTRAQ) labeling as indicated by the supplier. Remove a representative amount of each unlabeled sample for SRM assays.
8. Fractionate samples by SCX similar to **step 7** of Protocol 1, and submit SCX fraction to online reverse-phase LC-ESI-MS/MS for shotgun proteomics. The overall quantification results could be compared to isobaric tag labeling method with independent sample prepared for SRM method.

3.5 Analysis and Detection of Transitions

Although essentially any tandem MS instrument can be utilized for targeted quantification, triple-quadrupole and quadrupole-ion trap (Q-trap) MS are the most widely used instruments. The ability of the Q-trap instrument to acquire full scan product ion data with high mass accuracy allows this platform to be an alternative instrumentation for performing SRM quantification in addition to the more traditionally used triple-quadrupole mass spectrometer.

1. Verify, record, or tune the following parameters for SRM development: (1) retention time and scheduling intervals, (2) selectivity of transitions and possible interferences, (3) scan times, and (4) collision energies (*see* **Note 19**).
2. Verify the duration of the chromatography run to minimize the duration of chromatographic cycles (*see* **Note 20**).
3. Quantification of the protein: It was found that a relationship between the average MS signal of three most intense tryptic peptides per mole of protein is directly related to protein concentration according to the method of Silva et al. [14] (*see* **Note 21**).

4 Notes

1. It should be noted that information from genomic studies suggests that there is not always a direct correlation between mRNA and protein expression and thus the selection of proteotypic peptides depends on the use of various specialized software to generate the group of ion transitions necessary for the quantification of the protein of interest (Table 1).
2. Importantly, optimized SRM assays are successfully developed from empirical data, as neither the computational prediction of an optimal peptide set nor their optimal transitions particularly precise.
3. Skyline is the most popular user-friendly software to generate a list of peptide ions and their corresponding transition candidates [9]. Skyline program permits to import protein sequences from databases, allowing the construction of spectral libraries and definitions by the user, such as type of protease (trypsin, Lys-C), size and charge of the peptide precursor ion, types of fragments (*b* or *y*), and charge of product ions [9].

4. For beginners, many tutorials are available online for each step, from development to SRM assay using Skyline software, including features of software, both steps, initial and refinement of method, running of experiments, absolute quantification, retention time prediction, collision energy optimization, and spectral library exploration.
5. Skyline currently supports building libraries from different mass spectrometer vendors.
6. In the background, proteome may be spiked a protein unrelated to the species under study. We suggest chicken lysozyme for materials of human or mouse origin in the amount of 0.5–1.0 μg. Also the entire database in fasta format can be included to determine the specificity of proteotypic peptides chosen.
7. The most common enzyme used in proteomics is trypsin. It allows high-specificity cleavage in lysine and arginine residues. These amino acids are located at the C-terminus of the peptide which permits retention of proton in their side chains and enhances the stability of *y*-type ions commonly achieved by many mass spectrometers after collision-induced dissociation (CID or CAD).
8. The adequate identification of proteotypic peptides whose sequences are unique to each parent protein is very important. A preselection of ions is based on medium size of 8–25 amino acid residues, excluding the first 25 residues of each N-terminal due to possible limited proteolysis. The choice of proteotypic peptides should exclude missing cleavage of trypsin. Sites of known posttranslational modifications (e.g., phosphorylation, glycosylation, and acetylation) on peptides are avoided unless the assay is specifically targeted to the modified form. Additionally, peptides containing residues susceptible to artifact modifications during sample preparation, such as methionine (oxidation) and cysteine (carbamidomethylation and oxidation), aspartic acid–glycine pairs (deamidation), and N-terminal glutamine (pyroglutamic acid formation) and asparagine (deamidation) or histidine that confers multiple charges to product ions, should be avoided [10]. Select the protein sequences from SwissProt or NCBI database as a text or a fasta format, and submit it to Blast against the database which contains information about the proteome of organism of interest, and these will be used for SRM assay development.
9. When possible, it is highly recommended to do the selection of peptides based on two types of direct proteomics survey experiments (combined sources): (1) analysis of LC-MS/MS classical of sample digests and (2) using the approaches described above.
10. Optimization of transitions for peptide quantification in SRM should observe empirically the limitations of in vivo and in vitro protein modification, and experimentally it could be

tested by submitting at least 3–4 possible transitions of precursor ions and product ions, especially considering *y*-type ions. The *b* ion series often lose carbonyl group (−28 Da) to generate *a*-type ions which will compromise the sensitivity and accuracy of peptide quantification. It is important to select fragment ions with m/z values higher than precursor ions; it will avoid the detection of singly charged precursor ions that sometimes co-elute with precursor ions selected for SRM experiment.

11. RapiGest-based protocol facilitated quick and easy cell lysis, protein denaturation, digestion, and preparation of peptides for mass spectrometry analysis. This step avoids sample loss and makes the whole proteomic protocol easier and more robust. Multiple samples can be prepared in a 96-well plate, avoiding thus the common steps of detergent removal and lyophilization that might lead to sample loss and contribute to low reproducibility of analysis [11].
12. The depletion of abundant proteins may be necessary before the tryptic digestion, with the objective to improve the performance of SRM assay or shotgun proteomics. Removal of abundant plasma proteins allows the identification and quantification of low-abundance proteins.
13. Better results could be obtained using a solution of 10 % TCA in cold acetone during 18 h at −20 °C, followed by washing with chilled neat acetone.
14. An alternative to quench the excess of IAA is adding 12.5 mM final concentration N-acetyl-cysteine and incubating the reaction at 25 °C for 10 min.
15. The following precautions should be taken to minimize deterioration of peptides, such as oxidation of methionine and deamination of asparagine and glutamine, during storage and analysis: (1) peptides should be stored at −20 °C until use and then analyzed within a day; (2) tubes or plates should be sealed with silicone rubber sealing mats and kept at 10–15 °C during analysis.
16. The SRM protocol can be carried out in a single well of a 96-well plate, thus avoiding the common steps of detergent removal and lyophilization that might lead to sample loss and low reproducibility of analysis.
17. We use an HPLC dual-pump system to perform the SCX chromatography.
18. Alternatively, add 1 μL of freshly prepared 84 mM IAA solution (15.5 mg of IAA in 1 mL of ultrapure water or equivalent).
19. If multiple peptides per protein were detected, it is highly recommended to choose the two peptides with the highest SRM area and significantly different retention times. The typical median variation of SRM area in technical replicates should be

4–5 %, whereas the median biological variation before normalization can be ~15 % or ideally below 10 %.

20. An important consideration for throughput of SRM measurements is the duration of the chromatography run. For instance, using a 30-min gradient can lead to a total cycle time (including inter-sample wash) of 75 min. It would clearly be advantageous if the time could be reduced, allowing more samples to be run per day. Thus, SRM measurements in discrete time windows, an analysis mode commonly named scheduled SRM, may allow more SRMs to be brought closer together in elution time without sacrificing the required multiple measurements across each peak.
21. Precise and accurate quantification of proteins in SRM experiment via their tryptic peptides is usually a direct function of MS ion intensity statistics which are dependent on dwell time. Dwell time is defined in terms of how much time the mass spectrometer spent to detect an ion; longer dwell time helps to detect very-low-abundance peptides. There is no rule to apply, but dwell time that provides at least 12–15 data points can positively influence the detection and quantification of the peptide during SRM experiment.

Acknowledgments

AMCG is supported by Fundação de Amparo à Pesquisa do Estado de São Paulo (FAPESP) Post-Doctoral Fellowship (2012/16889-4). JCR is supported by Conselho Nacional de Desenvolvimento Científico e Tecnológico (CNPq) Researcher Fellowship (304627/2012-7). The work is supported by CNPq (grant 487105/2012-5 and 573754/2008-0) and FAPESP (grant 2011/07568-7).

References

1. Wilson R (2013) Sensitivity and specificity: twin goals of proteomics assays. Can they be combined? Expert Rev Proteomics 10(2): 135–149
2. Tu C, Rudnick PA, Martinez MY, Cheek KL, Stein SE, Slebos RJ, Liebler DC (2010) Depletion of abundant plasma proteins and limitations of plasma proteomics. J Proteome Res 9(10):4982–4991
3. Shushan B (2010) A review of clinical diagnostic applications of liquid chromatography-tandem mass spectrometry. Mass Spectrom Rev 29(6):930–944
4. Maiolica A, Jünger MA, Ezkurdia I, Aebersold R (2012) Targeted proteome investigation via selected reaction monitoring mass spectrometry. J Proteomics 75(12):3495–3513
5. Colangelo CM, Chung L, Bruce C, Cheung KH (2013) Review of software tools for design and analysis of large scale MRM proteomic datasets. Methods 61(3):287–298
6. Liebler DC, Zimmerman LJ (2013) Targeted quantitation of proteins by mass spectrometry. Biochemistry 52(22):3797–3806
7. Picotti P, Aebersold R (2012) Selected reaction monitoring-based proteomics: workflows,

potential, pitfalls and future directions. Nat Methods 9(6):555–566

8. Boja ES, Rodriguez H (2012) Mass spectrometry-based targeted quantitative proteomics: achieving sensitive and reproducible detection of proteins. Proteomics 12(8):1093–1110
9. MacLean B, Tomazela DM, Shulman N, Chambers M, Finney GL, Frewen B, Kern R, Tabb DL, Liebler DC, MacCoss MJ (2010) Skyline: an open source document editor for creating and analyzing targeted proteomics experiments. Bioinformatics 26(7):966–968
10. Bereman MS, MacLean B, Tomazela DM, Liebler DC, MacCoss MJ (2012) The development of selected reaction monitoring methods for targeted proteomics via empirical refinement. Proteomics 12(8):1134–1141
11. Drabovich AP, Pavlou MP, Dimitromanolakis A, Diamandis EP (2012) Quantitative analysis of energy metabolic pathways in MCF-7 breast cancer cells by selected reaction monitoring assay. Mol Cell Proteomics 11(8): 422–434
12. Anderson L, Hunter CL (2006) Quantitative mass spectrometric multiple reaction monitoring assays for major plasma proteins. Mol Cell Proteomics 5(4):573–588
13. Bradford MM (1976) A dye binding assay for protein. Anal Biochem 72:248–254
14. Silva JC, Gorenstein MV, Li GZ, Vissers JP, Geromanos SJ (2006) Absolute quantification of proteins by LCMSE: a virtue of parallel MS acquisition. Mol Cell Proteomics 5(1): 144–156

Chapter 21

Use of Universal Stable Isotope Labeling by Amino Acids in Cell Culture (SILAC)-Based Selected Reaction Monitoring (SRM) Approach for Verification of Breast Cancer-Related Protein Markers

Ning Qing Liu, Lennard J.M. Dekker, Martijn M. Van Duijn, and Arzu Umar

Abstract

Mass spectrometry-based proteomics facilitates high-throughput discovery of protein markers for diagnosis and treatment of breast cancer patients. Hundreds of putative prognostic and predictive markers are being identified every year, but only a very small proportion of them can be validated as clinically relevant markers. A quantitative and cost-efficient verification method is highly desirable to pick up real "nuggets" from the "sand." To fulfill these criteria, we previously introduced a stable isotope labeling by amino acids in cell culture (SILAC)-based selected reaction monitoring (SRM) approach for studying breast cancer-related protein markers. Here we describe a hands-on protocol of using this SILAC-SRM method for verification of breast cancer-related markers, which can also be used for verification of protein markers in other types of solid tumor tissues.

Key words Breast cancer, Proteomics, Biomarker verification, SILAC, SRM, Mass spectrometry

1 Introduction

Liquid chromatography hyphenated with mass spectrometry (LC-MS)-based proteomic techniques facilitate discovery of breast cancer-related protein markers [1]. Typically, LC-MS-based biomarker discovery can be classified into three stages: discovery, verification, and validation [2]. Well-established quantitative methods have been described for biomarker discovery and validation, including label-free shotgun proteomics-based discovery and targeted proteomics using absolute quantification (AQUA) for validation [3]. However, to date various techniques have been employed for biomarker verification.

Verification of breast cancer-related biomarkers requires precisely and reproducibly quantifying a large set of identified candidate markers at relatively low cost. Conventionally, labeling

Daniel Martins-de-Souza (ed.), *Shotgun Proteomics: Methods and Protocols*, Methods in Molecular Biology, vol. 1156, DOI 10.1007/978-1-4939-0685-7_21, © Springer Science+Business Media New York 2014

methods are often desirable for precise quantification. AQUA approach results in highly accurate and precise quantification by introducing stable isotope-labeled heavy peptides with identical chemical properties of the light counterparts and known absolute quantities [4]. This approach, although technically competitive, is little cost efficient for verification studies due to high cost of AQUA peptides. Alternatively, spot synthesis method has been introduced to reduce the cost [5]. However, special expertise is required to establish this method, which is not often available in many research laboratories. Stable isotope labeling by amino acids in cell culture (SILAC) technique simultaneously labels the entire proteome of cell lines, which can serve as a universal standard to quantify different protein markers [6]. SILAC method is also technically straightforward to establish in a biochemical or a molecular biological laboratory. Therefore, SILAC labeling could be an ideal method for biomarker verification.

To achieve reproducible quantification, a targeted MS strategy is frequently used [7]. Targeted MS methods, such as selected reaction monitoring (SRM), enable simultaneous measurement and quantification of multiple protein candidates in a high-throughput and highly reproducible manner [7]. Although shotgun or directed MS approaches have also been used for verification in some occasional cases [8–10], SRM appears to be a leading technique for targeted analysis in the context of intra- and inter-laboratory reproducibility [11].

Recently, we have described a SILAC-SRM approach to precisely quantify some proteins of the "focal adhesion" pathway from breast cancer tissues [8]. In this study, five "focal adhesion" proteins quantified with three peptides had coefficient of variations (CVs) within a range of 2.41 and 5.91 % in technical quadruplicates from a trace amount of breast cancer tissue lysate [8]. Therefore, SILAC-SRM appears to be a promising approach to verify a large set of clinically relevant protein markers.

In this chapter, we describe a hands-on protocol of our recently developed SILAC-SRM method for large-scale quantification of protein markers. This approach can be widely applied in the biomarker discovery field as well as functional proteomic studies for various solid tumor and other tissue materials.

2 Materials

2.1 Cell Culture and Harvest Materials

1. Breast cancer cell lines: Select 2–5 breast cancer cell lines to create a universal SILAC standard (*see* **Note 1**).
2. SILAC culture medium: Replace natural arginine (R) and lysine (K) by R10 and K8. Store at 4 °C.
3. Glutamax-I (100×): Store in 5 ml aliquots at room temperature.

4. Dialyzed fetal bovine serum (FBS): Store in 50 ml aliquots at –20 °C for short-term storage (≤1 year) and at –80 °C for long-term storage (>1 year).
5. Antibiotics solution (100 μg/ml penicillin G and 80 μg/ml streptomycin): Dissolve 500 μg penicillin G and 400 μg streptomycin in 5 ml sterilized water.
6. L-proline solution: Weigh 100 mg L-proline for 500 ml SILAC culture medium (final concentration: 200 mg/l) (*see* **Note 2**).
7. 6-well culture plates (inner diameter/I.D.: 3.48 cm) and culture flasks.
8. Cell lifters.
9. 15-ml Greiner tubes.
10. Ice-cold phosphate-buffered saline (PBS).
11. Dry ice.
12. Centrifuge.
13. Electronic weight balance.

2.2 Tissue Sample Preparation Reagents and Equipment

1. Fresh frozen breast cancer tissues: Store tumor tissues in liquid nitrogen (long term), at –80 °C (mid term), and in dry ice (short term, only prior to cryosectioning).
2. Cryostat.
3. TissueTek O.C.T. compound.
4. Lint-free wipes.
5. Dry ice.
6. Protein LoBind tubes, PCR clean (Eppendorf, Hauppauge, NY).
7. Pathological glass slides.
8. Pertex mounting medium (Histolab).
9. Tap water.
10. Hematoxylin.
11. Eosin.
12. Xylene.
13. Ethanol (technical grade): Prepare 50 and 70 % ethanol in tap water and 100 % ethanol for hydration and dehydration of tissue sections.
14. Electronic weight balance.

2.3 Protein Extraction Reagents and Equipment

1. LC-MS-grade water.
2. 50 mM ammonium bicarbonate (ABC) solution: Dissolve 20 mg ABC in 1 ml LC-MS-grade water (250 mM, 5×). Further dilute 250 mM ABC solution five times in LC-MS-grade water (50 mM, 1×). Store at 4 °C (stable up to 1 week) or –80 °C (stable up to 3 months).

3. 0.1 % RapiGest solution (w/v): Dissolve 1 mg RapiGest powder (Waters Corp., Milford, MA) in 1 ml 50 mM ABC solution. Store at 4 °C (stable up to 1 week).
4. Ultrasonics Disruptor Sonifier II (Model W-250/W-450, Branson Ultrasonics, Danbury, CT).
5. Bicinchoninic acid (BCA) assay reagents (Pierce Biotechnology, Rockford, IL): Prepare working reagent by mixing BCA reagent A and BCA reagent B in a ratio of 50:1 (v:v) before performing BCA assay (total volume of working reagent = number of samples × number of replicates × 200 μl).
6. Albumin standard (2 mg/ml): Construct a dilution series of albumin standard in 50 mM ABC solution (2,000, 1,500, 1,000, 750, 500, 250, 125, 25, and 0 μg/ml). Store at −20 °C (make a new dilution series every 3 months).
7. Plate reader with 570 nm filter.

2.4 Protein Digestion Reagents and Equipment

1. 100 mM dithiothreitol (DTT): Dissolve 15.4 mg DTT in 1 ml of 50 mM ABC solution. Stable for up to 10 h at 4 °C. For long-term storage, make 50-μl aliquots and store at −20 °C.
2. 300 mM iodoacetamide (IAA): Dissolve 9.25 mg IAA in 167 μl of 50 mM ABC solution. Keep in the dark and use immediately (*see* **Note 3**).
3. 50 mM acetic acid: Dissolve 75 μl acetic acid in 25 ml LC-MS-grade water (pH ~4).
4. 10 % trifluoroacetic acid (TFA) (v/v): Dissolve 1 ml TFA in 10 ml LC-MS-grade water (pH <2). Store the solution at a dry and cool place. Stable for up to 1 month.
5. 100 ng/μl MS-grade porcine-modified trypsin gold: Dissolve 100 μg lyophilized trypsin in 1 ml of 50 mM acetic acid. Store in 50 μl aliquots at −80 °C. Stable for up to 1 year.
6. 600 mM L-cysteine: Dissolve 7.3 mg L-cysteine in 100 μl of 50 mM ABC solution (*see* **Note 4**).
7. Thermomixer.

2.5 LC-MS/MS Analysis Components

1. Mobile phase: (a) Mobile phase A: composed of 97.9 % LC-MS-grade water, 2 % acetonitrile, and 0.1 % formic acid; (b) mobile phase B: composed of 80 % acetonitrile, 19.9 % LC-MS-grade water, and 0.1 % formic acid.
2. Nano-scale (n) LC system (Ultimate 3000, Dionex, Amsterdam, The Netherlands): (a) Trap column: C_{18}-particle-packed column (300-μm I.D. × 5 mm, 5-μm particle size, 100 Å pore size); (b) analytical column: reversed-phase C_{18}-particle-packed fused-silica capillary column (75-μm I.D. × 50 cm, 3-μm particle size); [3] nLC elution system.

3. High-resolution MS system: (a) Metal-coated nano-electrospray ionization (ESI) emitters (New Objective); (b) LTQ-Orbitrap-XL MS analyzer (Thermo Scientific).

2.6 LC-SRM-MS Analysis Components

1. Mobile phase: (a) Mobile phase A: composed of 99.9 % LC-MS-grade water and 0.1 % formic acid; (b) mobile phase B: composed of 99.9 % acetonitrile and 0.1 % formic acid.
2. Ultra performance LC (UPLC) system (Waters): (a) Trap column: Symmetry C_{18} nanoACQUITY column (180-μm I.D.×2 cm, 5-μm particle size); (b) analytical column: ACQUITY BEH300 C_{18} column (75-μm I.D.×20 cm, 1.7-μm particle size) (*see* **Note 5**).
3. MS analyzer (Waters): (a) Z-spray nanosource; [2] Xevo TQ-S triple-quadrupole (QQQ) mass spectrometer.

2.7 Data Analysis Pipeline

1. Human protein sequence database (downloaded from UniProt Knowledgebase).
2. MaxQuant software (downloaded from http://www.maxquant.org/) [12].
3. Skyline targeted proteomics software (downloaded from https://brendanx-uw1.gs.washington.edu/labkey/project/home/software/Skyline/begin.view?) [13].
4. Mascot Server and Mascot Daemon (Matrix Science Ltd).
5. LC-MS/MS dataset from breast cancer tissues.

3 Methods

This method can widely be used for biomarker verification of different solid tumors, which requires the use of cell lines from other types of cancers. In this chapter, we used breast cancer as an example due to the research focus of our laboratory. The method consists of the following sections: (1) cell culture and harvest, (2) tissue sample preparation, (3) protein extraction, (4) protein digestion, (5) LC-MS/MS analysis, (6) labeling check (MaxQuant analysis), (7) LC-SRM-MS analysis, and (8) SRM data analysis (Skyline analysis). The workflow of our pipeline is summarized in Fig. 1.

3.1 Cell Culture and Harvest

1. Preparation of SILAC medium: Add 10 % (v/v) dialyzed FBS, 1 % glutamax TM-I, 1 % antibiotics (100 μg/ml penicillin G and 80 μg/ml streptomycin), and 200 mg/l of L-proline into SILAC-labeled RPMI medium in which natural R and K are replaced by R10 and K8 (R10K8) (*see* **Note 2**).
2. Incorporation of R10K8 labels into breast cancer cell lines: Select 2–5 representative breast cancer cell lines, and maintain these cell lines in a 6-well plate for approximately ten doublings

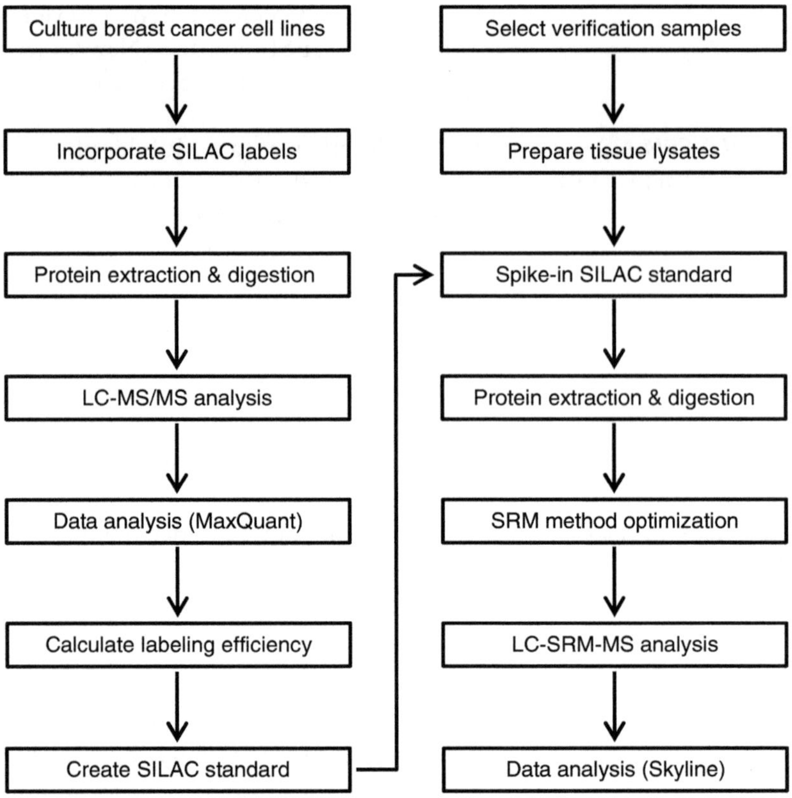

Fig. 1 A flow chart describing the procedure of developing SILAC-SRM assays to verify putative protein markers for breast cancer

prior to labeling check. Expand culturing of fully labeled cells (labeling efficiency >95 %) in culture flasks to collect enough SILAC standards for further experiments (*see* **Note 6**).

3. Harvest of SILAC standards: Pre-weigh equal numbers of 15-ml Greiner tubes as the number of cultured SILAC cell lines. Scrap cells using cell lifters, and suspend cells in 10 ml of ice-cold PBS. Add cell suspensions into 15-ml Greiner tubes. Use an additional 5 ml of ice-cold PBS to wash the culture flasks, and add the cell suspension into the same Greiner tubes. Centrifuge the Greiner tubes at 1,000 rpm for 5 min at 4 °C. Carefully remove supernatant, and immediately freeze cell pellets on dry ice. Weigh the Greiner tubes on a balance to roughly calculate the weight of each cell pellet. Store the cell pellets in −80 °C until further experiments.

3.2 Tissue Sample Preparation

1. Tumor tissue cryosectioning: Prior to cryosectioning, cool down the microtome system to desirable temperature (e.g., blade at −23 °C and tissue holder at −25 °C). Pre-weigh an equal number of protein LoBind tubes as the number of tumor tissues to be sectioned, and cool down these tubes on dry ice (*see* **Note 7**). Mount a tumor tissue on the holder with the aid

of TissueTek, but avoid fully embedding the tissue (*see* **Note 8**). Cut ten 10-μm sections of each tissue, and place these sections into a protein LoBind tube. Also collect few 5-μm sections on pathological glass slides before and after collecting ten 10-μm sections to inspect tumor percentage. Weigh tissue-containing LoBind tubes to calculate the approximate quantity of collected tissues (*see* **Note 7**).

2. Hematoxylin/eosin (HE) staining: Keep collected 5-μm sections of breast cancer tissues overnight at room temperature. Fix and rehydrate tissue sections consecutively in 70 and 50 % ethanol and tap water for 1 min respectively. Stain sections in hematoxylin for 30 s, and then blue the staining by washing tissue sections in running tap water for 5 min. Counterstain with eosin for 5 s, and subsequently rinse off excess staining in tap water. Dehydrate sections consecutively in 50, 70, and 2×100 % ethanol for 1 min and an additional 100 % ethanol wash for 5 min, followed by two times xylene washes for 1 min and a final xylene wash for 5 min. Place slides on a piece of tissue wipe, and add one drop of Pertex mounting medium on top of each section. Cover the sections with a cover slip without trapping air bubbles between tissue sections and cover slip. Let slides air-dry in the fume hood until Pertex mounting medium is solidified. Inspect the tissue sections under the microscope, and select those tissues with minimal 70 % tumor cell content for further experiments (*see* **Note 9**).

3.3 Protein Extraction

1. Preparation of protein lysates: Calculate quantity of SILAC cell lines and tissues by subtracting the weight of the empty Greiner/LoBind tube from the weight of the Greiner/LoBind tube containing SILAC cell lines or tissue sections. Add 0.1 % RapiGest solution to SILAC cell lines and tissue sections with a final concentration of 100 mg/ml (w/v). Vortex the mixture of SILAC cell lines and tissue sections and 0.1 % RapiGest for 10 s, followed by 1-min sonication at 70 % amplitude. Centrifuge the protein lysates at 14,000 rpm for 15 min at 4 °C. Transfer the supernatant to a new LoBind tube. Reserve 5 μl of protein lysates to test protein concentration. Store the rest of protein lysates in 20-μl aliquots at −80 °C.

2. Protein concentration assay: Dilute protein lysates 10 and 20 times 50 mM ABC solution (*see* **Note 10**). Pipette 10 μl of 2,000, 1,500, 1,000, 750, 500, 250, 125, 25, and 0 μg/ml BSA standard and 10× and 20× diluted protein lysates in duplicate in a 96-well plate. Add 200 μl of the working reagent to each well, and mix the plate on a plate shaker for 30 s. Cover the plate and incubate at 37 °C for 30 min. Measure the absorbance of each standard and sample at or near 562 nm on a plate reader. Calculate the concentration of each sample.

3. Sample dilution: Dilute each SILAC cell line and sample with 0.1 % RapiGest solution to a final concentration of 0.5 μg/μl. Aliquot each of the 10 μl tissue samples in protein LoBind tubes. For labeling check, keep each of the SILAC cell lines separately. To make a universal standard, mix equal amounts of different SILAC-labeled cell lines. For tissue samples, take one sample aliquot and mix it with the SILAC standard in 1:1 ratio. Store the samples at −80 °C until further experiments.

3.4 Protein Digestion

1. Protein denaturation: Preheat Eppendorf Thermomixer to 95 °C. Defrost the samples at room temperature, and immediately incubate samples at 95 °C for 5 min.
2. Protein reduction: Cool down Eppendorf Thermomixer to 60 °C. Add 100 mM DTT solution to the samples to reach a final concentration of 5 mM DTT. Briefly vortex and spin the samples down. Keep samples on ice until the Eppendorf Thermomixer has cooled down to 60 °C. Then incubate the samples at 60 °C and 650 rpm for 30 min. Cool down Eppendorf Thermomixer to 37 °C after protein reduction.
3. Protein alkylation: Add 300 mM IAA solution into the samples to reach a final concentration of 15 mM IAA. Briefly vortex and spin the samples down. Immediately place the samples in the dark for 30 min at room temperature.
4. Quench redundant IAA with L-cysteine (*see* **Note 3**): Add 600 mM L-cysteine solution into the samples to reach a final concentration of 30 mM IAA. Briefly vortex and spin the samples down. Immediately place the samples in the dark for 30 min at room temperature.
5. Trypsin digestion: Mix trypsin and the samples in a ratio of 1:50. Briefly vortex and spin the samples down. Incubate the samples at 37 °C and 650 rpm overnight.
6. Stop digestion, and precipitate RapiGest: Add 10 % TFA solution to the samples at a final concentration of 0.5 %. Briefly vortex and spin the samples down. Incubate the samples at 37 °C and 650 rpm for 30 min. Then centrifuge the samples at 14,000 rpm for 15 min. Transfer supernatant to LoBind tubes, and store them at −80 °C until LC-MS/MS or LC-SRM-MS analysis.

3.5 LC-MS/MS Analysis

1. Sample loading and washing: For each SILAC cell line, load 2 μl (equivalent to 1 μg protein) onto a trap column with 0.1 % TFA at a flow rate of 20 μl/min.
2. LC separation: Switch the trap column online to the analytical column. Gradually elute the peptide mixture out with a flow rate of 250 nl/min at 40 °C column temperature using the following binary gradient: start with 100 % mobile phase A to 25 %

mobile phase B over the first 120 min, and then increase mobile phase B to 50 % in the next 60 min.

3. MS measurement: Ionize eluted peptides using ESI. Acquire mass spectra over a mass-to-charge (m/z) ratio range of 400–1,800 at a resolving power of 30,000 at 400 m/z. Set target of automatic gain (AGC) at 10^6 ions and lock mass at 445.120025 u protonated with $((Si(CH_3)_2O)_6)$. Isolate top five intensive ions (AGC target set to 10^4 ions), and fragment these ions by collisionally activated dissociation applying 35 % normalized collision energy in the linear ion trap. Dynamically exclude parent ions within a mass window of ±5 ppm for dissociation for MS/MS fragmentation in the next 3 min or until the precursor ion intensity falls below a signal-to-noise of 1.5 for more than ten scans (early expiration). Acquire full and MS/MS scan partially and simultaneously in Orbitrap and linear ion trap, respectively.

3.6 Labeling Check

1. MaxQuant analysis: Analyze generated LC-MS/MS data in MaxQuant software. Set an initial search at a precursor mass window of 7 ppm. Select enzymatic cleavage rule of trypsin/P, and allow maximal two missed cleavage sites. Choose SILAC labels of R10 and K8 to analyze SILAC data. Select carbamidomethylation of cysteines as fixed modification while protein N-terminal acetylation, methionine oxidation, and heavy proline (P6) as variable modification (*see* **Note 2**). Construct the MS/MS file by extracting and submitting up to top eight peaks per 100 Da window to search against a concatenated forward and reverse versions of the human protein sequence database (UniProt Knowledgebase). Set a cutoff of global false discovery rate for peptide and protein identification at 0.01, and allow only peptides with ≥7 amino acid residues for identification. Require minimally one unique peptide for protein identification. Unselect "re-quantify" option for labeling check.
2. Calculate labeling efficiency: Open the "peptides.txt" file generated from MaxQuant analysis. Obtain unnormalized SILAC ratios from "Ratio H/L" columns. Compute labeling efficiency using the formula "1−1/average ratio" [14]. Only use SILAC-labeled cells to create the universal standard when the computed labeling efficiency is >95 %. In addition, check the number of peptides with P6 modifications. Inspect the labeling efficiency of R- and K-containing peptides separately when >2 % of peptides with P6 modification occur (indication of R-to-P conversion) (*see* **Note 2**).

3.7 LC-SRM-MS Analysis

1. Construction of MS/MS spectral library: Use LC-MS/MS dataset from discovery study or download publically available datasets generated from breast cancer tissues (*see* **Note 11**). Submit these

datasets to the Mascot Server, and search against a human protein sequence database (UniProt Knowledgebase) using similar settings as the MaxQuant search (*see* **Note 12**). Use search output ".dat" file to construct an MS/MS spectral library.

2. Selection of peptides for SRM analysis: Select peptides to develop SRM assays following the rules below: (a) unique to one protein; (b) fully tryptic; (c) 8–25 amino acid residues; (d) excluding first 25 amino acids at the N-terminus of proteins; (e) excluding ragged ends (tryptic peptides cleaved between R/K, K/R, R/R, and K/K); (f) excluding peptides with C, M, RP, and KP (*see* **Note 13**); (g) maximal one neutral loss; and (h) selecting 3–5 peptides per protein whenever possible (*see* **Note 14**).
3. Selection of SRM transitions: Select SRM transitions from spectral library using the following criteria: (a) precursor ions with two or three charges (*see* **Note 15**), (b) daughter ions are y ions with one or two charge(s) (*see* **Note 15**), (c) *m/z* range of precursor ions is between 50 and 1,500 Th, (d) mass tolerance of precursor ions is set at 0.055 Th, (e) mass tolerance of daughter ions is set at 0.5 Th, and (f) five most intense transitions per peptide are selected to test SRM performance (*see* **Note 16** and Fig. 2).
4. LC method: Inject 2 μl of the sample into the trap column and wash for 5 min at a flow rate of 8 μl/min with a solution of 99 % solvent A and 1 % solvent B. Switch the trap column online to the analytical column. Separate samples on the analytical column using a gradient increasing from 1 to 20 % solvent B in 60 min and in the next 30 min to 60 % solvent B at a flow rate of 300 nl/min (*see* **Note 5**).
5. ESI method: For the Xevo TQ-S QQQ mass spectrometer, set capillary voltage at 3.00 kV, cone voltage at 50 V, source offset at 50 V, and source temperature at 70 °C. Use appropriate settings for QQQ instruments from other vendors.
6. Optimization of SRM method: Measure a test sample on the Xevo TQ-S QQQ mass spectrometer using created methods. Initially test all the peptide precursors with both 2 and 3 charge states. Select the charge state with the most intense peak for further experiments. Optimize collision energy step by step through gradually increasing or decreasing 1 V from calculated default collision energy to obtain the desirable peak intensity (Fig. 3). Based on the initial screening, schedule retention time within a suitable window to obtain optimal sensitivity (*see* **Note 17** and Fig. 3). Based on the initial SRM screening, select three peptides per protein and three transitions per peptide (whenever possible, *see* **Notes 14** and **16**) for further measurements.
7. Large-scale measurements: Use optimized methods to measure a large set of samples. Check if there is a clear carryover effect

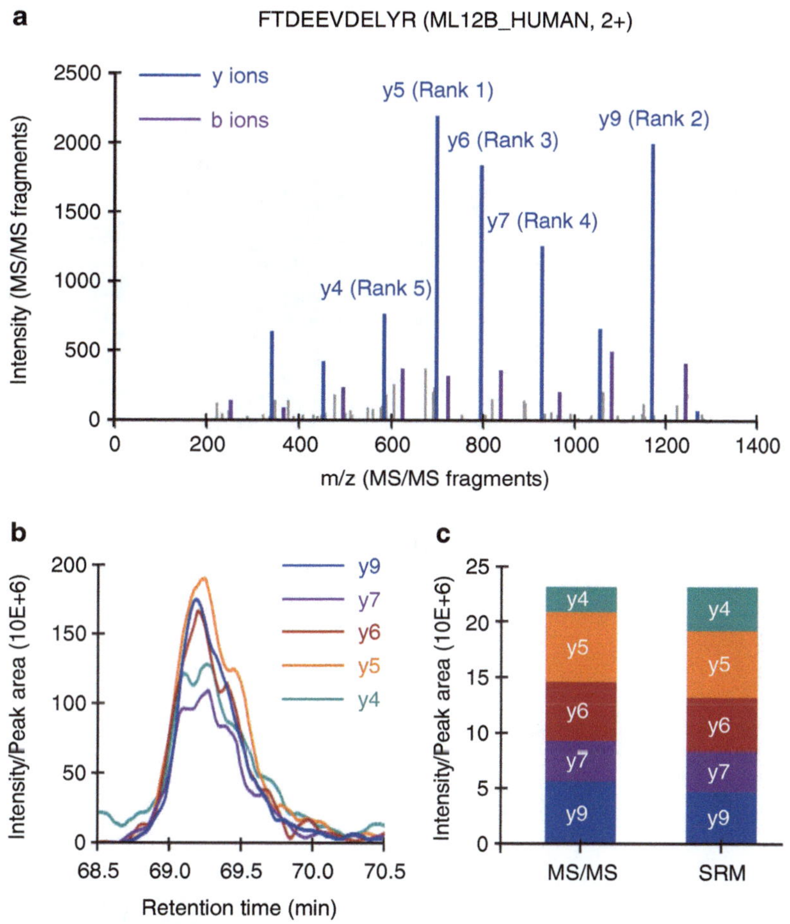

Fig. 2 Use of MS/MS spectral library to select robust transitions for SILAC-SRM assays. (**a**) An MS/MS spectrum of the peptide "FTDEEVDELYR" from the protein "ML12B_HUMAN." We selected the top five most intense y ions for development of SILAC-SRM assays. (**b**) Chromatograms of the transitions from these five y ions in SRM measurements. (**c**) A good concordance was observed between relative peak intensities in the MS/MS spectrum and peak areas measured by SRM assays

between different runs. If there is no clear carryover effect, no need to wash column in between different sample runs. Recalibrate retention time when a large shift in retention time is observed in between different runs (*see* **Note 17**).

3.8 SRM Data Analysis

1. Manual integration of incorrectly assigned peaks: When a peak is incorrectly assigned and integrated, manually assign and integrate it based on its transition and retention time information. Make sure to integrate light and heavy peaks over the same retention time range.

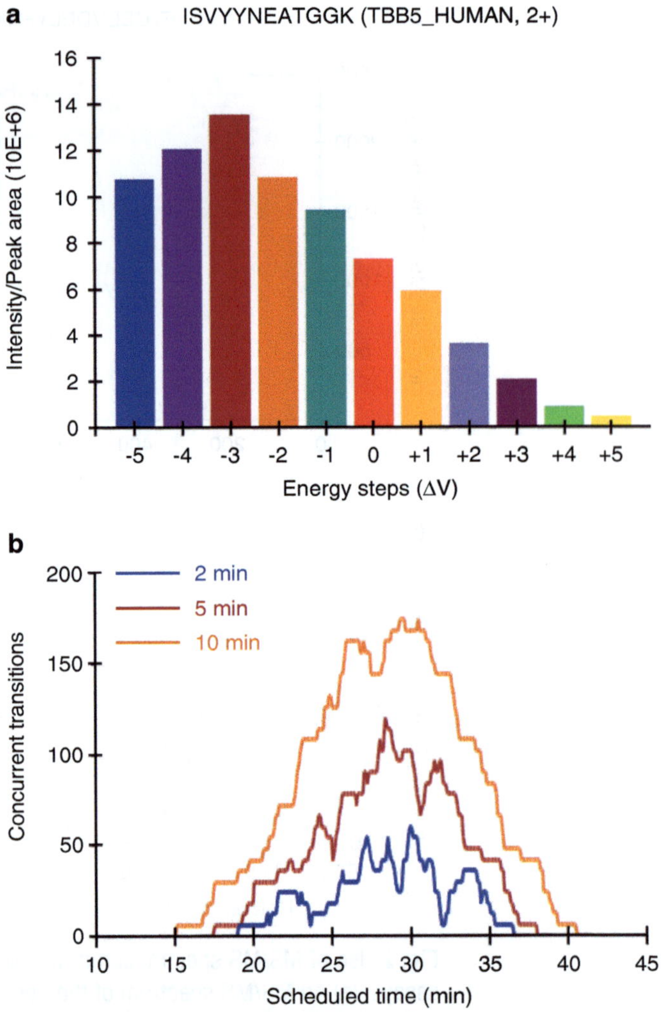

Fig. 3 Method optimization for SRM assays. (**a**) Optimization of collision energy for the peptide "ISVYYNEATGGK" from the protein "TBB5_HUMAN." The *red bar* is the collision energy predicted by the software (17.0 V plotted at 0). The optimal collision energy for this peptide can be achieved by decreasing 3 V (at −3) from the calculated default collision energy (*dark red bar*). (**b**) Number of concurrent transitions measured by the instrument using 2-, 5-, and 10-min window around previously measured unscheduled retention times

2. Peak quantification: Frequently use all three transitions of a peptide for quantification. Discard a transition when a severe interference is observed (*see* **Note 18** and Fig. 4).
3. Export of quantitative data: Export quantitative data into ".csv" file for further downstream data analysis.

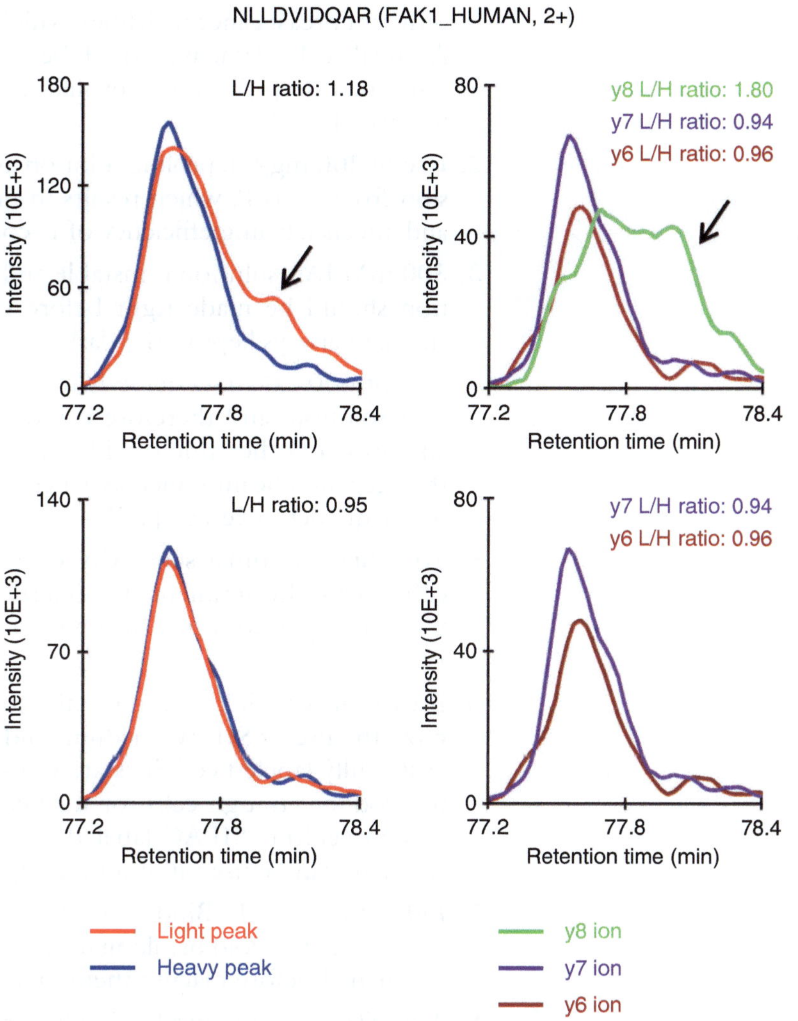

Fig. 4 Improvement of SILAC-SRM quantification in the presence of interference. Light peak of the peptide "NLLDVIDQAR" (FAK1_HUMAN) is interfered by a shoulder peak (*top left panel*), which is caused by the transition from y8 ions (*top right panel, green line*). This transition results in a different light/heavy (L/H) ratio (1.80) in comparison of the other two transitions (y7 = 0.94 and y6 = 0.96). After removing this transition, the shoulder peak disappears (*lower left panel*). This removal leaves two transitions with similar L/H ratios for quantification (*lower right panel*)

4 Notes

1. Generally, we select 2–5 breast cancer cell lines from different molecular subtypes (luminal, ErBB2/neu overexpression, basal and normal like [8]). Ideally, choice of cell lines should be based on the expression of the proteins of interest in certain cell lines. This, however, requires deep proteome profiling of

different breast cancer cell lines, which is currently not available. A small-scale database could be constructed from recently published deep proteome profiling of the NCI-60 cancer cell line panel [15].

2. Use of 200 mg/l L-proline solution avoids biochemical conversion from R to P, which results in unwanted P6 modification and affects labeling efficiency of R-containing peptides [16].
3. 300 mM IAA solution is instable and light sensitive. This solution should be made right before performing the alkylation step and always kept in the dark.
4. Use of 600 mM L-cysteine is to quench redundant 300 mM IAA solution, and therefore it reduces undesirable unspecific alkylation of other amino acids. From our previous experience, this step significantly increases peptide identification and protein sequence coverage [17].
5. To achieve optimal sensitivity of SRM measurements, short column and LC gradient are preferred in order to elute each of the single peptides in a short retention time interval (sharp LC peak).
6. Breast cancer cell lines can be labeled in a 6-well plate to minimize the use of SILAC medium and therefore reduce reagent cost. Fully labeled cell lines are transferred to 75- or 162-cm^2 to produce enough cells for further experiments. We recommend to culture SILAC-labeled cells in bulk so that the SILAC standard can be used in multiple experiments.
7. Lids of protein LoBind tubes become fragile under freezing condition. To avoid breaking lids, protein LoBind tubes should be opened before placing them on dry ice.
8. TissueTek is composed of polymers, which is incompatible with MS analysis. Trace amounts of TissueTek in tissue lysates can induce ion suppression and chemical background in MS measurements.
9. We recommend using tissues with minimally 70 % tumor cell content to reduce the influence of tumor heterogeneity on verification of protein markers.
10. When sufficient material is available, several dilutions of protein lysate can be made to ensure that at least one of the BCA measurements is in the range of the calibration curve.
11. MS/MS spectral library created from a large set of breast cancer tissues is desirable. An increased number of MS/MS spectra results in more choice for selecting candidate peptides and transitions for SRM assays.
12. Search results of MaxQuant (".res" file) are incompatible with Skyline software. Therefore, we usually use Mascot Server to generate ".dat" file in order to build MS/MS spectral library.

13. Peptides with carbamidomethylation on cysteine are sometimes used due to the lack of enough peptides for SRM assays, since this modification is relatively constant compared to other modifications.
14. We prefer to select more peptides per protein from the MS/MS spectral library. Some peptides may not have robust transitions in SRM assays, and therefore they should be excluded in finalized SRM methods.
15. Selection of charge state is based on the intensity of a peptide in each of the charge states. The most intense charge state should be selected to develop SRM assays.
16. Eventually, three transitions per peptide are required for a good SRM assay. However, more transitions can be used for initial test.
17. A good balance should be achieved in selecting the retention time window. A too narrow window results in failure of triggering a defined transition in some samples due to shift of retention time in between LC-SRM-MS runs. A too wide window leads to reduced sensitivity and therefore weakens the advantage of using a scheduled method. The retention time should be recalibrated when a large shift is observed in between different runs. The scheduled method can be modified based on shifted retention time.
18. A transition which is severely interfered by an unspecific transition should be discarded for quantification due to difficulty in estimating actual peak area from this transition.

Acknowledgements

This work was financially supported through the Netherlands Genomics Initiative (NGI) and the CTMM Breast CARE project 030–104.

References

1. Liu NQ, Braakman RBH, Stingl C et al (2012) Proteomics pipeline for biomarker discovery of laser capture microdissected breast cancer tissue. J Mammary Gland Biol Neoplasia 17:155–164
2. Rifai N, Gillette MA, Carr SA (2006) Protein biomarker discovery and validation: the long and uncertain path to clinical utility. Nat Biotechnol 24:971–983
3. Cima I, Schiess R, Wild P et al (2011) Cancer genetics-guided discovery of serum biomarker signatures for diagnosis and prognosis of prostate cancer. Proc Natl Acad Sci U S A 108: 3342–3347
4. Addona TA, Shi X, Keshishian H et al (2011) A pipeline that integrates the discovery and verification of plasma protein biomarkers reveals candidate markers for cardiovascular disease. Nat Biotechnol 29:635–643
5. Picotti P, Rinner O, Stallmach R et al (2010) High-throughput generation of selected reaction-monitoring assays for proteins and proteomes. Nat Methods 7:43–46

6. Geiger T, Cox J, Ostasiewicz P et al (2010) Super-SILAC mix for quantitative proteomics of human tumor tissue. Nat Methods 7:383–385
7. Domon B, Aebersold R (2010) Options and considerations when selecting a quantitative proteomics strategy. Nat Biotechnol 28:710–721
8. Liu NQ, Dekker LJM, Stingl C et al (2013) Quantitative proteomic analysis of microdissected breast cancer tissues: comparison of label-free and SILAC-based quantification with shotgun, directed, and targeted MS approaches. J Proteome Res 12:4627–4641
9. Jaffe JD, Keshishian H, Chang B et al (2008) Accurate inclusion mass screening: a bridge from unbiased discovery to targeted assay development for biomarker verification. Mol Cell Proteomics 7:1952–1962
10. Umar A, Kang H, Timmermans AM et al (2009) Identification of a putative protein profile associated with tamoxifen therapy resistance in breast cancer. Mol Cell Proteomics 8:1278–1294
11. Addona TA, Abbatiello SE, Schilling B et al (2009) Multi-site assessment of the precision and reproducibility of multiple reaction monitoring-based measurements of proteins in plasma. Nat Biotechnol 27:633–641
12. Cox J, Mann M (2008) MaxQuant enables high peptide identification rates, individualized p.p.b.-range mass accuracies and proteome-wide protein quantification. Nat Biotechnol 26:1367–1372
13. MacLean B, Tomazela DM, Shulman N et al (2010) Skyline: an open source document editor for creating and analyzing targeted proteomics experiments. Bioinformatics 26: 966–968
14. Geiger T, Wisniewski JR, Cox J et al (2011) Use of stable isotope labeling by amino acids in cell culture as a spike-in standard in quantitative proteomics. Nat Protoc 6:147–157
15. Moghaddas Gholami A, Hahne H, Wu Z et al (2013) Global proteome analysis of the NCI-60 cell line panel. Cell Rep 4:609–620
16. Bendall SC, Hughes C, Stewart MH et al (2008) Prevention of amino acid conversion in SILAC experiments with embryonic stem cells. Mol Cell Proteomics 7:1587–1597
17. Braakman RBH, Tilanus-Linthorst MMA, Liu NQ et al (2012) Optimized nLC-MS workflow for laser capture microdissected breast cancer tissue. J Proteomics 75: 2844–2854

Chapter 22

The Secretome Analysis by High-Throughput Proteomics and Multiple Reaction Monitoring (MRM)

Vitor M. Faça, Camila S. Palma, Daniele Albuquerque, Gabriela N.S. Canchaya, Mariana L. Grassi, Vera L. Epifânio, and Eduardo B. de Oliveira

Abstract

The secretome is a sub-proteome of great interest in several fields of biomedical sciences, especially as a source of diagnostics and therapeutic targets. Proteomics has been contributing significantly to elucidate the secretome of a great diversity of cells, tissues, and organisms, turning profiles of thousands of proteins a usual practice. After elucidation of long protein lists, targeted proteomics also plays important roles in accurate quantification and validation of such secreted proteins. Here we present detailed protocols to explore and quantify the secretome of cancer cells, even though this protocol can be employed to any kind of biological material.

Key words Secretome, SILAC, Cancer, Protein fractionation, Multiple reaction monitoring, Mass spectrometry

1 Introduction

The secretome comprises the subset of the proteome that is secreted by a cell, tissue, or organism. It is a very rich sub-proteome, since it contains proteins involved in signaling and communication, such as cytokines and chemokines involved in immune response, as well as proteases responsible for matrix degradation and remodeling such metalloproteases involved in cancer progression, therefore relevant to key biological processes [1, 2]. In addition, secreted proteins are of great biotechnological and medical relevance, since they can represent therapeutic products such as coagulation factors or diagnostic targets as the prostate-specific antigen (PSA) [3].

Profiling secretomes has been challenging, since it is very dynamic and susceptible to extracellular conditions, allowing posttranslational modification, degradation, or even fast excretion

Daniel Martins-de-Souza (ed.), *Shotgun Proteomics: Methods and Protocols*, Methods in Molecular Biology, vol. 1156, DOI 10.1007/978-1-4939-0685-7_22, © Springer Science+Business Media New York 2014

in a living system. Human plasma or serum can be considered the most complex and challenging secretomes to be profiled and characterized in detail. Not only the large number of components secreted by body cells but also the wide range of protein concentrations make these biological fluids still to be uncovered [4].

On the other hand, proteomics technology has evolved significantly in the last decade, especially in terms of instrumentation, allowing faster, more sensitive, and more accurate mass measurements [5]. This evolution drives comprehensive studies that are just recently capable to elucidate over thousands of proteins in complex mixtures such as secretomes. To our knowledge, protein fractionation plus state-of-the-art mass spectrometry has achieved the best results in terms of secretome coverage [6, 7]. In addition, bioinformatics workflows have also aided elucidating secretomes using better annotated database, appropriated data filtering, and improved prediction of secreted proteins [8, 9].

Among the well-succeeded secretome studies, those that used cells in culture have demonstrated the best results. Good examples can be represented by the characterization of the secretome of colon cancer stem cells [10] and from ovarian cancer cells [7]. In both studies, protein fractionation, respectively, by SDS-PAGE or by liquid chromatography were used prior to protein tryptic digestion and liquid chromatography coupled to mass spectrometry (LC-MS/MS) in order to decomplex sample and expand the list of protein identifications. In both cases, high-throughput proteomic workflows were important to elucidate hundreds of relevant proteins to colon and ovarian cancer. After that, the appropriate validation of proteins becomes the challenge. However, the lack of good immunoaffinity reagents for precise quantification of soluble proteins by ELISA, for example, hinders the clinical development of relevant proteins with potential biomedical application [11]. In that scenario, targeted proteomics, based on multiple reaction monitoring (MRM) by LC-MS/MS, has emerged as an important alternative for multiplex studies of targeting several different proteins and compatible with large sample sets [12].

Here we present our high-throughput and targeted proteomics detailed protocols for profiling and quantitation of secreted proteins in cancer cells maintained in culture. Nevertheless, this protocol is applicable to any kind of secretome.

2 Materials

Prepare fresh solutions using MilliQ water on the day of use and maintain at room temperature, unless indicated otherwise.

2.1 Cell Culture Components

1. Cell culture media: Dulbecco's modified Eagle medium (DMEM) supplemented with 10, 0.1, or 0 % fetal bovine serum (FBS) (*see* **Note 1**).
2. SILAC cell culture media: DMEM (Flex—Life Technologies) containing lysine $^{13}C_6$ (heavy lysine) supplemented with 10 % dialyzed FBS (*see* **Note 2**).
3. Tissue culture dishes for adherent cell culture (60.1 cm^2 area—TPP).

2.2 Protein Fractionation Components

1. Solvents for reversed-phase chromatography: Aqueous solvent (A)—5 % acetonitrile/95 % water/0.1 % of trifluoroacetic acid; organic solvent (B)—75 % acetonitrile/15 % isopropanol/10 % water/0.095 % trifluoroacetic acid.
2. Chromatography: Use a POROS Rl/10 column (Applied Biosystems—4.6 × 50 mm).

2.3 Sample Preparation Components for Proteomic Analysis

1. Bradford quantification kit.
2. Denaturation buffer: Urea 8 M, Tris–HCl 0.15 M, pH 8.5. Weight 1.5 g of urea and 454 mg of trizma base and dilute with 15 ml of water. Mix until solids are completely dissolved, and adjust pH with HCl (diluted 1:1 in water). Transfer to a 25 ml volumetric flask, and adjust volume.
3. Trypsin solution: Sequencing-grade modified trypsin (100 μg/ml). Resuspend one vial containing 20 μg of sequence grade modified trypsin (Promega) in 200 μl of 0.1 M ammonium bicarbonate solution (dissolve 395 mg of ammonium bicarbonate in 50 ml of water and filter through a 0.22 μm syringe filter).
4. Reducing and alkylation solutions: Dissolve 10 mg of dithiothreitol in 1 ml of 0.15 M Tris–HCl, pH 8.5 solution. Dissolve 10 mg of iodoacetamide in 0.5 ml 0.15 M Tris–HCl, pH 8.5 solution (*see* **Note 3**).
5. Solid-phase purification: OASIS HLB solid-phase-extraction columns (Waters). Equilibration solution is 95 % water/5 % acetonitrile/0.1 % of formic acid. Elution solution is 30 % water/70 % acetonitrile/0.1 % of formic acid.

2.4 High-Throughput Mass Spectrometry Components

1. Equipment: LTQ-ORBITRAP mass spectrometer (Thermo-Finnigan) coupled to a nanoflow chromatography system (Eksigent).
2. Chromatographic column: 25 cm long column (Picofrit 75 mm ID, New Objectives, packed in-house with MagicC18 resin).
3. Solvents for reversed-phase chromatography: Aqueous solvent (A)—5 % acetonitrile/95 % water/0.1 % of formic acid;

organic solvent (B)—95 % acetonitrile/5 % water/0.1 % formic acid. Bottled water, acetonitrile, and formic acid are obtained from Fisher Scientific.

2.5 Targeted Mass Spectrometry Components

1. Equipment: Tandem quadrupole Xevo-TQs mass spectrometer (Waters) coupled to a Class I Ultra Performance Chromatographic System (Waters).
2. Chromatographic column: ACQUITY UPLC HSS C18 Column, 1.8 μm particle size, 1 mm ID × 150 mm long (Waters).
3. Solvents for reversed-phase chromatography: Aqueous solvent (A)—5 % acetonitrile/95 % MilliQ water/0.1 % of formic acid; organic solvent (B)—99.9 % acetonitrile/0.1 % formic acid. Acetonitrile is of LC-MS grade and obtained from Fluka; formic acid is 99.9 % pure.

3 Methods

3.1 High-Throughput Profiling of Cancer Cell Lines

1. Cultivate cancer cell lines using standard cell culture conditions or according to user-adapted protocols (*see* **Note 4**).
2. Select one 60 cm^2 plate with cellular confluence of approximately 50 %, wash twice with sterile phosphate-buffered saline (PBS) solution preheated at 37 °C, and add 6 ml of the complete SILAC media (preheated at 37 °C) containing the heavy amino acid ($^{13}C_6$ Lys) and supplemented with 10 % dialyzed FBS provided with SILAC kit.
3. Perform the cellular expansion according to appropriate cell culture conditions up to five doubling cycles (*see* **Note 5**).
4. Remove cell media from all plates and replace with 6 ml of new SILAC media supplemented with 0.1 % dialyzed FBS. Keep cells in culture for an additional 48 h to collect cellular secretion.
5. Pool the cellular conditioned media containing secreted proteins, add immediately to complete protease inhibitor cocktail (1 tablet/10 ml of pooled media, Roche Diagnostics, Germany), and remove cells and debris first by centrifugation at 5,000 × *g* and then by filtration through a 0.22 mm filter.
6. Concentrate conditioned media to approximately 1 ml using centrifuge concentrators, with molecular weight cutoff of 3 kDa. Store solution at 4 °C until fractionation (*see* **Note 6**).
7. To the cells remaining attached to the plate, add 500 μL of a solution of 1 % (v/v) of the detergent *n*-octyl-glucoside (OG) in PBS, containing complete protease inhibitor cocktail (1 tablet/5 ml of solution, Roche Diagnostics, Germany). Collect cells using a cell scraper, and combine extracts. Additional cellular protein extraction is obtained by sonication

of solution using an ultrasound probe with alternating three cycles of 30-s sonication and ice bath. Centrifugate solution at 20.000 ×*g* to remove cellular debris. Store at 4 °C until processing.

8. Quantify concentrated conditioned media and cellular extract using Bradford Assay kit according to the manufacturer's instructions.
9. Aliquot 500 μg of both conditioned media and total cell extract, and perform protein reduction by adding 50 μL of reduction solution and maintaining the reaction at 37 °C for 1 h. Right after that, perform protein alkylation by adding 100 μL of alkylating solution and maintain the reaction at room temperature for an additional 1 h. Quench the reaction by direct injection in reversed-phase chromatography, used as fractionation step.
10. Fractionate the complex protein mixtures of conditioned media and total cell extract using reversed-phase chromatography. Perform separation at 2.7 ml/min using a linear gradient of 10–80 % of solvent B over a 30-min run. Collect fractions at a rate of three fractions/minute (*see* Subheading 2.2 and **Note 7**).
11. Pool protein fractions according to chromatographic features and lyophilize (*see* **Note 8**).
12. Perform in-solution trypsin digestion of pooled fractions by diluting samples in 30 μL of a solution containing 0.5 M urea, 0.1 M ammonium bicarbonate, and 3 % acetonitrile. Agitate vigorously, and add 10 μl of trypsin solution. Carry out digestion for 18 h or overnight at 37 °C. Trypsin enzymatic activity is quenched by addition of 1 μL of 10 % formic acid solution in water (*see* **Note 9**).
13. Centrifugate samples at 12,000 ×*g* for 15 min and transfer to mass spectrometry-compatible injection tubes.
14. Carry out the high-throughput LC-MS/MS data collection for each individual fraction obtained from conditioned media and total cell extract. Analyze samples over a 90-min linear gradient from 5 to 35 % of organic solvent at 350 nl/min (*see* Subheading 2.4 and **Note 10**).
15. Process LC-MS/MS files through data bank search and protein inference software (*see* **Note 11**).
16. Match the lists of proteins identified in the conditioned media (secretome) and proteins identified in the total cell extract, and select secreted proteins based on the higher value of enrichment obtained from the ratio of spectral counts observed in conditioned media profile/total cell extract profile. Table 1 presents a representative data obtained from ovarian cancer cells (*see* **Note 12**).

Table 1
Profiling of ovarian cancer cell secretome

				OVCAR3			CaOV3		
	IPI	Gene name	Description	Total extract counts	Secreted counts	Secreted / total extract	Total extract counts	Secreted counts	Secreted / total extract
1	IPI00180675	TUBA1A	Tubulin alpha-3 chain		156	505			
2	IPI00008580	SLPI	Antileukoproteinase 1 precursor		100	324		3	16
3	IPI00215965	HNRPA1	Heterogeneous nuclear ribonucleoprotein A1 isoform b		71	230	73	22	1
4	IPI00007257	CLSTN1	Alcadein alpha-1		65	211		54	270
5	IPI00027166	TIMP2	Metalloproteinase inhibitor 2 precursor		52	169		30	150
6	IPI00011695	PRSS2	Protease serine 2 isoform B		51	166			
7	IPI00297284	IGFBP2	Insulin-like growth factor-binding protein 2 precursor	1	102	165		9	46
8	IPI00032293	CST3	Cystatin C precursor		45	146		78	389
9	IPI00016112	PXDN	PREDICTED: Melanoma-associated gene		44	143		22	110
10	IPI00305380	IGFBP4	Insulin-like growth factor-binding protein 4 precursor		42	137		6	31
11	IPI00003590	QSOX1	Quiescin Q6, isoform a	1	78	126		18	91
12	IPI00018305	IGFBP3	Insulin-like growth factor-binding protein 3 precursor		37	120		3	16

13	IPI00290085	CDH2	Neural-cadherin precursor		35	114			
14	IPI00293881	COL5A2	COL5A2 protein	1	70	113			
15	IPI00021048	FER1L3	Splice isoform 1 of myoferlin		32	104	34		
16	IPI00022431	AHSG	Alpha-2-HS-glycoprotein precursor		31	101		3	16
17	IPI00007798	TRHDE	Thyrotropin-releasing hormone-degrading ectoenzyme		30	98			
18	IPI00553177	SERPINA1	Alpha-1-antitrypsin precursor		30	98	5	52	43
19	IPI00032464	KLK5	Kallikrein-5 precursor	1	58	94	2	59	98
20	IPI00010333	MDK	Midkine precursor		28	91		16	81
			Total # of protein	3,410	2,003		2,963	1,104	
			Normalization	1.000	0.310		1.000	0.201	
			Total counts	86,611	26,832		62,687	12,597	

Cell lines OVCAR3 and CaOV3 were profiled by high-throughput proteomics. The ratio of spectral counts of proteins detected in the secretome profile over spectral counts detected in total cell extract was used as an enrichment factor for secreted proteins. The list presents the top 20 proteins of OVCAR3 cell line and matching proteins from CaOV3 cell line. Data was adapted from Faca et al. [7]

3.2 Targeted Proteomic Analysis of Cancer Cell Line Secretome

1. Cultivate cancer cell lineages in 60 cm^2 plates in DMEM supplemented with 10 % FBS or according to the protocols for specific cells. Remove media containing FBS, and cultivate cells in serum-free medium for 48 h to collect conditioned medium (*see* **Note 13**).
2. Collect 1 ml of cellular conditioned media containing secreted proteins, and add immediately 5 μL of complete protease inhibitor cocktail (Sigma-Aldrich—Cat# P8340). Remove cells and debris first by centrifugation at 12,000 × *g* and then by filtration through a 0.22 mm filter (*see* **Note 14**).
3. Dry conditioned medium aliquot in speedvac.
4. Resuspend dried conditioned medium in 100 μl of denaturation buffer.
5. Add 900 μl of ice-cold acetone to precipitate proteins. Incubate samples at −20 °C for at a least 1 h with frequent vortexing (*see* **Note 15**).
6. Centrifuge precipitated samples for 30 min at 12,000 × *g* at 4 °C.
7. Discard supernatant, and allow the remaining acetone to evaporate at room temperature for 30 min in a vented hood or dry in speedvac.
8. Resuspend pellet in 100 μl denaturation buffer, and quantify the total proteins by Bradford protein assay (*see* **Note 16**).
9. Reduce the proteins cysteine residues by adding 5 μL of reduction solution and maintaining the reaction at 37 °C for 1 h. Right after that, perform protein alkylation by adding 10 μL of alkylating solution and maintain the reaction at room temperature for an additional 1 h.
10. Dilute samples to approximately 0.6 M by adding 1 ml of 100 mM Tris–HCl, pH 8.5 solution (*see* **Note 17**).
11. Perform in-solution trypsin digestion by adding 10 μl of trypsin solution to each digest to a final enzyme:protein ratio of 1:50 (w/w) and incubate for 2 h at 37 °C. Add more 5 μl of trypsin solution to each digest and incubate for an additional 16 h or overnight at 37 °C.
12. Desalt samples using solid-phase extraction as follows: (a) condition column with 1 ml of acetonitrile; (b) equilibrate column with 1.6 ml equilibration solution; (c) apply sample through the column; (d) wash column with 1.6 ml of equilibration solution; (e) elute peptides with 1.2 ml of elution solution; and (f) dry eluted peptides in speedvac. Samples can be stored dried at −80 °C until ready for MRM analysis.
13. Develop the MRM method for quantification of selected secreted proteins (*see* **Note 18**).

14. Reconstituting samples for MRM analysis with 50 μL of 3 % acetonitrile and 0.1 % formic acid. Vortex samples thoroughly, centrifugate for 15 min at 12.000 × *g*, and transfer the supernatant to a mass spectrometer-compatible injection vial.
15. Inject 10 μL of samples in triplicate through the LC-MS/MS system using a linear gradient of 5–30 % of organic solvent over 15 min at 130 μl/min in order to quantify targeted peptides by MRM (*see* Fig. 1, Subheading 2.5, and **Note 19**).

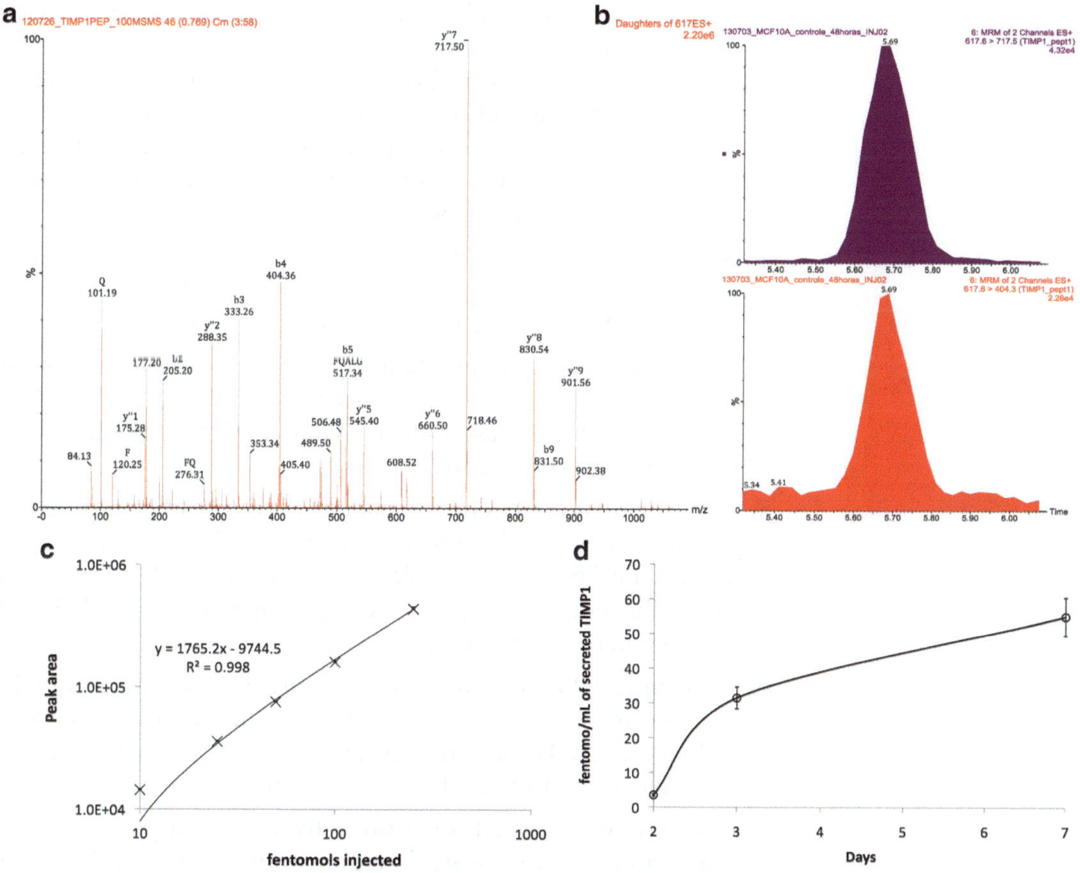

Fig. 1 Development of MRM assay for protein TIMP1. The peptide selected as an external standard for absolute quantification of TIMP1 protein has the sequence GFQALGDAADIR (*m/z* 617.5). (**a**) MS2 spectrum for TIMP1 peptide illustrating the higher intensity of fragment ions y7 (quantification ion) and b4 (confirmation ion). (**b**) MRM monitoring of quantification transition 617.5 > 715.5 and respective coeluting confirmation transition 617.5 > 404.3. (**c**) Calibration curve for TIMP1 peptide covering approximately two orders of magnitude. All injections were performed in triplicate, and the coefficients of variation for calibration points were all less than 4.3 %. (**d**) Absolute quantification of secretion of TIMP1 in MCF10A breast epidermal cell line indicating higher secretion rates at early days of culture. For these quantifications, 1 ml of conditioned media at days 2, 3, and 7 were collected and quantified

4 Notes

1. Cell culture in DMEM can be used for most common cancer cell lines. Adapt to appropriate cell culture media following ATCC (www.atcc.org) specifications if necessary.
2. There are currently in the market several kinds and manufacturers of culture media for SILAC experiments. User can choose among different base media types and isotope of amino acids used for labeling. We found that lysine isotope labeling ($^{13}C_6$-Lys) is convenient because of high proteome coverage in quantitative terms (>70 %) and it is simpler to process through standard proteomic pipelines such as Labkey Server [13] or Trans-Proteomic Pipeline [14]. The use of media without phenol red pH indicator is also recommended, since it is retained by reversed-phase chromatography, used in the described methods as fractionation step or for sample cleanup. Use the manufacturer's protocol for reconstitution and supplementation of SILAC culture media.
3. Always prepare fresh dithiothreitol and iodoacetamide solutions to guarantee efficient reduction and alkylation of cysteine residues. Do not store these solutions.
4. We performed experiments using standard conditions for the following cell lines: ovarian cancer (CaOV3, OVCAR3); lung cancer (A549); breast cancer (MCF7, MDA-MB-231); and pancreatic cancer (PANC1).
5. Upon cellular confluence, expand one culture plate to two new culture plates. At the end of five cycles of cellular doublings, you obtain 16 cell culture plates, with at a least 98 % incorporation of heavy lysine isotope. We expect approximately $2–5 \times 10^6$ cells per plate at this stage.
6. Concentrated conditioned media must be fractionated quickly in order to avoid precipitation. It is recommended to perform reduction, alkylation, and reversed-phase fractionation on the same day of conditioned media collection.
7. The system described for protein fractionation provides good separation and low protein retention into column. The same system can be used for all sorts of protein extracts, ranging from very soluble samples such as plasma or conditioned medias to membrane protein extractions. In the last case, the use of the detergent OG is a determinant to allow appropriate protein solubilization and adequate fractionation.
8. Pool reversed-phase factions in order to obtain approximately the same amount of protein in each fraction. Fractions containing intense peaks can be analyzed individually, while several fractions containing low-intensity peaks can be pooled together.

This strategy tries to maximize the number of identification in the following LC-MS/MS step. The number of pools is dependent on the time available for LS-MS/MS data collection. Routinely we use 15–20 fractions for deep secretome coverage.

9. For 500 μg of total protein fractionated, we estimate approximately 25 μg in each of the 20 pools formed. By adding 1 μg of trypsin to each pool we guarantee an enzyme:substrate ratio of 1:25.
10. Using the described conditions and instrumentation, it is expected that each run contains at least 300 good protein identifications. All fractions from conditioned media and total cell extracts are taken together; this procedure is expected to generate a list of confident identification containing more than 4,000 protein hits.
11. Acquired data can be automatically processed by the Computational Proteomics Analysis System—CPAS, nowadays called Labkey Server (www.labkey.org) [13]. Search data against the most recent version of the human IPI database or other appropriate human protein database. A fixed modification of 6.020129 mass units is added to lysine residues for database searching in order to account for incorporation of the heavy lysine isotope into cellular and secreted proteins. This way, lysine-containing peptides coming from FBS are automatically excluded. To estimate the significance of peptide and protein matches, we apply the tools PeptideProphet [15] and ProteinProphet [16]. Identifications with a PeptideProphet probability greater than 0.75 are selected and submitted to ProteinProphet to account for the protein inference problem. Overall, false discovery rates for such procedure are less than 5 %.
12. The spectral counting method is used to estimate protein secretion enrichment as previously described [7]. Briefly, the total number of spectral counts for each protein group output by ProteinProphet for secreted profile over the spectral count output for the total cell extract profile is used for a semiquantitative enrichment analysis. Each profile is normalized by the total number of counts in the entire experiment. The enrichment factor was calculated by the following expression: ((Csp/Nfsp) + 1)/((Ctep/Nftep) + 1), where C represents the counts of each protein in the conditioned media (sp) or total extract (tep) and Nf is the normalization factor represented by the sum of all spectral counts from conditioned media profile (sp) or total cell extract profile (tep). The addition of 1 to the counts was intended to take into account those proteins for which no observation was made in one of the profiles and represents the minimum enrichment factor for that particular protein.

13. We performed experiments using standard conditions for the following cell lines: ovarian cancer (CaOV3, OVCAR3); lung cancer (A549); breast cancer (MCF7, MCF10A, MDA-MB-231); and pancreatic cancer (PANC1, CAPAN2). The timing for conditioned media collection depends on the experiment purpose. In our case, we intended to compare concentration of few cancer-relevant secreted proteins among different cell types.
14. Note that in standard culture conditions, each 60 cm^2 culture plate contains a total of 6 ml of culture media; therefore, several consecutive removals of 1 ml (and replacement with the same fresh media) can be performed in time course experiments.
15. Protein precipitation is important for concentrating proteins and eliminating substances that interfere with downstream applications, such as the phenol red dye present in cell culture media, which interacts with downstream reversed-phased sample cleanup.
16. Quantification of total protein content in conditioned media is optional. The design of the experiment can take into account the number of cells in culture as a normalization factor to express the secretion rate of a given protein per number of cells and/or per hour.
17. Dilution of solution guarantees good trypsin activity, which is close to 100 % in urea solutions at concentration less than 1 M according to the manufacturer.
18. In order to develop an MRM method for target proteins we use the following workflow: (a) select tryptic peptides from the target protein containing from 10 to 20 amino acid; (b) rank these peptides by hydrophobicity index in order to appropriately distribute multiple peptides targeting multiple proteins in the chromatographic separation; (c) if possible, select peptides containing proline residues, which generate intense fragment peaks, and avoid peptides containing methionine or N-terminal glutamine, which suffer from in-source modifications; (d) synthesize standard peptides using solid-phase Fmoc chemistry, purify peptides by reversed-phase chromatography, and accurately quantify them by amino acid analysis; (e) acquire the MS2 spectra for standard peptides in the same instrument used for MRM analysis in order to identify the two most intense fragments, which will represent the quantification ion and the verification ion; and (f) construct a calibration curve with the standard peptides using the defined parameters. Refer also to the SRMatlas (www.srmatlas.org) for information regarding potential standard peptides previously observed in proteomics studies and suitable for MRM studies.
19. We use the Waters proprietary quantification tool (TargetLynx) in order to integrate, calibrate, and quantify samples analyzed

by our method. Based on our experience, the coefficient of variation for standard peptide injections is below 5 %, indicating the good reproducibility of the method. Additionally the limit of detection under the presented conditions reaches hundreds of femtomoles of injected peptides.

Acknowledgment

This work was supported by FAPESP (São Paulo State Research Foundation, Brazil), grant 2011/09740-1, and University of São Paulo grant 12.1.17598.1.3 for Research Support Centers.

References

1. Brown KJ, Formolo CA, Seol H, Marathi RL, Duguez S, An E, Pillai D, Nazarian J, Rood BR, Hathout Y (2012) Advances in the proteomic investigation of the cell secretome. Expert Rev Proteomics 9:337–345
2. Karagiannis GS, Pavlou MP, Diamandis EP (2010) Cancer secretomics reveal pathophysiological pathways in cancer molecular oncology. Mol Oncol 4:496–510
3. Stastna M, Van Eyk JE (2012) Secreted proteins as a fundamental source for biomarker discovery. Proteomics 12:722–735
4. Hanash SM, Pitteri SJ, Faca VM (2008) Mining the plasma proteome for cancer biomarkers. Nature 452:571–579
5. Cox J, Mann M (2011) Quantitative, high-resolution proteomics for data-driven systems biology. Annu Rev Biochem 80:273–299
6. Faca V, Pitteri SJ, Newcomb L, Glukhova V, Phanstiel D, Krasnoselsky A, Zhang Q, Struthers J, Wang H, Eng J, Fitzgibbon M, McIntosh M, Hanash S (2007) Contribution of protein fractionation to depth of analysis of the serum and plasma proteomes. J Proteome Res 6:3558–3565
7. Faça VM, Ventura AP, Fitzgibbon MP, Pereira-Faça SR, Pitteri SJ, Green AE, Ireton RC, Zhang Q, Wang H, O'Briant KC, Drescher CW, Schummer M, McIntosh MW, Knudsen BS, Hanash SM (2008) Proteomic analysis of ovarian cancer cells reveals dynamic processes of protein secretion and shedding of extra-cellular domains. PLoS One 3:e2425
8. Emanuelsson O, Brunak S, von Heijne G, Nielsen H (2007) Locating proteins in the cell using TargetP, SignalP and related tools. Nat Protoc 2:953–971
9. Petersen TN, Brunak S, von Heijne G, Nielsen H (2011) SignalP 4.0: discriminating signal peptides from transmembrane regions. Nat Methods 8:785–786
10. Emmink BL, Verheem A, Van Houdt WJ, Steller EJ, Govaert KM, Pham TV, Piersma SR, Borel Rinkes IH, Jimenez CR, Kranenburg O (2013) The secretome of colon cancer stem cells contains drug-metabolizing enzymes. J Proteomics 91:84–96
11. Hanash S, Taguchi A (2010) The grand challenge to decipher the cancer proteome. Nat Rev Cancer 10:652–660
12. Picotti P, Aebersold R (2012) Selected reaction monitoring-based proteomics: workflows, potential, pitfalls and future directions. Nat Methods 9:555–566
13. Rauch A, Bellew M, Eng J, Fitzgibbon M, Holzman T, Hussey P, Igra M, Maclean B, Lin CW, Detter A, Fang R, Faca V, Gafken P, Zhang H, Whiteaker J, Whitaker J, States D, Hanash S, Paulovich A, McIntosh MW (2006) Computational Proteomics Analysis System (CPAS): an extensible, open-source analytic system for evaluating and publishing proteomic data and high throughput biological experiments. J Proteome Res 5:112–121
14. Deutsch EW, Mendoza L, Shteynberg D, Farrah T, Lam H, Tasman N, Sun Z, Nilsson E, Pratt B, Prazen B, Eng JK, Martin DB, Nesvizhskii AI, Aebersold R (2010) A guided tour of the trans-proteomic pipeline. Proteomics 10:1150–1159
15. Keller A, Nesvizhskii AI, Kolker E, Aebersold R (2002) Empirical statistical model to estimate the accuracy of peptide identifications made by MS/MS and database search. Anal Chem 74:5383–5392
16. Nesvizhskii AI, Keller A, Kolker E, Aebersold R (2003) A statistical model for identifying proteins by tandem mass spectrometry. Anal Chem 75:4646–4658

Chapter 23

Preparation of Heteroelement-Incorporated and Stable Isotope-Labeled Protein Standards for Quantitative Proteomics

Anna Konopka, Nico Zinn, Christina Wild, and Wolf D. Lehmann

Abstract

A major obstacle for further development of quantitative proteomics is the lack of accurately quantified protein standards. The following protocol describes innovative methods for the production of stable isotope-labeled protein standards. Their production is achieved by cell-free protein synthesis, which enables simultaneous incorporation of selenomethionine and stable isotope-labeled amino acids. The selenium tag allows sensitive and accurate quantification by inductively coupled plasma mass spectrometry (ICP-MS). The stable isotope label allows internal standardization in mass spectrometry-based proteomics by electrospray ionization-tandem mass spectrometry (ESI-MS/MS). Both label types can be placed within a single protein RISQ standard (recombinant isotope-labeled and selenium quantified) or can be distributed over two types of related RSQ and RIQ standards for the same target protein (recombinant selenium quantified and recombinant isotope-labeled and quantified). The combination of cell-free synthesis as production method with ICP-MS and ESI-MS/MS as detection methods results in protein standards, which are quantified at an outstanding level of accuracy.

Key words Cell-free protein synthesis, Selenomethionine, Selenium incorporation, Stable isotope labeling, Protein standards, ICP-MS, ESI-MS, Absolute protein quantification

1 Introduction

Over the last decade mass spectrometry (MS)-based methods have been developed into powerful tools to study the proteome providing, so far the most accurate and reproducible quantitative data in this field [1–5]. Quantitative proteomics is a key technology in numerous innovative biological studies, such as biomarker discovery, interplay of cellular events at genomic and proteomic level, cellular signaling, and systems biology in general.

At the moment stable isotope labeling of proteins (often named isotope dilution strategy) is considered as *gold standard* for protein quantification. Typically, proteins are labeled with ^{13}C and/or ^{15}N isotopes on a particular amino acid, most commonly

Daniel Martins-de-Souza (ed.), *Shotgun Proteomics: Methods and Protocols*, Methods in Molecular Biology, vol. 1156, DOI 10.1007/978-1-4939-0685-7_23, © Springer Science+Business Media New York 2014

on lysine and arginine residues. Such labeling creates *heavy* analogs of corresponding non-labeled, *light* proteins. *Heavy* and *light* forms are characterized by highly identical chemical behavior during analytical proteomics workflows. Due to the mass difference between corresponding *light* and *heavy* protein forms, these can be easily distinguished in MS analysis. Majority of quantitative measurements of proteins are performed on peptides obtained by enzymatic digestion (*bottom-up* approach). In this strategy the intensity ratios of *light* and *heavy* peptides pairs are measured [6, 7].

Stable isotope labeling can be introduced either chemically, using stable isotope-labeled derivatization reagents, e.g., ICAT [8], iTRAQ [9], mTRAQ [10], TMTs [11], and ICPL [12], or by metabolic in vivo labeling, as, e.g., in the SILAC approach [13, 14]. Such differential labeling of proteins provides relative data, whereas to obtain absolute data, internal standards with known concentration have to be added. For this purpose the use of stable isotope-labeled analogs is the superior choice. Stable isotope-labeled peptide standards are broadly commercially available. Preparation of AQUA [15] and PASTA [16] peptide standards and QconCAT polypeptide standards [17] have been reported. They are added during or after proteolytic digestion implicating that only the downstream analytical steps are standardized. In contrast, stable isotope-labeled intact protein standards can be added at the very early point of sample processing, e.g., directly to a tissue lysate or body fluid. Thus, all target protein losses caused by the analytical workflow are standardized. For preparation of stable isotope-labeled intact protein standards, PSAQ [18], FlexiQuant [19], and RISQ [20] strategies are known.

All these approaches use the isotope dilution strategy, but they differ in the way of standard preparation and quantification. The analytical method applied for standard quantification has to be carefully considered, since the accuracy of standard quantification has strong influence on the trueness of the target protein quantification. At the moment the majority of standards are quantified either by colorimetric methods or by quantitative amino acid analysis. Both these methods quantify all peptides/proteins present in the sample, so a high purity of the standard protein is a prerequisite for accurate data. In practice, preparation of sufficiently pure peptide/protein standards is often laborious and connected with severe standard losses. Therefore, it is worthwhile to develop novel contamination-tolerant quantification strategies to ensure optimal quality of quantitative proteomic data. The RISQ (recombinant isotope-labeled and selenium quantified) methodology achieves this criterion. RISQ standards are full-length proteins generated by cell-free *E. coli* synthesis with incorporated selenomethionine (SeMet) and selected stable isotope-labeled amino acids, usually lysine and arginine. Selenium serves as a tag for accurate RISQ standard quantification by inductively coupled plasma mass spectrometry (ICP-MS), whereas $^{13}C/^{15}N$-labeled amino acids allow the use of a RISQ standard as internal standard for quantification of its target protein by liquid chromatography-electrospray ionization-mass spectrometry (LC-ESI-MS). The use of the cell-free protein synthesis as standard production method provides only one labeled protein, namely, that for which cDNA was added for protein expression. All other proteins present in the cell-free medium remain non-labeled. ICP-MS is characterized by very good precision (coefficient of variation < 5 %). ICP-MS analysis requires only 1–10 μg of protein for quantification (depending on the amount of SeMet in the protein). In contrast, the amino acid analysis method requires between 50 and 100 μg of purified protein [21]. Taken together, the combination of the cell-free protein synthesis as standard production method with ICP-MS as quantification method is an innovative and promising method for production of protein standards.

While RISQ protein standards can be quantified accurately by ICP-MS due to their exclusive selenium content, their chemical stability might be influenced by the reactivity of selenium. Therefore, RISQ approach was further developed into the RSQ and RIQ (recombinant selenium quantified and recombinant isotope-labeled and quantified) strategy. In the RSQ and RIQ approach the two types of labeling, SeMet and $^{13}C/^{15}N$, are separately positioned into two related standards for the same target protein. RSQ standard contains SeMet, and RIQ standard contains selected stable isotope-labeled amino acids.

In this chapter the detailed protocol for synthesis, purification, and accurate quantification of the full-length RSQ (optionally RISQ) and RIQ protein standards is presented. The workflow of their production and quantification is schematically shown in Fig. 1. The particular steps of this workflow are illustrated with corresponding examples of experiments performed for RSQ/RISQ and RIQ standards for their target proteins. The quality control experiments, which are essential for preparation of accurately quantified protein standards, are also described in detail.

2 Materials

2.1 cDNA Cloning: Preparation of the DNA Template for Cell-Free *E. coli* Protein Synthesis

1. pEXP1-DEST Gateway® Vector Kit, pEXP2-DEST Gateway® Vector Kit (Life Technologies™) (*see* **Notes 1** and **2**).
2. Gateway® Entry Clone encoding the target protein (Life Technologies™).
3. Gateway® LR Clonase™ II Enzyme Mix Kit (Life Technologies™) (*see* **Note 1**).
4. TE buffer, pH 8.0: 10 mM Tris (adjusted to pH 8.0 with HCl) and 1 mM EDTA in nuclease-free water.

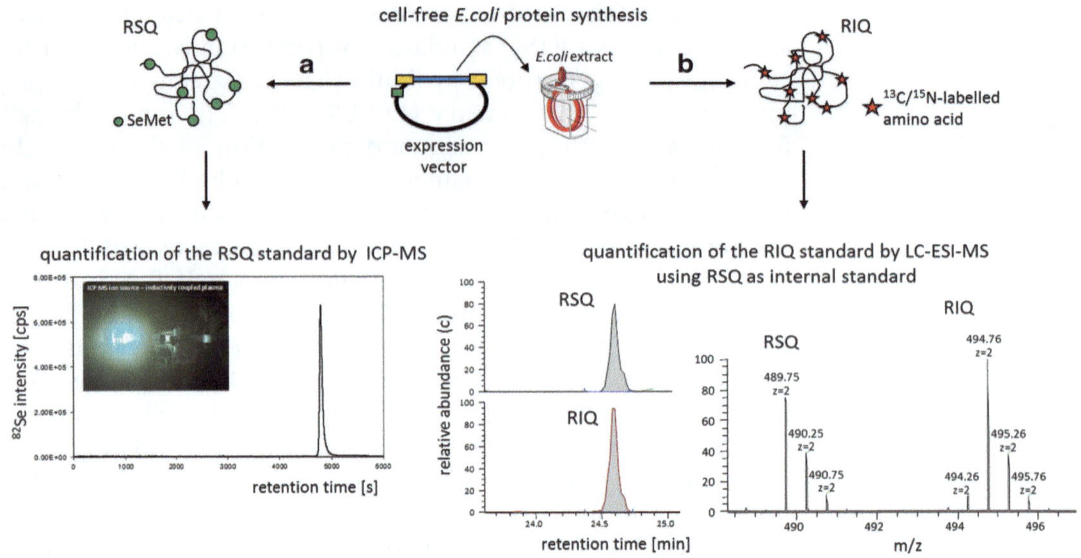

Fig. 1 Schematic workflow of preparation and quantification of RSQ and RIQ protein standards

5. One Shot® OmniMAX™ 2T1 Phage-Resistant Cells (Life Technologies™) (*see* **Note 3**).
6. LB medium prepared as follows: Dissolve 10 g tryptone, 5 g yeast extract, and 10 g NaCl in 1 L of distilled water. Adjust the pH to 7.0.
7. Petri dishes.
8. 37 °C incubator for bacterial culture (plates and suspension).
9. QIAprep® Spin Miniprep Kit (QIAGEN) (*see* **Note 4**).
10. QIAGEN Plasmid Maxi Kit (QIAGEN) (*see* **Note 4**).
11. 1.5 ml DNA low-binding microtubes.
12. Pipette filter tips.
13. Microtube centrifuge.
14. Spectrophotometer.

2.2 Cell-Free *E. coli* Synthesis of RSQ/RISQ and RIQ Protein Standards

1. RTS 500 Proteomaster *E. coli* HY Kit (5PRIME) (*see* **Note 5**).
2. RTS Amino Acid Sampler (5PRIME) (*see* **Note 5**).
3. Stable isotope-labeled amino acids: L-lysine-$^{13}C_6$-hydrochloride (99 atom % ^{13}C, 95 % CP) and L-arginine-$^{13}C_6$,$^{15}N_4$-hydrochloride (99 atom % ^{13}C, 99 atom % ^{15}N, 95 % CP) (Sigma-Aldrich) (*see* **Note 6**).
4. 1.5 ml DNase/RNase-free microtubes.
5. Thermomixer.
6. MTP Thermoblock equipped with an RTS 500 Adapter (5PRIME) (*see* **Note 5**).

2.3 Purification of Cell-Free Synthesized Protein Standards

1. Ni^{2+} immobilized metal ion chromatography spin kit.
2. SDS-PAGE gel 4–12 % 2-well.
3. SDS-PAGE running buffer, 20× concentrated, prepared as follows: Dissolve 62.8 g MOPS (acid), 72.6 g Tris (base), 10 g sodium dodecyl sulfate, and 6.5 g sodium bisulfite in 500 ml of distilled water. The pH should be between 8.2 and 8.3.
4. SDS-PAGE loading buffer, 4× concentrated, prepared as follows: Dissolve 4 g glycerol, 1.2 g triethanolamine, 0.93 g 6 N HCl, 0.4 g lithium dodecyl sulfate, 0.4 g Ficoll 400, 0.007 g EDTA disodium, 0.0025 g Brilliant Blue G250, and 0.0025 g phenol red in 10 ml of distilled water. The pH should be between 7.7 and 7.8.
5. Protein size standards, prestained.
6. Spin-X® Centrifuge Tube Filter, 0.22 μm Cellulose Acetate, RNase/DNase free, 2 ml (Costar, Corning Inc.).
7. Polyacrylamide gel electrophoresis cell.
8. Electro-eluter model 422 (Bio-Rad).
9. GeBAflex dialysis tubes, 0.2–15 ml with molecular weight cutoff 12–14 kDa (Gene Bio Application Ltd.).
10. Thermomixer.
11. Vacuum centrifuge.
12. pH meter.

2.4 Quantification of RSQ/RISQ Protein Standards by ICP-MS with Selenium Detection

1. Acetonitrile ULC/MS grade (*see* **Note 7**).
2. Water ULC/MS grade (*see* **Note 8**).
3. Selenium Standard for ICP TraceCERT®, 1,000 mg/L Se in nitric acid (Sigma-Aldrich).
4. Capillary LC-ICP-MS Element 2 system (Waters/Thermo Scientific).

2.5 Quantification of RIQ Protein Standards by nanoUPLC-ESI-MS

1. Water ULC/MS grade.
2. Acetonitrile ULC/MS grade.
3. SDS-PAGE gel 10–12 well.
4. SDS-PAGE running buffer (for composition *see* Subheading 2.3, **item 3**).
5. SDS-PAGE loading buffer (for composition *see* Subheading 2.3, **item 4**).
6. Polyacrylamide gel electrophoresis cell.
7. Coomassie Blue staining solution prepared as follows: Add 0.5 g Coomassie Brilliant Blue G250 to 500 ml of methanol (or ethanol), 400 ml of distilled water, and 100 ml of glacial acetic acid.

8. Trypsin (*see* **Note 9**).
9. Zip-Tip$_{C18}$ pipette tips (Millipore).
10. 1.5 ml protein low-binding microtubes.
11. Thermomixer.
12. Vacuum centrifuge.
13. 37 °C incubator.
14. nanoACQUITY UPLC-ESI-MS LTQ Orbitrap system (Waters/Thermo Scientific).

3 Methods

3.1 cDNA Cloning: Preparation of the DNA Template for Cell-Free *E. coli* Protein Synthesis

The DNA template preparation can be achieved using Gateway® recombination cloning technology provided by Life Technologies. This system allows fast, easy, and robust cloning that circumvents the traditional restriction enzyme-based cloning steps. The Gateway® Entry Clone contains cDNA from the gene of target protein flanked by *attL* sites. Using enzyme mixture called LR Clonase® this gene can be easily shuttled to the Gateway® Destination Vector containing *attR* sites to generate the Expression Vector. In Gateway® technology two destination vectors dedicated for cell-free *E. coli* protein expression are available: pEXP1-DEST Gateway® Vector and pEXP2-DEST Gateway® Vector. The pEXP1-DEST vector allows production of target protein as a fusion protein with N-terminal His-tag, whereas the pEXP2-DEST vector introduces a C-terminal His-tag. The destination vectors possess all elements required for DNA replication in *E. coli* (e.g., pUC origin), for cell-free *E. coli* protein transcription (e.g., bacteriophage T7 promoter), and for cell-free *E. coli* translation (e.g., ribosome-binding site, RBS). Gateway® LR Clonase® method ensures an extremely high yield of correct clones due to the antibiotic resistance gene included in the vectors and the presence of a ccdB cassette (lethal gene that targets DNA gyrase), which together are responsible for the growth inhibition of not properly recombined bacterial clones. Gateway® cloning involving destination vector and entry clone is schematically shown in Fig. 2.

3.1.1 Gateway® Cloning

The following recombination cloning steps are based on the instruction manual provided with LR Clonase™ II Enzyme Mix Kit.

1. In 1.5 ml tube mix 50–150 ng (in 1–7 μl) of entry clone with 150 ng (in 1–3 μl) of destination vector. Fill with TE buffer pH 8.0 to 8 μl.
2. Thaw on ice the LR Clonase™ II Enzyme Mix.
3. Add 2 μl of LR Clonase™ II Enzyme Mix to the reaction, and mix the solution well by vortexing. Spin shortly. Return the LR Clonase™ II Enzyme Mix to −20 °C for further storage.

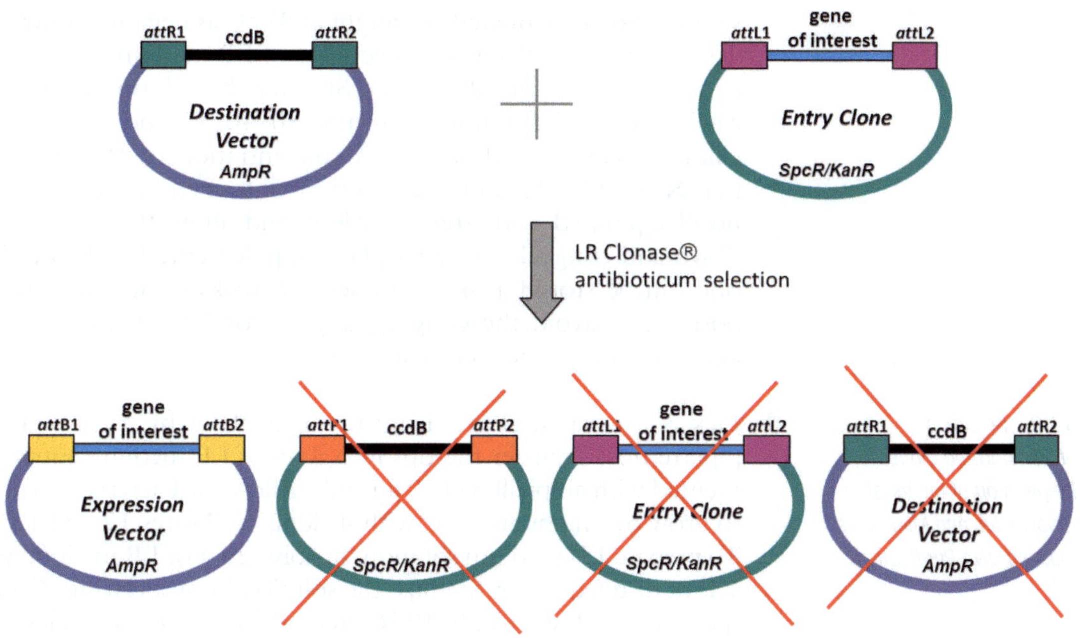

Fig. 2 Scheme of Gateway® LR Clonase recombination between the Gateway® Destination Vector and Gateway® Entry Clone. Only bacteria containing the correctly arranged expression vector are viable under the selected growth conditions; *Att* Gateway® attachments sites, *AmpR* ampicillin resistance gene, *SpcR* spectinomycin resistance gene, *KanR* kanamycin resistance gene, *ccdB* lethal gene that targets DNA gyrase

4. Incubate working reaction at 25 °C for 1 h.
5. Thaw on ice Proteinase K solution.
6. To terminate the reaction add 1 μl of Proteinase K solution to the working solution. Vortex shortly, and incubate the reaction solution at 37 °C for 10 min.
7. The reaction solution can be used for *E. coli* transformation within the next 24 h. For this purpose store the reaction solution at 4 °C.

3.1.2 E. coli Transformation with LR Clonase Recombination Mixture

1. Thaw on ice the competent cells: One Shot® OmniMAX™ 2T1 Phage-Resistant Cells (*see* **Note 3**).
2. Transform 1–3 μl of LR Clonase reaction mixture into 50 μl of competent cells, and incubate the mixture on ice for 30 min.
3. Heat-shock cells at 42 °C for 30 s.
4. Transfer cells again on ice.
5. Add 250 μl of pre-warmed (to 37 °C) SOC medium, and incubate cells at 37 °C for 1 h with shaking.
6. Plate 20, 50, and 100 μl of transformation mixture onto pre-warmed 37 °C agar plate containing ampicillin (100 μg/ml). Allow bacteria to grow overnight at 37 °C (*see* **Note 10**). One additional agar plate without plated transformation mixture

should also be incubated overnight at 37 °C as negative control. To prepare agar plates, dissolve 25 g of LB medium and 12 g of agar in 1 L of distilled water. Sterilize the solution by autoclaving. Cool the solution down to 30–40 °C, and add ampicillin to achieve final ampicillin concentration of 100 μg/ml (*see* **Note 11**). Spread the solution onto Petri dishes under hood equipped with sterile airflow, and allow them to cool down and congeal. The agar plates supplemented with antibiotic can be stored at 4 °C for several weeks in up-and-down position to avoid the congregating of condensed water onto agar which is prone to fungal growth.

3.1.3 Selection of Transformed Clones, Preparation of Bacterial Cryostocks, and Mini-scale Plasmid DNA Purification

1. Pick 3–5 isolated clones from the agar plates by using sterile pipette tips. Transfer the tips into 3 ml of LB medium supplemented with ampicillin (100 μg/ml). Allow the bacterial culture to grow overnight at 37 °C with shaking (*see* **Notes 12** and **13**). To prepare LB medium solution, dissolve 25 g of LB medium in 1 L of distilled water. Sterilize the solution by autoclaving. Cool the solution down to 30–40 °C, and add ampicillin to achieve a final ampicillin concentration of 100 μg/ml (*see* **Note 11**).
2. To prepare bacterial cryostocks of clones, which can be stored at −80 °C for further use, transfer 1 ml of each grown bacterial clone in LB medium into a cryotube and add 300 μl of glycerol (*see* **Note 14**). Mix the solution carefully by pipetting up and down in the tip several times. Freeze cryotubes at −80 °C.
3. Harvest the remaining 2 ml of bacterial cells suspended in LB medium by centrifugation at 6,000 × *g* in a tabletop centrifuge at room temperature for 10 min.
4. Purify the plasmid from each bacterial culture using a plasmid DNA Mini-Prep kit according to the protocol provided by the manufacturer (*see* **Note 4**). A purified plasmid DNA concentration and purity can be estimated by measuring absorbance at 260 and at 280 nm (*see* **Note 15**).
5. Verify the nucleotide sequence of purified expression vectors by DNA sequencing.

3.1.4 High-Scale Plasmid DNA Purification

1. Inoculate 100 ml of LB medium supplemented with ampicillin with bacteria clone (stored as cryostock) containing a Gateway® expression vector with gene of the target protein, which was verified by DNA sequencing.
2. Allow the bacterial culture to grow at 37 °C overnight in a shaking incubator (*see* **Note 12**).
3. Harvest the bacterial cells by centrifugation at 6,000 × *g* in a centrifuge at room temperature for 15 min.
4. Purify the plasmid from bacterial culture using a plasmid DNA Maxi-Prep kit according to the protocol provided by the

manufacturer (*see* **Note 4**). The purified plasmid DNA concentration and purity can be estimated by measuring the absorbance at 260 and 280 nm (*see* **Note 15**).

5. Store a purified plasmid DNA at −20 °C for further use.

3.2 Cell-Free E. coli Synthesis of RSQ/RISQ and RIQ Protein Standards

To produce the RSQ/RISQ and RIQ protein standards, a continuous-exchange cell-free *E. coli* protein synthesis system was applied. With this system the cell-free protein synthesis can be maintained for up to 24 h, which results in high protein yield (up to 6 mg/ml). The main advantage of cell-free compared to cell-based protein synthesis is the possibility to produce only one labeled protein. In cell-free extracts endogenous RNA polymerases are inhibited by adding of rifampicin, so that there is no need to remove endogenous DNA. Instead, specific bacteriophage RNA polymerases, e.g., T7, T3, or SP6, are utilized, which initiate transcription only from an externally delivered DNA template containing the respective phage RNA polymerase promoter site. The DNA template also contains the coding sequence of the protein to be expressed and some other regulatory elements required for cell-free protein expression. Moreover, the cell-free synthesis extract is depleted of natural canonical amino acids, which therefore have to be provided externally. In the provided amino acid mixture, selected canonical amino acids can be exchanged by corresponding labeled amino acids to incorporate desired labels into the newly synthesized protein. In this way, the protein of interest represents the only labeled protein in the reaction mixture, whereas all other proteins remain non-labeled.

The following procedure steps are based on the manufacturer's instruction manual provided with RTS 500 Proteomaster *E. coli* HY Kit.

3.2.1 Cell-Free Expression of RSQ Protein Standards

The RSQ standards contain SeMet incorporated quantitatively into protein structure at positions of methionine. Selenium incorporation is achieved by replacing canonical L-methionine by L-selenomethionine in the cell-free *E. coli* protein expression medium.

1. Reconstitute the lyophilizates of *E. coli* lysate, reaction mix, feeding mix, and amino acid mix without methionine in appropriate amounts of reconstitution buffer according to the manufacturer's instructions. Mix carefully all solutions by rolling. Do not vortex.
2. Prepare SeMet solution by dissolving 2.8 mg in 360 μl of reconstitution buffer at 30 °C. If the solution is to be stored longer than 1 day, add 0.2 mg of dithiothreitol.
3. To prepare the reaction solution for the RSQ standard expression, add 225 μl of reaction mix to the *E. coli* lysate, 270 μl of amino acid mix without methionine, 30 μl of freshly prepared

SeMet solution, and 10–15 μg (in 10–30 μl) of DNA expression vector containing the gene of the target protein. Mix the reaction solution carefully by rolling. Do not vortex.

4. To prepare the feeding solution for RSQ standard expression, add 2.65 ml of amino acid mix without methionine to the feeding mix and 300 μl of freshly prepared SeMet solution. Mix the feeding solution carefully by rolling. Do not vortex.
5. Fill the reaction chamber of RTS 500 CECF device with 1 ml of the reaction solution and the feeding chamber with 10 ml of the feeding solution according to the manufacturer's protocol.
6. Insert the filled RTS 500 CECF device into the Eppendorf Thermomixer using the RTS 500 Adapter, and start the protein synthesis at 30 °C with shaking speed of 990 rpm for 24 h.
7. Once the protein synthesis is finished, remove the reaction solution from the reaction chamber using a pipette equipped with capillary tip and transfer it into a fresh protein low-binding tube. Freeze the protein sample solution at –20 °C for further purification.

3.2.2 Cell-Free Expression of RIQ Protein Standards: Preparation of Labeled Amino Acid Mix

The RIQ standards usually contain stable isotope-labeled lysine and arginine, L-lysine-$^{13}C_6$ and L-arginine-$^{13}C_6$,$^{15}N_4$ (*see* **Note 6**). To achieve such kind of labeling, L-lysine and L-arginine with natural isotopic distribution of nitrogen and carbon are replaced by L-lysine-$^{13}C_6$ and L-arginine-$^{13}C_6$,$^{15}N_4$ in the cell-free *E. coli* protein expression medium. To prepare the labeled amino acid mix, the RTS Amino Acid Sampler Kit can be used, which contains stock solutions of 20 canonical L-amino acids in soluble form in separated vials.

Perform the following steps, which are based on the manufacturer's instruction documented in the manual provided with the RTS Amino Acid Sampler Kit.

1. Prepare 168 mM stock solution of L-lysine-$^{13}C_6$ and L-arginine-^{13}C ^{6},$^{15}N_4$ by dissolving an appropriate amount of each amino acid in the reconstitution buffer provided with RTS 500 Proteomaster HY Kit.
2. Treat the solutions of the remaining non-labeled amino acids from RTS Amino Acid Sampler at 37 °C for 20 min with vigorous shaking to ensure their complete dissolution. Mix 150 μl of each solution including labeled lysine and arginine to generate the labeled amino acid mix containing all 20 amino acids required for protein translation.

3.2.3 Cell-Free Expression of RIQ Protein Standards: RIQ Standard Expression in Cell-Free E. coli Extract

Reconstitute the lyophilizates of *E. coli* lysate, reaction mix, and feeding mix in appropriate amounts of reconstitution buffer according to the manufacturer's instructions. Mix carefully all solutions by rolling. Do not vortex.

1. To prepare the reaction solution for RIQ standard expression, to the *E. coli* lysate add 225 μl of reaction mix, 270 μl of labeled amino acid mix, and 10–15 μg (in 10–30 μl) of DNA expression vector containing the gene of the target protein. Mix the reaction solution carefully by rolling. Do not vortex.
2. To prepare the feeding solution for RIQ standard expression, to the feeding mix add 2.65 ml of labeled amino acid mix. Mix the feeding solution carefully by rolling. Do not vortex.
3. To complete the RIQ standard synthesis, follow the protocol **steps 5–7** of Subheading 3.2.1.

3.2.4 Cell-Free Expression of RISQ Protein Standards: Preparation of Labeled Amino Acid Mix Without Methionine

The RISQ standards contain stable isotope-labeled lysine (L-lysine-$^{13}C_6$) and arginine (L-arginine-$^{13}C_6$,$^{15}N_4$) (*see* **Note 6**) as well as SeMet. To achieve this, normal L-lysine and L-arginine in the *E. coli* protein expression medium are replaced by L-lysine-$^{13}C_6$ and L-arginine-$^{13}C_6$,$^{15}N_4$, respectively, and L-methionine is replaced by L-selenomethionine. To prepare the labeled amino acid mix without methionine perform the following steps, which are a combination of the procedure for synthesis of RSQ and RIQ standards.

1. Prepare 168 mM stock solution of L-lysine-$^{13}C_6$ and L-arginine-$^{13}C_6$,$^{15}N_4$ by dissolving an appropriate amount of each amino acid in the reconstitution buffer provided with RTS 500 Proteomaster HY Kit.
2. Incubate 17 (without methionine) non-labeled amino acid solutions from RTS Amino Acid Sampler at 37 °C for 20 min with vigorous shaking to ensure their complete dissolution. Mix 150 μl of each amino acid including labeled lysine and arginine to generate the labeled amino acid mix containing 19 amino acids.

3.2.5 Cell-Free Expression of RISQ Protein Standards: RISQ Standard Expression in Cell-Free E. coli Extract

Reconstitute the lyophilizates of *E. coli* lysate, reaction mix and feeding mix in appropriate amounts of reconstitution buffer according to the manufacturer's instructions. Mix carefully all solutions by rolling. Do not vortex.

1. Prepare SeMet solution by dissolving 2.8 mg in 360 μl of reconstitution buffer at 30 °C. If the solution is to be stored longer than 1 day, add 0.2 mg of dithiothreitol.
2. To complete the RISQ standard synthesis, follow the protocol **steps 3–7** of Subheading 3.2.1 using the prepared labeled amino acid mix without methionine instead of non-labeled amino acid mixture.

3.3 Purification of Cell-Free Synthesized Protein Standards

The cell-free synthesized proteins with His_6-tag can be purified using immobilized metal ion affinity chromatography (IMAC) methods since the string of six histidine residues specifically binds to immobilized divalent ions of nickel, cobalt, or copper.

If the protein standard is expressed as a native protein without any tags or IMAC purification is not accessible, protein purification can be achieved using polyacrylamide gel electrophoresis separation in combination with electro-elution from the polyacrylamide gel.

3.3.1 Ni^{2+}-IMAC Purification

The following steps are based on the instructions provided by the manufacturer together with the Ni-NTA Spin Kit. To enhance the purification efficiency, binding and washing steps are performed under denaturing conditions (*see* **Note 16**).

1. Prepare the binding buffer, wash buffer, and elution buffer of the following compositions:

   binding buffer—50 mM sodium phosphate buffer, 300 mM NaCl, 2 M thiourea, 6 M urea, pH adjusted to 7.4;

   washing buffer—50 mM sodium phosphate buffer, 300 mM NaCl, 2 M thiourea, 6 M urea, 30 mM imidazole, pH adjusted to 7.4;

   elution buffer—50 mM sodium phosphate buffer, 300 mM NaCl (optionally 2 M thiourea, 6 M urea), 300 mM imidazole, pH adjusted to 7.4 (*see* **Note 17**).
2. Incubate in 1:1 (v/v) ratio the cell-free *E. coli* reaction mixture with binding buffer for 30 min with vigorous shaking to solubilize the protein.
3. Equilibrate the IMAC spin column with 0.6 ml of binding buffer by centrifuging the spin column at 890 × *g* for 2 min. Repeat the equilibration step.
4. Filter the protein sample through 0.22 μm pore size filter to remove any remaining cellular debris just prior to the sample loading by centrifuging at 10,000 × *g* for 10 min.
5. Load up to 0.6 ml of the filtered sample onto the pre-equilibrated spin column (*see* **Note 18**). Centrifuge at 270 × *g* for 5 min (*see* **Note 19**). Reload the sample flow-through onto the spin column to increase the residence time between His_6-tagged target protein and immobilized metal ions to enhance the binding efficiency. Collect the final flow-through fraction for analysis by SDS-PAGE to check the binding efficiency. Store the fraction at −20 °C.
6. Wash the spin column three times with 0.6 ml of wash buffer to remove non-tagged proteins with no affinity to the immobilized metal ions by centrifuging the spin column at 890 × *g* for 2 min. Collect the wash fractions for analysis by SDS-PAGE to check the wash conditions. Store the fractions at −20 °C.
7. Elute the His_6-tagged target protein with 100–200 μl of elution buffer directly into a fresh microcentrifuge tube by centrifuging the spin column at 890 × *g* for 2 min. Repeat the elution step. Merge the second eluate with the first one. If higher concentrations are required, do not combine the two eluate fractions. Store the fractions at −20 °C.

8. To remove imidazole (and other low-molecular-weight compounds) from the IMAC-purified target protein sample, dialysis can be performed. Transfer the sample into 0.2–0.8 ml dialysis tube with MWCO 12–14 kDa. Dialyze the sample against 1 L of 10 mM ammonium bicarbonate over 18–24 h changing the ammonium bicarbonate equilibration solution every 4 h.
9. If required, concentrate the sample in a vacuum centrifuge.
10. Add acetonitrile to the target protein sample to achieve the final concentration of acetonitrile 10 % (v/v). Store at −20 °C for further use.

3.3.2 Electro-elution

1. Incubate in 1:1 (v/v) ratio the cell-free *E. coli* reaction mixture sample with 2 M thiourea and 6 M urea for 30 min with vigorous shaking to allow protein solubilization.
2. Filter the protein sample through a 0.22 μm pore size filter to remove any remaining cellular debris by centrifuging at 10,000 × *g* for 10 min.
3. Mix the filtered protein sample with 4× concentrated SDS-PAGE loading buffer to achieve 1x concentrated SDS-PAGE loading buffer and incubate at 70 °C with vigorous shaking for 10 min. Drive the sample to the bottom of the tube by short centrifugation.
4. Prepare 1 L of 1× concentrated SDS running buffer for polyacrylamide gel electrophoresis by diluting 20× concentrated SDS running buffer. Place the 2-well polyacrylamide gel into the electrophoresis cell, and fill the chamber with the 1× concentrated SDS running buffer.
5. Load the protein sample onto 2-well gel into the bigger well (ca. 450–500 μl for RunBlue SDS Gel 4–12 % 2-well). Load 5 μl of pre-stained protein size marker into the smaller well. Run the gel electrophoresis (e.g., under constant voltage of 120 V, 2 h).
6. Gently remove the gel from the plate, and submerse it in distilled water to wash out SDS running buffer. Do not stain the gel. Cut the gel region containing the target protein according to the reference bands of the pre-stained protein size marker.
7. Assemble the electro-elution cell as described in the instruction manual provided by the manufacturer.
8. Fragment the cutout gel into smaller pieces, approximately 2–3 mm^3 in size, and transfer the gel slices into the glass tubes of electro-elution cell. Fill the upper and lower chamber of the electro-elution cell with running buffer of 50 mM ammonium bicarbonate and 1 % sodium dodecyl sulfate.
9. Run the elution at 8–10 mA constant current per glass tube for 4–5 h.

10. Once the electro-elution is finished, disassemble the electro-elution chamber and transfer the solution containing the eluted target protein in a fresh microcentrifuge tube as described in the instruction manual provided by the manufacturer.
11. To remove SDS from the purified target protein sample, perform dialysis. Transfer the sample into 3–15 ml dialysis tube with MWCO 12–14 kDa. Dialyze the sample against 2 L of 10 mM ammonium bicarbonate over 18–24 h changing the ammonium bicarbonate equilibration solution every 4 h.
12. Concentrate the sample in a vacuum centrifuge (*see* **Note 20**).
13. Add acetonitrile to the target protein sample to achieve the final concentration of acetonitrile 10 % (v/v). Store at −20 °C for further use.

3.4 Quantification of RSQ/RISQ Protein Standards by ICP-MS with Selenium Detection

Purification of complex protein samples is a very laborious procedure, and often it is not feasible to obtain a really pure target protein sample. The presence of other proteins and peptides in the sample can bias the quantification of the target protein. The great advantage of cell-free compared to cell-based protein synthesis is the possibility to produce only a single labeled protein. In case of RSQ/RISQ protein standards only this labeled protein among all other proteins present in cell-free *E. coli* mixture contains selenium in the form of SeMet incorporated during cell-free synthesis. This fact ensures very accurate protein standard quantification by ICP-MS, which is independent of all other proteins/peptides present in the sample.

3.4.1 Quality Control: Verification of Selenium Incorporation into RSQ/RISQ Protein Standard

The verification of amino acid sequence coverage of the cell-free synthesized protein and labeling efficiency (SeMet incorporation in case of RSQ/RISQ proteins and stable isotope-labeled amino acid incorporation in case of RIQ proteins) can be achieved by mass spectrometric analysis after enzymatic digestion of the purified protein standard samples. An aliquot (5–10 μl) of purified RSQ/RISQ protein sample is enzymatically in-gel digested (*see* **Note 21**), and the purified peptide mixture is analyzed by nanoUPLC-ESI-MS/MS.

3.4.2 In-Gel Digestion

In-gel digestion procedure involves several steps such as SDS-PAGE, staining and destaining of protein bands, reduction of disulfide bonds of protein by treatment with dithiothreitol, alkylation of –SH groups with iodoacetamide, enzymatic in-gel protein digestion, and peptide extraction and purification.

1. Mix an aliquot (5–10 μl) of purified target protein sample with 4× concentrated SDS-PAGE loading buffer to achieve 1× concentrated SDS-PAGE loading buffer and incubate at 70 °C with vigorous shaking for 10 min. Drive the sample to the bottom of the tube by short centrifugation.

2. Prepare 1 L of 1× concentrated SDS running buffer for gel electrophoresis by diluting 20× concentrated SDS running buffer. Place the 10-well polyacrylamide gel into the electrophoresis cell, and fill the chamber with the 1× concentrated SDS running buffer.
3. Load 5 μl of pre-stained protein size marker. Load the incubated protein sample onto gel. Run the gel electrophoresis (e.g., under constant voltage of 120 V, 1–2 h).
4. Once the gel electrophoresis is finished, gently remove the gel from the plate and submerse the gel in enough staining solution to cover the gel. Leave the gel in staining solution for 1 h (optionally overnight) (*see* **Note 22**).
5. Wash the gel by agitation in distilled water changing the water three times every 10–15 min (*see* **Note 23**).
6. Cut out the target protein band from the gel as closely as possible with a sharp scalpel, and divide the gel band into smaller pieces of approx. 1 mm^3.
7. Place the gel pieces in a 1.5 ml plastic microcentrifuge tube, and drive the gel pieces to the bottom of the tube by short centrifugation.
8. Wash the gel pieces with 200 μl of water with vigorous shaking for 10 min.
9. Agitate with 1 ml of the 30 % acetonitrile/70 % 100 mM ammonium bicarbonate for 15 min with vigorous shaking (*see* **Note 24**). Discard the supernatant.
10. Agitate with 1 ml of the 50 % acetonitrile/0.1 % trifluoroacetic acid for 15 min with vigorous shaking. Discard the supernatant.
11. Dry the gel pieces in a vacuum centrifuge for about 10–15 min at room temperature.
12. Cover the gel pieces with 300 μl of freshly prepared 10 mM dithiothreitol, and drive the gel pieces to the bottom of the tube by short centrifugation (*see* **Note 25**).
13. Incubate the sample in a thermomixer at 56 °C for 45 min. Discard the supernatant.
14. Dry the gel pieces in a vacuum centrifuge for about 10–15 min at room temperature.
15. Add 300 μl of freshly prepared 55 mM iodoacetamide, and carry out the reaction for 30 min in the dark. Discard the supernatant (*see* **Notes 26** and **27**).
16. Agitate with 500 μl of 100 mM ammonium bicarbonate for 10 min with vigorous shaking. Discard the supernatant.
17. Agitate with 1 ml of acetonitrile for 15 min with vigorous shaking to dehydrate the gel pieces. When dehydrated, the gel

pieces will have a white color and will be significantly smaller in size. Discard the supernatant.

18. Dry the gel pieces in a vacuum centrifuge for about 10–15 min at room temperature.
19. Resuspend the trypsin in 40 mM ammonium bicarbonate/10 % acetonitrile to obtain the concentration of 20 μg/ml. Add 10–20 μl of the trypsin solution to the gel pieces, and keep the tube on the ice for 30 min to allow the trypsin penetration into gel (*see* **Note 28**).
20. Add 50 μl of 100 mM ammonium bicarbonate.
21. Wait for 15 min incubating the reaction tube on ice. The gel pieces should be completely covered with the liquid; if not, add again 20–50 μl of ammonium bicarbonate (*see* **Note 29**).
22. Carry out the digestion reaction overnight (12–18 h) at 37 °C.
23. Terminate the trypsin activity by adding 20–50 μl of 5 % formic acid to lower the pH of digestion mixture under 4 (*see* **Note 30**).
24. Carefully collect the supernatant, and transfer it into a fresh microcentrifuge tube.
25. Agitate the gel pieces with 50 μl of acetonitrile for 15 min.
26. Carefully collect the supernatant, and merge it with the previous one.
27. Agitate the gel pieces with 50 μl of 5 % formic acid for 15 min.
28. Carefully collect the supernatant, and merge it with the previous one.
29. Agitate the gel pieces with 50 μl of acetonitrile for 15 min.
30. Carefully collect the supernatant, and merge it with the previous one.
31. Reduce the volume of the extracted peptide solution in the vacuum centrifuge at room temperature to 2–3 μl.
32. Reconstitute the peptide sample with 20 μl of 2 % formic acid, and purify the sample using pipette tips with an immobilized reverse-phase C_{18} resin (Zip-Tip$_{C18}$).
33. Wash a fresh Zip-Tip$_{C18}$ pipette tip in 1 ml of 2 % formic acid/50 % acetonitrile by pipetting the liquid up and down three times. Equilibrate the Zip-Tip$_{C18}$ pipette tip in 1 ml of 2 % formic acid by pipetting the liquid up and down three times. Load the peptides onto the Zip-Tip$_{C18}$ resin by pipetting the peptide sample up and down twenty times. Expel the liquid. Wash the Zip-Tip$_{C18}$ in 300 μl of 2 % formic acid by pipetting the liquid up and down six times. Elute the peptides in 20 μl of 2 % formic acid/50 % acetonitrile by pipetting the solution up and down twenty times.

Reduce the volume of the purified peptide sample in the vacuum centrifuge at room temperature to dryness. Reconstitute in 5 μl of 0.1 % FA. Store the peptide sample at −20 °C for further analysis.

3.4.3 nanoUPLC-ESI-MS Analysis

The peptide sample can be separated and analyzed using reverse-phase liquid chromatography instrumentation coupled to a mass spectrometric system equipped with an ESI or a MALDI ion source. The presence of selenium in target RSQ/RISQ standard can be easily recognized by a characteristic isotopic pattern for selenium because of its several natural stable isotopes. To illustrate this, the simulated theoretical (*see* **Note 31**) and experimentally detected isotope pattern of doubly charged ion of the tryptic peptide EANNGSSPAGSLADA-SeM-SQK (derived from STAT5B RSQ standard; STAT5B; signal transducer and activator of transcription 5B) is shown in Fig. 3.

3.4.4 Quantification of RSQ/RISQ Protein Standards by Flow-Injection ICP-MS with Selenium Detection

Cell-free protein synthesis of RSQ/RISQ protein standards ensures that only one protein in the synthesis mixture has incorporated selenium in the form of SeMet. All other proteins originate from the cell-free *E. coli* extract are non-labeled. After protein synthesis the reaction mixture contains an excess of free SeMet, which has to be removed prior to an ICP-MS quantification of the RSQ/RISQ standard. Removal of free SeMet is achieved during RSQ/RISQ standard purification by using either Ni^{2+}-IMAC technology or electro-elution, both followed by dialysis (**steps 1** and **2**, Subheading 3.3). To demonstrate the presence of only a single high-molecular-weight species containing selenium, μLC-ICP-MS analysis can be performed. Such μLC-ICP-MS analysis is shown in Fig. 4 for the glutathione peroxidase 3 (GPx3) RSQ standard. The selenium μLC peak is characterized by a retention time typical for high-molecular-weight species, which confirms the absence of free SeMet. Thus, the RSQ/RISQ standard can be quantified by flow-injection ICP-MS using an inorganic selenium standard as reference (Fig. 4). To perform μLC-ICP-MS analysis, follow the procedure.

1. Add citric acid to the aliquot (5–10 μl) of purified RSQ/RISQ sample to obtain final citric acid concentration of 50 mM (*see* **Note 32**).
2. Run the μLC-ICP-MS analysis with selenium detection using the water–acetonitrile gradient from 3 % acetonitrile to 60 % acetonitrile containing 0.5 % isopropanol, 0.1 % formic acid, and 0.01 % heptafluorobutyric acid (*see* **Note 33**).
3. To quantify the RSQ/RISQ by flow-injection ICP-MS, perform the following steps:
   (a) Prepare the selenium calibration solutions ranging from 1 ppb to 1 ppm of selenium using Selenium Standard for

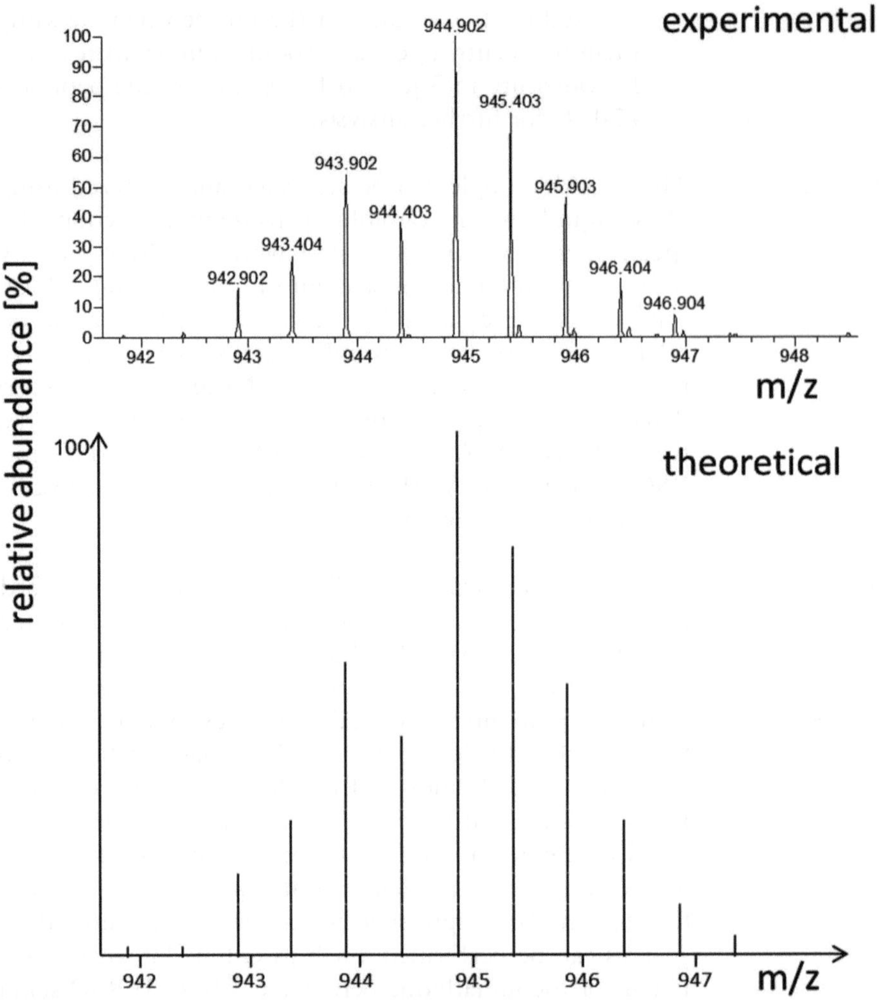

Fig. 3 Isotopic distribution pattern of the tryptic peptide EANNGSSPAGSLADA-SeM-SQK (derived from STAT5B RSQ standard); *upper panel*, observed isotope pattern for the doubly charged ion; *lower panel*, simulated isotope pattern for the doubly charged ion. MS analysis was performed using nanoACQUITY UPLC system equipped with 75 μm × 150 mm nanoACQUITY UPLC® 1.7 μm BEH130 C18 capillary column (Waters, Milford, USA) coupled online to an LTQ-Orbitrap XL instrument (Thermo Scientific, Bremen, Germany)

ICP TraceCERT®, 1,000 mg/L Se in nitric acid as stock solution.

(b) Analyze the inorganic selenium calibration solutions and the RSQ/RISQ sample by flow-injection ICP-MS with selenium detection.

(c) Based on the selenium calibration curve calculate the concentration of selenium in the RSQ/RISQ standard sample. To calculate the molar protein concentration of the RSQ/RISQ, the number of selenomethionine residues in the polypeptide chain must be taken into consideration.

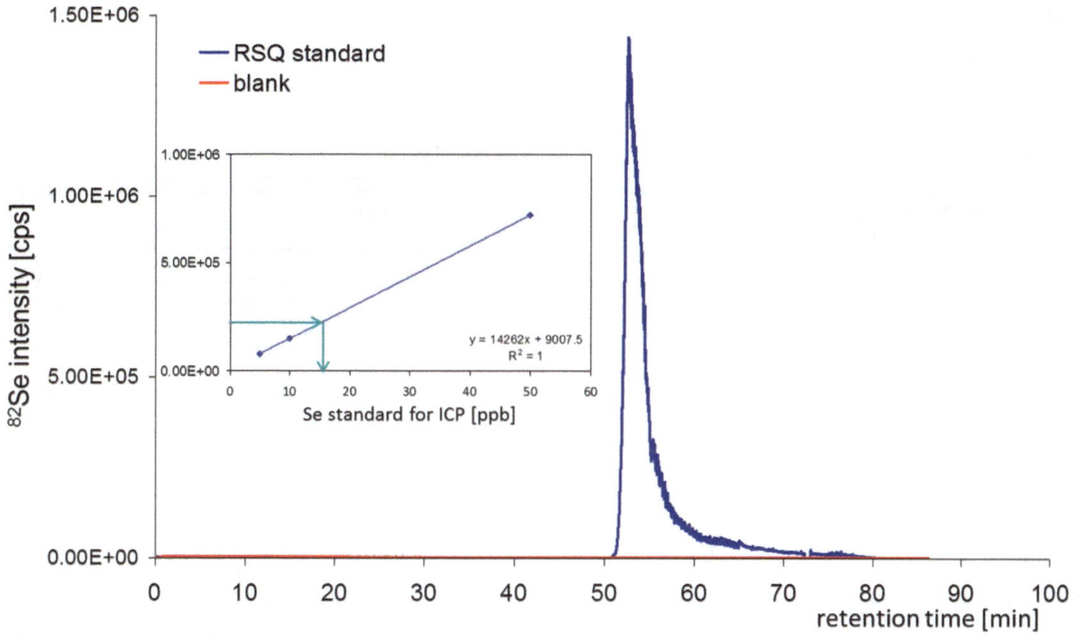

Fig. 4 Analysis of purified intact GPx3 RSQ standard by μLC-ICP-MS with ^{82}Se detection performed using a μLC system (Waters, Milford, MA) equipped with a 2.1 × 150 mm XBridge™ BEH300 C4 3.5 μm capillary column (Waters, Milford, MA) coupled online to a sector field ICP-MS Element 2 (Thermo, Bremen, Germany)

3.5 Quantification of RIQ Protein Standards by nanoUPLC-ESI-MS

3.5.1 Quality Control: Verification of Stable Isotope Label Incorporation into RIQ Standard

The stable isotope labeling efficiency of RIQ standards has to be carefully verified as incomplete labeling can affect the absolute quantification of target proteins. The labeling completeness verification can be achieved by nanoUPLC-ESI-MS analysis after enzymatic digestion of the purified RIQ protein standard sample. An aliquot (5–10 μl) of the purified RIQ protein sample is enzymatically in-gel digested (according to **step 1** of Subheading 3.4.1 of this protocol), and the peptide mixture is analyzed by nanoUPLC-ESI-MS. As an example, the labeling degree calculated for peptides from tryptic digestion of calmodulin-like protein 3 (Calml3) RIQ standard is above 98.9 % as illustrated in Fig. 5. All RIQ standards we have produced using cell-free RTS 500 Proteomaster *E. coli* HY Kit were characterized by a high stable isotope labeling degree above 98 %.

3.5.2 Accurate Quantification of RIQ Protein Standards by nanoUPLC-ESI-MS with RSQ as Internal Standard

RISQ standards are quantified via their selenium content and simultaneously function as internal standards for nanoUPLC-ESI-MS/MS via their incorporated stable isotope-labeled amino acids. While this concept is effective, since a single standard serves two functions, two points may be critically discussed. Firstly, the presence of selenium may reduce the standard lifetime due to the chemical reactivity of SeMet, and secondly, all methionine/SeMet peptide pairs are not available for relative quantification due to the

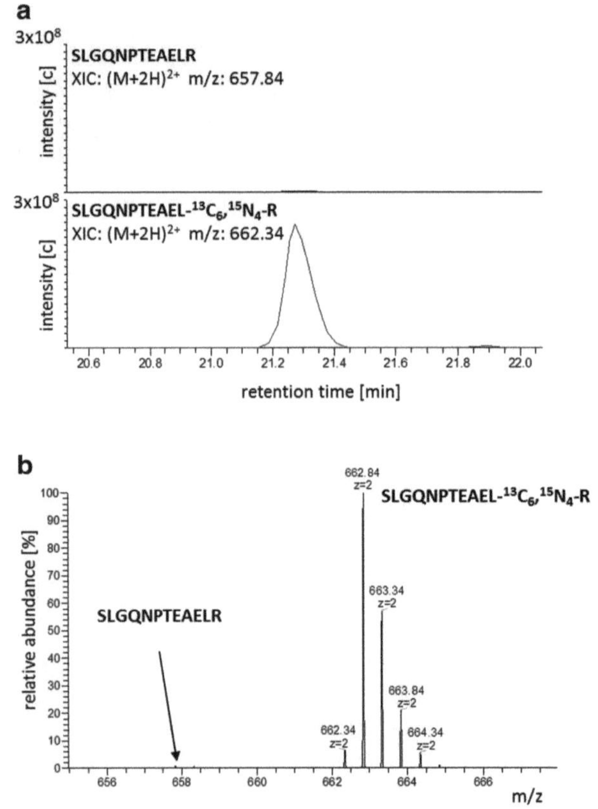

c

peptide number	peptide sequence	labelling degree [%]
37-53	EAFSLFDKDGDGCITTR	99.7
45-53	DGDGCITTR	99.1
54-60	ELGTVMR	99.1
61-72	SLGQNPTEAELR	99.0
73-80	DMMSEIDR	99.1
81-97	DGNGTVDFPEFLGMMAR	99.1
98-109	KMKDTDNEEEIR	99.8
98-113	KMKDTDNEEEIREAFR	99.8
99-109	MKDTDNEEEIR	99.4
99-113	MKDTDNEEEIREAFR	99.9
101-109	DTDNEEEIR	99.0
101-113	DTDNEEEIREAFR	99.7
114-129	VFDKDGNGFVSAAELR	99.8
118-129	DGNGFVSAAELR	99.0
135-149	LGEKLSDEEVDEMIR	99.0
139-149	LSDEEVDEMIR	99.1
150-166	AADTDGDGQVNYEEFVR	98.9

Fig. 5 Control of the degree of labeling of a ^{13}C,^{15}N-labeled RIQ protein standard by nanoUPLC-LC-ESI-MS

chemical difference between both species. This reduces the number of *light/heavy* peptide pairs. Both limitations can be eliminated by placing the two types of labels in two standards. The selenium-labeled standard is named RSQ (recombinant selenium quantified), while the stable isotope-labeled standard is named RIQ (recombinant isotope-labeled and quantified). The quantified RSQ standard is applied for quantification of the corresponding RIQ standard by nanoUPLC-ESI-MS. An aliquot of RSQ standard is mixed with the aliquot of RIQ standard. This mixture is subjected to enzymatic digestion (according to **step 1** of Subheading 3.4.1 of this protocol), and obtained peptides are analyzed by nanoUPLC-ESI-MS. The peptide fragments form non-labeled (*light*, coming from the RSQ standard) and labeled (*heavy*, coming from the RIQ standard) pairs and *light/heavy* ratios (L/H) are calculated. The principle of such quantification is schematically shown in Fig. 6. Pairs formed by *heavy* and *light*+SeM peptides cannot be used for relative quantification due to their chemical difference.

As an example, a typical nanoUPLC-ESI-MS trace observed in the elution of *light/heavy* peptide pair not containing SeMet or methionine, respectively, and the corresponding MS spectrum

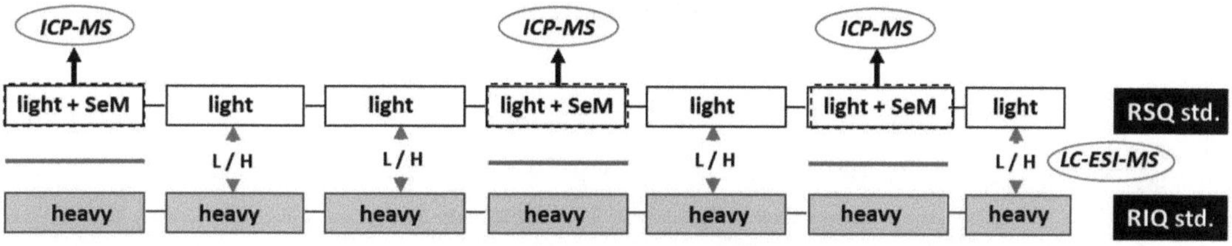

Fig. 6 Quantification transfer from a selenium-quantified RSQ standard to an RIQ standard, containing only stable isotope-labeled amino acids. The quantification transfer is performed by nanoUPLC-ESI-MS/MS on the peptide level (bottom-up approach)

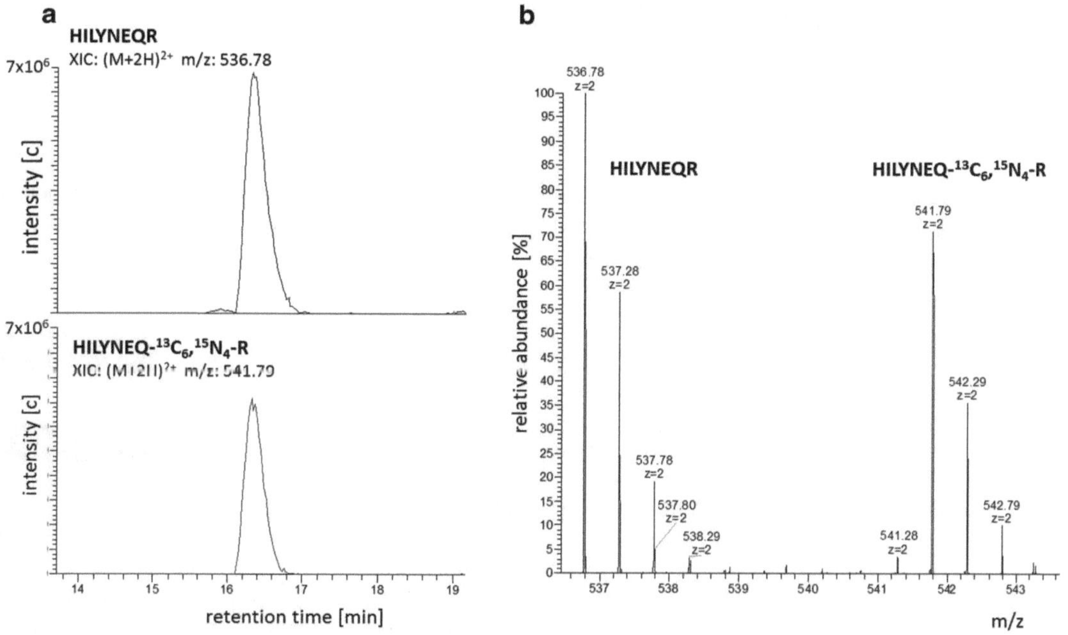

Fig. 7 Example of nanoUPLC-ESI-MS trace observed in the elution of *light*/*heavy* peptide pair not containing selenomethionine or methionine, respectively; (**a**) extracted ion chromatograms for $(M+2H)^{2+}$ ions of *light* and *heavy* form of tryptic peptide HILYNEQR (derived from STAT5A); (**b**) the corresponding mass spectrum averaged over the complete LC peak

averaged over the complete LC peak for STAT5A RSQ and RIQ standards is presented in Fig. 7. The extracted ion chromatograms of doubly charged ions show virtual co-elution of *light* and *heavy* peptides.

3.6 Use of RIQ Protein Standards for Absolute Quantification of Target Protein

Once the RIQ standard is quantified, it can be applied as an internal standard for quantification of the target protein by nanoUPLC-ESI-MS. An aliquot of RIQ standard is added to the sample containing the target protein, e.g., cell lysate. 1D polyacrylamide gel electrophoresis of a sample containing STAT5A RIQ standard added to the cell lysate after immunoprecipitation step is presented in Fig. 8 as an example.

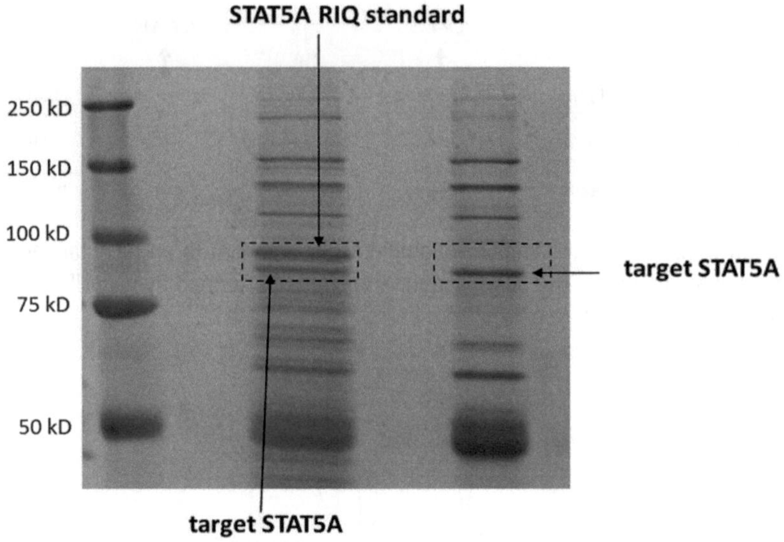

Fig. 8 1D PAGE for STAT5A quantification using a STAT5A RIQ standard; first line: protein size marker; second line: cell lysate with RIQ standard added after immunoprecipitation; third line: cell lysate without RIQ standard after immunoprecipitation

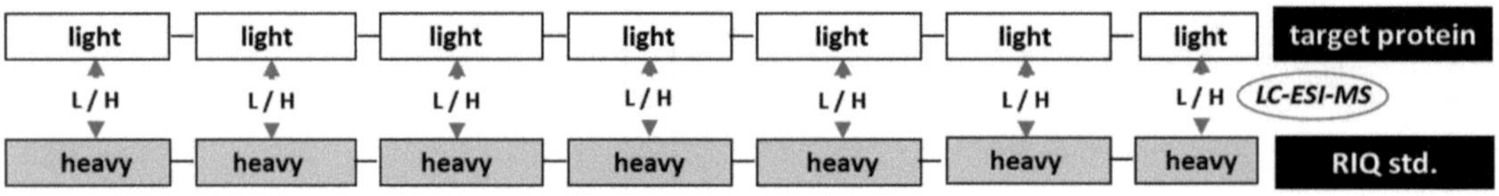

Fig. 9 Principle of quantification of a target protein by nanoUPLC-ESI-MS using its RIQ standard

A gel band containing both the target protein and its RIQ standard is cut, subjected to in-gel digestion (according to Subheading 3.4.2 of this protocol), and obtained peptides are analyzed by nanoUPLC-ESI-MS. The peptide fragments form non-labeled (*light*, coming from protein to be quantified) and labeled (*heavy*, coming from the RIQ standard) pairs, and L/H ratios are calculated. The principle of such quantification is schematically shown in Fig. 9. As shown, all peptide pairs can be used for L/H ratio calculation, which increases the accuracy of target protein quantification.

4 Notes

1. Alternatively, RTS pIVEX His-tag, 2nd Generation Vector Set (5PRIME) can be used together with conventional cloning steps. This kit contains two vectors, which are suitable for cell-free *E. coli* protein translation: the pIVEX 2.3d vector destined for production of target protein with C-terminal His_6-tag and the

pIVEX 2.4d vector suitable for introduction of an N-terminal His_6-tag. For protein expression without any tags, the pIVEX 2.3d vector can be used in combination with addition of a stop codon at the end of the target protein cDNA. The pIVEX 2.4d vector contains a factor Xa restriction protease cleavage site downstream to the His_6-tag nucleotide sequence so that the His_6-tag can be cleaved after protein synthesis by appropriate protease action. Both RTS pIVEX vectors contain a multicloning site (MSC), which allows incorporation of the cDNA of target protein into expression vector using traditional restriction enzyme-based cloning strategy.

2. pEXP2-DEST Gateway® Vector was modified to incorporate an ATG initiation codon and an RBS with optimal spacing for proper protein translation.
3. Any competent cells with transformation efficiency of $>10^8$ transformants/μg can be used.
4. We have used Mini-Prep and Maxi-Prep plasmid DNA purification kits from QIAGEN, but similar kits are commercially available also from other manufacturers.
5. We have used RTS 500 Proteomaster *E. coli* HY Kit from 5PRIME, but alternative cell-free protein expression systems are also commercially available from Promega, QIAGEN, and Thermo Scientific.
6. In principle, any stable isotope-labeled canonical amino acid can be incorporated into protein during cell-free protein synthesis. It is convenient to use stable isotope-labeled lysine and arginine, when trypsin protein digestion is performed, since trypsin cleaves on the C-terminal side of lysine and arginine residues. Amino acid with other amounts of ^{13}C and ^{15}N labels can be used as well (e.g., L-lysine-$^{13}C_6$,$^{15}N_2$, L-lysine-$^{15}N_2$, and/or L-arginine-$^{13}C_6$). Care must be taken by data evaluation if stable isotope labeling provides a small mass shift, e.g., 2 Da when using only L-lysine-$^{15}N_2$ labeling. In such a case isotopic patterns for non-labeled and labeled peptides are not completely separated and correction factors must be applied. Stable isotope-labeled amino acids are also available from Cambridge Isotope Laboratories.
7. To estimate the presence of other proteins in the purified RSQ/RISQ sample (purification efficiency), sulfur isotope can be monitored in addition to selenium during ICP-MS analysis. To minimize sulfur background signal, use of acetonitrile with low sulfate content (<10 μg/kg) is recommended.
8. To estimate the presence of other proteins in the purified RSQ/RISQ sample (purification efficiency), sulfur isotope can be monitored in addition to selenium during ICP-MS analysis. To minimize sulfur background signal, use of water

(ion chromatography grade) with low sulfate content (<1 μg/kg) is recommended.

9. Trypsin is available from various manufacturers. We have used Trypsin Gold Mass Spectrometry grade (Promega). Other proteases such as LysC or AspN can be applied as well.
10. Avoid too long bacteria-growing times because of the risk of creation of secondary colonies. Such colonies impede proper selection of individual clones from the plate.
11. Do not add the ampicillin to the solution with temperature above 60 °C to avoid the degradation of ampicillin at high temperature.
12. Do not allow to grow bacteria longer than 17 h. Growth time longer than 17 h is not recommended since the bacterial cells may begin to lyse and lose plasmid, so plasmid purification yield may be reduced.
13. For bacterial culture use a tube or a flask with a volume of at least four times of volume of the culture to ensure enough oxygen supply for growth.
14. Be careful during pipetting of pure glycerol. Glycerol is a very viscous liquid and therefore needs time to be pulled into the pipette tip.
15. DNA concentration estimation can be achieved by measuring absorbance at 260 nm (A_{260}) (adjusted for turbidity measured at 320 nm) by using a spectrophotometer equipped with a UV lamp. The calculation is based on the relationship that an A_{260} equals 1.0 for a pure dsDNA solution of 50 μg/ml. To ensure the highest measurement accuracy, the A_{260} readings should be within the instrument's linear range (generally 0.1–1). To evaluate DNA purity, measurement at 280 nm is recommended. At this wavelength aromatic amino acids absorb. The ratio A_{260}/A_{280} is often used as purity calculation. Good-quality DNA solution should have A_{260}/A_{280} ratio higher than 1.8.
16. It is recommended to analyze the aliquots of all purification fractions (flow-through, wash, and elution) by SDS-PAGE to check the efficiency of IMAC purification process.
17. The elution step can also be performed under denaturing conditions in the presence of thiourea and urea in the elution buffer. To remove thiourea and urea from the protein sample, dialysis can be performed.
18. Check the binding capacity of applied IMAC column, and according to that adjust the volume of loaded protein sample. If required, the cell-free synthesized target protein sample can be purified using more than one IMAC columns.
19. Do not exceed 270 × g to ensure efficient binding.

20. Use of volatile ammonium bicarbonate buffer allows easy reduction of sample volume in a vacuum centrifuge. Protein sample containing SDS is prone to foaming during treatment in a vacuum centrifuge. Therefore, SDS must be removed beforehand to avoid sample losses.
21. Alternatively, in-solution digestion can also be performed according to the instruction manual provided with an applied protease.
22. To stain low-abundance protein bands, colloidal Coomassie Blue can be applied, which is characterized by better sensitivity compared to classical Coomassie Blue.
23. Wash solution of 25 % methanol in water can also be applied.
24. Ensure that destaining process in solution 30 % acetonitrile/70 % 100 mM ammonium bicarbonate is complete. If the blue color of gel pieces is still visible, repeat the incubation in 30 % acetonitrile/70 % 100 mM ammonium bicarbonate. Incomplete destaining will decrease the digestion efficiency.
25. The reduction of disulfide bonds can be omitted; however, lower protein sequence coverage in nanoUPLC-ESI-MS analysis may be observed.
26. The alkylation of –SH groups can be omitted; however, lower protein sequence coverage in nanoUPLC-ESI-MS analysis may be observed.
27. IAA treatment exceeding 30 min is not recommended because of the risk of unspecific alkylation reactions.
28. Add trypsin in a protease:protein ratio of 1:100 to 1:20 (w/w).
29. Check the pH of the digestion mixture. It should be in the range of 7–9 at which pH trypsin achieves its maximal activity.
30. Check the pH of the inhibited digestion mixture. Trypsin is reversibly inactivated at pH lower than 4.
31. A theoretical isotope pattern of labeled peptides can be calculated using online-available software, e.g., Isotope Pattern Calculator v4.0 or Sheffield ChemPuter Isotope Pattern Calculator.
32. Addition of citric acid improves the protein/peptide recovery from the reverse-phase chromatography columns. Citric acid is an effective complexing agent for trivalent ions. Iron(III) is present in solvents and/or analytical columns and may bind proteins/peptides, thus decreasing their LC recovery. Citric acid effectively complexes Fe^{3+} ions and removes them from analytical columns.
33. Another LC gradient can be applied depending on the LC system available.

Acknowledgements

We are thankful to L. Adler, M. Schilling, and U. Klingmüller for numerous valuable discussions and to M. Boehm and D. Winter for support in nanoUPLC-ESI-MS analysis. We thank also W. Konopka for training and support in molecular biology techniques. We gratefully acknowledge funding by the SBCancer network in the Helmholtz Alliance on Systems Biology and the German Federal Ministry of Education and Research (BMBF)-funded MedSys-Network LungSys and the Virtual Liver Network. This research was undertaken within the project HLT05 (EMRP Researcher Grant). The EMRP is jointly funded by the EMRP participating countries within EURAMET and the European Union.

References

1. Ong SE, Foster LJ, Mann M (2003) Mass spectrometric-based approaches in quantitative proteomics. Methods 29:124–139
2. Olsen JV, Blagoev B, Gnad F, Macek B, Kumar C, Mortensen P, Mann M (2006) Global, in vivo, and site-specific phosphorylation dynamics in signaling networks. Cell 127:635–648
3. Yates JR, Ruse CI, Nakorchevsky A (2009) Proteomics by mass spectrometry: approaches, advances and applications. Annu Rev Biomed Eng 11:49–79
4. Monetti M, Nagaraj N, Sharma K, Mann M (2011) Large-scale phosphosite quantification in tissues by a spike-in SILAC method. Nat Methods 8:655–685
5. Bantscheff M, Lemmer S, Savitski MM, Kuster B (2012) Quantitative mass spectrometry in proteomics: critical review update from 2007 to the present. Anal Bioanal Chem 404:939–965
6. Duncan MW, Aebersold R, Caprioli RM (2010) The pros and cons of peptide-centric proteomics. Nat Biotechnol 28:659–664
7. Domon B, Aebersold R (2010) Options and considerations when selecting a quantitative proteomics strategy. Nat Biotechnol 28:710–721
8. Gygi SP, Rist B, Gerber SA, Turecek F, Gelb MH, Aebersold R (1999) Quantitative analysis of complex protein mixtures using isotope-coded affinity tags. Nat Biotechnol 17:994–999
9. Ross PL, Huang YLN, Marchese JN, Williamson B, Parker K, Hattan S, Khainovski N, Pillai S, Dey S, Daniels S, Purkayastha S, Juhasz P, Martin S, Bartlet-Jones M, He F, Jacobson A, Papin DJ (2004) Multiplexed protein quantitation in Saccharomyces cerevisiae using amine-reactive isobaric tagging reagents. Mol Cell Proteomics 3:1154–1169
10. Mertins P, Udeshi ND, Clauser KR, Mani DR, Patel J, Ong SE, Jaffe JD, Carr SA (2012) iTRAQ labeling is superior to mTRAQ for quantitative global proteomics and phosphoproteomics. Mol Cell Proteomics 11:M111.014423. doi:10.1074/mcp.M111.014423
11. Thompson A, Schafer J, Kuhn K, Kienle S, Schwarz J, Schmidt G, Neumann T, Hamon C (2003) Tandem mass tags: a novel quantification strategy for comparative analysis of complex protein mixtures by MS/MS. Anal Chem 75:1895–1904
12. Lottspeich F, Kellermann J (2011) ICPL labeling strategies for proteome research. Methods Mol Biol 753:55–64
13. Ong SE, Blagoev B, Kratchmarova I, Kristensen DB, Steen H, Pandey A, Mann M (2002) Stable isotope labeling by amino acids in cell culture, SILAC, as a simple and accurate approach to expression proteomics. Mol Cell Proteomics 1:376–386
14. Ong SE, Mann M (2006) A practical recipe for stable isotope labeling by amino acids in cell culture (SILAC). Nat Protoc 1:2650–2660
15. Kirkpatrick DS, Gerber SA, Gygi SP (2005) The absolute quantification strategy: a general procedure for the quantification of proteins and posttranslational modifications. Methods 35:265–273
16. Zinn N, Hahn B, Pipkorn R, Schwarzer D, Lehmann WD (2009) Phosphorus-based absolutely quantified standard peptides for quantitative proteomics. J Proteome Res 8:4870–4875
17. Brownridge P, Holman SW, Gaskell SJ, Grant CM, Harman VM, Hubbard SJ, Lanthaler K, Lawless C, O'cualain R, Sims P, Watkins R, Beynon RJ (2011) Global absolute quantification of a proteome: challenges in the deployment

of a QconCAT strategy. Proteomics 11: 2957–2970

18. Picard G, Lebert D, Louwagie M, Adrait A, Huillet C, Vandenesch F, Bruley C, Garin J, Jaquinoda J, Brun V (2012) PSAQ™ standards for accurate MS-based quantification of proteins: from the concept to biomedical applications. J Mass Spectrom 47:1353–1363
19. Singh S, Springer M, Steen J, Kirschner MW, Steen H (2009) FLEXIQuant: a novel tool for the absolute quantification of proteins, and the simultaneous identification and quantification of potentially modified peptides. J Proteome Res 8:2201–2210
20. Zinn N, Winter D, Lehmann WD (2010) Recombinant isotope labeled and selenium quantified proteins for absolute protein quantification. Anal Chem 82:2334–2340
21. Lebert D, Dupuis A, Garin J, Bruley C, Brun V (2011) Production and use of stable isotope-labeled proteins for absolute quantitative proteomics. Methods Mol Biol 753:93–115

Part IX

Posttranslational Modifications

Chapter 24

One-Source Peptide/Phosphopeptide Ratio Standards for Accurate and Site-Specific Determination of the Degree of Phosphorylation

Martin E. Boehm, Bettina Hahn, and Wolf D. Lehmann

Abstract

Reversible protein phosphorylation is a key mediator for intracellular signal transduction. Here we describe an innovative method for the production of pairs of peptide standards designed for quantitative mass spectrometry. These standard pairs can be used for site-specific analysis of the degree of phosphorylation of proteins in a bottom-up approach. The method starts from an isotopically labeled phosphopeptide analogue of the analyte phosphopeptide and ends up with a labeled peptide/phosphopeptide ratio standard in which the molar ratio between the phosphorylated and the unphosphorylated form is exactly defined. The signals of the ratio standard are used to standardize the corresponding analyte signals. This compensates for differences in LC recovery or ionization efficiency between the phosphorylated and unphosphorylated forms. The method can also be extended to quantitative analysis of multisite phosphorylation in a single peptide, which is exemplified for the presence of two phosphorylation sites. Peptide/phosphopeptide ratio standards exhibit high ratio accuracy, since ratio adjustment is performed by volumetric operations only.

Key words One-source peptide/phosphopeptide ratio standards, Degree of phosphorylation, Phospho-form abundance, Relative quantification

1 Introduction

Reversible protein phosphorylation is one of the most important regulation principles for protein functions such as enzymatic activity, protein–protein interaction, and signal transduction [1, 2]. Therefore, quantitative protein phosphorylation analysis is essential for the understanding of the regulation mechanisms and their dynamics as well as to characterize signaling cascades and networks. Quantitative protein phosphorylation analysis is mostly performed by a combination of enzymatic digestion and tandem mass spectrometry. In bottom-up proteomics, numerous quantitative methods have been developed over the last years. Among these are methods for global [3, 4] or targeted phosphorylation site analysis (for review *see* ref. 5). Targeted phosphorylation site analysis can be

Daniel Martins-de-Souza (ed.), *Shotgun Proteomics: Methods and Protocols*, Methods in Molecular Biology, vol. 1156, DOI 10.1007/978-1-4939-0685-7_24, © Springer Science+Business Media New York 2014

performed standard free [6] and by using internal standardization. The principle of targeted approaches is often the use of absolutely quantified internal standard peptides for both the phosphorylated and the non-phosphorylated form [7, 8]. In this concept, the peptide/phosphopeptide ratio is influenced by two quantification steps, since both species have to be quantified separately. The typical error of peptide quantification by amino acid analysis is 10–30 %. Here we describe a protocol for the production of a peptide/phosphopeptide ratio standard without absolute quantification steps. Carrying just the error of volumetric mixing, this method provides a high accuracy for degree of phosphorylation determination of proteins by internal standardization on the peptide level [9]. An extension of this method is possible to quantify the individual phospho-forms of a peptide carrying two phosphorylated sites. In this case one has to provide a unique isotopically labeled phosphopeptide for each of the three different phospho-forms [10].

2 Materials

Use always ultrapure reagents with analytical grade. To use this method it is necessary to have a purified isotopically labeled phosphopeptide. We recommend to use surface-optimized pipette tips and vials for standard production and sample processing. For quality control of the standard phosphopeptide and for control of the completeness of the dephosphorylation step we recommend static nanoESI-MS.

2.1 Standard Production

1. Surface-optimized pipette tips (10 μl, 200 μl) and vials (1.5 ml).
2. Accurate pipettes (10 μl/100–200 μl/1,000 μl).
3. Purified stable isotope-labeled phosphopeptide solution with a concentration of about 2 μmol/μl prepared on gravimetric basis.
4. Antarctic phosphatase and its reaction buffer (*see* **Note 2**).
5. 50 % aqueous glycerol solution.
6. Parafilm.
7. Thermomixer (36 °C/65 °C).
8. Solid-stem thermometer (65 °C).
9. Dilution buffer: 5 % acetonitrile and 0.1 % formic acid in water.

2.2 Quality Control

1. C18 micro bed ZipTip.
2. ZipTip buffer1: 50 % acetonitrile and 2 % formic acid in water.
3. ZipTip buffer2: 2 % formic acid in water.
4. Gold-coated borosilicate spray needles.
5. MS system with static electrospray ionization (*see* **Note 1**).

3 Methods

3.1 Standard Production (Scheme See Fig. 1)

1. Control the phosphopeptide solution for absence of the non-phosphorylated form (*see* **Note 3**).
2. Produce a 50 % glycerol solution.
3. Take a 1.5 ml vial (vial 1) and pipette 160 μl water, 20 μl 10× antarctic phosphatase buffer, and 20 μl of the controlled phosphopeptide solution into the vial.
4. Adjust a pipette to 95 μl, and use a 100 μl tip to mix the solution from **step 3**. Use the same tip to transfer 95 μl of this solution to two 1.5 ml vials (vials 2 and 3) (Fig. 1a).
5. Pipette 5 μl of 50 % glycerol into vial 2 and 5 μl antarctic phosphatase solution into vial 3—gently mix the solutions and spin down (*see* **Note 4**) (Fig. 1b).
6. Close vials 2 and 3 with parafilm and incubate for 2 h at 37 °C (*see* **Note 5**).
7. Put vials 2 and 3 for 5 min in a thermomixer set to 65 °C to inactivate antarctic phosphatase (*see* **Note 6**) (Fig. 1c).
8. Prepare two 1.5 ml vials, vial 4 with 960 μl and vial 5 with 980 μl of dilution buffer.
9. Remixing (1:1) and first dilution step (1:50): Following incubation of vials 2 and 3, cooling down, short mixing, and spinning down, remove the parafilm and pipette from each of the vials 20 μl to vial 4, then mix vial 4, and spin down (*see* **Note 7**) (Fig. 1d).

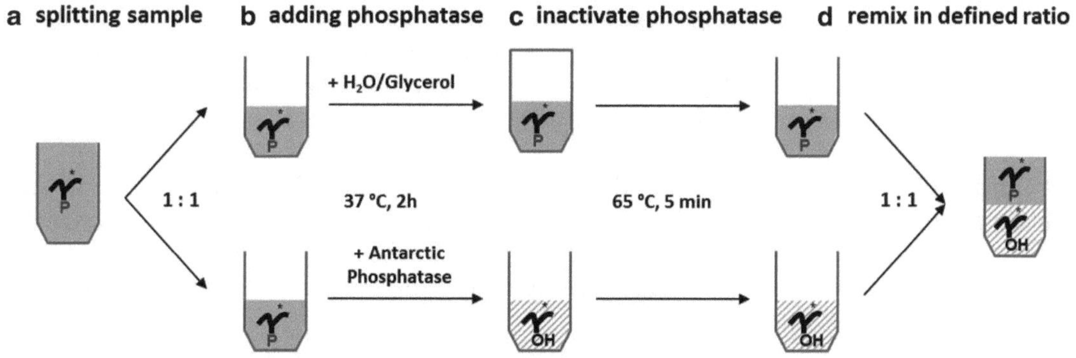

Fig. 1 Scheme of OS ratio standard production. Starting from an isotopically labeled phosphopeptide you perform steps (a–d) and end up with a solution containing peptide and phosphopeptide at a defined molar ratio. This solution can then be used as internal peptide/phosphopeptide ratio standard for LC-MS measurements to determine site-specific degrees of phosphorylation in a very robust and accurate way

10. Second dilution step (1:2,500): Transfer 20 μl of vial 4 to vial 5, mix vial 5, and spin down. Start with 2 μmol/μl of phosphopeptide solution, and follow these two dilution steps to obtain an 80 fmol/μl one-source ratio standard (*see* **Note 8**).

3.2 Quality Control of Produced Standard

1. Use the residual solution from vial 2 and vial 3 to measure if the phosphopeptide in vial 2 is really 100 % phosphorylated and if the peptide in vial 3 is actually completely dephosphorylated. Before analysis desalt the solutions by using ZipTip as described below (*see* **Note 9**).
2. Take a micro bed ZipTip, wash it in 100 μl ZipTip buffer 1, and equilibrate the tip in 100 μl ZipTip buffer2 by pipetting 10 μl of this solvent up and down several times.
3. Transfer 2 μl from vial 2 (phosphorylated standard peptide) in 10 μl ZipTip buffer 2, provided in a new vial (vial 6). Take the prepared ZipTip and enrich the phosphopeptide of vial 6 by pipetting 10 μl sample up and down ten times.
4. Wash the ZipTip in an extra vial with 25 μl ZipTip buffer 2 by pipetting up and down four times. Then elute the phosphopeptide which is still adsorbed on your ZipTip in 6 μl ZipTip buffer 1 by pipetting up and down six times (*see* **Note 10**).
5. Measure the desalted sample with nanoESI direct injection. Therefor, fill the solution in a gold-coated borosilicate spray capillary and acquire mass spectra over a mass range covering all relevant charge states of your peptide and phosphopeptide.
6. Check the spectrum for the ratio between phosphorylated and non-phosphorylated form. In this sample (derived from vial 2/vial 6) there should be the phosphorylated form only. In practice, signals for the non-phosphorylated form should be below 1 % compared to the phospho-form (*see* Fig. 2a).
7. Repeat **steps 3–6**. This time use 2 μl of vial 3 for desalting. The spectra of this standard fraction should show a completely dephosphorylated standard peptide. The dephosphorylation extent should be ≥99 % (Fig. 2b) (*see* **Note 11**).

3.3 Standard Addition to Sample

Add 0.5–20 μl of the diluted peptide/phosphopeptide ratio standard to the sample. The standard should be added to the samples as early as possible. We recommend to pipet the standard directly into your digest at the end of the incubation time (*see* **Note 12**).

3.4 Data Evaluation

1. Look for the signals of your analyte and standard. For each of them there should be a non-phosphorylated and a phosphorylated pair of heavy and light peptides. pTyr peptides often have a reduced and pSer/pThr peptides an enhanced LC retention time of about 0–3 min compared to the corresponding non-phosphorylated form (Fig. 3a). Look for the most abundant charge state of each, the phosphorylated and the

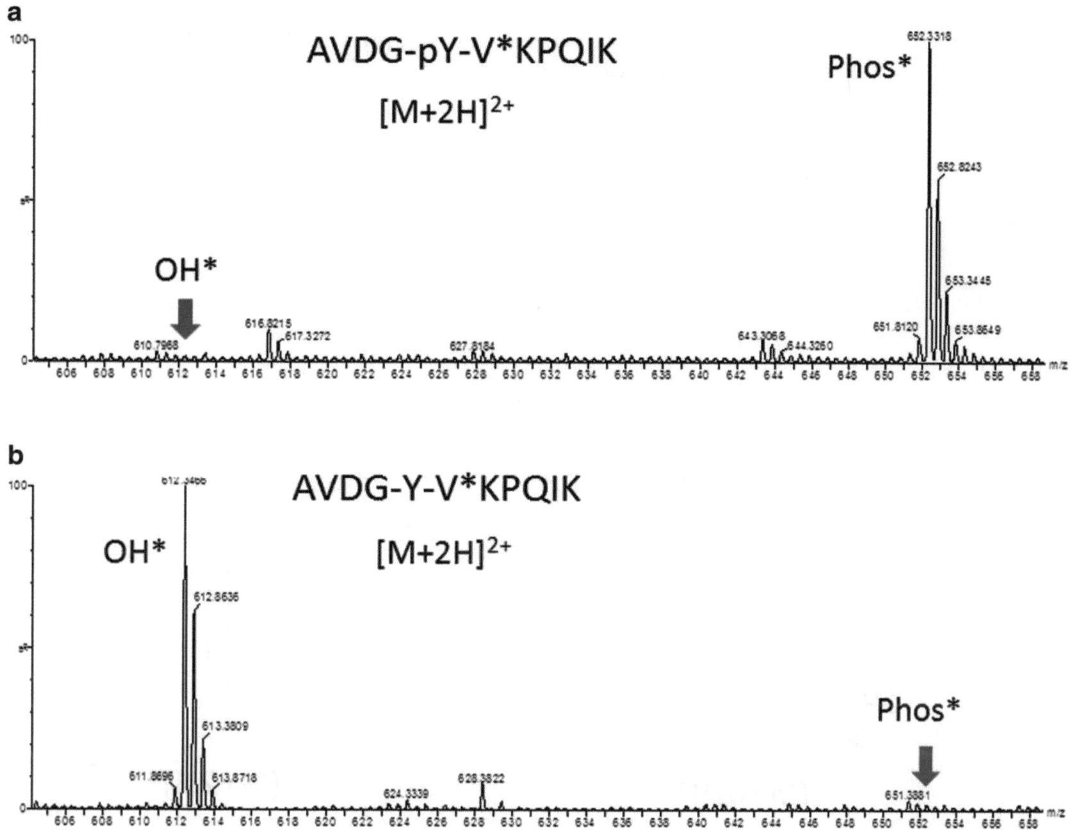

Fig. 2 Control of standard preparation steps by static nanoESI-MS; spectrum A shows the completely phosphorylated peptide before phosphatase treatment; spectrum B shows the spectrum of the completely dephosphorylated peptide following treatment with antarctic phosphatase

non-phosphorylated peptide (charge states 2 or 3 are observed most frequently) (Fig. 3b, c).

2. Then extract the LC peak areas for the appropriate charge states of the non-phosphorylated analyte peptide (OH), non-phosphorylated standard peptide (OH*), phosphorylated analyte peptide (Phos), and phosphorylated standard peptide (Phos*) (*see* **Note 13**).

3. Calculate the ratio (R_n) of the normalized intensity of the phosphopeptide signal over that of the peptide signal according to formula 1. The mixing ratio R_{mix} represents a correction factor if other ratios than 1 were put into practice in the remixing step of the ratio standard. R_{mix} is equal to the ratio of the applied volume of phosphorylated peptide and volume of dephosphorylated peptide in the remixing step (**step 8**, Subheading 3.1):

$$R_n = \frac{\text{OH}^* \times \text{Phos}}{\text{OH} \times \text{Phos}^*} \times R_{\text{mix}} \quad (1)$$

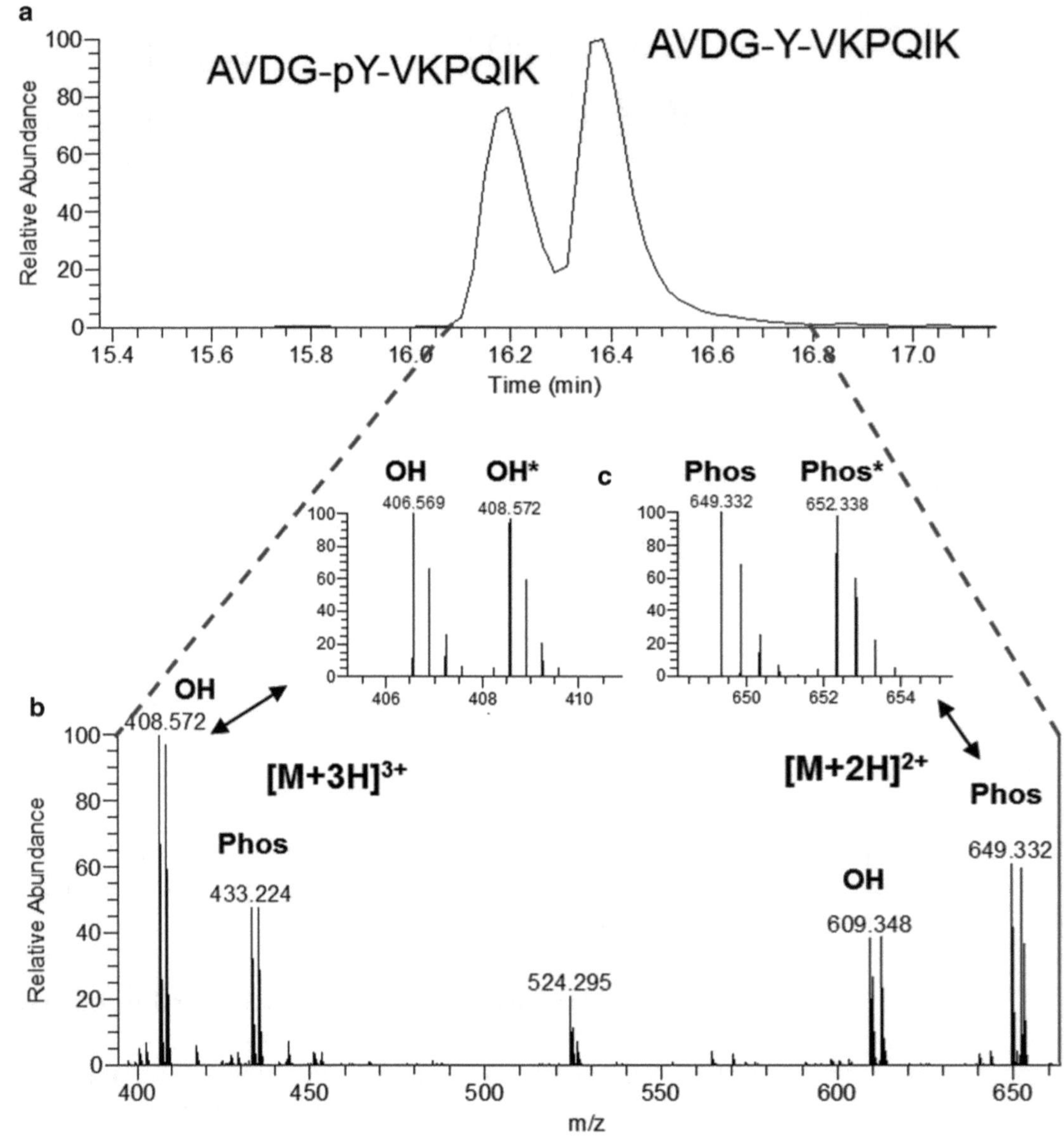

Fig. 3 Degree of phosphorylation determination for STAT5 by one-source ratio standardization. The analyte to standard ratio is 1:1, and the degree of STAT5 phosphorylation is 50 %; (**a**) selected ion chromatogram shows the UPLC separation of the pTyr-peptide and its non-phosphorylated counterpart, with the pTyr-peptides eluting earlier; (**b**) the different charge distribution between peptide and phosphopeptide, exhibiting a shift to lower charge states for the phosphopeptide; (**c**) the most intense signals used to calculate the degree of phosphorylation are shown

4. Insert the R_n value into formula 2 to calculate the degree of phosphorylation.

$$\text{Degree of phosphorylation}\,[\%] = \frac{R_n}{1+R_n} \times 100\,[\%] \qquad (2)$$

3.5 Extension to Peptides Containing Two Phosphorylation Sites

1. Provide three differently labeled standard phosphopeptides representing the three phospho-forms of interest (two singly and one doubly phosphorylated form) (e.g., Fig. 4). The first step is to control the quality of your labeled phosphopeptides and to adjust the concentrations of the stock solutions. This can be performed by analyzing a defined mix of all three peptides (*see* **Note 14**).
2. For each phospho-form perform **steps 1–9** of Subheading 3.1.
3. Control each peptide and phosphopeptide solution as described in Subheading 3.2.
4. Add the diluted standard for each phospho-form to the sample as described in Subheading 3.3 (*see* **Note 15**).
5. For data evaluation intensities of ten signals are extracted from the MS data set. For an example *see* ERK2 phospho-form abundance measurement displayed in Fig. 5. For each phospho-form the charge state with the best signal intensity should be selected, as done in this example (*see* **Note 13**).

protein	target	protease	peptide-sequence	ref.
STAT5A/B_human STAT5A_mouse	pY-694/699	trypsin	A V* D G pY V K P Q I K	[12, 13]
STAT5B_mouse	pY-699	trypsin	A A D G pY V* K P Q I K	
STAT6	pY641	AspN+LysC	D G R G pY V* P A T I K	[9, 12]
ERK1	pT-202, pY-204	trypsin	I A* D P E* H D H T G F L pT E Y V A* T R I A* D P E H D H T G F L T E pY V A* T R I A* D P E H D H T G F L pT E pY V A T R	[10]
ERK2	pT185, pY187	trypsin	V A* D P D H D H T G F* L pT E Y V A* T R V A* D P D H D H T G F L T E pY V A* T R V A* D P D H D H T G F L pT E pY V A T R	[10]
p38 alpha	pT180, pY182	trypsin	H T D* D E M pT G Y V A* T R H T D* D* E M T G pY V A* T R H T D D E M pT G pY V A* T R	
p38 beta	pT180, pY182	trypsin	Q A D E* E M pT G Y V A* T R Q A D E*E* M T G pY V A* T R Q A D E E M pT G pY V A* T R	

A* = $^{13}C_3,^{15}N$; D* = $^{13}C_4,^{15}N$; E* = $^{13}C_5,^{15}N$; F* = $^{13}C_6$; V* = $^{13}C_5,^{15}N$

Fig. 4 Examples for the design of one-source ratio standards for peptides containing one or two phosphorylation sites. STAT5 and STAT6 are examples for singly phosphorylated peptides. ERK1, ERK2, p38alpha, and beta are examples for two phosphorylation sites. In the latter case, three phospho-forms of the peptide occur. Each phospho-form requires an individual standard with individual mass shift so that the non-phosphorylated form of each pair can be detected individually. The STAT6 peptide standard is designed for the combined use of AspN and LysC, all other examples represent purely tryptic peptides. Peptide labeling is described in detail on the bottom of the table

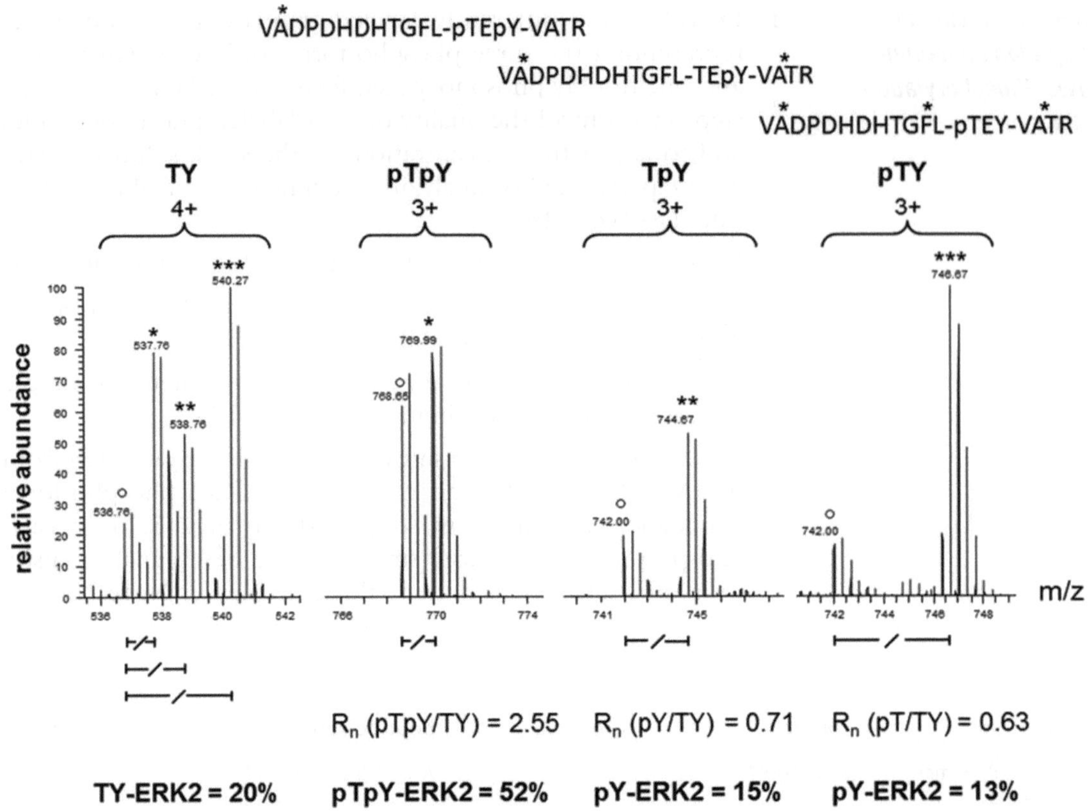

Fig. 5 Phospho-form abundance analysis by ratio standards at the activation motif of ERK2 with two closely adjacent phosphorylation sites. To standardize the three phospho-forms, three differently labeled one-source ratio standards are required. The selected standard phosphopeptides are displayed. Ten signal intensities are extracted to calculate first normalized intensities (R_n) and then the relative abundance of each phospho-form. In this example the most intense ions for each phosphorylation state were selected. These are the $[M + 4H]^{4+}$ ions for the non-phosphorylated and the $[M + 3H]^{3+}$ ions for the phosphorylated peptides (ref. 10, Fig. 2)

6. With the help of formulae 3–5, the extracted intensities are first converted into normalized intensity ratios (R_n):

$$R_{n_{pTpY/TY}} = \frac{[TY^{*}] \times [pTpY^{\circ}]}{[TY^{\circ}] \times [pTpY^{*}]} \tag{3}$$

$$R_{n_{pY/TY}} = \frac{[TY^{**}] \times [pY^{\circ}]}{[TY^{\circ}] \times [pY^{**}]} \tag{4}$$

$$R_{n_{pT/TY}} = \frac{[TY^{***}] \times [pT^{\circ}]}{[TY^{\circ}] \times [pT^{***}]} \tag{5}$$

7. To calculate the relative abundance of each phospho-form these R_n ratios are inserted in formulae 6–9:

$$TY[\%] = \frac{1}{1 + R_{n_{pTpY/TY}} + R_{n_{pY/TY}} + R_{n_{pT/TY}}} \times 100 \tag{6}$$

$$pTpY[\%] = \frac{R_{n_{pTpY/TY}}}{1 + R_{n_{pTpY/TY}} + R_{n_{pY/TY}} + R_{n_{pT/TY}}} \times 100 \tag{7}$$

$$pY[\%] = \frac{R_{n_{pY/TY}}}{1 + R_{n_{pTpY/TY}} + R_{n_{pY/TY}} + R_{n_{pT/TY}}} \times 100 \tag{8}$$

$$pT[\%] = \frac{R_{n_{pT/TY}}}{1 + R_{n_{pTpY/TY}} + R_{n_{pY/TY}} + R_{n_{pT/TY}}} \times 100 \tag{9}$$

4 Notes

1. Direct injection by static nanoESI is recommended, but it is also possible to use (nano)LC-MS to check the standards. Since quality control of the peptide and the phosphopeptide standard requires a dynamic range of at least 100:1 (Fig. 2), inject a sufficiently large amount of sample. In case LC-MS is used for control, addition of a metal ion-mobilizing additive, such as citrate, is recommended to avoid selective loss of phosphopeptides [11].
2. For dephosphorylation we used antarctic phosphatase from New England Biolabs (M0289S), but antarctic phosphatases from other suppliers may also work.
3. If there is too much dephosphorylated peptide (>1 %) in the sample derived from vial 2, you may remove the non-phosphorylated species by phosphopeptide enrichment. For ratio standard production rather work with high concentrations (>1 μmol/μl) to avoid selective adsorption effects which would influence the accuracy of the ratio standard. We normally use 5–10 % acetonitrile for the peptide/phosphopeptide stock solutions and for subsequent dilution steps, since this may reduce peptide adsorption to the tube wall and pipette tips. If the amount of recombinant phosphopeptide is limited the standard production can also be started with lower concentrations. In this case you can also reduce phosphatase amount and adjust standard dilution in **steps 8** and **9** accordingly.
4. The antarctic phosphatase is stored at −20 °C. Put the phosphatase as well as the 50 % glycerol 15 min before use on +4 °C to ensure that both have the same temperature for pipetting. The antarctic phosphatase from New England

Biolabs is diluted in 50 % glycerol; if you want to use phosphatases from another supplier, adjust the glycerol concentration to that one.

5. To be on the safe side, gently mix and spin down the solutions every 30 min to ensure that phosphopeptides which are possibly adsorbed on the tube wall get in contact with the phosphatase.
6. Use a control vial filled with a few 100 μl water and a solid-stem thermometer to check if the temperature in your samples is really 65 °C.
7. If you want to use this standard to analyze degrees of phosphorylation in a wide range, e.g., 10–80 % analyte phosphorylation, it is recommended to remix in a 1:1 ratio. If you analyze more extreme degrees of phosphorylation (below 10 % or over 90 %), adjust your remix ratio so that your peptide/phosphopeptide ratio standards reflect the situation in your analyte. If standard ratios other than 1:1 are applied, this has to be corrected in the final calculations with the R_{mix} correction factor (*see* Subheading 3.4, **step 1**).
8. The concentration of your standard should be adjusted to your analyte concentration. Your analyte concentration is mainly dependent on the number of cells in your cell lysate, the copy number per cell of your target protein, and the quality of your antibody for IP. If cellular concentration of the target protein is known, the optimal amount of ratio standard can be roughly estimated.
9. It is recommended to desalt the solutions before the control measurement. Salts from impurities of peptide synthesis and buffer/glycerol from the dephosphorylation step may lead to a signal loss in nanoESI-MS.
10. If you want to control the standards by LC/MS, it is recommended to use metal ion-mobilizing additives [11]. The volume of ZipTip elution buffer can be reduced down to 3 μl. The eluate can then be diluted with 10 mM citrate to the peptide concentration needed for injection to your LC/MS system. To ensure retention on C18 column add enough citrate solution to keep the overall acetonitrile content of your sample below 4 %. To enhance the positive effect of citrate, rinse your HPLC system by a citrate blank injection.
11. For safety reasons, control the sample derived from vial 2 (untreated aliquot) for the absence of the dephosphorylated form. Control the sample derived from vial 3 (dephosphorylated aliquot) for the absence of the phosphorylated form. If you detect more than 1 % of this form, you can repeat this step with more phosphatase and/or longer incubation time.

12. The best accuracy is obtained if the standard is given to the samples in analyte/standard ratios of around 1, but ratios down to 0.05 are also feasible.
13. The charge distributions of the native and labeled analyte peptide are equal. However, peptide and phosphopeptide may show different charge state distributions. Typically, phosphopeptides show a shift to lower charge states compared to corresponding peptide. For calculation of the degree of phosphorylation, it is not necessary to use the same charge state for peptide and phosphopeptide, so that normally the most abundant signals can be selected.
14. Beyond the use for mono-phosphorylated peptides, this standardization method can in addition be applied for analyte peptides containing two (or even more) phosphorylation sites. To be able to standardize the signal intensities of all phosphoforms it is necessary to prepare individual one-source ratio standards for each form. For the analysis of two phosphorylation sites within one peptide, three differentially labeled phosphopeptides are required. The number of phosphorylation sites (n) is connected with the number of required standards (x) by the formula $x=2n+1$.
15. It is possible to adjust the standard addition of each phosphoform to its expected abundance. If a tenfold higher amount of doubly phosphorylated compared to singly phosphorylated analyte is expected, you can add a tenfold higher amount of doubly phosphorylated standard to the sample.

Acknowledgements

We are indebted to U. Klingmüller and their group members and gratefully acknowledge financial support of the BMBF (Bundesministerium für Bildung und Forschung) via the projects SBCancer and LungSys.

References

1. Marks F, Klingmüller U, Müller-Decker K (2008) Cellular signal processing: an introduction to the molecular mechanisms of signal transduction. Garland Science, New York
2. Preisinger C, von Kriegsheim A, Matallanas D, Kolch W (2008) Proteomics and phosphoproteomics for the mapping of cellular signalling networks. Proteomics 8:4402–4415
3. Olsen JV, Blagoev B, Gnad F, Macek B, Kumar C, Mortensen P, Mann M (2006) Global, in vivo, and site-specific phosphorylation dynamics in signaling networks. Cell 11:635–648
4. Lehmann WD (2010) Protein phosphorylation analysis by electrospray mass spectrometry. RSC Publishing, Cambridge
5. Boehm ME, Seidler J, Hahn B, Lehmann WD (2012) Site-specific degree of phosphorylation in proteins measured by liquid chromatography-electrospray mass spectrometry. Proteomics 12:2167–2178

6. Seidler J, Adal M, Kübler D, Bossemeyer D, Lehmann WD (2009) Analysis of autophosphorylation sites in the recombinant catalytic subunit alpha of cAMP-dependent kinase by nano-UPLC-ESI-MS/MS. Anal Bioanal Chem 7:1713–1720
7. Zinn N, Hahn B, Pipkorn R, Schwarzer D, Lehmann WD (2009) Phosphorus-based absolutely quantified standard peptides for quantitative proteomics. J Proteome Res 8: 4870–4875
8. Kirkpatrick DS, Gerber SA, Gygi SP (2005) The absolute quantification strategy: a general procedure for the quantification of proteins and post-translational modifications. Methods 35:265–273
9. Hahn B, Böhm M, Raia V, Zinn N, Möller P, Klingmüller U, Lehmann WD (2011) One-source peptide/phosphopeptide standards for accurate phosphorylation degree determination. Proteomics 11:490–494
10. Hahn B, D'Alessandro LA, Depner S, Waldow K, Boehm ME, Bachmann J, Schilling M, Klingmüller U, Lehmann WD (2013) Cellular ERK phospho-form profiles with conserved preference for a switch-like pattern. J Proteome Res 12:637–646
11. Seidler J, Zinn N, Haaf E, Boehm ME, Winter D, Schlosser A, Lehmann WD (2011) Metal ion-mobilizing additives for comprehensive detection of femtomole amounts of phosphopeptides by reversed phase LC-MS. Amino Acids 41:311–320
12. Raia V, Schilling M, Böhm M, Hahn B, Kowarsch A, Raue A, Sticht C, Bohl S, Saile M, Möller P, Gretz N, Timmer J, Theis F, Lehmann WD, Lichter P, Klingmüller U (2011) Dynamic mathematical modeling of IL13-induced signaling in Hodgkin and primary mediastinal B-cell lymphoma allows prediction of therapeutic targets. Cancer Res 71: 693–704
13. Bachmann J, Raue A, Schilling M, Böhm ME, Kreutz C, Kaschek D, Busch H, Gretz N, Lehmann WD, Timmer J, Klingmüller U (2011) Division of labor by dual feedback regulators controls JAK2/STAT5 signaling over broad ligand range. Mol Syst Biol 7:516

Chapter 25

Quantitative Glycoproteomics for N-Glycoproteome Profiling

Sheng Pan

Abstract

Glycoproteomics, the global analysis of glycoproteins, is a subfield of proteomics with high biological and clinical relevance. Glycoproteomics integrates knowledge and techniques drawing from glycotechnology, separation science, mass spectrometry, and bioinformatics to support systematic identification and quantification of glycoproteins in a complex biological sample. In this chapter, we describe a shotgun proteomics-based glycoproteomics platform that provides precise mapping of protein N-glycosylation sites and quantitative assessment of glycosylation site occupancy on a global scale.

Key words Glycoproteomics, Proteomics, Glycosylation, Glycoprotein, Mass spectrometry, Hydrazide chemistry, Lectin, Glycosylation site occupancy, Quantification

1 Introduction

Glycosylation is considered among the most heterogenic and complex protein posttranslational modifications (PTMs), involving not only the complexity of glycoprotein constituent but also the intricacy within glycoprotein molecule. Protein glycosylation may be altered due to changes in cellular physiological state resulted from diseases or different biological conditions. Aberrant glycosylation can involve structure alteration of glycan moiety as well as change on glycosylation site occupancy that results in hypo- or hyper-glycosylation on a specific glycosylation site [1, 2]. While glycomics analysis focuses on characterizing the structure of glycan moieties attached to proteins [3], the emerging technology of glycoproteomics emphasizes identification of protein glycosylation sites and further quantification of change in glycosylation site occupancy.

A comprehensive quantitative glycoproteomics analysis of a complex biological system requires an integrated platform orchestrating different techniques. While sample preparation strategy for different sample types varies, a typical glycoproteomics pipeline includes the following modules: enrichment of glycoproteins or

Daniel Martins-de-Souza (ed.), *Shotgun Proteomics: Methods and Protocols*, Methods in Molecular Biology, vol. 1156, DOI 10.1007/978-1-4939-0685-7_25, © Springer Science+Business Media New York 2014

glycopeptides, mass spectrometric analysis, and bioinformatics data processing. In addition, for a quantitative study, stable isotope labeling techniques are commonly applied providing differential mass tags to represent samples from different origins, enabling quantitative mass spectrometric analysis. Several studies have reported using quantitative glycoproteomics techniques to investigate glycoproteome of a variety of specimens, including tissues, cells, plasma or serum, and other bodily fluids [4]. In most of these studies, either hydrazide chemistry [5, 6] or lectin affinity [7–9] was used for glyco-enrichment, and various quantitative methods were applied to facilitate mass spectrometric analysis, including iTRAQ [10–12], succinic anhydride [6], N-acetoxy-succinimide [13], 2-nitrobenzenesulfenyl [14], ^{18}O labeling [15], and SILAC [16].

Recently, we have established a robust, sensitive, and cost-effective glycoproteomics approach for quantitative global profiling of N-glycopeptides [17]. The method allows precise mapping of N-glycosylation sites and comparative quantification of the abundance of formerly N-linked glycopeptides using high-resolution mass spectrometry. Using this approach, quantitative glycoproteomics method can be established to investigate N-glycoproteome of a variety of biological sample and clinical specimens. As demonstrated in Fig. 1, equal amount of proteins from each sample of comparison (e.g., disease vs. healthy) are alkylated and digested into peptides, which are separately labeled with different stable isotopes using formaldehyde-based dimethyl labeling [18]. The differentially labeled peptides from the compared samples are combined, and a small portion of the combined sample is taken for quantitative global protein profiling so that core protein quantification information can be obtained. The majority of the combined sample is used for glycoproteomics analysis and subjected to glycopeptide enrichment using either hydrazide chemistry-based solid-phase extraction or multi-lectin affinity column. After cleavage of N-glycans using PNGase F enzyme, the labeled, de-glycosylated N-glycopeptides are collected and subjected to LC MS/MS analysis. The obtained MS data is searched against protein database for peptide/protein identification. N-glycosylation sites can be mapped precisely using consensus sequence of Asn-X-Ser/Thr (X—any amino acid except proline) in which asparagine is converted to aspartic acid following enzyme cleavage, introducing a mass difference of 0.9840 Da. The quantification of an N-glycopeptide is achieved using the signal intensity ratio between the heavy and light isotopic forms of the de-glycosylated N-glycopeptides. With the availability of quantitative information in both an N-glycopeptide abundance and its core protein expression, the changes on the corresponding N-glycosylation site occupancy can be estimated as follows: fold change in N-glycosylation site occupancy = fold change in N-glycopeptide abundance/fold change in core protein expression.

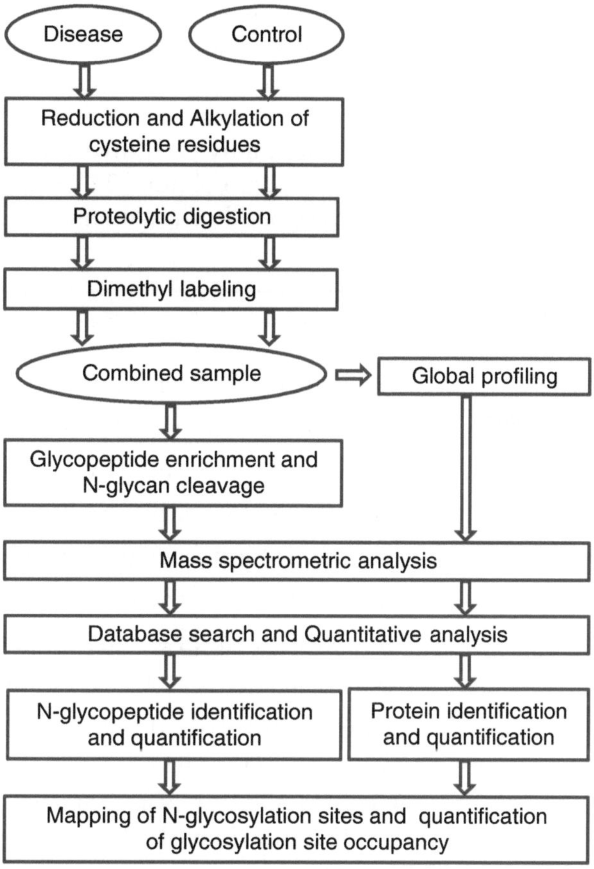

Fig. 1 Quantitative glycoproteomics work flow

A variety of sample types, such as tissue (including snap-frozen or paraffin-embedded formalin-fixed (PEFF) tissues), cell lysate/secretome, plasma/serum, and other bodily fluids, can be readily analyzed using this approach for glycoproteomics profiling. Multiple proteases can be used to enhance the identification of glycopeptides when trypsin digestion does not generate peptides with suitable mass range for MS detection. Using formaldehyde-based dimethyl labeling, the dimethyl groups are tagged on the N-terminal and lysine residues of a peptide, conceptually covering the whole proteome for differential isotopic labeling for quantitative analysis. The combination of differentially labeled samples prior to glycopeptide enrichment significantly reduces the variations that may be introduced when individual samples are processed separately. One advantage of enriching N-glycopeptides for mass spectrometry analysis is to reduce the possibility of false identification of N-glycosylation sites due to deamidation of asparagine. Using solid-phase enrichment, glycopeptides are retained on the solid phase, and the majority of the non-glycopeptides, including deamidated peptides, is washed off during the sample preparation process and is not included in the mass spectrometric analysis. While hydrazide chemistry-based solid-phase extraction and lectin affinity columns have been the most versatile and widely used approaches in glycoproteomics studies, other mechanisms, such as boronic acid [19], size-exclusion chromatography [20], hydrophilic interaction [21], and a graphite powder micro-column [22], can also be integrated into the pipeline for glycopeptide enrichment.

2 Materials

2.1 Proteomics Sample Preparation

1. T-PER® Tissue protein extraction reagent (Thermo Scientific).
2. Protease Inhibitor Cocktail Kit.
3. Glycoprotein standard: Yeast invertase 2 (Sigma-Aldrich).
4. 10 mM DL-dithiothreitol (DTT) solution.
5. 25 mM iodoacetamide solution.
6. 100 % (w/v) trichloroacetic acid (TCA) solution.
7. Sequencing-grade modified trypsin.
8. Formaldehyde, 37 wt% solution in water.
9. Formaldehyde-d2, 98atom %D, 20 wt% solution in deuterium oxide (Isotech).
10. 3 M sodium cyanoborohydride solution in 100 mM sodium acetate (*see* **Note 1**).
11. 10 % (v/v) trifluoroacetic acid (TFA) solution (*see* **Note 2**).
12. C18 loading buffer: 0.1 % (v/v) TFA in ultrapure water.
13. C18 eluting buffer: 85 % (v/v) acetonitrile (LC/MS grade) and 0.1 % TFA in ultrapure water.
14. BCA Protein Assay Kit.
15. Nano Drop spectrophotometer (Thermo Scientific).

2.2 Hydrazide Chemistry-Based Solid-Phase Glyco-Enrichment

1. Coupling buffer: 0.1 M sodium acetate and 150 mM sodium chloride, pH 5.5.
2. Oxidation solution: 250 mM sodium (meta)periodate.
3. 500 mM sodium sulfite solution.
4. Resin washing solution: 80 % (v/v) acetonitrile and 0.1 % TFA.
5. UltraLink hydrazide resins (Thermo Scientific).
6. PNGase F (New England BioLabs).

2.3 Multi-lectin Affinity Column for Glyco-Enrichment

1. Binding buffer: 1 mM calcium chloride, 1 mM manganese (II) chloride, 1 mM magnesium chloride, and 0.5 M sodium chloride in 25 mM Tris, pH 7.4.

2. Elution buffer 1: 200 mM α-methyl mannoside, 200 mM α-methyl glucoside, 100 mM L-fucose, 250 mM lactose, and 250 mM N-acetyl-D-glucosamine in binding buffer.
3. Elution buffer 2: 200 mM acetic acid.

2.4 LC MS/MS Analysis

1. LTQ-Orbitrap hybrid mass spectrometer (Thermo Fisher Scientific) coupled with a nano-flow HPLC (Eksigent Technologies). Other high-resolution tandem mass spectrometer and nano-flow HPLC may be employed.
2. 1.5 cm trap column (IntegraFrit 100 μm, New Objective) packed with Magic C18AQ resin (5 μm, 200 Å particles; Michrom Bioresources).
3. 27 cm analytical column (PicoFrit 75 μm, New Objective) packed with Magic C18AQ resin (5 μm, 100 Å particles; Michrom Bioresources).
4. Buffer A: Water with 0.1 % formic acid.
5. Buffer B: Acetonitrile with 0.1 % formic acid.

3 Methods

3.1 Preparation of Tissue Lysate

1. Place fresh frozen tissue (100–150 mg) into T-per with addition of Protease Inhibitor Cocktail and homogenize completely using homogenizer. Incubate the lysate on ice for 15 min (*see* **Note 3**).
2. Transfer the homogenized lysate into 1.5 ml microcentrifuge tubes. Centrifuge sample for 15 min at 4 °C at 13,000 × *g* to pellet the debris. Collect the supernatant into a new microcentrifuge tube, and keep the sample on ice.
3. Determine protein concentration using BCA Protein Assay Kit and Nano Drop spectrophotometer.
4. Make aliquots of each sample and store in −80 °C.

3.2 Proteomics Sample Preparation: Cysteine Alkylation, Trypsin Digestion, and Dimethyl Labeling

1. Preparation of glycoprotein standard: Dissolve 500 μg of yeast invertase-2 glycoprotein standard into 250 μl of ultrapure water and heat at 90–95 °C for 10 min.
2. 1,000 μg of tissue lysate was used per comparison group (e.g., control group vs. diseased group), add 100 μg of the nonhuman glycoprotein standard, and dilute each sample to 300 μl with 50 mM ammonium bicarbonate (pH 7.4–8.4).
3. Reduce cysteine residues by adding 10 mM DTT into sample and incubating the sample at 50 °C for 1 h.
4. Alkylate cysteine residues by adding 25 mM IDA into sample and incubating the sample at room temperature for 30 min in the dark.

5. Add ¼ volume of 100 % (w/v) TCA solution into the sample, and incubate the sample on ice for 10 min.
6. Centrifuge the sample at 14,000 × *g* for 5 min, and discard supernatant.
7. Wash pellet with 200 μl of ice-cold acetone, centrifuge at 14,000 × *g* for 5 min, discard supernatant, and repeat the washing step one more time.
8. Let the pellet air-dry until most of the acetone has evaporated, and then resuspend the protein in 300 μl 50 mM sodium bicarbonate (pH 7.4–8.4).
9. Trypsin is added at 1:50 trypsin-to-sample ratio. Add half the amount of trypsin to the samples and incubate for 2 h at 37 °C.
10. Add the rest of the trypsin to the sample and incubate overnight (16–18 h) at 37 °C.
11. Spin down briefly to collect all digested sample, and adjust the pH to 5.5 with 10 % acetic acid.
12. Add stable isotopic labeling agents to each sample for comparison as follows. Light labeling: add 5.4 μl of 37 wt% formaldehyde. Heavy labeling: add 10 μl of 20 wt% formaldehyde-D_2.
13. Immediately add 10 μl of freshly prepared 3 M sodium cyanoborohydride to the samples.
14. Vigorously vortex the samples and incubate at RT for 15 min. Combine the H/D-labeled samples.
15. Adjust sample to <pH 3 by adding 10 % TFA carefully (*see* **Note 4**).
16. Purify sample using C18 purification column according to the manufacturer's instruction.
17. Dry down sample using SpeedVac and store at −20 °C.

3.3 Glycopeptide Enrichment (Hydrazide Chemistry or Lectin Affinity)

3.3.1 Hydrazide Chemistry-Based Solid-Phase Extraction

1. Dissolve combined sample in 300 μl of the coupling buffer.
2. Add freshly prepared 250 mM sodium meta-periodate to a final concentration of 10 mM. Incubate at RT for 1 h in the dark with gentle rotation.
3. Add freshly prepared 500 mM sodium sulfite to a final concentration of 20 mM and incubate at RT for 10 min with gentle rotation.
4. Centrifuge 400 μl of UltraLink hydrazide resin slurry for 2 min at 1,000 × *g*. Aspirate and discard the supernatant.
5. Wash resin with 1 ml of the coupling buffer, centrifuge for 2 min at 1,000 × *g*, aspirate, and discard the supernatant. Repeat the process three more times.

6. Resuspend washed resin in 300 μl coupling buffer. Add sample and incubate at RT for 10 h with gentle rotation.
7. Wash the resin with 1 ml of freshly prepared resin washing solution, centrifuge for 2 min at 1,000 ×*g*, aspirate, and discard the supernatant.
8. Wash the resin five more times with 1 ml 1× PBS, pH 7.4.
9. Resuspend the resin in 250 μl 1× PBS, pH 7.4. Add 2 μl PNGase F, and incubate the sample for 6 h at 37 °C to cleave glycopeptides.
10. Centrifuge the sample for 2 min at 1,000 ×*g*, and collect the supernatant as de-glycosylated N-glycopeptides.
11. Rinse the resin with 250 μl 1× PBS, pH 7.4, centrifuge, and combine the supernatant to the saved de-glycosylated N-glycopeptides.
12. Purify de-glycosylated N-glycopeptides using C18 purification column according to the manufacturer's instruction.
13. Dry down the peptides and store in −20 °C until mass spectrometric analysis.

3.3.2 Multi-lectin Affinity Column

1. Prepare the multi-lectin capturing column using Costar Spin-X Centrifuge Tube Filters. For every 1,000 μg peptides, add and mix the following lectin agarose bead slurries in the filter tube: 125 μl agarose-bound Concanavalin A, 125 μl agarose-bound Aleuria Aurantia Lectin, 300 μl agarose-bound Elderberry Bark Lectin, and 100 μl agarose-bound Germ Agglutinin. Centrifuge at 800 ×*g* at RT for 5 min, and discard the flow through (*see* **Note 5**).
2. Add 500 μl of the binding buffer, centrifuge at 800 ×*g* at RT for 5 min, and discard the flow through. Repeat.
3. Load the sample onto the column, incubate at RT for 1 h with gentle rotation, and centrifuge at 800 ×*g* at RT for 5 min.
4. Reload the sample onto the column, incubate at RT for 30 min with gentle rotation, and centrifuge at 800 ×*g* at RT for 5 min. Repeat.
5. Wash the column with 500 μl of the binding buffer three times.
6. Add 500 μl of the elution buffer 1, incubate for 15 min, centrifuge at 800 ×*g* at RT for 5 min, and collect the eluate. Repeat.
7. Add 500 μl of the elution buffer 2, incubate for 30 min, centrifuge at 800 ×*g* at RT for 5 min, and collect the eluate. Repeat.
8. Combine all four eluates, and purify the captured peptides using C18 purification column according to the manufacturer's instruction.

9. Resuspend the sample in 50 μl 1× PBS, pH 7.4, add 2 μl PNGase F, and incubate for 6 h in 37 °C.
10. Purify the de-glycosylated N-glycopeptides with C18 column. Collect the glycan flow through for glycan analysis if needed.
11. Elute and dry down the peptides and store in −20 °C until mass spectrometric analysis.

3.4 Mass Spectrometric Analysis

The sample is analyzed using a high-resolution LTQ-Orbitrap hybrid mass spectrometer coupled with a nano-flow HPLC. The sample is first loaded onto a 1.5 cm trap column packed with Magic C18AQ resin with buffer A at a flow rate of 3 μl/min. The peptide sample is then separated by a 27 cm analytical column packed with Magic C18AQ resin followed by mass spectrometric analysis. A 90-min LC gradient is used as follows: 5–7 % buffer B versus buffer A over 2 min, then to 35 % over 90 min, then to 50 % over 1 min, hold at 50 % for 9 min, change to 95 % over 1 min, hold at 95 % for 5 min, drop to 5 % over 1 min, and recondition at 5 %. The flow rate for the peptide separation is 300 nL/min. For MS analysis, a spray voltage of 2.25 kV is applied to the nanospray tip. The mass spectrometric analysis is performed using data-dependent acquisition with a m/z range of 400–1,800, consisting of a full MS scan in the Orbitrap followed by up to five MS/MS spectra acquisitions in the linear ion trap using collision-induced dissociation (CID). Other mass spectrometer parameters include the following: isolation width 2 m/z, target value 1e4, collision energy 35 %, and max injection time 100 ms. Lower abundance peptide ions are interrogated by using dynamic exclusion (exclusion time 45 s, exclusion mass width −0.55 m/z low to 1.55 m/z high). Charge state screen is used, allowing for MS/MS of any ions with identifiable charge states +2, +3, and +4 and higher.

3.5 Data Analysis

Raw machine output files of MS/MS spectra are converted to mzXML files and searched with X!Tandem [23] configured with the k-score scoring algorithm [24], against a combined human protein database with the addition of yeast invertase 2. The search parameters are as follows: enzyme: trypsin; maximum missed cleavages: 1; fixed modifications: carboxyamidomethylation on cysteine and light dimethylation on N-terminus and lysine; variable modifications: oxidation on methionine, difference between light and heavy dimethylation on N-terminus and lysine, and enzymatic conversion of asparagine to aspartic acid; and parent monoisotopic mass tolerance: 2.5 Da. Peptide identifications are assigned probability by PeptideProphet [25] (*see* **Note 6**). Relative quantitation of light and heavy peptide abundance is performed with Xpress [26]. Proteins present in sample are inferred using ProteinProphet [27].

4 Notes

1. 3 M sodium cyanoborohydride solution is freshly prepared in 100 mM sodium acetate right before use.
2. TFA reacts with oxygen-containing plastic, so it is recommended to use glass syringe and container to handle and store TFA solution.
3. When preparing more than one lysate at a time, wash homogenizer blades thoroughly with water and lysis buffer before preparing the next sample.
4. Due to the presence of multiple chemicals in sample, pH may not settle for several minutes after addition of TFA. Also, the reaction may produce gas and cause the lid to pop/explode, so leave the lid open to let the gas escape.
5. Different combinations of lectins can be used for glycopeptide enrichment depending on study design.
6. It is recommended that only peptides with a PeptideProphet score ≥ 0.95 are used for quantitative analysis.

Acknowledgment

This work was supported in part with federal funds from the National Institutes of Health under grants K25CA137222 and R21CA149772.

References

1. Jones J, Krag SS, Betenbaugh MJ (2005) Controlling N-linked glycan site occupancy. Biochim Biophys Acta 1726:121–137
2. Zhao YY, Takahashi M, Gu JG, Miyoshi E, Matsumoto A, Kitazume S, Taniguchi N (2008) Functional roles of N-glycans in cell signaling and cell adhesion in cancer. Cancer Sci 99:1304–1310
3. Wuhrer M (2013) Glycomics using mass spectrometry. Glycoconj J 30:11–22
4. Pan S, Chen R, Aebersold R, Brentnall TA (2011) Mass spectrometry based glycoproteomics—from a proteomics perspective. Mol Cell Proteomics 10:R110
5. Liu T, Qian WJ, Gritsenko MA, Camp DG, Monroe ME, Moore RJ, Smith RD (2005) Human plasma N-glycoproteome analysis by immunoaffinity subtraction, hydrazide chemistry, and mass spectrometry. J Proteome Res 4:2070–2080
6. Zhang H, Li XJ, Martin DB, Aebersold R (2003) Identification and quantification of N-linked glycoproteins using hydrazide chemistry, stable isotope labeling and mass spectrometry. Nat Biotechnol 21:660–666
7. Kaji H, Saito H, Yamauchi Y, Shinkawa T, Taoka M, Hirabayashi J, Kasai K, Takahashi N, Isobe T (2003) Lectin affinity capture, isotope-coded tagging and mass spectrometry to identify N-linked glycoproteins. Nat Biotechnol 21:667–672
8. Qiu R, Regnier FE (2005) Use of multidimensional lectin affinity chromatography in differential glycoproteomics. Anal Chem 77: 2802–2809
9. Yang Z, Hancock WS (2004) Approach to the comprehensive analysis of glycoproteins isolated from human serum using a multi-lectin affinity column. J Chromatogr A 1053:79–88

10. Jung K, Cho W, Regnier FE (2009) Glycoproteomics of plasma based on narrow selectivity lectin affinity chromatography. J Proteome Res 8:643–650
11. Lee HJ, Na K, Choi EY, Kim KS, Kim H, Paik YK (2010) Simple method for quantitative analysis of N-linked glycoproteins in hepatocellular carcinoma specimens. J Proteome Res 9:308–318
12. Lei Z, Beuerman RW, Chew AP, Koh SK, Cafaro TA, Urrets-Zavalia EA, Urrets-Zavalia JA, Li SF, Serra HM (2009) Quantitative analysis of N-linked glycoproteins in tear fluid of climatic droplet keratopathy by glycopeptide capture and iTRAQ. J Proteome Res 8: 1992–2003
13. Wilson CR, Regnier FE, Knapp DW, Raskin RE, Andrews DA, Hooser SB (2008) Glycoproteomic profiling of serum peptides in canine lymphoma and transitional cell carcinoma. Vet Comp Oncol 6:171–181
14. Ueda K, Katagiri T, Shimada T, Irie S, Sato TA, Nakamura Y, Daigo Y (2007) Comparative profiling of serum glycoproteome by sequential purification of glycoproteins and 2-nitrobenzensulfenyl (NBS) stable isotope labeling: a new approach for the novel biomarker discovery for cancer. J Proteome Res 6:3475–3483
15. Liu Z, Cao J, He Y, Qiao L, Xu C, Lu H, Yang P (2010) Tandem 18O stable isotope labeling for quantification of N-glycoproteome. J Proteome Res 9:227–236
16. Wollscheid B, Bausch-Fluck D, Henderson C, O'Brien R, Bibel M, Schiess R, Aebersold R, Watts JD (2009) Mass-spectrometric identification and relative quantification of N-linked cell surface glycoproteins. Nat Biotechnol 27:378–386
17. Pan S, Tamura Y, Chen R, May D, McIntosh MW, Brentnall TA (2012) Large-scale quantitative glycoproteomics analysis of site-specific glycosylation occupancy. Mol Biosyst 8: 2850–2856
18. Hsu JL, Huang SY, Chow NH, Chen SH (2003) Stable-isotope dimethyl labeling for quantitative proteomics. Anal Chem 75: 6843–6852
19. Sparbier K, Koch S, Kessler I, Wenzel T, Kostrzewa M (2005) Selective isolation of glycoproteins and glycopeptides for MALDI-TOF MS detection supported by magnetic particles. J Biomol Tech 16:407–413
20. Alvarez-Manilla G, Atwood J III, Guo Y, Warren NL, Orlando R, Pierce M (2006) Tools for glycoproteomic analysis: size exclusion chromatography facilitates identification of tryptic glycopeptides with N-linked glycosylation sites. J Proteome Res 5:701–708
21. Hagglund P, Bunkenborg J, Elortza F, Jensen ON, Roepstorff P (2004) A new strategy for identification of N-glycosylated proteins and unambiguous assignment of their glycosylation sites using HILIC enrichment and partial deglycosylation. J Proteome Res 3:556–566
22. Larsen MR, Hojrup P, Roepstorff P (2005) Characterization of gel-separated glycoproteins using two-step proteolytic digestion combined with sequential microcolumns and mass spectrometry. Mol Cell Proteomics 4:107–119
23. Craig R, Beavis RC (2004) TANDEM: matching proteins with tandem mass spectra. Bioinformatics 20:1466–1467
24. Keller A, Eng J, Zhang N, Li XJ, Aebersold R (2005) A uniform proteomics MS/MS analysis platform utilizing open XML file formats. Mol Syst Biol 1:2005
25. Keller A, Nesvizhskii AI, Kolker E, Aebersold R (2002) Empirical statistical model to estimate the accuracy of peptide identifications made by MS/MS and database search. Anal Chem 74:5383–5392
26. May D, Law W, Fitzgibbon M, Fang Q, McIntosh M (2009) Software platform for rapidly creating computational tools for mass spectrometry-based proteomics. J Proteome Res 8:3212–3217
27. Nesvizhskii AI, Vitek O, Aebersold R (2007) Analysis and validation of proteomic data generated by tandem mass spectrometry. Nat Methods 4:787–797

Chapter 26

A Practical Recipe to Survey Phosphoproteomes

William C. Edelman, Kelsey M. Haas, Joanne I. Hsu, Robert T. Lawrence, and Judit Villén

Abstract

The field of cellular signaling is fueled by the discovery of novel protein phosphorylation events. Phosphoproteomics focuses on the large-scale identification and characterization of serine, threonine, and tyrosine phosphorylation of proteins. Phosphopeptide enrichment followed by mass spectrometry has emerged as the most powerful technique for unbiased, discovery-driven analysis by offering high sensitivity, resolution, and speed. Methods for mass spectrometry-based phosphoproteomics analysis have improved substantially over the last decade, making the discipline more approachable to the broader scientific community. Herein we describe the status of the field of phosphoproteomics and provide a robust workflow covering the major aspects of large-scale phosphorylation analysis from phosphopeptide enrichment via IMAC to data analysis.

Key words Phosphorylation, Phosphoproteomics, Cellular signaling, Mass spectrometry, IMAC

1 Introduction

The reversible phosphorylation of serine, threonine, and tyrosine is a chief regulatory mechanism within the cell. Phosphorylation can rapidly alter a protein's structure, enzymatic activity, and interactions with other proteins, and thus plays the lead role in cellular signaling events. Phosphorylation-dependent signaling pathways are ubiquitous, extending from cell surface receptors and cytosolic kinases to thousands of substrates throughout every subcellular domain. Signaling pathways drive a myriad of cellular functions including cell motility, division, growth, metabolism, gene expression, and apoptosis. At present, more than 200,000 distinct phosphorylation sites have been reported, offering a vast regulatory landscape primed for further exploration [1]. In this chapter, we review mass spectrometry-based phosphoproteomics technology and its applications, and provide a protocol to comprehensively survey the phosphorylated species of a biological sample.

Daniel Martins-de-Souza (ed.), *Shotgun Proteomics: Methods and Protocols*, Methods in Molecular Biology, vol. 1156, DOI 10.1007/978-1-4939-0685-7_26, © Springer Science+Business Media New York 2014

Phosphoproteomic applications are well suited to answer many fundamental questions in signal transduction research. At one extreme, phosphoproteomics can be used to power signaling network analyses. For example, Bodenmiller et al. systematically profiled the yeast kinome, determining the relationship between 97 kinases and 27 phosphatases and their substrates [2]. In another study, the phosphorylation patterns amongst a panel of nine mouse tissues were examined, revealing a diverse signaling landscape across nearly 36,000 phosphorylation sites [3]. At the other extreme, phosphoproteomics is used to identify *bona fide* substrates of a particular protein kinase. This can be accomplished by genetically or pharmacologically manipulating kinases prior to analysis, as was the case with work on Cdk1 [4], Polo and Aurora mitotic kinases [5], and mTOR [6, 7]. Additional methods have also been developed to profile kinase activities by in vitro assays performed on complex peptide mixtures [8–11] or to globally ascertain phosphorylation site stoichiometry [12]. Finally, one of the most fascinating applications is the large-scale characterization of dynamic biological processes. Mass spectrometry-based phosphoproteomics was used to quantify the phosphorylation of more than 20,000 sites throughout mitosis [13] and, more recently, to dynamically quantify more than 37,000 distinct sites during the adipocyte insulin response [14]. These studies offer only a glimpse of the potential applications of phosphoproteomics. We anticipate that the diversity of uses will grow as the technique becomes a more routine feature of the biologist's toolbox.

As a discipline, phosphoproteomics began in earnest around a decade ago when phosphorylated peptides were first enriched using immobilized metal affinity chromatography (IMAC) followed by liquid chromatography-tandem mass spectrometry (LC-MS/MS) analysis to identify 383 phosphorylation sites in yeast [15]. Since then, the technology has improved exponentially, combining optimized enrichment and fractionation strategies, superior mass spectrometers with higher sensitivity, speed and resolution, and statistically robust computational algorithms. It is now achievable for a non-expert with modest resources to profile more than 5,000 phosphorylation sites in a single experiment.

To transform phosphoproteomics into a mainstream technology, researchers had to overcome three main challenges. First, phosphorylated species spanned six to seven orders of magnitude in cellular protein abundance and varied in phosphorylation stoichiometry. Second, coverage of protein sequences was incomplete. Enzymatic digestion of proteins generates peptides of diverse lengths. Peptides that are too short cannot be uniquely mapped to a protein, and those that are too long or highly charged pose a challenge for mass spectrometry analysis. Third, MS/MS fragmentation of phosphopeptides was poor because it generates a phosphate neutral loss fragment dominant over sequence-informative b- and y-type ions.

The dynamic range and abundance problem has mostly been addressed by incorporating into the proteomic workflow an enrichment step that selects for phosphorylated species. This enrichment is typically performed at the peptide level and can target the entire phosphoproteome or just a specific subset. Enrichment methods using immunoprecipitation, chromatographic separation, or metal ion affinity reagents prior to mass spectrometry have been successfully used in phosphoproteomics (Table 1). Peptide immunoprecipitation uses antibodies against a phosphorylation motif (e.g., phosphotyrosine) to isolate peptides containing the motif [16]. IMAC enrichment exploits phosphate's coordination to metal cations (e.g., Fe^{3+}, Ga^{3+}) to purify phosphorylated peptides from non-phosphorylated [15, 17, 18]. Similarly, enrichment with metal oxides (e.g., TiO_2, ZrO_2) relies on a Lewis acid–base interaction between the oxide and the phosphate group [19, 20]. All of these methods start with a peptide complex mixture to obtain a phosphopeptide complex mixture. Alternatively, chromatographic techniques can simultaneously fractionate complex peptide mixtures and enrich for phosphopeptides. Strong cation-exchange (SCX) chromatography separates peptides on the basis of solution charge. At acidic pH, a phosphate group contributes to the peptide solution charge with a negative charge, facilitating the separation between phosphopeptides and their non phosphorylated counterparts [21]. SCX chromatography has received considerable attention and is commonly combined with the enrichment methods exposed above for deep phosphoproteome coverage [22–25]. Electrostatic repulsion hydrophilic interaction chromatography (ERLIC, also called eHILIC and ion-pair normal phase) uses high organic concentration on an anion-exchange column to promote the hydrophilic interaction of phosphopeptides with the column while decreasing the usual electrostatic repulsion to the stationary phase [26].

Proteome coverage can be increased by performing separate digestions with different enzymes [27]. This strategy has been mainly applied to single-protein phosphorylation analysis, given the added workload and sample amount requirements.

To alleviate poor MS/MS fragmentation of phosphopeptides by collision-induced dissociation (CID), several mass spectrometry data acquisition methods have been developed (Table 2). Neutral loss-dependent MS^3 [21] and multistage activation (also known as pseudo-MS^3) [28] rely on extended CID fragmentation schemes on the phosphate loss product ion to obtain sequence-informative fragment ions. However, with increased sensitivity of current ion trap mass analyzers, the neutral loss peak dominance is not such a burden and may no longer necessitate MS^3 approaches, which, in fact, are slower and less sensitive than classic CID MS/MS fragmentation [29]. Alternatively, electron transfer dissociation (ETD) [30] induces chemical fragmentation of the peptide sequence, generating c- and z-type ions without affecting amino acid side chains, and has thus been successfully applied to the analysis of peptides

Table 1
Phosphopeptide enrichment methods

Enrichment method	Scale	Enrichment bias	Co-purified peptides	Throughput	Cost	Applications
Metal ion affinity (IMAC)	0.1–0.5 mg	Multiply phosphorylated	Acidic	High	$	Global phosphorylation
Immuno-affinity purification	5–50 mg	Phosphorylation motif	–	Medium	$$$	Phosphorylation motif
Acid–base interaction (TiO_2, ZrO_2)	0.1–0.5 mg	Multiply phosphorylated	Acidic	High	$	Global phosphorylation
SCX chromatography	1–10 mg	Peptides of low solution charge	Acetylated, protein C-terminal	Low	$$	Global phosphorylation
Precipitation with divalent cations (Ca^{2+}, Ba^{2+})	0.2–1 mg	Acidic phosphopeptides	Acidic	High	$	Global phosphorylation, compatible with detergents
SCX + IMAC/TiO_2	1–10 mg	–	–	Low	$$	Global phosphorylation at deep coverage
ERLIC	1–10 mg	Multiply phosphorylated	Sialylated glycopeptides	Low	$$	Global phosphorylation

Commonly used phosphopeptide enrichment methods. We indicate the recommended amount of peptide starting material, their potential bias as to which phosphopeptides are enriched, and other (non-phospho) peptides that might co-purify. To guide the selection of one method over another, we also indicate the throughput level, the relative cost per sample and the recommended applications

Table 2
MS/MS fragmentation methods used in phosphoproteomics

MS fragmentation method	Advantages	Disadvantages	Best used
CID MS/MS	• High sensitivity • Fast scan times	• Intense neutral loss product	With large-scale proteomics experiments
ETD MS/MS	• High peptide sequence coverage • No neutral phosphate loss • Precise site localization	• Limited availability on instrumentation • Poor fragmentation of peptides with $z=2$	With highly charged peptides
HCD MS/MS	• Low noise spectra • High mass precision • Covers low m/z fragment ions • Precise site localization	• Lower sensitivity than CID MS/MS • Slower scan times	With large-scale proteomics experiments
MS^3 and pseudoMS^3	• High sensitivity • Resolves problematic neutral phosphate loss	• Lower sensitivity than CID MS/MS • Slower scan times	With phosphopeptide samples of lower complexity (e.g., single-protein analysis)
Precursor ion scanning	• High sensitivity • Sequencing only when a phosphopeptide is detected	• Requires switching between negative and positive modes • Slow	For determining whether a peptide is phosphorylated

Mass spectrometry acquisition strategies and fragmentation methods used in phosphoproteomics. A list of advantages and disadvantages for each method is provided, as well as their recommended applications. Some of these methods may be combined on some instrument configurations

with labile modifications, including phosphorylation [31]. Higher energy collisional dissociation (HCD) is yet another fragmentation method that has been applied to phosphopeptide analysis [32, 33]. HCD produces b- and y-type ions, and fragment analysis is performed at high mass precision on a mass analyzer that permits the detection of low m/z fragments. Consequently, HCD is one of the best fragmentation methods for unambiguously localizing phosphorylation sites, in particular near peptide termini. Overall, any of these fragmentation methods performed on today's mass spectrometers is suitable for phosphorylation analysis. However, given its high speed and sensitivity, CID seems the best choice for analyzing highly complex phosphopeptide mixtures, despite the neutral loss problem.

Apart from the automation of sample preparation and LC-MS/MS analysis, phosphoproteomics has achieved a high level of routine automation in the realm of spectral data interpretation. Adequate interpretation of phosphopeptide spectra requires the correct localization of a peptide phosphorylation site within the peptide sequence. To localize a phosphorylation site, the presence of site-determining ions in the fragmentation spectrum is critical. Several algorithms (e.g., Ascore [34], PTM score [23], phosphoRS [35], and LuciPHOr [36]) have been developed to provide a confidence metric of correct site localization, all revolving around the idea of scoring matches to site-determining ions.

Paramount to understanding cellular signaling is the ability to compare the phosphoproteomes between two or more biological conditions. Phosphoproteomics via LC-MS/MS enables quantitative comparisons between controls and genetic mutants, drug treatments, and disease states. Quantitative proteomic methods have been reviewed elsewhere within this book.

In the last 10 years, proteomics researchers devoted much effort to developing technology to analyze phosphoproteomes, from methods for phosphopeptide enrichment to LC-MS/MS data acquisition, to automated data interpretation and statistics. Currently, robust and automated phosphoproteomics workflows are available, setting a prime time for global cellular signaling research. It is certain that phosphoproteomics research within the next years will make important contributions in deciphering kinase–substrate relationships, assembling signaling networks, and overall expanding the protein phosphorylation knowledge base.

The protocol presented is a starting point for phosphoproteomics experiments using enrichment of phosphopeptides via IMAC and analysis by LC-MS/MS. On a practical note, this protocol is robust and portable and can be combined with a wide variety of experimental pipelines and sample types.

2 Materials

2.1 Solid-Phase Extraction for Peptide Desalting on a Cartridge

1. C18 solid-phase extraction cartridges.
2. Vacuum manifold.
3. pH paper strips.
4. Sample acidification solution: 10 % trifluoroacetic acid (TFA).
5. Column wash solution: 100 % acetonitrile (ACN).
6. Column equilibration and sample wash solution: 0.1 % TFA.
7. Sample wash solution: 0.5 % acetic acid (AA).
8. Column wash and sample elution solution: 50 % ACN and 0.5 % AA.
9. Elution tubes: 2 mL microcentrifuge tubes or 15 mL conical tubes, depending on elution volume.
10. Vacuum concentrator equipped with cold trap and vacuum pump.

2.2 Solid-Phase Extraction for Peptide Microscale Desalting on a Stage Tip

1. Stage tip assembly components: Empore C18 polymer disks (3M), plastic pipette tips (200 μL size), blunt-end cannula with any rod that snugly fits into the cannula shaft.
2. Liquid passing mechanism: Syringe or centrifuge.
3. Stage tip wash solution: 100 % methanol (MeOH).
4. Stage tip wash solution: 100 % ACN.
5. Stage tip wash and sample elution solution: 70 % ACN and 1 % formic acid (FA).
6. Stage tip equilibration and sample wash solution: 1 % FA.
7. Sample resuspension solution: 10 % FA, or 1 % FA (*see* Subheading 3.2, **step 4**).
8. Vacuum concentrator equipped with cold trap and vacuum pump.

2.3 IMAC

1. Ni-NTA magnetic agarose beads (5 % suspension, Qiagen).
2. Metal stripping solution: 40 mM EDTA pH 8.
3. Metal loading solution: 100 mM $FeCl_3$.
4. Wash solution: Deionized water.
5. Wash and sample resuspension solution: 80 % ACN, 0.1 % TFA.
6. Phosphopeptide elution solution: 70 % ACN, 1 % NH_4OH, 29 % H_2O.
7. Acidifying solution: 10 % FA.
8. Shaker.
9. Magnet bar.
10. 0.2 mL microcentrifuge tubes.
11. Stage tip assembly components: Empore C18 polymer disks, plastic pipette tips (200 μL size), blunt-end cannula with any rod that snugly fits into the cannula shaft.
12. Liquid passing mechanism: Syringe or centrifuge.
13. Stage tip wash solution: 100 % MeOH.
14. Stage tip wash solution: 100 % ACN.
15. Vacuum concentrator equipped with cold trap and vacuum pump.

2.4 LC-MS/MS

1. Nano-liquid chromatography system consisting of autosampler and pumps.
2. Analytical column: Capillary column fabricated from fused-silica capillary tubing (50–100 μm inner diameter) 15–40 cm in length and packed with 3 μm C18 reversed-phase material (*see* **Note 1**).
3. Pre-column: Capillary pre-column of the same inner diameter and packing material as the analytical column, 1–2 cm in length.

4. Mass spectrometer: LTQ Orbitrap Velos (Thermo Fisher Scientific).
5. LC solvents: 0.1 % FA in water and 0.1 % FA in ACN.
6. LC-MS/MS sample resuspension solution: 3 % ACN, 4 % FA.

3 Methods

The IMAC protocol has been adapted from Ficarro et al. [37], and it is designed for the enrichment of phosphopeptides from digested whole-cell protein extract. In our hands, IMAC enrichment alone can yield up to 8,000 unique phosphopeptide identifications. Optionally, to obtain deeper phosphoproteome coverage, we suggest including SCX fractionation in the experimental workflow.

To prepare the peptide sample prior to IMAC, we recommend supplementing the lysis buffer with protease inhibitors (e.g., leupeptin and aprotinin, or a protease inhibitor cocktail) to prevent protein degradation, and phosphatase inhibitors (e.g., sodium β-glycerophosphate, sodium pyrophosphate, sodium orthovanadate, calyculin A, and/or okadaic acid) to prevent protein dephosphorylation. We also recommend avoiding the use of detergents or eliminating them prior to digestion. Enzymatic digestion is typically conducted with trypsin or Lys-C. While sufficient peptide data can be collected from samples digested with one enzyme, we have found that dividing the protein sample for use with two different enzymes increases the number of identified phosphorylation sites. We are providing two equivalent methods for sample cleanup via solid-phase extraction, depending on the scale of the experiment. We recommend using solid-phase extraction cartridges (Subheading 3.1) to desalt peptide sample amounts between 40 μg and 1 g and stage tip desalting (Subheading 3.2) for sample amounts less than 40 μg. An overview of the method described in this chapter is provided in Fig. 1.

3.1 Solid-Phase Extraction for Peptide Desalting on a Cartridge

1. Acidify the peptide sample to pH 2 by adding 10 % TFA to achieve a final concentration of 0.1–0.5 %. Verify pH using a pH strip (*see* **Note 2**).
2. To desalt the peptide sample via C18 solid-phase extraction cartridge, choose a cartridge size with packing material about 20 times the total mass of the peptides (*see* **Note 3**). Each peptide sample should receive its own cartridge, unless combining chemically labeled samples. Insert the cartridges into the knob slots of the vacuum manifold for use.
3. Wash the cartridge C18 polymer: Fill the cartridges to the top (approximately five column volumes) with the designated solution, open the knob valves, and run the liquid through the cartridge, repeating as indicated (*see* **Note 4**). Use this technique

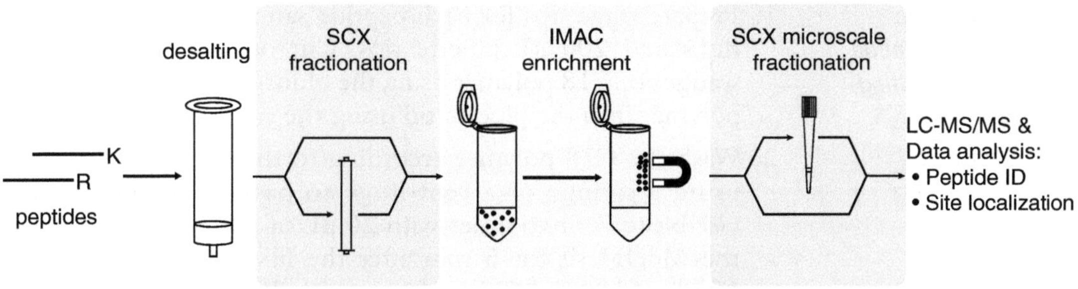

Fig. 1 Experimental procedure for phosphopeptide enrichment and analysis by mass spectrometry. Prior to enrichment, peptides derived from enzymatic digestion must be desalted on a C18 solid-phase extraction cartridge or a C18 stage tip (*see* Subheadings 3.1 and 3.2). Phosphopeptide enrichment is performed by immobilized metal affinity chromatography (IMAC), using a magnetic Fe^{3+}-coated solid support (shown on the *green rectangle*) (Subheadings 3.3 and 3.4). Phosphopeptides are analyzed by LC-MS/MS to identify their sequence and their site/s of phosphorylation (Subheadings 3.5 and 3.6). IMAC enrichment alone can yield up to 8,000 unique phosphopeptide identifications. To achieve deeper phosphoproteome coverage, it is advisable to incorporate SCX fractionation into the workflow. This optional SCX chromatography step (shown on *grey rectangles*) provides peptide separation and partial phosphopeptide enrichment. SCX fractionation can be performed before IMAC on a column format or after IMAC on a stage tip format, depending on starting sample amounts and proteome complexity (*see* **Note 13**) (Color figure online)

for the following sequence of washes: three washes with 100 % ACN and one wash with 50 % ACN and 0.5 % AA. Washing can be done at a fast flow-through speed.

4. Equilibrate the C18 cartridge to peptide sample conditions by passing through three cartridge volumes of 0.1 % TFA. Equilibration can also be done at fast flow-through speed.
5. Load samples into the designated cartridges using the above technique. Adjust the vacuum manifold pressure to a slow flow-through speed to ensure efficient peptide binding to the C18 polymer (*see* **Note 5**).
6. Wash the peptides bound to the polymer three times with 0.1 % TFA and once with 0.5 % AA. Washing can be done at a fast flow-through speed.
7. Use the above technique to elute the peptides from the polymer at slow flow-through speed with five column volumes of 50 % ACN and 0.5 % AA, placing a fresh microcentrifuge tube or conical tube to collect the peptide eluate from this step.
8. Reserve a 20 μg aliquot of each desalted peptide sample in a vial insert for analysis by mass spectrometry to verify sample integrity. Divide the remaining peptide samples into 1 mg aliquots in fresh microcentrifuge tubes for later use in IMAC.
9. Dry all aliquots to completion in the vacuum concentrator. Resuspend the 20 μg aliquot in 20 μL of 3 % ACN and 4 % FA and analyze by LC-MS/MS (*see* Subheading 3.5). Store dried 1 mg aliquots at −20 °C.

3.2 Solid-Phase Extraction for Peptide Microscale Desalting on a Stage Tip

1. Prepare stage tips for each peptide sample with Empore C18 disks and 200 μL pipette tips. Cut out four small circular wedges of C18 polymer using the blunt-end cannula. Push the polymer into the pipette tip using the rod (*see* **Note 6**).
2. Wash the C18 polymer according to the following sequence, using a syringe or a centrifuge to pass the liquid through (*see* **Note 7**): two times with 20 μL of 100 % MeOH, letting the MeOH sit for 5 min after the first addition; once with 20 μL of 100 % ACN; and once with 20 μL of 70 % ACN and 1 % FA.
3. Equilibrate the polymer by passing 20 μL of 1 % FA two times through the column.
4. Prepare the sample for stage tip desalting. If the sample is in solution containing ACN, add 10 μL of 10 % FA and dry the sample with the vacuum concentrator for about 10 min to evaporate the ACN. When the sample is dried down, resuspend in 20 μL of 1 % FA.
5. Load the peptide sample on the stage tip micro-column by passing the peptide solution through the polymer (*see* **Note 8**).
6. Wash the sample by passing 20 μL of 1 % FA twice through the column.
7. Elute the sample into an MS insert by passing 20 μL of 70 % ACN and 1 % FA.
8. Dry the sample to completion in the vacuum concentrator. Store at −20 °C.

3.3 Preparation of Fe^{3+} Magnetic Beads for IMAC Phosphopeptide Enrichment

1. Use 1 mL of Qiagen Ni-NTA magnetic agarose beads (5 % suspension).
2. Wash beads three times with 1 mL of deionized water turning the test tube several times over between washes. To remove supernatant pull beads to the side of the tube with the help of a magnet bar and aspirate liquid with a pipette.
3. Incubate the beads for 30 min in 1 mL 40 mM EDTA pH 8.
4. Repeat the three washes with 1 mL of deionized water.
5. Load the beads with Fe^{3+} by incubating them with 1 mL of 100 mM $FeCl_3$ for 30 min.
6. Wash the beads twice with 1 mL of deionized water.
7. Wash the beads three times with 1 mL 80 % ACN and 0.1 % TFA (*see* **Note 9**).
8. Resuspend the beads in 1 mL 80 % ACN and 0.1 % TFA to make a 5 % suspension.

3.4 IMAC Phosphopeptide Enrichment with Fe^{3+} Magnetic Beads

1. Transfer the Fe^{3+}-loaded bead slurry to microcentrifuge tubes. Use 1 μL of the 5 % bead suspension per 10 μg of peptides (*see* **Note 10**). Pipette out the liquid from the slurry to isolate the beads.
2. Dissolve the lyophilized peptide sample in 80 % ACN and 0.1 % TFA at 1 μg/μL, and transfer the solution to the microcentrifuge tubes containing the beads. Incubate the peptides with the beads on a shaker for 30 min.
3. During the incubation time, construct C18 stage tips, which will later be used to filter out the magnetic beads during phosphopeptide elution (*see* **Note 11**). To construct a stage tip, press the blunt-end cannula into a two-layer sheet of Empore C18 and push the excised disk into the end of a 200 μL pipette tip until a snug fit is achieved. Wash the C18 disk by passing through 20 μL of MeOH, followed by 20 μL of 100 % ACN. Add 60 μL of 10 % FA to a 2 mL microcentrifuge tube, and suspend the stage tip in the tube using a centrifuge adapter.
4. After the 30-min bead incubation, wash the phosphopeptide-bound beads three times with three volumes of 80 % ACN and 0.1 % TFA to remove any non-phosphorylated peptides that are not bound to the beads.
5. Elute the phosphopeptides from the beads by adding three volumes of 70 % ACN, 1 % NH_4OH, and 29 % water. Incubate for 2 min, and transfer the eluent containing phosphopeptides to stage tips for filtering (*see* **Note 12**). To filter off any remaining beads, push the liquid through the stage tip into the microcentrifuge tube containing 10 % FA using a syringe or a centrifuge.
6. Lyophilize the phosphopeptide elution with the vacuum concentrator.
7. To desalt by stage tip, *see* Subheading 3.2.

As an optional procedure, SCX chromatography can be conducted before or after IMAC to obtain deeper coverage of the phosphoproteome. Fractionation is recommended for highly complex peptide samples, such as those derived from mammalian cell proteomes. SCX chromatography separates peptides based on their charge state [21], and, because of phosphate's negative charge, it provides partial phosphopeptide enrichment (*see* **Note 13**).

3.5 LC-MS/MS

The LC-MS/MS method will depend on the instrument model and configuration. Here, we provide an example of a 120-min data-dependent acquisition method using the Easy nanoLC (Proxeon) nano-liquid chromatography system coupled to an LTQ-Orbitrap Velos mass spectrometer (Thermo).

1. *High-performance liquid chromatography (HPLC) method*: LC gradient should be set to yield maximum phosphopeptide separation. For tryptic digest phosphopeptide samples, use a gradient of 10–30 % ACN in 0.1 % FA delivered at ~250 nl/min over 100 min. End the gradient with a high percentage of ACN (≥90 % for 10 min) to wash the column of remaining peptides and then a low percentage of ACN (5 % for 10 min) to re-equilibrate the column for subsequent runs (*see* **Note 14**).
2. *Mass spectrometry acquisition method*: For each cycle, schedule one full MS scan in the Orbitrap followed by 20 MS/MS scans after CID fragmentation (*see* **Note 15**). These scans target the 20 most intense ions present in the MS scan. Use dynamic exclusion to prevent the selection of precursors that have been fragmented on the preceding 30 s (*see* **Note 16**). Exclude precursors of charge 1+ or unknown charge.
3. *Load sample onto column and run LC-MS/MS method*: Resuspend the dried phosphopeptide sample in 3 % ACN and 4 % FA at 1 μg/μL, and load 1 μg of sample onto the analytical column for LC-MS/MS analysis following the acquisition method described above.

3.6 Data Analysis: Peptide Identification and Site Localization

Peptide spectra can be analyzed using a variety of database search algorithms to identify matching peptide sequences (e.g., Sequest [38], Mascot [39], X!Tandem [40], OMSSA [41]).

1. Regardless of the search algorithms used, there are key parameter specifications to include as follows: the digestive enzyme used, constant modifications (cysteine carboxymethylation), variable peptide modifications (methionine oxidation and serine/threonine/tyrosine phosphorylation), and, in quantitative experiments, isotope label modifications. The complete, sequenced protein database for the organism of study should also be supplied when performing the search. These can be accessed on a variety of servers; one commonly used is the UniProt protein knowledge base.
2. Searching against concatenated target/decoy databases (for example, using as a decoy all peptide sequences written in reverse) allows estimation of the false discovery rate (FDR) for the peptide data set [42]. The final list of peptide identifications is filtered to eliminate false identifications based on an FDR cutoff, typically set at 1 %.
3. Most search algorithms do a poor job in confidently identifying the correct phosphorylation positional isomer for a phosphopeptide. Thus, it is important to use a phosphorylation site localization algorithm to address this limitation. There are a number of phosphorylation site localization algorithms, such as Ascore [34], SLoMo [43], phosphoRS [35], and LuciPHOr [36].

Most of these algorithms focus on site-determining ions (i.e., those that are unique to a particular localization) and provide a site localization confidence score. In general, they are more accurate in determining the precise position of phosphorylation on a given peptide than database search algorithms.

4. Finally, when the phosphoproteomic experiment involves different experimental conditions, quantification algorithms provide a way to extract peak intensities and calculate relative peptide abundances.
5. Proteomic software packages such as MaxQuant [44], Scaffold [45], and Proteome Discoverer (Thermo Scientific) integrate tools for peptide identification, site localization, and peptide relative quantification.

4 Notes

1. Analytical column length and bead particle size will depend upon the maximum pressure allowed by the HPLC. Longer columns and smaller beads enhance peptide separation but result in higher HPLC pressures.
2. The purposes of acidifying the peptide samples are to inactivate the enzyme used in digestion and to increase binding to the C18 solid-phase extraction material.
3. Total mass of the peptide sample is estimated after cell lysis/protein extraction and before protein reduction, alkylation, and digestion, commonly by bicinchoninic acid method.
4. For all steps using the solid-phase extraction cartridges, be sure to close the vacuum manifold knob valves before the liquid in the cartridge runs past the polymer. Do not let the polymer dry out during any step.
5. An additional, optional step to maximize the quantity of column-bound peptides is to reload the eluent through the column. This requires separately collecting the eluent during the first sample loading step.
6. Do not force the polymer wedge into the pipette tip or it will be difficult to pass liquid through the polymer. Insert the polymer so that it is resting in place but not tightly compacted in the pipette tip. Incubation with methanol in the next step will cause the polymer to expand; thus, the wedges will become more snugly fit.
7. If using a syringe to push the liquid through, pass the liquid relatively quickly through the C18 polymer and take care to avoid dislodging the polymer from the pipette tip. If using a centrifuge, spin the pipette tip in a microcentrifuge tube at 700–1,000 rpm for 30–90 s, depending on how quickly the

liquid passes through the polymer. In either case, do not let the polymer dry out; retain a small amount of liquid above the polymer. These stipulations apply to all wash, equilibration, sample load, and elution steps, unless further specified.

8. For sample load and elution steps, pass the liquid slowly by syringe or centrifuge about 500 rpm for approximately 2 min.
9. If the beads are to be stored for future use, wash and resuspend them in 30 % ethanol. At the time of the experiment, wash the beads three times with an 80 % ACN and 0.1 % TFA solution.
10. Optimal ratio of peptide to beads varies across biological samples. The values provided are a good starting point, but we recommend optimizing this for each sample.
11. It is important to filter out the magnetic beads that accidentally ended up in the phosphopeptide elution, since they could cause damage to the HPLC during sample analysis.
12. The basic environment assists the detachment of the phosphopeptides from the beads. However, be sure to limit the time of phosphopeptide exposure in the basic environment to 3 min to prevent side reactions involving the phosphate group. The purpose of the formic acid is to neutralize the pH of the elution.
13. Peptide fractionation by SCX chromatography: SCX chromatography can be performed prior to IMAC enrichment using an SCX column and an HPLC elution gradient to obtain high-resolution peptide separation [21]. SCX fractionation can also be performed in a solid-phase extraction cartridge format for a rapid but lower resolution fractionation [46]. Alternatively, SCX can be conducted after IMAC in microscale solid-phase extraction format by utilizing a stage tip packed with an SCX disk [47]. This method is recommended for less complex samples, such as yeast proteomes, or for smaller sample amounts. Stage tip fractionation can be done alone or in conjunction with peptide desalting, in which case the stage tip is constructed with a C18 polymer disk stacked on top of an SCX disk. The advantages of either solid-phase extraction SCX format are that they are simpler, less expensive, and less technical to operate than the SCX chromatography HPLC system. However, SCX chromatography produces higher resolution fractions and allows a greater number of obtainable peptide fractions and total phosphopeptide identifications compared to the SCX disk stage tips. For inclusion of these optional procedures in the phosphopeptide enrichment protocol refer to previous literature [29, 47].
14. For mammalian samples or other highly complex samples, a longer acquisition time will improve peptide separation and increase the depth of phosphoproteome coverage. These parameters are flexible and may depend upon sample complexity, column material, and mass spectrometer scanning speed.

15. Peptides scheduled for isolation and sequencing can be fragmented by several methods, depending on the mass spectrometer used. These include CID, ETD, and HCD. CID frequently suffers from the neutral loss of phosphate as dominant fragment in the spectrum [21, 29]. ETD reduces the risk of phosphate loss during fragmentation and is optimally suited for high charge states. HCD enables higher mass accuracy and the observation of lower *m/z* ions, yet is a slower fragmentation method than CID.
16. Dynamic exclusion time should be adjusted to match chromatographic peak widths.

Acknowledgements

The authors would like to thank Danielle Swaney, Ricard Rodriguez-Mias, and Daniel Hernandez for their helpful advice. This work was supported in part by NIH/NCI grant R00CA140789 and an Ellison Medical Foundation New Scholar Award (to J.V.). W.C.E. is a recipient of an NSF Graduate Research Fellowship.

References

1. Hornbeck PV, Kornhauser JM, Tkachev S, Zhang B, Skrzypek E, Murray B, Latham V, Sullivan M (2012) PhosphoSitePlus: a comprehensive resource for investigating the structure and function of experimentally determined post-translational modifications in man and mouse. Nucleic Acids Res 40:D261–D270
2. Bodenmiller B, Wanka S, Kraft C et al (2010) Phosphoproteomic analysis reveals interconnected system-wide responses to perturbations of kinases and phosphatases in yeast. Sci Signal 3:rs4
3. Huttlin EL, Jedrychowski MP, Elias JE, Goswami T, Rad R, Beausoleil SA, Villen J, Haas W, Sowa ME, Gygi SP (2010) A tissue-specific atlas of mouse protein phosphorylation and expression. Cell 143:1174–1189
4. Holt LJ, Tuch BB, Villen J, Johnson AD, Gygi SP, Morgan DO (2009) Global analysis of Cdk1 substrate phosphorylation sites provides insights into evolution. Science 325:1682–1686
5. Kettenbach AN, Schweppe DK, Faherty BK, Pechenick D, Pletnev AA, Gerber SA (2011) Quantitative phosphoproteomics identifies substrates and functional modules of Aurora and Polo-like kinase activities in mitotic cells. Sci Signal 4:rs5
6. Yu Y, Yoon S-O, Poulogiannis G et al (2011) Phosphoproteomic analysis identifies Grb10 as an mTORC1 substrate that negatively regulates insulin signaling. Science 332:1322–1326
7. Hsu PP, Kang SA, Rameseder J et al (2011) The mTOR-regulated phosphoproteome reveals a mechanism of mTORC1-mediated inhibition of growth factor signaling. Science 332:1317–1322
8. Cutillas PR, Khwaja A, Graupera M, Pearce W, Gharbi S, Waterfield M, Vanhaesebroeck B (2006) Ultrasensitive and absolute quantification of the phosphoinositide 3-kinase/Akt signal transduction pathway by mass spectrometry. Proc Natl Acad Sci U S A 103:8959–8964
9. Yu Y, Anjum R, Kubota K, Rush J, Villen J, Gygi SP (2009) A site-specific, multiplexed kinase activity assay using stable-isotope dilution and high-resolution mass spectrometry. Proc Natl Acad Sci U S A 106:11606–11611
10. Kubota K, Anjum R, Yu Y et al (2009) Sensitive multiplexed analysis of kinase activities and activity-based kinase identification. Nat Biotechnol 27:933–940
11. Xue L, Wang W-H, Iliuk A, Hu L, Galan JA, Yu S, Hans M, Geahlen RL, Tao WA (2012) Sensitive kinase assay linked with phosphoproteomics for identifying direct kinase substrates. Proc Natl Acad Sci U S A 109:5615–5620
12. Wu R, Haas W, Dephoure N, Huttlin EL, Zhai B, Sowa ME, Gygi SP (2011) A large-scale method to measure absolute protein phosphorylation stoichiometries. Nat Methods 8:677–683

13. Olsen JV, Vermeulen M, Santamaria A et al (2010) Quantitative phosphoproteomics reveals widespread full phosphorylation site occupancy during mitosis. Sci Signal 3:ra3
14. Humphrey SJ, Yang G, Yang P, Fazakerley DJ (2013) Dynamic adipocyte phosphoproteome reveals that Akt directly regulates mTORC2. Cell Metab 17:1009–1020
15. Ficarro SB, McCleland ML, Stukenberg PT, Burke DJ, Ross MM, Shabanowitz J, Hunt DF, White FM (2002) Phosphoproteome analysis by mass spectrometry and its application to Saccharomyces cerevisiae. Nat Biotechnol 20:301–305
16. Rush J, Moritz A, Lee KA, Guo A, Goss VL, Spek EJ, Zhang H, Zha X-M, Polakiewicz RD, Comb MJ (2005) Immunoaffinity profiling of tyrosine phosphorylation in cancer cells. Nat Biotechnol 23:94–101
17. Andersson L, Porath J (1986) Isolation of phosphoproteins by immobilized metal (Fe3+) affinity chromatography. Anal Biochem 154:250–254
18. Posewitz MC, Tempst P (1999) Immobilized gallium(III) affinity chromatography of phosphopeptides. Anal Chem 71:2883–2892
19. Sano A, Nakamura H (2004) Titania as a chemo-affinity support for the column-switching HPLC analysis of phosphopeptides: application to the characterization of phosphorylation sites in proteins by combination with protease digestion and electrospray ionization mass spectrometry. Anal Sci 20: 861–864
20. Pinkse MWH, Uitto PM, Hilhorst MJ, Ooms B, Heck AJR (2004) Selective isolation at the femtomole level of phosphopeptides from proteolytic digests using 2D-NanoLC-ESI-MS/MS and titanium oxide precolumns. Anal Chem 76:3935–3943
21. Beausoleil SA, Jedrychowski M, Schwartz D, Elias JE, Villen J, Li J, Cohn MA, Cantley LC, Gygi SP (2004) Large-scale characterization of HeLa cell nuclear phosphoproteins. Proc Natl Acad Sci U S A 101:12130–12135
22. Gruhler A, Olsen JV, Mohammed S, Mortensen P, Faergeman NJ, Mann M, Jensen ON (2005) Quantitative phosphoproteomics applied to the yeast pheromone signaling pathway. Mol Cell Proteomics 4:310–327
23. Olsen JV, Blagoev B, Gnad F, Macek B, Kumar C, Mortensen P, Mann M (2006) Global, in vivo, and site-specific phosphorylation dynamics in signaling networks. Cell 127:635–648
24. Villen J, Beausoleil SA, Gerber SA, Gygi SP (2007) Large-scale phosphorylation analysis of mouse liver. Proc Natl Acad Sci U S A 104: 1488–1493
25. Villen J, Gygi SP (2008) The SCX/IMAC enrichment approach for global phosphorylation analysis by mass spectrometry. Nat Protoc 3:1630–1638
26. Alpert AJ (2008) Electrostatic repulsion hydrophilic interaction chromatography for isocratic separation of charged solutes and selective isolation of phosphopeptides. Anal Chem 80:62–76
27. Swaney DL, Wenger CD, Coon JJ (2010) Value of using multiple proteases for large-scale mass spectrometry-based proteomics. J Proteome Res 9:1323–1329
28. Schroeder MJ, Shabanowitz J, Schwartz JC, Hunt DF, Coon JJ (2004) A neutral loss activation method for improved phosphopeptide sequence analysis by quadrupole ion trap mass spectrometry. Anal Chem 76:3590–3598
29. Villen J, Beausoleil SA, Gygi SP (2008) Evaluation of the utility of neutral-loss-dependent MS3 strategies in large-scale phosphorylation analysis. Proteomics 8:4444–4452
30. Syka JEP, Coon JJ, Schroeder MJ, Shabanowitz J, Hunt DF (2004) Peptide and protein sequence analysis by electron transfer dissociation mass spectrometry. Proc Natl Acad Sci U S A 101:9528–9533
31. Chi A, Huttenhower C, Geer LY, Coon JJ, Syka JEP, Bai DL, Shabanowitz J, Burke DJ, Troyanskaya OG, Hunt DF (2007) Analysis of phosphorylation sites on proteins from Saccharomyces cerevisiae by electron transfer dissociation (ETD) mass spectrometry. Proc Natl Acad Sci U S A 104:2193–2198
32. Nagaraj N, D'Souza RCJ, Cox J, Olsen JV, Mann M (2010) Feasibility of large-scale phosphoproteomics with higher energy collisional dissociation fragmentation. J Proteome Res 9:6786–6794
33. Jedrychowski MP, Huttlin EL, Haas W, Sowa ME, Rad R, Gygi SP (2011) Evaluation of HCD- and CID-type fragmentation within their respective detection platforms for murine phosphoproteomics. Mol Cell Proteomics 10:M111.009910
34. Beausoleil SA, Villen J, Gerber SA, Rush J, Gygi SP (2006) A probability-based approach for high-throughput protein phosphorylation analysis and site localization. Nat Biotechnol 24:1285–1292
35. Taus T, Köcher T, Pichler P, Paschke C, Schmidt A, Henrich C, Mechtler K (2011) Universal and confident phosphorylation site localization using phosphoRS. J Proteome Res 10:5354–5362
36. Fermin D, Walmsley SJ, Gingras A-C, Choi H, Nesvizhskii AI (2013) LuciPHOr: algorithm for phosphorylation site localization with false

localization rate estimation using target-decoy approach. Mol Cell Proteomics. doi:10.1074/mcp.M113.028928

37. Ficarro SB, Adelmant G, Tomar MN, Zhang Y, Cheng VJ, Marto JA (2009) Magnetic bead processor for rapid evaluation and optimization of parameters for phosphopeptide enrichment. Anal Chem 81:4566–4575
38. Eng JK, McCormack AL, Yates JR (1994) An approach to correlate tandem mass spectral data of peptides with amino acid sequences in a protein database. J Am Soc Mass Spectrom 5:976–989
39. Perkins DN, Pappin DJ, Creasy DM, Cottrell JS (1999) Probability-based protein identification by searching sequence databases using mass spectrometry data. Electrophoresis 20:3551–3567
40. Craig R, Cortens JP, Beavis RC (2004) Open source system for analyzing, validating, and storing protein identification data. J Proteome Res 3:1234–1242
41. Geer LY, Markey SP, Kowalak JA, Wagner L, Xu M, Maynard DM, Yang X, Shi W, Bryant SH (2004) Open mass spectrometry search algorithm. J Proteome Res 3:958–964
42. Elias JE, Gygi SP (2007) Target-decoy search strategy for increased confidence in large-scale protein identifications by mass spectrometry. Nat Methods 4:207–214
43. Bailey CM, Sweet SMM, Cunningham DL, Zeller M, Heath JK, Cooper HJ (2009) SLoMo: automated site localization of modifications from ETD/ECD mass spectra. J Proteome Res 8:1965–1971
44. Cox J, Mann M (2008) MaxQuant enables high peptide identification rates, individualized p.p.b.-range mass accuracies and proteome-wide protein quantification. Nat Biotechnol 26:1367–1372
45. Searle BC (2010) Scaffold: a bioinformatic tool for validating MS/MS-based proteomic studies. Proteomics 10:1265–1269
46. Dephoure N, Gygi SP (2011) A solid phase extraction-based platform for rapid phosphoproteomic analysis. Methods 54:379–386
47. Rappsilber J, Mann M, Ishihama Y (2007) Protocol for micro-purification, enrichment, pre-fractionation and storage of peptides for proteomics using StageTips. Nat Protoc 2:1896–1906

Chapter 27

Quantitation of the Phosphoproteome Using the Library-Assisted eXtracted Ion Chromatogram (LAXIC) Strategy

Justine V. Arrington, Liang Xue, and W. Andy Tao

Abstract

Phosphorylation is a key posttranslational modification that regulates many signaling pathways, but quantifying changes in phosphorylation between samples can be challenging due to its low stoichiometry within cells. We have introduced a mass spectrometry-based label-free quantitation strategy termed LAXIC for the analysis of the phosphoproteome. This method uses a spiked-in synthetic peptide library designed to elute across the entire chromatogram for local normalization of phosphopeptides within complex samples. Normalization of phosphopeptides by library peptides that co-elute within a small time frame accounts for fluctuating ion suppression effects, allowing more accurate quantitation even when LC-MS performance varies. Here we explain the premise of LAXIC, the design of a suitable peptide library, and how the LAXIC algorithm can be implemented with software developed in-house.

Key words Phosphoproteomics, Label-free quantitation, LAXIC, Quantification software, Peptide library

1 Introduction

The goal of shotgun proteomics is the large-scale identification and quantitation of proteins using a bottom-up approach that employs liquid chromatography coupled with mass spectrometry (LC-MS). The identity and quantity of various proteins can help distinguish diseased from physiologically healthy tissues or provide insights into cellular processes. However, many cellular functions are in fact regulated by signal transduction pathways [1]. In particular, protein phosphorylation controls a large number of cellular processes and its dysregulation is implicated in a variety of illnesses including cancers [2, 3], neurological diseases [4], and immune system disorders [5]. Thus, "shotgun phosphoproteomics" seeks to map and quantify protein phosphorylation in large scale to elucidate system-wide signaling networks.

Daniel Martins-de-Souza (ed.), *Shotgun Proteomics: Methods and Protocols*, Methods in Molecular Biology, vol. 1156, DOI 10.1007/978-1-4939-0685-7_27, © Springer Science+Business Media New York 2014

Improvements in LC-MS technology, phosphopeptide enrichment strategies, and analysis software have advanced the field of phosphoproteomics, so thousands of phosphorylation events can be analyzed in a single experiment [6–8]. However, quantitation of these phosphorylation events can still be challenging. Quantitative phosphoproteomics typically follows one of the two strategies: stable isotope labeling or label-free quantitation. Stable isotope labeling includes metabolic labeling through stable isotope labeling of amino acids in cell culture (SILAC) and chemical labeling with multiplexed isobaric tags for relative and absolute quantification (iTRAQ) or isotope-coded affinity tags (ICAT) [9–11]. Although these labeling techniques have been used extensively, they require additional sample preparation or chemical reactions and may not be compatible with all sample types. Because of its greater compatibility with different sample and instrument types, label-free quantitation provides an attractive alternative to isotope labeling.

Label-free quantitation can be accomplished through comparison of ion intensities [12], but a major pitfall of this approach is variation in LC-MS performance from run to run [13]. Ideally, error due to LC-MS variations can be corrected by the use of spiked-in standards. For instance, absolute quantification peptide technology (AQUA) employs isotope-labeled synthetic peptides as internal standards [14]. This technique may be difficult to apply in global-level studies because a vast number of reference peptides are needed. To forgo isotope labeling, some have proposed using peptides from an endogenous housekeeping protein as a standard [15]. This approach eliminates the need for a spiked-in standard, but finding a protein that is consistently expressed across all samples can be problematic. Therefore, many label-free methods spike in a standard to improve normalization. In particular, Waters produces MassPREP™ internal standard peptides and phosphopeptides for spiking in after enzymatic digestion of the sample. It is also possible to spike in a known amount of a standard protein before digestion [16], but phosphoproteomic analyses often utilize phosphopeptide enrichments that may introduce error with this method. After LC-MS analysis, sample peptides are either normalized to the total peak intensity of the spiked-in standards [17] or compared before the application of a normalization factor that is determined as the median [18] or average ratio [19]. Spectral counting is another popular label-free quantitation method because it is simple and easy to apply to large data sets, but spectral counts can be low for phosphopeptides, leading to inaccurate quantitation of phosphoproteins [20–22].

To address the issue of LC variability across runs, we present a novel label-free quantitation strategy named Library-Assisted eXtracted Ion Chromatogram (LAXIC) [23]. As opposed to using a single peptide as an internal standard, LAXIC uses a synthetic peptide library that is designed to elute consecutively across the entire LC gradient with m/z values that fall within the normal scanned mass range of typical experiments. In addition to this optimized library, LAXIC employs a local normalization method. Raw intensity ratios for library peptides (*see* Fig. 1) are first calculated by dividing the peak intensity of a certain peptide in one LC-MS run by its intensity in a reference LC-MS run:

$$\mathrm{NF}(t) = \frac{\mathrm{Intensity\,I1}(t)}{\mathrm{Intensity\,I2}(t)} \quad (1)$$

This provides a normalization factor (NF) for a particular time t. Next, LAXIC generates a weighted normalization factor for each sample peptide based upon the two closest eluting library peptides (*see* Fig. 1). If a sample peptide elutes at time x, and the closest library peptides elute at times i and j, then the normalization factor is calculated as

$$\mathrm{NF}_x = \mathrm{NF}_i \left|\frac{x-j}{i-j}\right| + \mathrm{NF}_j \left|\frac{x-i}{i-j}\right| \quad (2)$$

where NF_i and NF_j are the peak intensity ratios of the two closest library peptides. The quantitation ratio for the phosphopeptide eluting at retention time x is then calculated using its raw peak intensity ratio and the weighted normalization factor as

$$\mathrm{LAXIC\ ratio} = \frac{\mathrm{Raw\ ratio}}{\mathrm{NF}_x} \quad (3)$$

This accounts for varying ion suppression effects across the chromatogram and for variations in LC-MS performance. Here, we describe the LAXIC workflow used for quantitation of phosphopeptides identified by high-resolution LC-MS and note the important considerations for designing the synthetic peptide library.

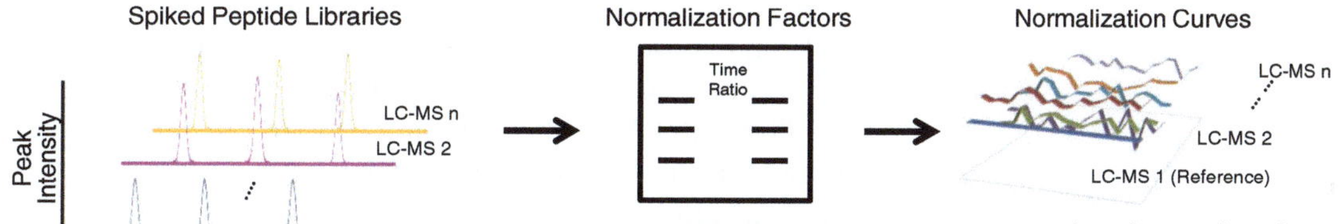

$$NF(t) = \frac{Intensity\ I1(t)}{Intensity\ I2(t)} \qquad NF_x = NF_i \cdot \left|\frac{x-j}{i-j}\right| + NF_j \cdot \left|\frac{x-i}{i-j}\right|$$

Fig. 1 Normalization by spiked-in peptide library. The same amount of a synthetic peptide library is spiked into each sample for analysis, including the control. Once the chromatograms of the samples and control have been aligned, LAXIC compares the library peptide peak intensities in the samples to those in the control to generate raw intensity ratios. Each library peptide ratio is associated with the retention time of that peptide. The library is designed to elute consecutively across the entire chromatogram, and each sample phosphopeptide can be normalized using two co-eluting library peptides

2 Materials

All solutions should be prepared with ultrapure water and analytical grade reagents. Unless otherwise noted, solutions are created using reagents from Sigma (St. Louis, MO). Ultrapure water can be prepared by purifying deionized water to a sensitivity of 18 MΩ cm at 25 °C with a Barnstead Nanopure system (Thermo Fisher, Waltham, MA).

1. Lyophilized peptide samples (0.5–2 μg each) from an enzymatic digest: The samples should have a low salt concentration (*see* **Note 1**) and should be stored at −80 °C to preserve phosphorylation until needed for analysis.
2. Solution of 0.1 % formic acid in ultrapure water for sample resuspension: This can also serve as the mobile phase for liquid chromatography. The elution buffer for liquid chromatography comprises 0.1 % formic acid (buffer A) in LC-grade acetonitrile (buffer B).
3. Synthetic peptide library for spiking into samples prior to LC-MS analysis: Prepare 1 mL of a library stock solution by dissolving the lyophilized library (*see* **Notes 2** and **3**) in ultrapure water to a concentration of 1 pmol/μL in a low-binding microcentrifuge tube. Dilute the stock to 2.5 fmol/μL in 0.1 % formic acid. Store the solutions at −20 °C.
4. A liquid chromatography system coupled to a mass spectrometer: Our lab employs an Eksigent NanoLC Ultra 2D HPLC system (Eksigent, CA) coupled to an LTQ-Orbitrap Velos (Thermo Fisher). Reverse-phase chromatography is achieved using a C18 column packed in-house with 5 μm particle size C18 Magic resin (Michrom, Auburn, CA; 75 μm i.d. with a bed length of 30 cm).
5. An alignment file produced from the LC-MS data for each sample that matches peptide retention times and m/z values in the sample to those in the control.
6. A FASTA database file for the samples and a separate FASTA file containing the peptide sequences found in the synthetic library: The FASTA databases should utilize Universal Protein Resource (UniProt) accession numbers.
7. A computer with Internet access.

3 Methods

1. Prepare the peptide samples for analysis by directly resuspending the samples in 10 μL of 0.1 % formic acid containing the combinatorial library at a concentration of 2.5 fmol/μL (*see* **Note 4**).

Vortex the samples for 10 s, spin down briefly, and vortex again. Centrifuge the samples at 16,100×*g* for 10 min, and carefully transfer the supernatant to an LC-MS vial.

2. Complete the analysis of peptide samples via high-resolution LC-MS. A linear gradient of buffer A in buffer B is run over 60 or 90 min at a flow rate of 300 nL/min. The Orbitrap mass spectrometer is run in the data-dependent mode with a scan range of *m/z* 300–1,700 and a resolution of 60,000 at *m/z* 400. The top 20 most abundant ions in the MS scan are selected for MS/MS with an exclusion time of 15 s.
3. Create an alignment file for each sample in which the peptides of the sample are matched to the corresponding peptides of a control by retention time and *m/z* (*see* **Note 5**).
4. Generate two spreadsheets or tab-delimited text files for each sample and the control. One file should contain a list of all library peptides identified with columns for sequence, peptide spectrum match ambiguity, protein group accession, modifications, *m/z* value, charge, and retention time. The second file should list the same information for all identified sample (non-library) peptides (*see* **Notes 6** and 7).
5. Gather all of the spreadsheets, alignment files, and FASTA databases. It is important that the sample spreadsheet and alignment file names preceding the file extensions match exactly. The library files should have the same names as the samples followed by "-library" before the file extension.
6. Access the LAXIC software through an Internet browser at *thr.genomics.purdue.edu/~xue2/LAXIC.html* (*see* Fig. 2).
7. Under "STEP 1" upload one spreadsheet containing the library peptides found in the control in the first box. In the next box, upload the library spreadsheets for all of the remaining samples.
8. Under "STEP 2" upload one spreadsheet containing the peptides found in the control in the first box. In the second box, upload the peptide spreadsheets for all of the remaining samples. In the third box, upload all of the sample alignment files.
9. Under "STEP 3" adjust the mass accuracy tolerance and retention time tolerance values. The defaults are set to 5 ppm and 1 min, respectively.
10. Under "STEP 4" upload the sample FASTA file in the first box. In the next box, upload the library FASTA file.
11. Under "STEP 5" provide a user or a job name, and click the "Submit Query" button.
12. The quantitation may take a few minutes depending upon the server load. When a "Done!" notification appears, retrieve the quantitation results by clicking on one of the result links.

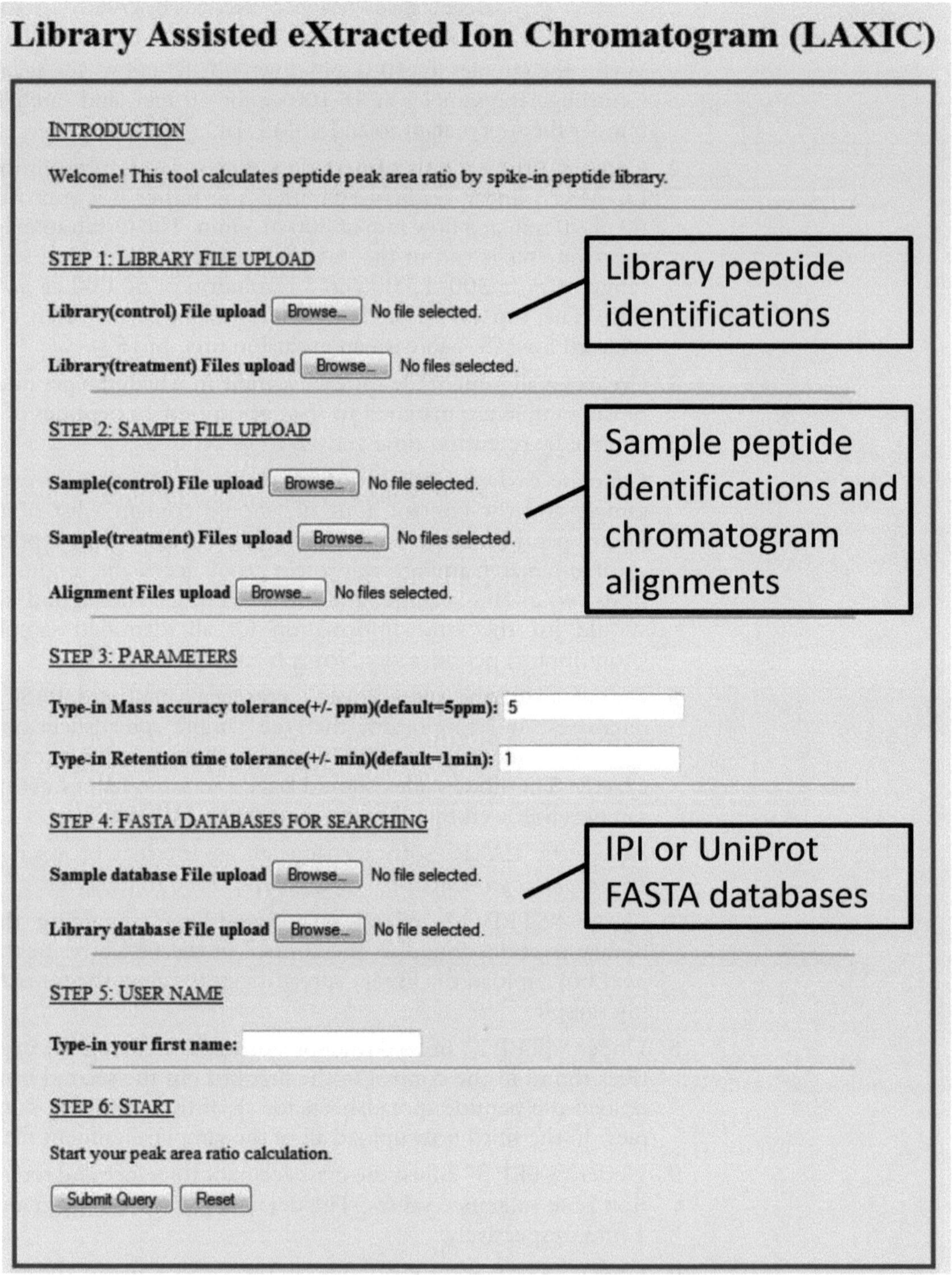

Fig. 2 Setup of the LAXIC analysis. The first files to be uploaded are the spreadsheets or tab-delimited text files containing the sequences, PSM ambiguity, protein accession numbers, modifications, *m*/*z* values, charge, and retention times of the library peptides identified in the control and treated samples. The file for the control is entered in a *separate box* from the sample files. The second step is the upload of the spreadsheets or the text files containing the same information for the sample phosphopeptides identified in the control and treated samples. Again, the software accepts one file for the control in a *separate box* from the other sample files. The chromatogram alignment files should be entered in *one box* followed by the upload of one FASTA database for the samples and a second FASTA database containing the library peptide sequences

If the browser times out, the analysis will still be completed, but a "Done!" notification may not appear. If this happens, simply click on the result links to check the status of the quantitation.

13. The LAXIC output provides links to the calculated quantitation ratios. The "Result of the Peptide Quantification" link gives the phosphopeptide sequences identified in the control with their modifications, protein accession numbers, and corresponding protein names. Most importantly, this link provides the LAXIC peptide abundance ratios for each peptide that was quantified at least once. The "Count" column indicates the number of samples in which a particular peptide sequence was quantified. "NA" in a sample ratio column indicates that the software was unable to quantify a particular peptide in that sample. The "Result of the Protein Quantification" link provides the abundance ratios for the proteins identified in the samples along with their accession numbers. Lastly, the "Result of the Normalization Factors" link lists the peak apex retention times and peak intensity ratios for identified library peptides (*see* Eq. 1). These ratios are used to calculate the weighted normalization factors used to adjust sample phosphopeptide peak intensity ratios (*see* Eqs. 2 and 3). It may be useful to check these ratios for indication of instrument performance; for instance, wildly differing ratios may result from LC-MS variance, while ratios consistently above or below one may indicate systematic error.
14. The quantitation results found under all three links discussed above may be saved as text files and then imported into a spreadsheet program for easier manipulation.

4 Notes

1. We typically desalt peptide samples using C18 NuTips (Glygen) or Sep-Pak C18 columns (Waters) and perform any desired enrichments for modified peptides before spiking in the synthetic peptide library for quantitation. The desalting procedure consists of conditioning the column with methanol followed by a solution of 0.1 % trifluoroacetic acid (TFA) in 80 % acetonitrile. The organic solvent is rinsed out of the column with two washes of 0.1 % TFA. Next, the sample is loaded on the column, and the column is washed three times with 0.1 % TFA. Finally, the desalted sample is eluted with 0.1 % TFA in 80 % acetonitrile and dried to completion in a centrifuge vacuum concentrator.
2. Synthetic peptide libraries may be purchased from various vendors, but our library was generated in-house through standard solid-phase synthesis. Describing the synthesis of the peptide

library in detail is beyond the scope of this chapter, but solid-phase synthesis with phenylacetamidomethyl (PAM) resin beads should generally begin with arginine or lysine residues to produce tryptic-like peptides. The peptide is lengthened from the C-terminus using *N*-tertbutoxycarbonyl (Boc)-protected amino acids (Midwest Biotech, Indianapolis, IN) and a coupling reagent, 3-(diethoxy-phosphoryloxy)-3H-benzo [d][1,2,3]triazin-4-one (DEPBT, National Biochemicals, Twinsburg, OH). When two amino acids need to be incorporated at a single position, the PAM beads can be dried and divided equally by mass before the next coupling reaction. *See* refs. 24, 25 for further details.

3. The peptide library should be created according to certain recommendations. (a) Distribution of the peptides should cover the entire chromatogram, and gaps in retention time between sequentially eluting library peptides should be minimized. (b) The number of peptides and amount of library used should be controlled to maintain LC-MS sensitivity while allowing accurate quantitation by LAXIC. (c) Amino acids that may have modifications are not recommended to be included in peptide sequences. Based upon these major criteria, we spike in a library that contains a total of 160 unique peptide sequences. The hydrophobicity of the peptides covers a wide range; the hydrophilic peptides are used as a model for phosphopeptides, while the more hydrophobic peptides mimic the behavior of non-phosphorylated peptides.
4. The library amount may be adjusted according to the number of peptide sequences within the library to maintain LC-MS sensitivity, provided that the same amount is spiked into each sample of a given experiment. The amounts given here are based upon a library of 160 unique peptides.
5. There are a variety of ways to accomplish the task of alignment, but we find the OpenMS Proteomics Pipeline Assistant (TOPPAS) easiest to implement [26]. First, the peptide peaks in the mass spectra are converted to a series of features using the FeatureFinderCentroided node; this generates a FeatureXML file for each sample. Next, the FeatureXML files for each sample are compared to those of the control using the MapAlignerPoseClustering node; this node aligns the retention times of the peptides in the samples to those found in the control. Finally, the FeatureLinkerUnlabeled node is used to group peptides found in the samples to the control. The TOPPAS output is a ConsensusXML file that contains the peptide groupings for each sample. Depending on the software installed, this output file may have an "Unknown" file extension, but it can still be recognized by LAXIC.

6. Our group uses the SEQUEST algorithm built into ProteomeDiscoverer 1.4 (Thermo Fisher) for peptide/protein identification and the PhosphoRS 3.0 node [7] for phosphorylation localization. Thus, to generate the necessary spreadsheets, we export the SEQUEST search results as a Microsoft Excel file directly from ProteomeDiscoverer. After exporting the data, we open and re-save the Excel file or transfer it to another computer before entering it into the LAXIC software. Otherwise, LAXIC may display an error indicating that the file cannot be read because it is in use by another program (ProteomeDiscoverer). LAXIC accepts both ".xls" and ".xlsx" Excel formats.

7. If software other than ProteomeDiscoverer is used to generate the spreadsheets, it may be necessary to rename certain columns. LAXIC requires the following column headings: Sequence, PSM Ambiguity, Protein Group Accessions, Modifications, m/z (Da), Charge, and RT (min). If peptide spectrum match ambiguity is unavailable, create a column titled "PSM Ambiguity" and fill in "Unambiguous" for each peptide sequence. Column heading capitalization must follow the conventions indicated above.

Acknowledgments

This project has been funded in part by an NSF CAREER award CHE-0645020 (W.A.T.) and by National Institutes of Health GM088317.

References

1. Hunter T (2000) Signaling—2000 and beyond. Cell 100:113–127
2. Blume-Jensen P, Hunter T (2001) Oncogenic kinase signaling. Nature 411:355–365
3. Pawson T (2004) Specificity in signal transduction: from phosphotyrosine-SH2 domain interactions to complex cellular systems. Cell 116:191–203
4. Abou-Sleiman PM, Muqit MMK, Wood NW (2006) Expanding insights of mitochondrial dysfunction in Parkinson's disease. Nat Rev Neurosci 7:207–219
5. Nag K, Chaudhary A (2009) Mediators of tyrosine phosphorylation in innate immunity: from host defense to inflammation onto oncogenesis. Curr Signal Trans Ther 4:76–81
6. Cantin GT, Shock TR, Park SK, Madhani HD, Yates JRI (2007) Optimizing TiO2-based phosphopeptide enrichment for automated multidimensional liquid chromatography coupled to tandem mass spectrometry. Anal Chem 79:4666–4673
7. Taus T, Kocher T, Pichler P, Paschke C, Schmidt A, Henrich C, Mechtler K (2011) Universal and confident phosphorylation site localization using phosphoRS. J Proteome Res 10:5354–5362
8. Olsen JV, Blagoev B, Gnad F, Macek B, Kumar C, Mortensen P, Mann M (2006) Global, in vivo, and site-specific phosphorylation dynamics in signaling networks. Cell 127:635–648
9. Matsuoka S, Ballif BA, Smogorzewska A, McDonald ERIHKE, Luo J, Bakalarski CE, Zhao Z, Solimini N, Lerenthal Y, Shiloh Y, Gygi SP, Elledge SJ (2007) ATM and ATR substrate analysis reveals extensive protein networks responsive to DNA damage. Science 316:1160–1166

10. Trinidad JC, Thalhammer A, Specht CG, Lynn AJ, Baker PR, Schoepfer R, Burlingame AL (2008) Quantitative analysis of synaptic phosphorylation and protein expression. Mol Cell Proteomics 7:684–696
11. Li L, Steen H, Gygi SP (2003) Protein profiling with cleavable isotope-coded affinity tag (cICAT) reagents: the yeast salinity stress response. Mol Cell Proteomics 2:1198–1204
12. Ong SE, Mann M (2005) Mass spectrometry-based proteomics turns quantitative. Nat Chem Biol 1:252–262
13. Kim YJ, Zhan P, Feild B, Ruben SM, He T (2007) Reproducibility assessment of relative quantitation strategies for LC-MS based proteomics. Anal Chem 79:5651–5658
14. Gerber SA, Rush J, Stemman O, Kirschner MS, Gygi SP (2003) Absolute quantification of proteins and phosphoproteins from cell lysates by tandem MS. Proc Natl Acad Sci U S A 100:6940–6945
15. Tabata T, Sato T, Kuromitsu J, Oda Y (2007) Pseudo internal standard approach for label-free quantitative proteomics. Anal Chem 79:8440–8445
16. Han CL, Chen JS, Chan EC, Wu CP, Yu KH, Chen KT, Tsou CC, Tsai CF, Chien CW, Kuo YB, Lin PY, Yu JS, Hsueh C, Chen MC, Chan CC, Chang YS, Chen YJ (2011) An informatics-assisted label-free approach for personalized tissue membrane proteomics: case study on colorectal cancer. Mol Cell Proteomics 10:M110.003087
17. Bondarenko PV, Chelius D, Shaler TA (2002) Identification and relative quantitation of protein mixtures by enzymatic digestions followed by capillary reversed-phase liquid chromatography-tandem mass spectrometry. Anal Chem 74: 4741–4749
18. Wang W, Zhou H, Lin H, Roy S, Shaler TA, Hill LR, Norton S, Kumar P, Anderle M, Becker CH (2003) Quantification of proteins and metabolites by mass spectrometry without isotopic labeling or spiked standards. Anal Chem 75:4818–4826
19. Zhang R, Barton A, Brittenden J, Huang JT, Crowther D (2010) Evaluation for computational platforms of LC-MS based label-free quantitative proteomics: a global view. J Proteomics Bioinf 3:260–265
20. Xie F, Liu T, Qian W, Petyuk VA, Smith RD (2011) Liquid chromatography-mass spectrometry-based quantitative proteomics. J Biol Chem 286:25443–25449
21. Zybailov B, Mosely AL, Sardiu ME, Coleman MK, Florens L, Washburn MP (2006) Statistical analysis of membrane proteome expression changes in Saccharomyces cerevisiae. J Proteome Res 5:2339–2347
22. Mueller L, Brusniak M, Mani DR, Aebersold R (2008) An assessment of software solutions for the analysis of mass spectrometry based quantitative proteomics data. J Proteome Res 7:51–61
23. Xue L, Wang P, Wang L, Renzi E, Radivojac P, Tang H, Arnold R, Zhu J, Tao WA (2013) Quantitative measurement of phosphoproteome response to osmotic stress in Arabidopsis based on Library-Assisted eXtracted Ion Chromatogram (LAXIC). Mol Cell Proteomics 12(8):2354–2369
24. Bohrer BC, Li YF, Reilly JP, Clemmer DE, DiMarchi RD, Radivojac P, Tang H, Arnold RJ (2010) Combinatorial libraries of synthetic peptides as a model for shotgun proteomics. Anal Chem 82:6559–6568
25. Lebl M, Krchnak V (1997) Synthetic peptide libraries. Method Enzymol 289:336–392
26. Junker J, Bielow C, Bertsch A, Sturm M, Reinert K, Kohlbacher O (2012) TOPPAS: a graphical workflow editor for the analysis of high-throughput proteomics data. J Proteome Res 11:3914–3920

Chapter 28

Fast, Efficient, and Quality-Controlled Phosphopeptide Enrichment from Minute Sample Amounts Using Titanium Dioxide

Clarissa Dickhut, Sonja Radau, and René P. Zahedi

Abstract

Bottom-up mass spectrometry (MS) is still the method of choice for analyzing protein phosphorylation. However, the low stoichiometry of phosphorylation, especially in highly complex samples, renders the specific enrichment of phosphopeptides prior to analysis inevitable. In recent years, specific phosphopeptide enrichment strategies combined with high-performance liquid chromatography (HPLC)-MS (LC-MS) provided researchers deeper insights into the phosphorylation networks of biological systems.

Here, we describe two protocols for the enrichment of phosphopeptides from biological samples using titanium dioxide (TiO_2) resins, enabling the handling of small sample amounts (<20 μg of protein) as well as larger sample amounts (up to the milligram range), depending on the scientific issue to be solved.

Furthermore, we imply quality control steps during sample preparation to ensure the reproducibility and reliability of the phosphoproteomic findings.

Key words Phosphopeptide enrichment, Phosphoproteomics, Protein phosphorylation, Titanium dioxide

1 Introduction

Metal oxides, especially titanium dioxide, are known for their high affinity for organic phospho-moieties [1], rendering these resins convenient for the enrichment of phosphorylated peptides from biological samples. However, unspecific binding of mostly acidic peptides was a limitation in the first setups, which used online TiO_2 chromatography as the first dimension before LC-MS analysis [2, 3]. In 2005, Larsen et al. improved the specificity of phosphopeptide enrichment by the addition of dihydroxybenzoic acid to the loading step, leading to the competitive inhibition of non-phosphopeptide binding [4]. In recent years, various TiO_2-based protocols have

Clarissa Dickhut and Sonja Radau contributed equally to this work.

Daniel Martins-de-Souza (ed.), *Shotgun Proteomics: Methods and Protocols*, Methods in Molecular Biology, vol. 1156, DOI 10.1007/978-1-4939-0685-7_28, © Springer Science+Business Media New York 2014

been published, particularly for off-line methods using TiO_2 in combination with LC-MS approaches [5–8]. However, many protocols have the drawback that they are tested only for a particular kind of sample and require large starting protein amounts. As, e.g., samples for clinical research, such as purified blood cells or cancer tissue, are mostly only accessible in limited amounts, we describe an efficient protocol for phosphopeptide enrichment from small sample amounts (<20 µg) using micro-columns, which is based on established TiO_2 protocols [9, 10]. We demonstrate that the use of self-made micro-columns increased the identification of phosphopeptides by 40 % compared to batch mode.

Furthermore, we provide a second protocol for the enrichment of phosphopeptides from larger protein amounts employing batch mode, which is convenient up to the milligram range, faster than micro-column-based strategies and easy to parallelize for large-scale experiments. Both protocols show a high specificity for phosphopeptides of >85 %.

As a thorough and reproducible sample preparation is critical for the success of phosphoproteomic studies, we recommend the implication of quality control steps during sample preparation. These steps imply monitoring of proteolytic digestion by monolithic chromatography [11] as well as addition of synthetic, non-endogenous phosphopeptides into the digested sample prior to enrichment, enabling the dedicated validation of the reproducibility and recovery of the phosphopeptide enrichment step in different replicates.

2 Materials

Prepare all solutions using ultrapure deionized water. Store all reagents at 4 °C, if not indicated otherwise. Use LC-MS or UPLC grade for high-performance liquid chromatography (HPLC) solvents.

2.1 Cell Culture and Lysis

1. Cell medium: Dulbecco's modified Eagle's medium (DMEM) with L-glutamine supplemented with 10 % fetal bovine serum and penicillin/streptomycin.
2. Cell line: HeLa S3 cells (DSMZ), grow under standard conditions in 75 cm^2 cell culture flask (*see* **Note 1**).
3. Cell detachment: Solution of trypsin/ethylenediaminetetraacetic acid (EDTA) 1:250 and Phosphate buffered saline (PBS).
4. Cell lysis buffer: 1 % sodium dodecyl sulfate (SDS), 150 mM NaCl, 50 mM Tris–HCl, pH 7.8 (adjust pH with HCl) plus phosphatase inhibitor cocktail phosSTOP (Roche) (*see* **Note 2**).
5. Buffer for hydrolysis of nucleic acids: Benzonase (purity >99 %) plus 2 mM $MgCl_2$.
6. Determination of protein concentration: Bicinchoninic acid (BCA) protein assay.

2.2 Carbamidomethylation and Ethanol Protein Precipitation

1. Reduction buffer: 2 M dithiothreitol (DTT) stock, can be stored at −40 °C.
2. Alkylation buffer: 1 M iodoacetamide (IAA, *see* **Note 3**).
3. Ethanol precipitation: 100 % cold ethanol, store at −40 °C before use.

2.3 Protein Digestion

1. Solubilization buffer: 4 M guanidine hydrochloride (GuHCl) in 50 mM triethylammonium bicarbonate (TEAB), pH 7.8 (adjust pH with HCl).
2. Digestion buffer: Dilute solubilization buffer with 50 mM TEAB, pH 7.8 to 0.2 M GuHCl.
3. Proteolytic enzyme: Sequencing-grade modified trypsin, lyophilized. Dissolve lyophilized trypsin in an appropriate volume of digestion buffer and add the trypsin solution to the sample.
4. Stop digestion: 10 % trifluoroacetic acid (TFA) (*see* **Note 4**).

2.4 Digestion Control by Monolithic Reversed-Phase Chromatography

1. HPLC: UltiMate 3000 rapid separation liquid chromatography (RSLC) system (Thermo Fisher Scientific) or similar nano-HPLC system.
2. Monolithic columns: Monolithic trap and main column (e.g.: PepSwift monolithic trap column 200 μm × 5 mm and PepSwift monolithic capillary column 200 μm × 5 cm (Thermo Fisher Scientific)).
3. Solvent A: 0.1 % TFA.
4. Solvent B: 0.08 % TFA, 84 % acetonitrile (ACN).

2.5 Synthetic Peptides for Quality Control

1. Spike in: Approximately 500 fmol of synthetic phosphopeptides. Use either peptides without homology to human sequences or stable isotope-coded peptides (*see* **Note 5**).

2.6 Titanium Dioxide Phosphopeptide Enrichment

1. Adsorption material: Titanium dioxide beads, 5 μm (Titansphere, GL Sciences).
2. Loading buffer: 80 % ACN, 5 % TFA, and 1 M glycolic acid (*see* **Note 6**).
3. Washing buffer 1: 80 % ACN, 1 % TFA.
4. Washing buffer 2: 10 % ACN, 0.1 % TFA.
5. Elution buffer: 1 % ammonium hydroxide, pH 11.3.
6. Depletion of beads: Glass microfiber filters.
7. Acidification: 100 % formic acid (FA) and 10 % TFA.

2.7 Sample Purification by Solid-Phase Extraction

1. Solid phase: C18 desalting tips.
2. RP washing buffer: 0.1 % TFA.
3. RP elution buffer: 60 % ACN, 0.1 % TFA.

2.8 Reversed-Phase Chromatographic Separation and Mass Spectrometry

1. HPLC: UltiMate 3000 nano RSLC system (Thermo Fisher Scientific) or a similar nano-HPLC system.
2. RP columns: C18 trap and main column (e.g.: Acclaim C18 PepMap100 Nano-Trap column 100 μm × 2 cm and Acclaim PepMap100 C18 column 75 μm × 50 cm (Thermo Fisher Scientific)).
3. HPLC loading buffer: 0.1 % TFA.
4. HPLC solvent A: 0.1 % FA.
5. HPLC solvent B: 0.1 % FA, 84 % ACN.
6. Mass spectrometer: Q-Exactive mass spectrometer (Thermo Scientific) or a similar instrument with high mass accuracy and high resolution.

2.9 Data Interpretation

1. Data analysis software: Proteome Discoverer (version 1.3, Thermo Scientific).
2. Search algorithm: Mascot algorithm (version 2.4.1, Matrix Science).

3 Methods

3.1 Cell Culture and Lysis

1. Grow HeLa cells in 75 cm^2 cell culture flask in DMEM with L-glutamine, 10 % fetal bovine serum, and penicillin/streptomycin at 37 °C in a humidified atmosphere with 5 % CO_2 to a confluence of 70–80 %.
2. Remove medium, and wash cells once with 5 ml of PBS buffer. Detach cells using 2 ml of trypsin/EDTA solution and incubate at 37 °C for 3–5 min. Add 8 ml of fresh medium to inactivate trypsin and transfer cell suspension to a falcon tube. Pellet cells by centrifugation (max. 300 × *g*), and wash cell pellet one to two times with PBS.
3. Lyse cells in the cell lysis buffer (*see* **Note 7**), using an ultrasonic bath for up to 30 s and mechanic forces (such as pipetting or mixing) until the lysate is clear (*see* **Note 8**).
4. Add benzonase (1 μL per 200 μL lysis buffer) plus 2 mM $MgCl_2$ and incubate at 37 °C for 30 min. Benzonase will hydrolyze DNA, reducing the viscosity of the sample (*see* **Note 9**).
5. Clarify supernatant, if necessary, by centrifugation at 10,000 × *g* for 10 min at 4 °C. At this step the supernatant can either be processed further or stored at −80 °C.
6. Determine protein concentration by the BCA protein assay according to the manufacturer's instructions. For reliable results measure at least three different dilutions of the sample as well as all standard dilutions and prepare everything in triplicate (*see* **Note 10**).

3.2 Sample Preparation: Carbamidomethylation, Protein Purification, and Digestion

1. For reduction of disulfide bonds add 10 mM DTT to the clarified cell lysate and incubate at 56 °C for 30 min. Further, alkylate free sulfhydryl groups using 30 mM IAA at room temperature in the dark for another 30 min. Always prepare IAA freshly (*see* **Note 3**).
2. Add 9 volumes of cold ethanol to 1 volume of protein solution for protein precipitation. Incubate at −40 °C for 1 h, and pellet the proteins by centrifugation at 20,000 × *g* at 4 °C for 30 min. Discard supernatant carefully.
3. Solubilize protein pellet in the solubilization buffer containing 4 M GuHCl. For digestion dilute the sample to a final concentration of 0.2 M GuHCl using 50 mM TEAB (*see* **Note 11**). Take an aliquot of 0.5–1 μg for digestion control.
4. Use a trypsin-to-protein ratio of 1:20 (w/w), and incubate sample at 37 °C overnight.
5. Add 10 % TFA to a final concentration of 1 % to quench trypsin.
6. Perform a digestion control using a monolithic RP separation on an Ultimate 3000 HPLC (*see* **Note 12** and Fig. 1). For each

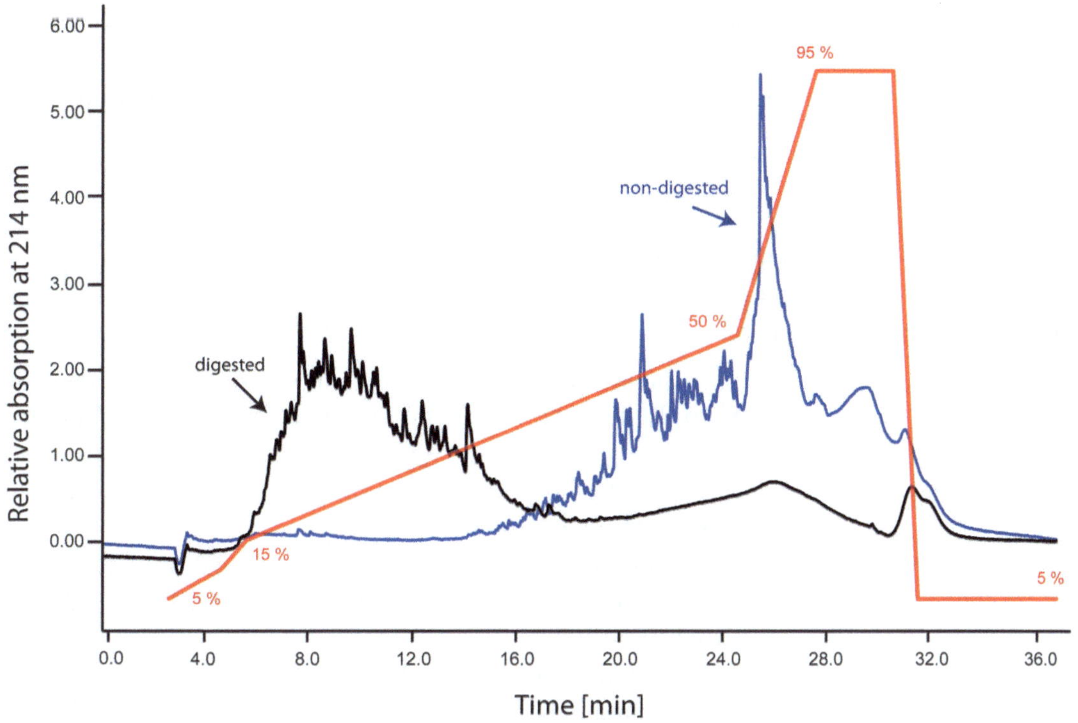

Fig. 1 Monolithic digestion control of HeLa samples. The separation of digested and non-digested samples on a monolithic column provides a fast and straightforward digestion control. Peptides elute in the early part of the gradient, proteins at higher retention times. The samples were separated using a binary gradient (solvent A: 0.1 % TFA, solvent B: 0.08 % TFA, 84 % acetonitrile) reaching from 15 to 50 % solvent B in 20 min

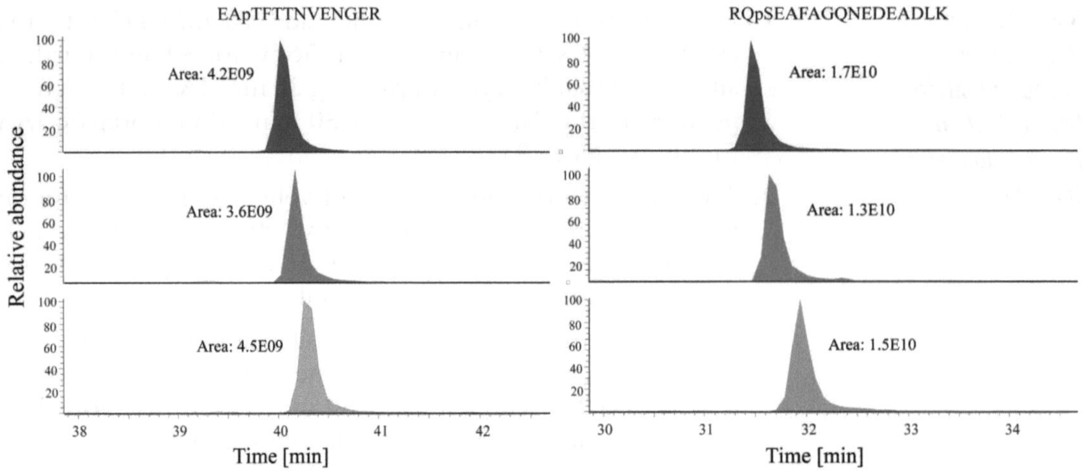

Fig. 2 Quality control of phosphopeptide enrichment. Prior to enrichment, synthetic phosphopeptides were spiked into each replicate. Here, the extracted ion chromatogram areas of two spiked-in peptides are given as an example (EApTFTTNVENGER and RQpSEAFAGQNEDEADLK). By comparing the peak areas between replicates, the reproducibility of the enrichment step can be evaluated. The standard deviations should be below 30 %

sample, take 0.5–1 μg before (use aliquot from **step 3**) and after digestion and dilute each to 15 μL (loading volume) with 0.1 % TFA (solvent A).

7. Prior to TiO_2 enrichment spike approximately 500 fmol of the synthetic phosphopeptides into the digested sample to control reproducibility and efficiency of the phosphopeptide enrichment (*see* Fig. 2).

3.3 TiO_2 Phosphopeptide Enrichment in Batch Mode

This protocol is based on a TiO_2 phosphopeptide enrichment protocol published by Martin Larsen and co-workers [9]. The following procedure is convenient especially for sample amounts above 20 μg.

1. Dry the sample completely in an Eppendorf tube under vacuum. Redissolve the sample in 1 ml of TiO_2 loading buffer.
2. Adjust the amount of TiO_2 beads to the amount of sample. Use a bead-to-peptide ratio of 6:1, e.g., 600 μg TiO_2 beads to 100 μg HeLa lysate (*see* **Note 13**).
3. Incubate the peptide–bead suspension for 10 min on a vortex mixer at low speed. Then, centrifuge at 3,000 × *g* for 30 s to pellet the beads.
4. Incubate the supernatant with fresh TiO_2 beads a second time as described in **step 3**. Use a bead-to-peptide ratio of 3:1.
5. Incubate the supernatant with fresh TiO_2 beads a third time as described in **step 3**. Use a bead-to-peptide ratio of 1.5:1 (*see* **Note 14**).

6. Pool the beads of all incubation steps in one Eppendorf tube using 100 μL of loading buffer. Centrifuge as above to pellet the beads, and discard the supernatant.
7. Wash the beads with 100 μL of TiO_2 washing buffer 1, mix well using a vortex mixer for 15 s, centrifuge as above to pellet the beads, and discard supernatant carefully. Repeat this step using 100 μL of washing buffer 2.
8. Elute the phosphopeptides from the beads using 100 μL of TiO_2 elution buffer, mix shortly at high speed, and incubate the sample on a vortex mixer at low speed for at least 10 min.
9. Pellet the beads by centrifugation at 3,000 × *g* for 1 min. Transfer the eluate carefully to a new Eppendorf tube (*see* **Note 15**).
10. Wash the beads with 30 μL of TiO_2 elution buffer, and combine the supernatant with the eluate from **step 9**.
11. Acidify eluate prior to the solid-phase extraction (SPE) using 2 μL of 10 % TFA and 8 μL of 100 % FA (*see* **Note 16**).

3.4 TiO_2 Phosphopeptide Enrichment in Micro-columns

This protocol is convenient especially for sample amounts below 20 μg. A thorough comparison of the two described phosphopeptide enrichment protocols, batch mode and micro-column enrichment, clearly showed the advantage of using micro-columns for small sample amounts (*see* **Note 17** and Fig. 3).

1. Dry the sample completely in an Eppendorf tube under vacuum. Redissolve the sample in 1 ml of TiO_2 loading buffer (*see* **Note 6**).
2. Adjust the amount of TiO_2 beads to the amount of sample. Use a bead-to-peptide ratio of 6:1 (*see* **Note 13**).

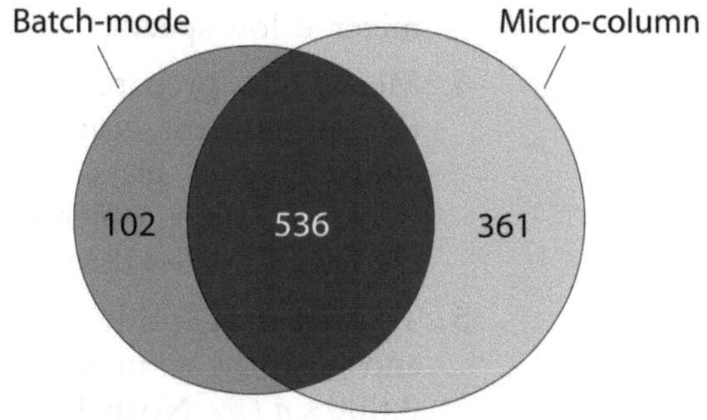

Fig. 3 Comparison of batch mode and micro-column phosphopeptide enrichment for 10 μg of starting material. For 10 μg of starting material about 40 % more phosphopeptides (average from triplicate enrichments with phosphoRS probabilities >75 %) could be identified using micro-columns for the enrichment, when compared to batch mode. Therefore, we recommend the use of micro-columns for small sample amounts (<20 μg)

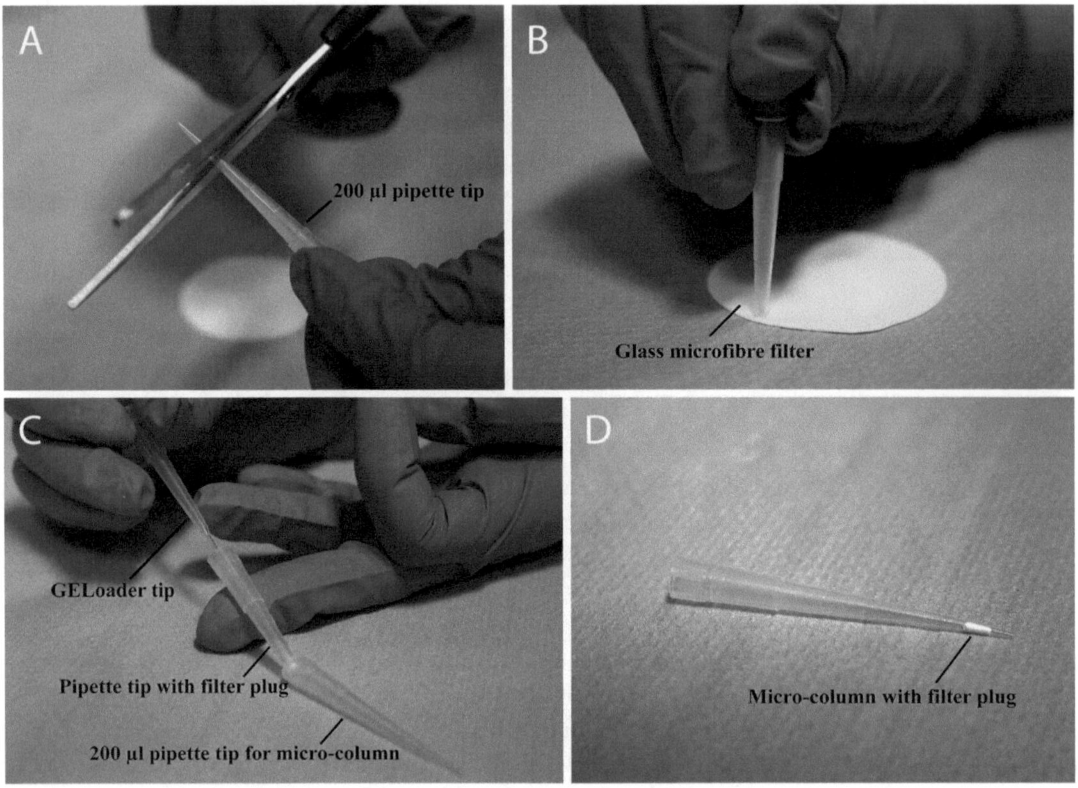

Fig. 4 Instructions for building a micro-column. For a micro-column, two 200 μL pipette tips, a GELoader tip, and a glass microfiber filter are required. (**a**) First, approximately 5 mm of a 200 μL pipette tip are cut from the bottom and then (**b**) the shortened tip is used to excise a plug of the glass microfiber filter. (**c**) The plug is placed into a new 200 μL pipette tip with the help of a GELoader tip. (**d**) Finally, a micro-column with filter plug is built, which can be loaded with TiO_2 beads

3. Incubate the peptide–bead suspension for 10 min on a vortex mixer at low speed.
4. Meanwhile, prepare the micro-column for the phosphopeptide enrichment, using a conventional 2–200 μL pipette tip and a plug of glass microfiber filter paper [12]. This plug works as a filter, which retains the TiO_2 beads, while it allows the liquid to pass through (*see* Fig. 4).
5. Transfer the peptide–bead suspension onto the micro-column, and push the solution through the filter by centrifugation at 3,000 × *g* (*see* **Note 18**). As the volume of the tip is limited, repeat this step until the complete suspension has passed the filter.
6. Keep flow-through and incubate with fresh TiO_2 beads a second time. Use a bead-to-peptide ratio of 3:1, and load the suspension on the same micro-column to combine both incubation steps.

7. For complex samples keep flow-through and incubate with fresh TiO_2 beads a third time. Use a bead-to-peptide ratio of 1.5:1, and load the suspension on the same micro-column to combine all incubation steps (*see* **Note 14**).
8. Wash the TiO_2 micro-column with 100 μL of loading buffer, and mix beads by pipetting carefully. Discard the loading buffer by pushing the solution through the micro-column using centrifugation at 3,000 × *g*.
9. Repeat **step 8** twice. Use 100 μL of TiO_2 washing buffer 1 and 100 μL of washing buffer 2 consecutively.
10. Incubate the TiO_2 beads with 100 μL of elution buffer for at least 10 min at room temperature (*see* **Note 19**), and collect the phosphopeptides in a clean vial by centrifugation of the eluate through the micro-column at 3,000 × *g*.
11. Furthermore, wash the micro-column with 30 μL of elution buffer and afterwards with 2 μL of 30 % ACN to elute peptides bound to the filter material (*see* **Note 20**).
12. Acidify eluate prior to SPE using 2 μL of 10 % TFA and 8 μL of 100 % FA (*see* **Note 16**).

3.5 Desalting of Enriched Samples by Solid-Phase Extraction

1. Cut 5 mm from the top of a 2–200 μL pipette tip, and use this tip as an adapter for the C18 tip. The extension of the C18 tip enables desalting of samples with volumes above 10 μL in one step.
2. Activate the RP material by pipetting 40 μL of 100 % ACN with the prolonged tip. Discard solution after one passage.
3. Equilibrate the RP material with 40 μL of 0.1 % TFA as described in **step 2**.
4. Load the acidified phosphopeptide sample onto the RP column by pipetting the sample slowly up and down (*see* **Note 21**). To minimize sample loss, repeat this procedure at least three times.
5. Wash the RP column with 40 μL of 0.1 % TFA to remove residual salts, etc. from the sample.
6. Elute phosphopeptides from the RP material into a new Eppendorf tube using 40 μL of 60 % ACN, 0.1 % TFA.
7. Dry the eluted peptides completely under vacuum.
8. For LC-MS analysis, reconstitute the dried sample in 15 μL of 0.1 % TFA (HPLC loading buffer).

3.6 LC-MS Analysis of the Enriched Phosphopeptide Sample

1. Nano-LC-MS/MS was performed on a Q-Exactive mass spectrometer online coupled to a nano-HPLC system. The described setup serves as an example and can be modified according to the facilities of the laboratory as well as the amount and complexity of the sample.
2. In brief, the samples were loaded onto an RP trap column with 0.1 % TFA (HPLC loading buffer) at a flow rate of 20 μL/min,

followed by separation on a 50 cm RP main column using a binary gradient (HPLC solvents A: 0.1 % FA and B: 0.1 % FA, 84 % ACN) from 5 to 42 % solvent B at a flow rate of 250 nL/min in 90 min, followed by 4 min at 95 % B.

3. The Q-Exactive was operated in a data-dependent acquisition mode acquiring MS survey scans at a resolution of 70,000 and a target value of 3×10^6 ions with a maximum fill time of 120 ms, followed by MS/MS scans of the 15 most abundant ions (top 15), using 5×10^4 ions as the target value and a maximum fill time of 250 ms. Spectra were acquired with a normalized collision energy of 27 and a dynamic exclusion of 12 s, selecting only precursor ions with charge states between +2 and +5 (*see* **Note 22**).

3.7 Data Analysis

1. Mass spectrometry data were searched against the human Uniprot database (August 07, 2012; 20,232 target sequences) using the Proteome Discoverer software.
2. In the Proteome Discoverer software raw data were processed using the Spectrum Selector node for precursor ion selection with default settings.
3. Peptide identification was performed using the Mascot search algorithm. Search parameters were as follows: trypsin cleavage specificity with a maximum of two missed cleavages, oxidation of methionine as well as phosphorylation of serine, threonine, and tyrosine as variable and carbamidomethylation of cysteine as fixed modification, precursor mass tolerance of 10 ppm, and fragment mass tolerance of 0.02 Da.
4. The Peptide Validator node was used for the assessment of the false discovery rate (FDR) on the peptide spectrum match (PSM) level by searching against a randomized decoy database.
5. Furthermore, the PhosphoRS node (version 3.1, [13]) was implemented in the workflow. For each PSM, this algorithm conducts a dedicated phosphorylation site localization and calculates probabilities for correct assignments. Here, only phosphopeptides with phosphoRS probabilities >75 % were considered (*see* **Note 23**).
6. The following filter criteria were applied to the processed dataset: high confidence corresponding to an FDR <1 % on the PSM level and peptide search engine rank of 1.

4 Notes

1. The cultivation of one 75 cm^2 cell culture flask of HeLa cells yields approximately 8×10^6 cells (grown to 70–80 % confluence), which corresponds to approximately 2–3 mg of protein.

Any other adherent or suspension cell line is also appropriate, though few parameters such as the peptide-to-bead ratio might need optimization due to the potentially different phosphorylation level of cell lines.

2. The addition of phosphatase inhibitors prevents the dephosphorylation of proteins by endogenous phosphatases during sample preparation, increasing yield, reliability, and reproducibility of the phosphopeptide analysis.
3. IAA is unstable and light sensitive; therefore, prepare the 1 M IAA stock freshly to minimize hydrolysis and keep in the dark.
4. As concentrated TFA is highly corrosive, wear protective clothing (lab coat, goggles, and gloves) and work under a chemical fume hood.
5. To prevent any overlap with endogenous phosphopeptides check sequence homology using protein BLAST (http://blast.ncbi.nlm.nih.gov).
6. As the loading step is essential for the success of the phosphopeptide enrichment, we recommend preparing the TiO_2 loading buffer freshly each time.
7. The volume of the lysis buffer has to be adapted to the size of the cell pellet. Use approximately 200 μL lysis buffer for the lysis of 2–3 mg of HeLa cells. If the cell lysate remains cloudy, add more lysis buffer until the solution is completely clear.
8. If the lysis buffer was stored at 4 °C, warm it prior to usage as SDS precipitates in the cold.
9. Prolong incubation times, if sample is still viscous. For large sample volumes carefully use microtip sonication (15 W output with two bursts of 30 s each) to disrupt DNA and cool the sample on ice during sonication.
10. For the determination of the protein concentration perform a colorimetric assay, such as BCA protein assay, Bradford, or Lowry. Accurate determination is important for subsequent phosphopeptide enrichment, as both the amount of trypsin as well as the amount of TiO_2 beads depend on the protein amount.
11. Though sequencing-grade modified trypsin retains part of its activity under denaturing conditions up to 2 M GuHCl (*see Promega Usage Information*), dilute the lysis buffer to a final concentration of 0.2 M GuHCl prior to protein digestion to ensure full enzyme activity and high cleavage specificity.
12. In contrast to C18 RP chromatography, the monolithic system enables the detection of peptides and proteins in a single LC run and is more robust against detergents. Compared to digestion controls using SDS-PAGE and silver staining, the monolithic separation is 5–10× more sensitive and faster.

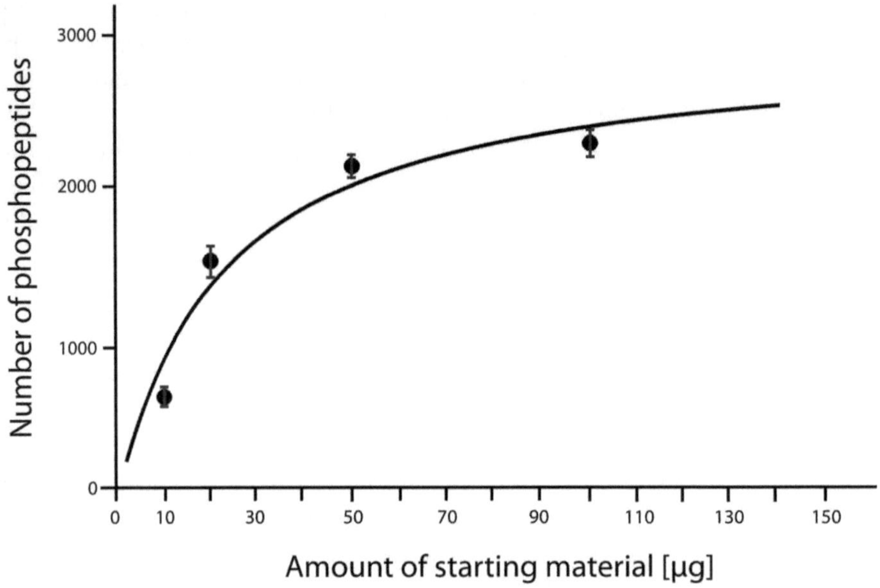

Fig. 5 Correlation between the amount of starting material and the number of identified phosphopeptides. With the described protocol already 70 different phosphopeptides (phosphoRS probability >75 %) can be identified from as little as 1 µg of starting material. Starting from 50 µg a maximum of 2,000 identified phosphopeptides is achieved, notably with only 140 min of LC-MS analysis time (enrichment workflow was conducted in triplicate). To increase the amount of identified phosphopeptides, it is necessary to either prolong the LC-MS gradient or fractionate the sample after the enrichment step (e.g., with HILIC [16])

In addition, the UV traces allow for a better comparison of digestion reproducibility [11].

13. Although the bead-to-peptide ratio depends on the complexity of the starting material, the described ratios have proven to be suitable for a broad variety of samples (*see* Fig. 5).
14. For complex samples a third incubation step is recommended, as approximately 5 % more phosphorylated peptides can be identified with this additional step.
15. Despite the removal of residual beads during the adjacent desalting step, try to transfer as few beads as possible using a GELoader tip.
16. To prevent loss of the phosphate moiety due to β-elimination under highly alkaline conditions [14], directly acidify the phosphopeptide sample after the elution step.
17. Lower identification rates using the batch mode approach are probably due to the loss of beads during the washing steps. The tiny amount of beads used for small sample amounts (e.g., 60 µg beads for 10 µg starting material) is often not visible as a solid pellet and thus unanticipated partial removal might occur.

18. Use a centrifugation tip adapter to fix the micro-column in an Eppendorf tube during centrifugation.
19. Rotate micro-column manually from time to time to allow an efficient elution of phosphopeptides from the beads.
20. Do not increase the volume of ACN in this step, as higher organic amounts will reduce the binding affinity of the peptides to the reversed-phase material during SPE.
21. Perform this step slowly and carefully, allowing the phosphopeptides to bind to the RP material.
22. This study was conducted using higher energy collision dissociation (HCD), yet collision-induced dissociation (CID) is also appropriate for phosphoproteomic studies, yet CID is prone to more profound neutral losses of phosphoric acid [15]. As fragmentation of phosphopeptides tends to generate MS/MS spectra of lower quality compared to non-phosphorylated peptides, an MS instrument with high sensitivity and high mass accuracy is recommended.
23. To select only phosphopeptides containing confident phosphorylation sites, export the processed peptide list from Proteome Discoverer to Microsoft Excel. In Excel, a ready-to-use macro, provided by the Mechtler group (http://ms.imp.ac.at), can be applied to the dataset, which determines the number of confident phosphorylation sites for each peptide dependent on a chosen minimum value for the phosphoRS probability score and furthermore provides the position of the phosphorylation sites within the protein sequences.

References

1. Matsuda H, Nakamura H, Nakajima T (1990) New ceramic titania—selective absorbent for organic phosphates. Anal Sci 6:911–912
2. Pinkse MW, Uitto PM, Hilhorst MJ et al (2004) Selective isolation at the femtomole level of phosphopeptides from proteolytic digests using 2D-NanoLC-ESI-MS/MS and titanium oxide precolumns. Anal Chem 76(14):3935–3943
3. Sano A, Nakamura H (2004) Chemo-affinity of titania for the column-switching HPLC analysis of phosphopeptides. Anal Sci 20(3): 565–566
4. Larsen MR, Thingholm TE, Jensen ON et al (2005) Highly selective enrichment of phosphorylated peptides from peptide mixtures using titanium dioxide microcolumns. Mol Cell Proteomics 4(7):873–886
5. Engholm-Keller K, Birck P, Storling J et al (2012) TiSH–a robust and sensitive global phosphoproteomics strategy employing a combination of TiO2, SIMAC, and HILIC. J Proteomics 75(18):5749–5761
6. Molina H, Horn DM, Tang N et al (2007) Global proteomic profiling of phosphopeptides using electron transfer dissociation tandem mass spectrometry. Proc Natl Acad Sci U S A 104(7):2199–2204
7. Wu J, Shakey Q, Liu W et al (2007) Global profiling of phosphopeptides by titania affinity enrichment. J Proteome Res 6(12): 4684–4689
8. Kettenbach AN, Gerber SA (2011) Rapid and reproducible single-stage phosphopeptide enrichment of complex peptide mixtures: application to general and phosphotyrosine-specific phosphoproteomics experiments. Anal Chem 83(20):7635–7644
9. Palmisano G, Parker BL, Engholm-Keller K et al (2012) A novel method for the simultaneous enrichment, identification, and quantification of phosphopeptides and sialylated glycopeptides applied to a temporal profile of mouse brain development. Mol Cell Proteomics 11(11):1191–1202

10. Thingholm TE, Jorgensen TJ, Jensen ON et al (2006) Highly selective enrichment of phosphorylated peptides using titanium dioxide. Nat Protoc 1(4):1929–1935
11. Burkhart JM, Schumbrutzki C, Wortelkamp S et al (2012) Systematic and quantitative comparison of digest efficiency and specificity reveals the impact of trypsin quality on MS-based proteomics. J Proteomics 75(4): 1454–1462
12. Beck F, Lewandrowski U, Wiltfang M et al (2011) The good, the bad, the ugly: validating the mass spectrometric analysis of modified peptides. Proteomics 11(6):1099–1109
13. Taus T, Kocher T, Pichler P et al (2011) Universal and confident phosphorylation site localization using phosphoRS. J Proteome Res 10(12):5354–5362
14. Meyer HE, Hoffmann-Posorske E, Korte H et al (1986) Sequence analysis of phosphoserine-containing peptides. Modification for picomolar sensitivity. FEBS Lett 204(1):61–66
15. Loroch S, Dickhut C, Zahedi RP et al (2013) Phosphoproteomics – more than meets the eye. Electrophoresis 34(11):1483–1492
16. Engholm-Keller K, Hansen TA, Palmisano G et al (2011) Multidimensional strategy for sensitive phosphoproteomics incorporating protein prefractionation combined with SIMAC, HILIC, and TiO(2) chromatography applied to proximal EGF signaling. J Proteome Res 10(12):5383–5397

Chapter 29

Quantifying Small Molecule-Induced Changes in Cellular Protein Expression and Posttranslational Modifications Using Isobaric Mass Tags

Isabelle Becher, Maria Fälth Savitski, and Marcus Bantscheff

Abstract

Proteomics enables the comprehensive analysis of cellular perturbations induced by bioactive small molecules and contributes to our understanding of the mechanisms by which drugs elicit their activity in disease situations. Here we describe a quantitative proteomics approach to study dose-dependent changes in protein expression and posttranslational protein modifications in human promyelocytic leukemia cells in response to inhibition of histone deacetylases by Vorinostat. The method employs isobaric mass tags (tandem mass tags, TMT) to enable the multiplexed quantitative analysis of up to six samples and antibodies directed against acetylated lysine residues for immunoenrichment of TMT-encoded acetylated peptides.

Key words Posttranslational modifications, Protein abundance, Isobaric labeling, Acetylome, Mass spectrometry

1 Introduction

Acetylation of lysine residues is a reversible posttranslational modification of proteins, and it is involved in diverse cellular processes, such as metabolism, chromatin remodeling, cell cycle, nuclear transport, and actin nucleation [1–4]. Acetylation of histone proteins was one of the first protein posttranslational modifications reported [5]. Recently, novel tools for immunoaffinity-based enrichment of acetylated proteins and peptides have enabled the discovery that acetylation is a major posttranslational modification in the human proteome [4, 6]. Histone deacetylases (HDACs) enzymatically catalyze the removal of acetyl groups from lysine residues of histone and nonhistone proteins. HDACs are grouped into different classes, where class I, IIa, IIb, and IV are Zn^{2+}-dependent enzymes and class III HDACs, commonly referred to as sirtuins, require NAD^+. With the exception of HDAC8, class I HDACs are organized in megadalton co-repressor complexes [7].

Daniel Martins-de-Souza (ed.), *Shotgun Proteomics: Methods and Protocols*, Methods in Molecular Biology, vol. 1156, DOI 10.1007/978-1-4939-0685-7_29, © Springer Science+Business Media New York 2014

Vorinostat (suberoylanilide hydroxamic acid, SAHA) is a rather unselective HDAC inhibitor and an approved drug for treatment of cutaneous T-cell lymphoma (CTCL, [8]). The exact mechanism by which the drug exerts its anticancer activity is unknown. Proteomics can contribute to our understanding of drug activity as it enables the study of compound-induced changes of protein abundance and posttranslational modifications. For example, a recent study indicated that out of more than 3,000 acetylation sites identified in the human proteome only approximately 10 % were shown to be regulated upon treatment with Vorinostat [4]. Such data can also give guidance for the development of more class and isoform-selective HDAC inhibitors [9–12], which are believed to exhibit fewer side effects.

Recent advances in instrumentation [13] and chromatographic strategies [14] enabled the characterization of the human proteome at unprecedented analytical depth. In recent studies up to 10,000 proteins were identified in a human cancer cell line [15, 16] representing half of the protein-encoding genes [17] and an even higher fraction of the actually transcribed genes in this cell line. For simpler eukaryotes such as yeast full proteome coverage has been achieved in single-run analysis omitting extensive, time-consuming fractionation [18]. Protein activity and signal transduction within biological pathways are regulated by dynamic posttranslational modification of proteins further adding to the complexity of the proteome. In recent years, several tools have been made available enabling the comprehensive enrichment of peptides carrying modifications such as phosphorylation and acetylation [6, 19]. For example acetyl-lysine-directed antibodies facilitated the detection and quantification of 3,600 acetylation sites on 1,750 proteins after on-cell treatment with two HDAC inhibitors [4]. Further, a robust quantitative dimension has been added to proteomics workflows allowing researchers to study dynamic cellular processes. These methods are typically either based on stable isotope labeling or rely on signal intensities in the mass spectrometer or surrogate measures thereof to infer on protein abundance (label-free methods) [20]. Popular stable isotope-based methods include metabolic labeling with stable isotope-encoded amino acids (SILAC [21]), chemical approaches such as dimethyl labeling [22], and isobaric mass tags (tandem mass tags, TMT [23], iTRAQ [24]). In contrast to the majority of quantitative proteomics approaches, isobaric mass tags do not enable relative quantification of intact peptide ion signals but quantification is based on reporter ion signals generated upon fragmentation of peptides labeled with different stable isotope-encoded variants of the reagents. The method can be applied on any cell line or tissue sample without requiring major adaptation of experimental procedures and allows the multiplexed analysis of up to ten samples in a single analysis [25, 26].

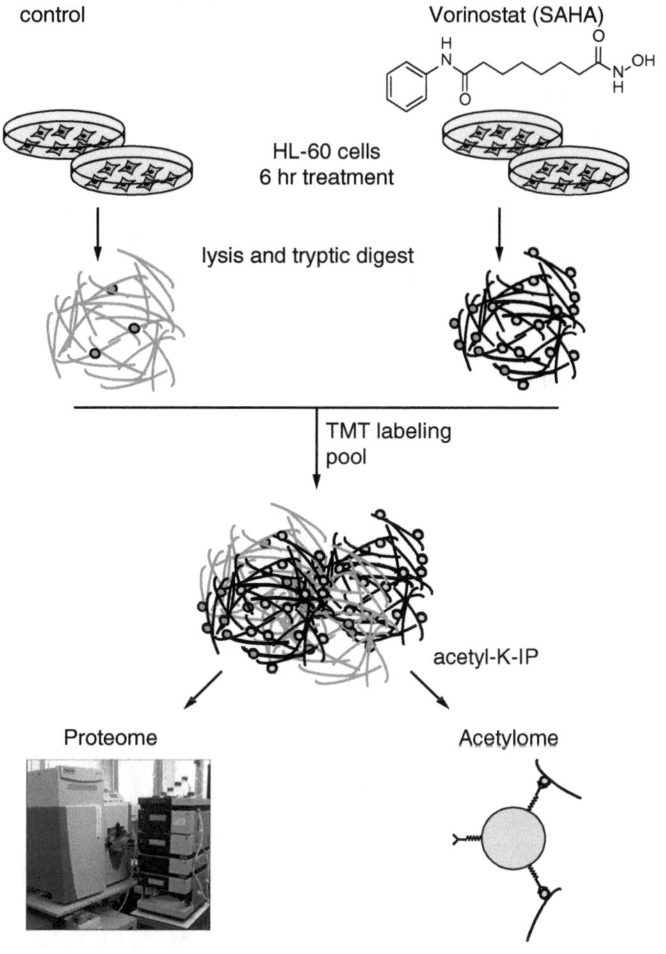

Fig. 1 Experimental scheme: HL-60 cells are treated for 6 h either with vehicle (DMSO) or 5 μM Vorinostat. Proteins are extracted using high-salt urea lysis and tryptic digested, and peptides are labeled with tandem mass tags (TMT). After pooling, peptides are incubated with acetyl-lysine antibody beads, and bound peptides are eluted and subjected to LC-MS/MS analysis. An aliquot of the non-bound fraction is used for analysis of protein abundances

Here we present an isobaric tag-based protocol for measuring changes in proteome abundance and global acetylation in response to cellular treatment with an HDAC inhibitor. Cells were incubated with Vorinostat for 6 h, and extracted proteins were digested with trypsin. Peptide pools were labeled with TMT followed by antibody enrichment of acetylated lysine-containing peptides. LC-MS/MS of the non-bound fraction of this enrichment step allowed to study compound-induced changes in protein abundance (Fig. 1), whereas analysis of the acetylated peptide pool enabled identification of Vorinostat treatment-induced acetylation sites in a dose-dependent fashion. For all steps a peptide pre-fractionation was performed using a high-pH reversed-phase chromatography approach to increase coverage of acetylome as well as proteome [27].

2 Materials

2.1 Cell Culture and Treatment

1. Cell culture medium with supplements as required: For example HL-60 cells.
2. Phosphate-buffered saline (PBS).
3. Dissolve small molecule in DMSO in stock concentration, store in aliquots at −20 °C, avoid thaw–freeze, e.g., Vorinostat (SAHA, Selleckchem). For a selection of different HDACi and their selectivity consider reference 7.

2.2 Cell Lysis and Protein Digestion, Isobaric Labeling

1. Urea lysis buffer: 20 mM 4-(2-hydroxyethyl)-1-piperazineethanesulfonic acid (HEPES), 1 mM β-glycerophosphate, 2.5 mM sodium pyrophosphate, 8 M urea. Optional: 1 mM sodium vanadate, 25 mM sodium fluoride (NaF), 400 nM trichostatin A (TSA, supplier Selleckchem), 10 mM nicotinamide.
2. Trypsin, mass spectrometry grade
3. Ultrasonic homogenizer, ice.
4. 1 M Dithiothreitol (DTT), 1 M iodoacetamide (IAA), aqueous stock solutions stored at −20 °C.
5. Gel electrophoresis equipment, precast NuPAGE 4–12 % Bis-Tris gel (Invitrogen).
6. Bradford protein assay.
7. C18 cleanup columns, 10 % aqueous trifluoroacetic acid (TFA) solution; 0.1 % aqueous TFA solution; solution containing 0.1 % TFA, 2 % ACN in water; solution containing 0.1 % TFA, 50 % ACN, 49.9 % water.
8. Isobaric labeling reagent, TMT reagents, e.g., TMT6plex (Thermo Fisher Scientific).
9. 5 % (v/v) aqueous hydroxylamine solution.

2.3 Enrichment of Modified Peptides

1. Anti-acetyl-lysine antibody beads (Immunechem).
2. Immunoprecipitation (IP) buffer: 50 mM 3-(*N*-morpholino) propanesulfonic acid (MOPS), 10 mM sodium phosphate, 50 mM sodium chloride (NaCl), pH 7.2.
3. Microcentrifuge.
4. Elution buffer: 0.1 % aqueous TFA solution.

2.4 Chromatographic Pre-fractionation of Peptides

1. Bioinert HPLC or UPLC system with fraction collector.
2. Solvents for fractionation: A: 1.25 % ammonia solution in water, B: 70 % ACN, 1.25 % ammonia solution.
3. Base-stable C18 column 15 cm × 1 mm (e.g., Waters X-Bridge, 3.5 μm).
4. Speedvac.

2.5 LC-MS/MS Analysis

1. Nano-flow HPLC system, pre-column and separation column.
2. ESI source.
3. Mass spectrometer: Fast scanning, e.g., Q Exactive or Orbitrap Elite (Thermo Fisher Scientific).
4. Buffer A: 0.1 % formic acid in water.
5. Buffer B: 70 % ACN/29.9 % water/0.1 % formic acid.
6. Reversed-phase column 50 cm long, 75 μm inner diameter.

2.6 Data Processing and Analysis

1. Mascot (Matrix Science) was used as a search engine for identification of peptides and proteins. Recommended search parameter settings include the variable modifications: TMT on lysine and N-terminus of peptide, acetylation on lysine and N-terminus of protein, and oxidation of methionine; and fixed modification: carbamidomethylation of cysteines (due to alkylation after reduction with DTT).
2. The statistical programming language R has been used for data analysis and visualizations [http://www.r-project.org/].

3 Methods

3.1 Cell Culture and Cell Treatment

1. Grow HL-60 cells in a spinner flask to 80 % confluence; reduce FCS to 1 % before incubation with inhibitor.
2. Add 5 μM Vorinostat to 2×10^8 HL-60 cells (e.g., 20 ml cell culture), to control treatment add same volume of DMSO (max. 0.2 %) to control flask, incubate for 6 h, and perform treatment with $n = 2$; therefore, in total four samples are generated (*see* **Note 1**).
3. After incubation time wash cells with cold PBS and either freeze the cell pellet in liquid nitrogen and store at −80 °C or continue with cell lysis.

3.2 Cell Lysis and Protein Digestion

1. Prepare urea lysis buffer (*see* **Notes 2** and **3**).
2. Add 2 ml of lysis buffer to fresh/frozen cell pellet.
3. Pipette slurry up and down to resuspend pellet (*see* **Note 4**).
4. Sonicate on ice with three bursts of 10 s each, and cool sample between bursts.
5. Clear the supernatant by centrifugation in an ultracentrifuge with 140,000 × *g* at 4 °C. Transfer supernatant to fresh 50 mL conical tube.
6. Protein reduction and alkylation: Add 4.5 mM DTT to lysate, incubate for 45 min at room temperature, add 10 mM IAA to lysate, and incubate for 30 min at room temperature in the dark (*see* **Note 5**).

7. Dilute urea to 2 M end concentration with 20 mM HEPES buffer before protein assay and digestion.
8. Determine protein concentration using Bradford assay: usually 1–2 mg/ml could be expected (*see* **Notes 6** and 7).
9. Trypsin is dissolved in HEPES buffer (1 mg/mL), and 20 μg is added to 1 mg of lysate. Proteolysis is carried out overnight at ambient temperature (*see* **Note 8**).
10. Dissolve labeling reagent in ACN according to the manufacturer's recommendation, and add 2.5 mg of labeling reagent to 1 mg of digested protein. Incubate samples for 1 h at room temperature (*see* **Note 9**). Stop excess reagent by adding 5 % hydroxylamine and incubate for 15 min. Samples with different labels are combined.
11. To remove excess label reagent and urea a C18 cleanup of the pooled peptides is performed to enable proper antibody enrichment of acetylated peptides. Therefore peptides are acidified with a 10 % TFA stock solution to a final concentration of 1 % TFA and incubated for 30 min. Sample is centrifuged, and supernatant is transferred to new tube. C18 columns are pre-wet with ACN and equilibrated with 0.2 % ACN/0.1 % TFA. Add sample and wash column with 0.1 % TFA and 0.1 % TFA in 2 % ACN. Peptides are eluted with 50 % ACN and 1 % TFA.
12. Eluted peptides are lyophilized.

3.3 Enrichment of Modified Peptides

1. Dissolve 6 mg of pooled peptides in 1 mL of IP buffer.
2. Wash 50 μL anti-acetyl antibody beads three times with 500 μL IP buffer.
3. Add peptide sample to antibody beads and incubate overnight at 4 °C.
4. Settle antibody beads at 500×*g* in a microcentrifuge. Collect supernatant, and wash beads two times with IP buffer and two times with H_2O, before elution buffer is added. Peptides are eluted from beads with two volumes of 0.1 % TFA.
5. Eluted peptides are dried in vacuo and resuspended in 0.1 % formic acid. 5–10 % of the sample is measured in advance; the rest of the sample is used for further fractionation to improve acetylome coverage.

3.4 Pre-fractionation of Peptides Using Basic RP-LC (for Acetylome as Well as Proteome Analysis)

For protein expression analysis 100–150 μg of the non-bound fraction of the acetyl-lysine immunoprecipitation experiment is fractionated using high-pH reversed-phase liquid chromatography. Also, the antibody-enriched acetyl-peptides are fractionated before LC-MS/MS analysis to increase the identification of acetylation sites.

Peptides are separated using a 120-min gradient, with solvent B increasing from 3 to 60 % within 71 min. Fractions are collected every 120 s (or 60 s if less fractions are pooled) in a meandering scheme. Flow rate is 40 μl/min, and UV detection is performed at 214 nm.

3.5 LC-MS/MS Analysis

The following instrument settings are described for the currently very common Q Exactive. Similar instrumental setups are also suited.

1. The sample is injected into a nano-LC system coupled to Q Exactive mass spectrometer (Thermo). Peptides are separated on a custom-made 50 cm × 75 μM (inner diameter) reversed-phase columns at 40 °C. Gradient elution is performed from 2 % acetonitrile to 40 % acetonitrile in 0.1 % formic acid over 2–4.5 h at a flow rate of 350 μl/min.
2. Intact peptides are detected at 70,000 resolution, with a target value of 3e6. For MS2 resolution is set to 35,000 and precursors are fragmented using 33 % normalized collision energy. Internal calibration was performed using the ion signal from $(Si(CH_3)_2O)_6H)^+$ at m/z 445.1200252.

3.6 Quantitative Analysis of Acetylated Peptides

1. For identification of peptides the following settings for the search engine (Mascot) are used: 4 ppm mass tolerance for peptide precursor, 0.02 Da for MS/MS fragment ions, and maximum three missed cleavages (includes "miss-cleaved" acetylated lysines). Modifications include TMT on N-terminus and lysine, acetylation on lysine, oxidation on methionine, and carbamidomethylation on cysteines (fixed).
2. Spectra acquired in HCD often contain the acetyl-lysine immonium ion (m/z 126.09) which can be used to verify identified acetylated peptides.
3. Acetylation of lysines must be localized internally in the peptide sequence.
4. H-Score is calculated for PTM data [28]. The H-score is based on Mascot identification and further assignment of fragments to peptide and its posttranslational modification, where each fragment increases the H-score.
5. Thresholds applied: Mascot peptide score >10, H-score >5, S2I (signal-to-interference ratio, [29]) >0.5, and P2T (precursor abundance-to-threshold ratio, [29]) >4. Spectra not fulfilling these criteria are filtered out from further analysis.
6. All spectra matching the same peptide sequence and modification are aggregated, and the number of spectra for each such pair is recorded. The aggregation is done by summing up the sum ion area of the reporter ions in the spectra. The sum ion area is defined as reporter ion intensity multiplied by the acquisition time of the tandem mass spectrum.
7. For each peptide the fold change is calculated by dividing the aggregated sum ion area of the SAHA-treated sample with the

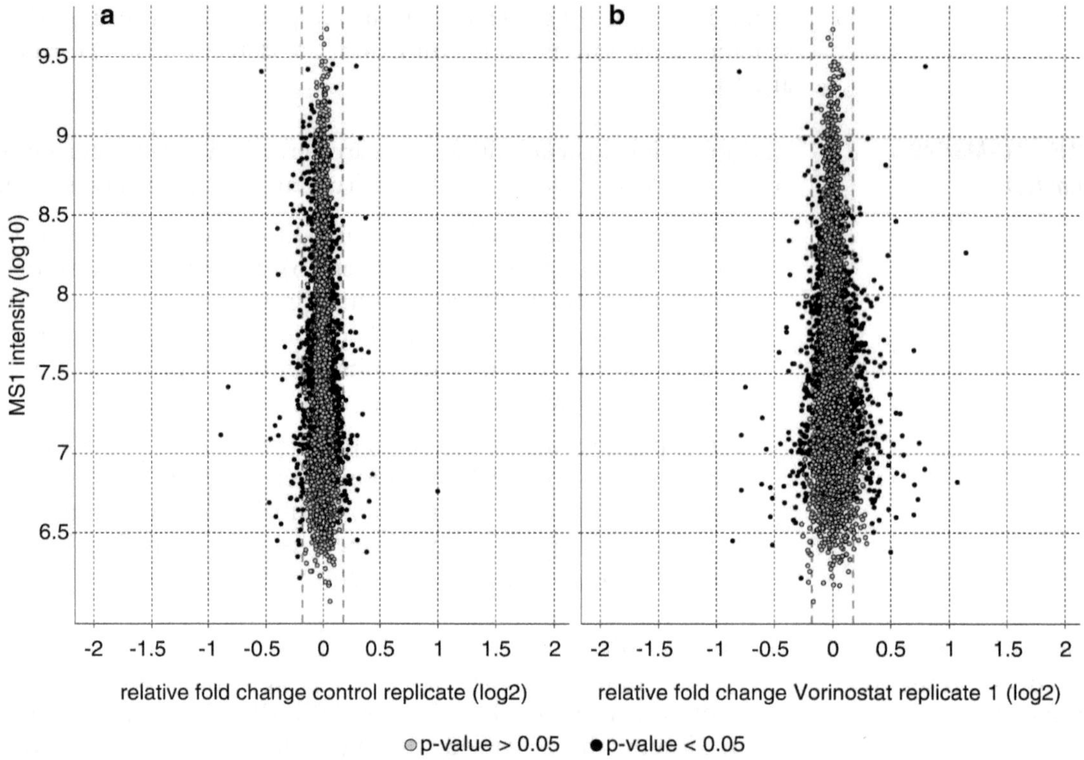

Fig. 2 Distribution of relative fold changes (relative to control and log2 transformed) of (**a**) a replicate of vehicle-treated samples and (**b**) Vorinostat treatment (replicate 1), plotted against MS1 intensity (*y*-axis, log10 transformed, based on three most abundant peptides). *Dashed vertical lines* indicate 20 % deviation. *Black circles* represent proteins with a *p*-value below 0.05

DMSO-treated control sample. The fold change is then log2 transformed.

8. Calculation of *p*-values: Peptides are binned according to the number of peptide-to-sequence matches obtained; each bin should contain at least 300 peptides. For each bin the robust standard deviation is calculated by using the 15.87 and 84.13 percentiles. Using the fold changes of the peptides in the bin a normal distribution is calculated with the robust standard deviations. A *p*-value for each peptide fold change in the bin is calculated using the normal distribution [30]. To control the false discovery rate *p*-values are corrected for multiple testing by the Benjamini–Hochberg method [31].

9. A peptide is considered significantly regulated if the multiple testing-adjusted *p*-value is smaller or equal to 0.05. A fold change cutoff can also be applied. The fold change threshold depends on data quality and needs to be adjusted accordingly. For a tight distribution, a fold change of even 20 % could be significant (Figs. 2 and 3).

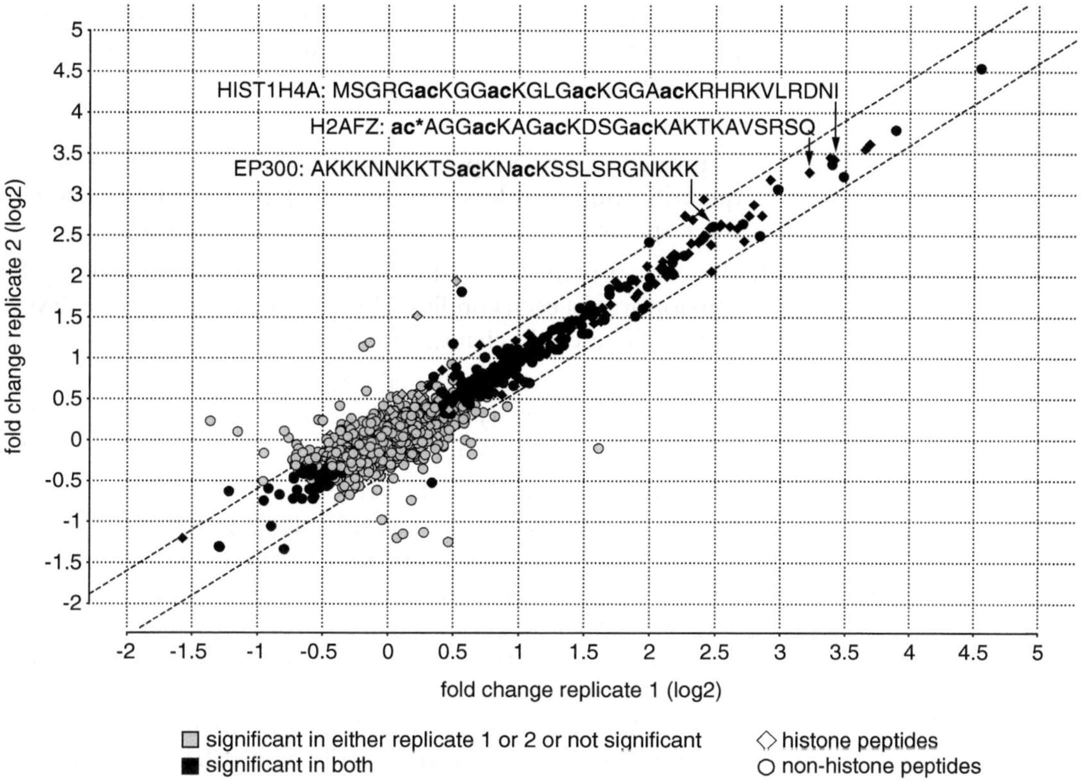

Fig. 3 Reproducibility of treatment with 5 μM Vorinostat: Replicate 1 is plotted against replicate 2, fold changes were calculated versus vehicle control and log2 transformed. *Dashed lines* indicate 50 % deviation. *Light grey shapes* represent acetylation site without significance or being only significant for one of both replicates. *Black shapes* represent acetylation sites that are significant in both replicates. *Rhombs* indicate histone peptides, *circles* indicate nonhistone peptides. Three example peptides are labeled, for which strong increase (more than eightfold) was observed upon Vorinostat treatment: EP300, histone H2A, and histone H4

10. The analysis of the proteome should also be considered for PTM analysis. Changes in PTM level could also be false positives as a strong protein change also increases or decreases the PTM level of this protein. Hence, the combination of both protein and PTM data should lead to the identification of true effects.

3.7 Quantitative Analysis of Protein Expression Changes

From the non-bound fraction of the acetyl-lysine enrichment ~100–150 μg of peptides are used for further protein expression analysis. The sample is fractionated using basic RPLC into 96 fractions which are then pooled to 17 fractions; recently this procedure was described in detail [32]. These are analyzed for protein abundance changes due to (1) changes in protein expression or (2) differences in sample handling before pooling (*see* **Note 10**).

1. For identification of proteins the following settings for the search engine (Mascot) are used: 4 ppm mass tolerance for peptide precursor, 0.02 Da for MS/MS fragment ions, and

maximum three missed cleavages. Modifications include TMT label on N-terminus (fixed) and lysine (variable), oxidation on methionine (variable), and carbamidomethylation on cysteines (fixed).

2. A proper identified protein required the identification and quantification of at least two unique peptides, based on at least three spectra.
3. For each protein the fold change is calculated by dividing the sum ion area of each condition by the sum ion area of the DMSO control. These fold changes are then log2 transformed.
4. For each protein the MS1 abundance is calculated, which is based on the intensity of the three most abundant (top3) peptides of the protein [33, 34].
5. The statistical analysis of the protein expression data is done in the same way as for the peptide data. Proteins are binned according to the number of spectra used for quantification, with at least 300 proteins per bin. For each bin the robust standard deviation is calculated by using the 15.87 and 84.13 percentiles. Using the fold changes of the proteins in the bin a normal distribution is calculated with the robust standard deviations. A p-value for each protein fold change in the bin is calculated using the normal distribution [30]. To control the false discovery rate p-values are corrected for multiple testing by the Benjamini–Hochberg method [31].

4 Notes

1. Cell number to be used: We suggest using 2×10^8 cells for cell treatment and lysis, which will yield approximately 5 mg protein. This amount is enough to perform several experiments. We typically use 1 mg of this sample for each technical replicate.
2. Urea buffer: Only use freshly prepared urea buffer, and do not heat sample to prevent modification of lysines. To control this, a variable modification can be defined for the database search engine to check for artificial lysine and N-terminal carbamylation.
3. Protease inhibitors and HDAC inhibitors in buffer: Cells are lysed in 8 M urea. This high concentration of denaturizing agent is supposed to also inactivate proteases. Additionally the lysis is followed by a proteolytic digestion which would also be inhibited by serine protease inhibitors in the lysate buffer. It should be considered to add phosphatase inhibitors as well as inhibitors preventing enzymes to remove modification of interest, e.g.,

1 mM sodium vanadate, 40 nM TSA, and 10 mM nicotinamide to inhibit HDAC- and sirtuin-dependent deacetylation.

4. If pipetting is difficult due to high viscosity of the sample continue with sonication.
5. Do not heat the sample for reduction with DTT to prevent carbamylation of lysine residues.
6. It is possible to adjust protein concentration in all samples to same concentration. This would have the advantage that all samples are handled with the same volume for the following steps, which makes reproducibility easier (which is required for quantitative analysis).
7. Western Blot analysis: After cell treatment/cell lysis it is recommended to perform a Western Blot analysis to check for changes in protein abundance and/or changes in modification levels of samples. This serves as control for proper cell treatment before extensive analysis of data is done. Here, a pan-acetyl-lysine antibody (e.g., Immunechem) could be used.
8. Before proceeding with the protocol the achieved digestion efficiency should be checked. 25 μg of input (lysate) and the same amount of digest are loaded on SDS-PAGE to check protein degradation with Coomassie Blue staining. Additionally a control by mass spectrometry could be performed to check reproducibility of the digestion between samples (e.g., missed cleavages, carbamylation of K- and N-terminus). If digestion efficiency is too low (e.g., <70 % of complete digested peptides) fresh trypsin should be added and another incubation step should be carried out.
9. Before label reagent is stopped and samples are pooled it is recommended to check labeling efficiency by mass spectrometry. Therefore, 1 μg of labeled peptides is measured in a single LC-MS/MS run. TMT(K) and TMT(N-term) are searched as variable modifications. The percentage of completely labeled peptides should be above 90 % to proceed with sample pooling. If labeling efficiency is too low, a re-labeling should be performed by adding fresh label reagent.
10. Sequential analysis of samples: It is possible to re-use NBF for, e.g., phospho-enrichment or ubiquitin analysis (but additional requirements for cell lysis).

Acknowledgements

We would like to thank Stephanie Melchert, Thilo Werner, Birgit Duempelfeld, and Toby Mathieson for contributions to the development of the presented protocol and Frank Weisbrodt for help with the figures.

References

1. Johnson ES, Kornbluth S (2012) Life, death, and the metabolically controlled protein acetylome. Curr Opin Cell Biol 24(6):876–880
2. Yang T et al (2012) Acetylation of histone deacetylase 1 regulates NuRD corepressor complex activity. J Biol Chem 287(48): 40279–40291
3. Liu Y et al (2012) Modulation of histone deacetylase 6 nuclear import and tubulin deacetylase activity through acetylation. J Biol Chem 287(34):29168–29174
4. Choudhary C et al (2009) Lysine acetylation targets protein complexes and co-regulates major cellular functions. Science 325(5942):834–840
5. Phillips DM (1963) The presence of acetyl groups of histones. Biochem J 87:258–263
6. Kim SC et al (2006) Substrate and functional diversity of lysine acetylation revealed by a proteomics survey. Mol Cell 23(4):607–618
7. Bantscheff M et al (2011) Chemoproteomics profiling of HDAC inhibitors reveals selective targeting of HDAC complexes. Nat Biotechnol 29(3):255–265
8. Marks PA, Breslow R (2007) Dimethyl sulfoxide to vorinostat: development of this histone deacetylase inhibitor as an anticancer drug. Nat Biotechnol 25(1):84–90
9. Balasubramanian S et al (2008) A novel histone deacetylase 8 (HDAC8)-specific inhibitor PCI-34051 induces apoptosis in T-cell lymphomas. Leukemia 22(5):1026–1034
10. Butler KV et al (2010) Rational design and simple chemistry yield a superior, neuroprotective HDAC6 inhibitor, tubastatin A. J Am Chem Soc 132(31):10842–10846
11. Olson DE et al (2013) Discovery of the first histone deacetylase 6/8 dual inhibitors. J Med Chem 56(11):4816–4820
12. Methot JL et al (2008) Exploration of the internal cavity of histone deacetylase (HDAC) with selective HDAC1/HDAC2 inhibitors (SHI-1:2). Bioorg Med Chem Lett 18(3):973–978
13. Kelstrup CD et al (2012) Optimized fast and sensitive acquisition methods for shotgun proteomics on a quadrupole orbitrap mass spectrometer. J Proteome Res 11(6):3487–3497
14. Angel TE et al (2012) Mass spectrometry-based proteomics: existing capabilities and future directions. Chem Soc Rev 41(10):3912–3928
15. Nagaraj N et al (2011) Deep proteome and transcriptome mapping of a human cancer cell line. Mol Syst Biol 7:548
16. Beck M et al (2011) The quantitative proteome of a human cell line. Mol Syst Biol 7:549
17. International Human Genome Sequencing Consortium (2004) Finishing the euchromatic sequence of the human genome. Nature 431(7011):931–945
18. Mann M et al (2013) The coming age of complete, accurate, and ubiquitous proteomes. Mol Cell 49(4):583–590
19. Olsen JV et al (2006) Global, in vivo, and site-specific phosphorylation dynamics in signaling networks. Cell 127(3):635–648
20. Bantscheff M et al (2012) Quantitative mass spectrometry in proteomics: critical review update from 2007 to the present. Anal Bioanal Chem 404(4):939–965
21. Ong SE et al (2002) Stable isotope labeling by amino acids in cell culture, SILAC, as a simple and accurate approach to expression proteomics. Mol Cell Proteomics 1(5):376–386
22. Boersema PJ et al (2009) Multiplex peptide stable isotope dimethyl labeling for quantitative proteomics. Nat Protoc 4(4):484–494
23. Thompson A et al (2003) Tandem mass tags: a novel quantification strategy for comparative analysis of complex protein mixtures by MS/MS. Anal Chem 75(8):1895–1904
24. Ross PL et al (2004) Multiplexed protein quantitation in Saccharomyces cerevisiae using amine-reactive isobaric tagging reagents. Mol Cell Proteomics 3(12):1154–1169
25. McAlister GC et al (2012) Increasing the multiplexing capacity of TMTs using reporter ion isotopologues with isobaric masses. Anal Chem 84(17):7469–7478
26. Werner T et al (2012) High-resolution enabled TMT 8-plexing. Anal Chem 84(16):7188–7194
27. Kruse U et al (2011) Chemoproteomics-based kinome profiling and target deconvolution of clinical multi-kinase inhibitors in primary chronic lymphocytic leukemia cells. Leukemia 25(1):89–100
28. Savitski MM et al (2010) H-score, a mass accuracy driven rescoring approach for improved peptide identification in modification rich samples. J Proteome Res 9(11):5511–5516
29. Savitski MM et al (2011) Delayed fragmentation and optimized isolation width settings for improvement of protein identification and accuracy of isobaric mass tag quantification on Orbitrap-type mass spectrometers. Anal Chem 83(23):8959–8967
30. Cox J, Mann M (2008) MaxQuant enables high peptide identification rates, individualized p.p.b.-range mass accuracies and proteome-wide protein quantification. Nat Biotechnol 26(12):1367–1372

31. Benjamini Y, Hochberg Y (1995) Controlling the false discovery rate: a practical and powerful approach to multiple testing. J R Stat Soc Ser B Methodol 57:289–300
32. Mertins P et al (2013) Integrated proteomic analysis of post-translational modifications by serial enrichment. Nat Methods 10(7): 634–637
33. Grossmann J et al (2010) Implementation and evaluation of relative and absolute quantification in shotgun proteomics with label-free methods. J Proteomics 73(9):1740–1746
34. Silva JC et al (2006) Absolute quantification of proteins by LCMSE: a virtue of parallel MS acquisition. Mol Cell Proteomics 5(1): 144–156

Part X

Mass Spectrometry for Protein Analysis

Chapter 30

Analysis of Protein Structure by Cross-Linking Combined with Mass Spectrometry

Evgeniy V. Petrotchenko, Karl A.T. Makepeace, Jason J. Serpa, and Christoph H. Borchers

Abstract

Cross-linking combined with mass spectrometry is a powerful technique to study protein structure. Here, we present an optimized protocol for the preparation, processing, and analysis of a protein sample cross-linked with isotopically coded, affinity-enrichable, and CID-cleavable cross-linker CyanurBiotinDimercapto PropionylSuccinimide using LC/ESI-MS/MS on a Thermo Scientific Orbitrap mass spectrometer.

Key words Chemical cross-linking, Structural proteomics, Isotope-labeled

1 Introduction

Structural proteomics approaches are valuable tools for studying proteins, particularly in cases where the exact mechanisms of protein conformational changes or the structures of proteins and protein complexes cannot be elucidated by traditional structural biology techniques such as X-ray crystallography or NMR methods. Structural proteomics, which is a combination of protein chemistry methods (such as limited proteolysis, surface modification, hydrogen/deuterium exchange, chemical cross-linking, and affinity labeling) with mass spectrometry, can provide novel insights into the three-dimensional structures of proteins and protein complexes.

Cross-linking combined with mass spectrometry is being gradually accepted as a technique which enables detailed structural characterization of proteins and protein complexes [1–6]. Intrinsic challenges of the method, which have prevented its widespread use, are being progressively addressed by improvements in the capabilities of mass spectrometric instrumentation, by the development of new cross-linking reagents and techniques, and by the development of specialized software tools for processing mass spectrometric cross-linking data.

Daniel Martins-de-Souza (ed.), *Shotgun Proteomics: Methods and Protocols*, Methods in Molecular Biology, vol. 1156, DOI 10.1007/978-1-4939-0685-7_30, © Springer Science+Business Media New York 2014

The current experimental paradigm in structural cross-linking analysis is to enzymatically digest or otherwise fragment cross-linked proteins and then to identify the cross-linked peptides (cross-links) by mass spectrometry, thus determining the sites of cross-linking and providing information about protein conformation and protein–protein interaction sites. Unfortunately, inter-peptide cross-links, which provide information on pairs of cross-linked residues, are minor constituents in the complex digestion peptide mixture and need to be enriched for confident detection and identification. This can be accomplished by gel filtration [7], by strong cation-exchange chromatography [8], or by using affinity-tagged cross-linking reagents [9–16]. Unequivocal detection of the cross-links can be achieved by isotopic coding of the cross-linking reagents [17]. Another complication in the analysis of the inter-peptide cross-links is their combinatorial nature. To overcome these hurdles, cleavable cross-linking reagents can be employed [18–21].

Here, we describe an application of our isotopically coded, affinity-tagged, and CID-cleavable cross-linker CyanurBiotin DimercaptoPropionylSuccinimide (CBDPS) (Fig. 1) for protein cross-linking analysis using LC/ESI-MS/MS on an Orbitrap mass spectrometer. This combination of reagent and high-resolution instrument enables confident detection and identification of inter-peptide cross-links [22].

A typical structural cross-linking experiment consists of several steps: preliminary cross-linking titration, protein cross-linking reaction, optional separation of the cross-linked products, digestion of the cross-linked proteins, and enrichment, followed by LC/ESI-MS/MS analysis of the cross-linked peptides, which includes data acquisition, analysis, and interpretation. Each of these steps is described in detail below.

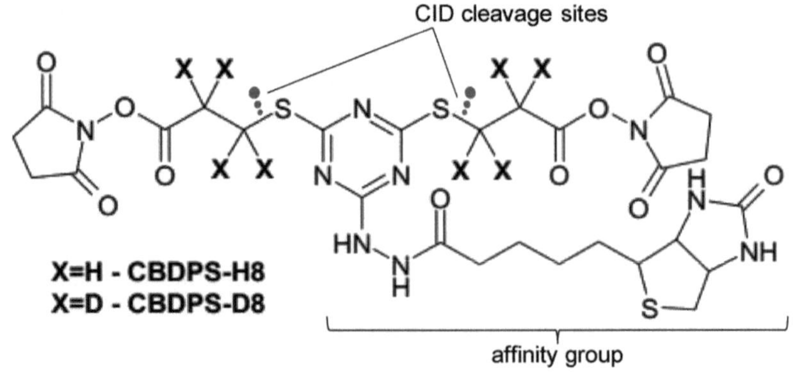

Fig. 1 Structure of CBDPS-H8/D8. The *dotted arrows* indicate CID cleavage sites. The *bracket* denotes the affinity biotin group

This combination of an advanced cross-linking reagent, selective cross-link purification strategy, optimized mass spectrometric techniques, and software tailored to specifically analyze and assign inter-peptide cross-links creates a powerful analytical toolkit for structural proteomics cross-linking applications.

2 Materials

Prepare all solutions using HPLC-grade water and analytical grade reagents. Prepare and store all reagents at room temperature (unless otherwise indicated).

Please note that other reagents and suppliers may be substituted and may give equally good results, but we have had no experience with their use. In each case where a specific manufacturer is given, please recognize that the phrase "or the equivalent" is implied.

2.1 Cross-Linking Reaction Materials

1. Protein sample dissolved in PBS pH 7.4 or a similar LC/MS-compatible buffer that does not contain primary amines (*see* **Note 1**).
2. Sodium phosphate dibasic (Na_2HPO_4), 0.2 M aqueous solution.
3. CBDPS-H8/D8, 1:1 molar ratio mixture of CBDPS-H8 and CBDPS-D8 (Creative Molecules, Inc).
4. Dimethyl sulfoxide (DMSO), ACS reagent grade, ≥99.9 %.
5. Ammonium bicarbonate (NH_4HCO_3) 100 mM, pH 8.4.
6. Proteinase K solution, 20 mg/mL.
7. Trypsin sequencing grade, 0.5 mg/mL.
8. 4-(2-Aminoethyl)benzenesulfonyl fluoride hydrochloride (AEBSF).
9. Compact Reaction Columns (CRC) (Affymetrix).
10. CRC Filters, Lower filter with 35 μm pore size (Affymetrix).
11. Pierce® Monomeric Avidin Agarose, capacity: >1.2 mg biotinylated BSA/ml resin, supplied as a 50 % aqueous slurry, with 0.02 % sodium azide (Thermo Scientific).
12. Ammonium acetate ($NH_4CH_3CO_2$), 100 and 500 mM solutions.
13. Trifluoroacetic acid, Formic acid, LC/MS grade, 0.1 % solution (v:v) in HPLC-grade water.
14. 50:50 acetonitrile:water (v:v), both HPLC grade, containing 0.1 % trifluoroacetic acid, eluant additive for LC/MS grade.
15. Formic acid, Formic acid, LC/MS grade, 0.1 % solution (v:v) in HPLC-grade water.
16. Bond-Breaker TCEP Solution, Neutral pH (Thermo Scientific).

2.2 Reversed-Phase High-Performance Liquid Chromatography and Mass Spectrometry Instrumentation and Materials

1. HPLC instrument, such as an Easy-nLC II (Thermo Fisher Scientific).
2. Trapping column: 100 μm ID, 360 μm OD trap column packed with Magic C18AQ (Bruker-Michrom), 100 Å, 5 μm pore size (prepared in-house).
3. Analytical column: 75 μm ID, 360 μm OD analytical column packed (in-house) with Magic C18AQ, 100 Å, 5 μm pore size with IntegraFrit (New Objective).
4. Acetonitrile, HPLC grade.
5. Water, HPLC grade.
6. Formic acid, eluent additive for LC/MS.
7. Mobile phases: Phase A: 0.1 % (v/v) formic acid and 2 % (v/v) acetonitrile in an aqueous water solution, and phase B: 0.1 % (v/v) formic acid in acetonitrile.
8. PicoTip™ Emitter SilicaTip™ (FS360-20-10-N-20-6.35CT) (New Objective).
9. Axygen® 96-Well Flat Top Polypropylene PCR Microplate, No Skirt, Clear, Nonsterile (Tewksbury, MA).
10. Axygen® AxyMat™ PCR Microplate Sealing Mat for 96-Well Plates (Tewksbury, MA).
11. Thermo LTQ Orbitrap Velos with Nanospray Flex Ion Source (Thermo Electron Corporation) (*see* **Notes 2** and **3**).

2.3 Data Acquisition, Analysis, and Interpretation

1. Xcalibur Xtract (Thermo Scientific), *see* **Note 4**.
2. Thermo File Converter (Thermo Scientific), *see* **Note 5**.
3. Proteome Discoverer (Thermo Scientific).
4. ICC-CLASS [23, 24] (Creative Molecules Inc., www.creative-molecules.com, *see* **Note 3**).
5. PyMOL [25] (Schrödinger, Inc., www.pymol.org).

3 Methods

The protocol described below pertains to cross-linking analysis of a protein mixture wherein proteins of interest are present in the reaction buffer in relatively pure form and the protein sequences are known.

Carry out all procedures at room temperature (21 °C) unless otherwise specified.

3.1 Cross-Linking

The purpose of a cross-linking reaction is to trap proteins of interest in a specific conformational state and to produce covalent linkages between amino acid residues, thus providing distance information between them (*see* **Note 6**). It is recommended to first

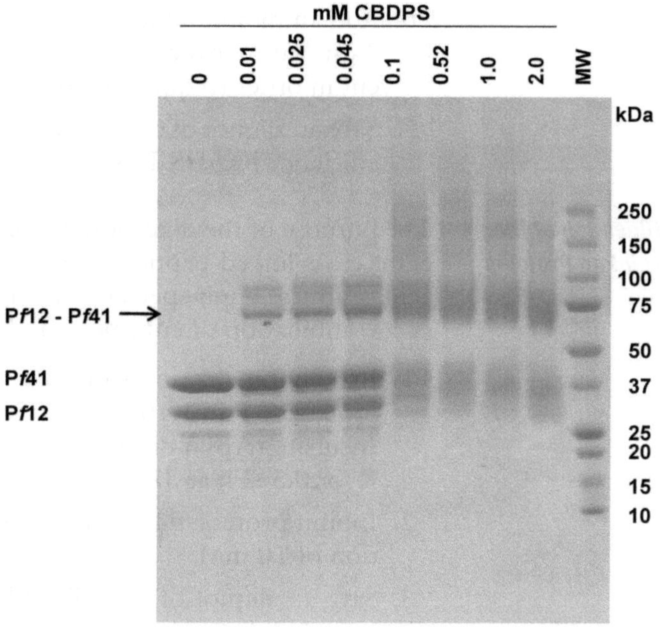

Fig. 2 Titration of a cross-linking reaction. Based on the titration results, a concentration of 0.025 mM CBDPS was chosen for the actual cross-linking reaction. Adapted from [29], with permission. Note that this figure was originally published in the Journal of Biological Chemistry. Tonkin ML, Arredondo SA, Loveless BC, Serpa JJ, Makepeace KAT, Sundar N, Petrotchenko EV, Miller LH, Grigg ME, Boulanger MJ, Structural and biochemical characterization of *Plasmodium falciparum* 12 (*Pf*12) reveals a unique interdomain organization and the potential for an antiparallel arrangement with Pf41. *J Biol Chem* 2013; 288: 12805-12817. © The American Society for Biochemistry and Molecular Biology

titrate the amount of cross-linker needed to achieve a desirable cross-linking reaction yield (Fig. 2) (*see* **Note 7**).

After these titration experiments, the cross-linking of more than 20 μg of the protein sample can be then performed. Aliquots from each step of a protein cross-linking experiment should be assessed by SDS-PAGE: proteins before cross-linking, after cross-linking, and after enzymatic digestion.

1. Protein sample preparation: Amine-containing buffers (Tris, glycine, ammonium salts, etc.) should be avoided for amine-reactive cross-linking reactions. The pH should be 7–9 (*see* **Note 8**); adjust with sodium phosphate dibasic solution if necessary.
2. Prepare CBDPS reagent at a concentration of 100 mM in DMSO immediately prior to carrying out the cross-linking reaction (*see* **Notes 9** and **10**).
3. Add CBDPS to sample (*see* **Note 11**).
4. Mix briefly.
5. Incubate at room temperature (21 °C) for 30 min.

6. Add ammonium bicarbonate to cross-linked sample in at least 2× molar excess compared to the amount of cross-linker present in order to quench cross-linking reaction.
7. Save an aliquot of cross-linked sample for verifying cross-linking efficiency by SDS-PAGE.

3.2 Digestion of the Cross-Linked Proteins

The purpose of the digestion of the cross-linked proteins is to generate cross-linked peptides which will contain cross-linked amino acid residues. High-specificity or nonspecific proteolytic enzymes, or a combination of enzymes, can be used (*see* **Note 12**).

1. In-solution digestion of cross-linked sample: Add trypsin or proteinase K in 1:10 w/w to cross-linked sample (*see* **Note 13**). Incubate trypsin digestion mixture for 12–18 h and proteinase K for 0.5–2 h at 37 °C.
2. Inhibit proteolytic enzymes: Add AEBSF to a final concentration of 10 mM.
3. Save an aliquot of cross-linked and digested sample for verification of digestion efficiency.

3.3 Check of the Cross-Linking Reaction and Digestion by SDS-PAGE

Equal aliquots, by protein mass, of non-cross-linked sample, sample after cross-linking reaction, sample after digestion, and a control sample containing the enzyme used for digestion are analyzed by SDS-PAGE to confirm cross-linking yield and completeness of the digestion (*see* **Note 14**).

3.4 Affinity Enrichment of Cross-Linked Peptides

The purpose of affinity enrichment of the cross-links is to eliminate overwhelming the sample with non-cross-linked peptides, which otherwise will interfere with the detection and identification of the cross-links. The CBDPS cross-linker carries a biotin tag, and therefore CBDPS cross-links can be enriched by avidin (*see* **Note 15**).

1. Prepare compact reaction columns containing monomeric avidin-agarose beads. The amount of beads needed can be estimated based on the binding capacity of the sorbent and total amount of the biotinylated cross-linking reagent in the reaction mixture (*see* **Note 16**).
2. To equilibrate, wash beads with 3× bead volume of 100 mM ammonium acetate.
3. Repeat **step 1** two more times.
4. Load sample onto beads and incubate at ambient room temperature for 10 min.
5. Push sample volume through column, collect, and reload onto beads.
6. Repeat two more times, and collect flow through when finished.
7. Wash beads with 3× bead volume of 100 mM ammonium acetate solution.

8. Wash beads with 3× bead volume of 500 mM ammonium acetate solution.
9. Wash beads with 3× bead volume of 100 mM ammonium acetate solution.
10. Wash beads with 3× bead volume of HPLC-grade water.
11. Repeat **step 10** two more times.
12. To elute, acidify beads with 3× bead volume of 0.1 % TFA solution.
13. Elute from beads with 3× bead volume of 0.1 % TFA in 50 % acetonitrile:water.
14. Combine elution with acidification flow through from **steps 12** to **13**.
15. Lyophilize combined eluates to reduce the volume down to the capacity of the HPLC injection loop.
16. Raise to pH 4–9 with 100 mM ammonium bicarbonate.
17. To reduce, add TCEP to give a final concentration of 10 mM.
18. Acidify with 0.1 % formic acid solution.

3.5 Reversed-Phase HPLC

Even after affinity enrichment, the mixture of the cross-linked peptides is still very complex and requires a separation for the individual cross-links to be detected. For this we use online nano-flow capillary HPLC coupled to an LTQ Orbitrap Velos mass spectrometer.

1. Inject samples onto the trapping column and wash for 15 min with 0.1 % formic acid solution.
2. Separate peptides with the analytical column over an ACN gradient. Set the gradient profile to 4–40 % B from 0 to 60 min, 40 to 80 % B from 60 to 62 min, and 80 % B from 62 to 70 min.

3.6 Mass Spectrometry

Mass spectrometry is used for determining the precise and accurate mass of the cross-links and the precise masses of their fragments. These measured masses are used for the cross-link assignments and to confirm their identification. The mass spectrometric experiment thus consists of detecting the cross-links in the MS spectrum and fragmenting them by MS/MS.

The isotopic coding of the CBDPS cross-linking reagent produces a specific mass spectrometric signature in the cross-linked peptides. When used as 1:1 molar ratio mixture of light and heavy isotopic forms, each cross-linked peptide will appear in the mass spectrum as a doublet of signals separated by 8.05 Da (mass difference corresponding to eight hydrogen atoms in the light form having been substituted by eight deuterium atoms in the heavy form). Both the light and heavy cross-link forms need to be selected for MS/MS fragmentation. On an Orbitrap instrument, data-dependent acquisition of the MS/MS spectra from the cross-link precursor ions can be performed in three different ways using Thermo's acquisition software: the Top N (TN) method, the Mass Tags (MT) method, and the inclusion-list (IL) method. In the TN method, a predefined number (N) of the most intense peaks are selected for MS/MS acquisition after each MS scan; in the MT method, the instrument acquisition software detects doublets of peaks in the MS scan which have a specified mass difference and automatically selects the light and heavy monoisotopic peaks of the doublet as precursor masses for MS/MS acquisition; in the IL method the mass list of doublets (containing the monoisotopic masses of both light and heavy isotopic forms of the cross-links, along with their retention times) can be loaded as an inclusion list for the acquisition of the MS/MS spectra. We use the MT method as a standard approach (described below), and, if sufficient sample amounts and machine time are available, we then perform consecutive runs in the IL mode using an inclusion list derived from the first MT run.

1. In the "Thermo Xcalibur Instrument Setup" window, select or create a new method that you wish to use for acquisition of your data, and ensure that "Dependent scan" is enabled for all scan events except for "Scan Event 1."
2. Ensure that "Scan Ranges" are set to 400.00 for "First Mass (m/z)" and 2,000.00 for "Last Mass (m/z)."
3. Set the resolution at 60,000 for scan 1 and 30,000 for all subsequent data-dependent scans.
4. Click on the "Settings…" button next to "Dependent scan" on any one of the scan events where you have it enabled.
5. In the "Data Dependent Settings" window go to Global>Global and ensure that "Use m/z values as masses" is not enabled.
6. In the "Data Dependent Settings" window, go to Global>Dynamic Exclusion and set "Repeat count" to 2, "Repeat duration (s)" to 15, "Exclusion list size" to 500, "Exclusion duration" to 60.0, "Exclusion mass width" to "Relative to reference mass (ppm)," "Low" to 10.000, "High" to 10.000, and "Early expiration" not enabled.
7. In the "Data Dependent Settings" window, go to Global>Mass Tags and enable Mass Tags. Set "Partner intensity range (%)" to 50.0–100.0. Fill "Mass Delta" table with expected DXs for charge states 2–5 (i.e., 4.02511, 2.68340, 2.01255, and 1.61004).
8. In the "Data Dependent Settings" window, go to Segment> Current Segment and ensure that "Most intense if no Parent Masses found" is enabled. "Predict ion injection time" and "Enable preview mode for FTMS master scans" are enabled.
9. In the "Data Dependent Settings" window go to Segment> Charge State and ensure that "Enable charge state screening" is enabled, "Enable monoisotopic precursor selection" is enabled, "Charge state rejection" is enabled, and "Reject

charge states" is enabled for charge states of 1 and unassigned charge states.

10. For each data-dependent scan event (scan event 2 and greater) go to the "Data Dependent Settings" window and go to Scan Event>Current Scan Event. Ensure that "Minimum signal threshold (counts)" is set at minimum 10,000 (*see* **Note 17**), "Mass determined from scan event" is set at 1, "Same MS order as referenced scan event" is enabled, and "Nth most intense ion" is set at 1 minus the scan event number (e.g., scan event 2 will be set to 1, scan event 3 will be set to 2, and so on).
11. For each data-dependent scan event go to the "Data Dependent Settings" window and go to Scan Event>Mass Tags. Ensure that "Perform scan on: Both partners" is selected.
12. For each data-dependent scan event go to the "Data Dependent Settings" window and go to Scan Event>Activation. Set "Activation type" to CID, "Default charge state" to 2, "Isolation width (*m/z*)" to 2.0, "Normalized collision energy" to 35.0, and "Activation time" to 0.100.
13. Save any changes that may have been made.

3.7 Data Analysis

The purpose of the data analysis is to extract information from the acquired raw mass spectrometric data for the assignment and identification of the cross-links. We have developed a complete set of programs for the processing of mass spectrometric data obtained with isotopically coded CID-cleavable cross-linkers such as CBDPS [24] (*see* **Notes 18** and **19**). Here, we present a data analysis workflow for the data acquired using the MT method. For the analysis of data obtained using TN and IL acquisition methods, *see* **Note 20**. An example of the resultant table of the inter-protein cross-links is shown in Fig. 3.

1. MGFs: Generate MGF files from .raw files using Proteome Discoverer Daemon (v. 1.3).
2. Mass Tags precursor pairs merge: Use "MGF Mass Tags spectra merge L H to LH.exe" to generate light and heavy forms of MS/MS spectra merged in MGF file, which will be used as "MGF File" input for "DX MSMS Match ESI.exe" program.
3. Precursor list extraction: Use "MGF to Precursor list.exe" with merged MGF file as input to generate a total list of all precursors in merged MGF. Output .txt file will serve as DXMSMS Match ESI "DX mass list file" input.
4. To perform the DXMSMS Match ESI cross-link assignment, first load protein sequences in FASTA format. Modify sequences if necessary, and click "Update." Select CBDPS as cross-linker (*see* **Note 21**).

Scan Data					Prediction		Peptide 1							Peptide 2							Scoring						
Deconvoluted MS Peak (MH+) (Light)	MS Intensity (Light)	MS Retention Time (min)	Original Precursor (m/z) (Light)	z	Theoretical (MH+) (Light)	ppm	Protein (1)	AA# Start (1)	AA# End (1)	Modified AA# (1)	AA -1 (1)	Peptide Sequence (1)	AA +1 (1)	Protein (2)	AA# Start (2)	AA# End (2)	Modified AA# (2)	AA -1 (2)	Peptide Sequence (2)	AA +1 (2)	Fragment Score	Intensity Score	401-1	401-2	401-3	401-4	401 Total
1465.68517	1510265	20.33	733.34650	2	1465.6861	0.6	*Pf*12	28	30	30	S	VTK	L	*Pf*41	285	289	288	Q	VLHKN	K	133	9	1	1	1	1	4
1772.82761	1014019	38.53	886.91772	2	1772.8281	0.3	*Pf*12	43	45	45	V	TIK	C	*Pf*41	258	266	259	Q	GKPGELVGF	K	125	29	1	1	1	1	4
1293.63718	7111886	17.12	431.88428	3	1293.6377	0.4	*Pf*12	75	77	75	K	KIK	E	*Pf*41	286	288	288	V	LHK	N	225	23	1	1	1	1	4
1651.75401	926302	27.32	826.38092	2	1651.7542	0.1	*Pf*12	109	114	111	M	PEKPIY	C	*Pf*41	286	288	288	V	LHK	N	114	36	1	1	1	1	4
1260.55357	612906	16.92	420.85641	3	1260.5547	0.9	*Pf*12	193	195	193	V	KAH	A	*Pf*41	127	129	127	M	KHL	K	325	23	1	1	1	1	4
1388.64868	534052	12.90	463.55478	3	1388.6497	0.7	*Pf*12	193	195	193	V	KAH	A	*Pf*41	127	130	127	M	KHLK	G	320	42	1	1	1	1	4
1982.89133	312441	35.28	991.94958	2	1982.8921	0.4	*Pf*12	224	226	225	A	GKN	E	*Pf*41	10	18	13	C	DFTKEKYLL	S	125	14	1	1	1	1	4
1821.73461	2221569	30.35	911.37122	2	1821.7353	0.4	*Pf*12	224	229	225	A	GKNENL	E	*Pf*41	10	14	13	C	DFTKE	K	122	15	1	1	1	1	4
2339.06077	114661	38.20	1170.03430	2	2339.0617	0.4	*Pf*12	224	229	225	A	GKNENL	E	*Pf*41	10	18	13	C	DFTKEKYLL	S	88	18	1	1	1	1	4
1249.52683	2197969	27.20	625.26733	2	1249.5275	0.5	*Pf*12	224	226	225	A	GKN	E	*Pf*41	15	17	15	E	KYL	L	125	18	1	1	1	1	4
1362.61071	2700420	33.43	681.80927	2	1362.6116	0.6	*Pf*12	224	226	225	A	GKN	E	*Pf*41	15	18	15	E	KYLL	S	150	26	1	1	1	1	4
1605.69663	1115278	32.28	803.35223	2	1605.6971	0.3	*Pf*12	224	229	225	A	GKNENL	E	*Pf*41	15	17	15	E	KYL	L	136	24	1	1	1	1	4
1707.78635	524344	19.53	854.39709	2	1707.7876	0.7	*Pf*12	224	229	225	A	GKNENL	E	*Pf*41	127	130	127	M	KHLK	G	125	19	1	1	1	1	4
2078.96825	480907	17.47	1039.98804	2	2078.9681	-0.1	*Pf*12	224	232	225	A	GKNENLENK	L	*Pf*41	127	130	127	M	KHLK	G	123	27	1	1	1	1	4
1651.71323	741391	25.23	826.36053	2	1651.7138	0.3	*Pf*12	227	232	232	N	ENLENK	L	*Pf*41	127	129	127	M	KHL	K	143	13	1	1	1	1	4

Fig. 3 Example of the Excel table output showing the assigned and validated inter-protein cross-links. Adapted from [29], with permission. Note that this Excel table was originally published in the Journal of Biological Chemistry. Tonkin ML, Arredondo SA, Loveless BC, Serpa JJ, Makepeace KA T, Sundar, N, Petrotchenko EV, Miller LH, Grigg ME, Boulanger MJ, Structural and biochemical characterization of *Plasmodium falciparum* 12 (*Pf*12) reveals a unique interdomain organization and the potential for an antiparallel arrangement with Pf41. *J Biol Chem* 2013; 288:12805–12817. © The American Society for Biochemistry and Molecular Biology

5. Search parameters:

Peptide selection parameters:	
Digest sites:	KR (trypsin), all (proteinase K)
Including CL site:	Not enabled (trypsin), enabled (proteinase K)
Missed digest sites:	All
Cross-linked sites:	K or KYST (*see* **Note 22**)
Dead ends only:	Not enabled
Intra-peptide only:	Not enabled
Parameters for matching cross-link masses:	
Systematic error:	0 ppm
Mass tolerance±:	2 ppm
Parameters for locating and matching MS/MS spectrum in MGF file:	
Precursor tolerance±:	2 ppm
Retention time tolerance±:	1 s
Fragment tolerance±:	10 ppm

6. Load DX mass list file.
7. Load merged MGF file.
8. "Run" analysis. Found matches will be pasted in the "Crosslink" box and saved in the output text file.
9. Validation: Found matches with at least three CID cleavage products found in MS/MS spectra (each designated as "1" in output box), percent of b- and y- fragment ions matched >50 %, and percent of total spectrum signal intensity matched >10 % can be manually validated using MS/MS Spectrum View window (*see* **Note 23**). Select cross-link in the output box, and press "Spectrum" button. Matched cleavage products and fragment ions will be highlighted in red.

3.8 Data Interpretation

The result of the cross-linking analysis is a set of cross-linked pairs of amino acid residues for a given protein or protein complex. With a known span of the cross-linker applied these data can be used as distance constraints for the studied protein structures. Depending on the task, applications of the constraints may vary as determining three-dimensional structure of the proteins, detecting conformational changes, determining topology of the multi-subunit protein assemblies, determining mutual orientation of the proteins in the complexes, etc. Accordingly, varying software tools could be used as programs for ab initio structure modeling, molecular dynamics simulations, docking, etc. Here, as a common step in data interpretation process, we describe an example for how to visualize obtained intra- and inter-protein cross-links for the protein heterodimer using PyMOL software (Fig. 4).

1. Load the .pdb files for your protein(s) of interest in PyMOL.
2. Show sequence pane (Display>Sequence).
3. Represent all proteins with cartoon (.pdb object in right panel>S>as>cartoon).
4. Select all cross-linked residues (can be done by clicking on the residues in the sequence pane).
5. Show side chains for all cross-linked residues ("(sele)" in right panel>S>side chain>sticks).
6. To draw linear distances between cross-linked residues go to "Wizard>Measurement." You can now click on the two atoms that are cross-linked to create a distance measure between them (*see* **Note 24**).

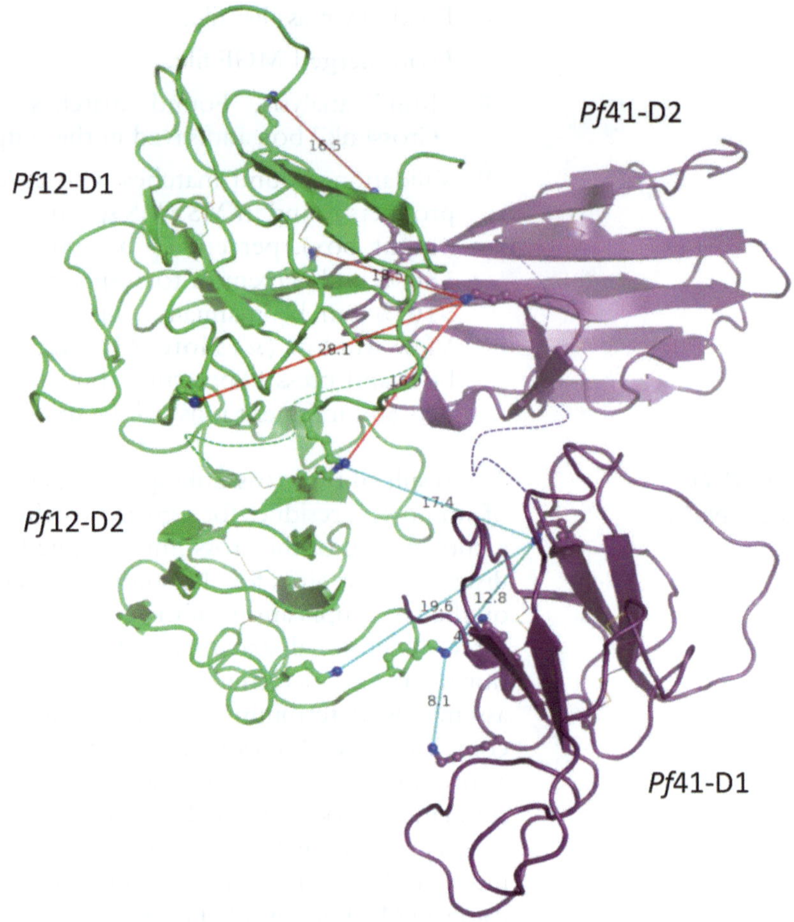

Fig. 4 Example of a PyMOL session image with the inter-protein cross-links found mapped onto the protein heterodimer structure. Cross-linking data was obtained and adapted from [29], with permission. Note that this figure was originally published in the Journal of Biological Chemistry. Tonkin ML, Arredondo SA, Loveless BC, Serpa JJ, Makepeace KAT, Sundar N, Petrotchenko EV, Miller LH, Grigg ME, Boulanger MJ, Structural and biochemical characterization of *Plasmodium falciparum* 12 (*Pf*12) reveals a unique interdomain organization and the potential for an antiparallel arrangement with Pf41. *J Biol Chem* 2013; 288:12805–12817. © The American Society for Biochemistry and Molecular Biology

4 Notes

1. If the protein sample is stored in an incompatible buffer, then dialysis against a suitable buffer (i.e., which does not contain primary amines) will need to be performed prior to cross-linking.
2. Any mass spectrometer that has high mass accuracy in both the MS and MSMS modes may be used as an alternative. However,

the current version of the ICC-CLASS software, which is used for the doublet identification, is designed for use with the Thermo Orbitrap instrument.

3. The current version of the ICC-CLASS software, which is used for the doublet identification, may be compatible with other instrument software but has been *designed* for use with the Thermo Orbitrap instrument and its associated software, Xcalibur Xtract.
4. Software capable of producing a deconvoluted mass list of MS spectra from ".raw" may be used as an alternative to Xcalibur Xtract. The comment regarding the ICC-CLASS software, as stated in **Note 3**, applies here as well.
5. Software capable of producing a ".txt" file containing information (that includes deconvoluted MH+, peak intensity, and retention time) for each deconvoluted MS spectra may be used as an alternative to Thermo File Converter. The comment regarding the ICC-CLASS software, as stated in **Note 3**, applies here as well.
6. Several considerations need to be taken into account: protein and cross-linking reagent concentrations need to be optimized to achieve satisfactory yield while avoiding nonspecific cross-linking or over-cross-linking. The time of the cross-linking reaction should correspond to the time scale within which changes may take place in the protein system. Temperature of the reaction mixture should be in accordance with the protein stability data. Protein complexes should be used at concentrations not lower than their K_d values.
7. Titration of the cross-linking reaction consists of performing several test reactions under similar conditions with increasing amounts of cross-linker and visualizing the cross-linking efficiency by SDS-PAGE. Conditions where cross-linked protein complexes can be seen, but where nonspecific cross-linking products are absent, should be chosen for the cross-linking reaction.
8. We typically adjust the pH of the mixture to 8.0–8.5 with the addition of 0.2 M Na_2HPO_4.
9. CBDPS stocks should be prepared immediately prior to use, as the NHS ester moieties will rapidly hydrolyze in neutral aqueous solution resulting in decreased cross-linking efficiencies.
10. The DMSO concentration should be kept to less than 5 % v/v in the cross-linking reaction mixture to minimize distortion of protein structure.
11. Typically, the cross-linker concentration for the cross-linking reaction will be in the range of 0.01–1.0 mM. It is best to determine the ideal concentration to use for the cross-linking reaction

from a titration experiment, and this concentration will depend on the nature of the sample and the extent of cross-linking which the experimenter desires.

12. The choice of which proteolytic enzymes to use will depend on the complexity of the protein system under study. We typically perform both trypsin and proteinase K digestions in parallel. Tryptic digestion produces longer peptides and cleaves at specific residues. Proteinase K, in contrast, produces short peptides cut at random sites, thereby resulting in multiple but related cross-linked species for each pair of cross-linked residues [26].
13. In-gel and out-gel digests may also be performed to specifically isolate a particular cross-linked species for analysis [27, 28].
14. Complete digestion should not leave any stained "smear" on the gel. A single band representing the proteolytic enzyme may be visible.
15. We use monomeric avidin, which appears to have a lower affinity than the tetrameric forms and leads to more efficient elution of the bound biotinylated cross-links.
16. Typically, in order to minimize nonspecific binding, we use 10- to 50-fold less sorbent than would be required for an equimolar amount of sorbent to total biotinylated cross-linking reagent used in the cross-linking reaction.
17. We have observed the best results for Mass Tags doublet-pair acquisition and MS/MS spectrum quality when "Minimum signal threshold (counts)" is set at 30,000 for affinity-enriched CBDPS cross-linked samples. A lower threshold may be necessary for samples that are not affinity enriched because cross-linked precursor doublets will be observed at lower intensities than in enriched samples.
18. The main program for the analysis of LC/ESI-MS/MS data is called DXMSMS Match, and several other supporting programs within the ICC-CLASS software suite are used to create the DMSMS Match input files. DXMSMS Match calculates the possible peptide cross-links for a given precursor mass, based on the known protein sequence, searches the corresponding merged MS/MS spectrum (the combined spectra from the light and heavy forms) for predicted fragment ions, and assigns non-probabilistic scores based on the number of matched peaks.
19. All output files are tab-delimited text files and can be opened and processed in Microsoft Excel.

20. Generate a deconvoluted .raw file using Xcalibur Xtract with the following parameters:

    Scan filter: FTMS + c NSI Full ms [400.00–2,000.00]

    Generate masses mode: MH+

    Extract what: Monoisotopic masses

    Time range: This will depend on the length of your run and where in the time range the cross-linker-modified peptides appear. Specify a "Begin Time" and "End Time" that include these modified peptides.

    Mass range: "Low Mass:" 400.00; "High Mass:" 2000.00

    S/N threshold: 10

    Generate a .txt file from the deconvoluted .raw file using Thermo File Converter with the following parameters:

    Source data type: Xcalibur Files *.raw

    Destination data type: Text Files *.txt

    Generate a mass list .txt file from the deconvoluted .txt file using Mass_List_from_Xtract_txt.exe (an ICC-CLASS program).

    Generate a doublet only containing mass list ("DX masslist") .txt file from the masslist .txt file using DX_ESI_LCMS.exe (an ICC-CLASS program).

    Generate a filtered DX masslist .txt file from the DX masslist .txt file using DX_ESI_LCMS_Mass_List_Filter.exe (an ICC-CLASS program).

    Use the MGF created in **step 1** as the DXMSMS Match ESI input for **step 6** (do not merge the MGF file).

    Use the filtered DX masslist .txt file as the DXMSMS Match ESI input for **step 7**.

    All parameters detailed in **step 5** can remain the same with the exception of "Retention time tolerance±" which should be changed to "60" s.

21. Mip can be modified to search for other common peptide modifications, e.g., oxidation or methylation. Mclrest1 and 2 are the mass additions to the peptide masses for the cleavage products. These parameters are more fully explained on the product information sheet for CBDPS-H8/D8 from Creative Molecules Inc.

22. NHS ester reactivity and cross-linking on tyrosine, serine, and threonine side chains may be observed. YST cross-linking is generally far less likely to be observed than K cross-linking when using the conditions described in this protocol. You may want to initially search your data set for K–K cross-links before considering Y, S, and T cross-linking because the analysis

is less complicated and is likely to yield the majority of the cross-linking results.

23. For a detailed description of the output refer to DX MSMS Match user guide (available at www.creativemolecules.com). During manual inspection of merged MS/MS spectra we verify that cross-linker-non-containing fragments are represented as single peaks and cross-linker-containing fragments are represented as doublets of peaks. For cleavage product peaks (displayed next to cross-linked peptide sequences in the MS/MS Spectrum View window) we are looking for heavy peaks that are +4 Da. For b- and y-ion fragments that are cross-linker modified we are looking for heavy peaks that are +8 Da. We also note that a significant proportion of total spectrum intensity may be due to ($M+H-H_2O$) peaks of z equal to +2 or greater that may account for a low total spectrum signal intensity match % value.
24. This process can be performed using a script to facilitate drawing of multiple cross-link distance measures. For example, executing "dist CBDPS_K30-K63, /Pf12_D1//A/30/NZ, /Pf41_D2//B/63/NZ" will create a distance measure object called "CBDPS_K30-K63" that will show a dotted line and labeled distance in ångströms between *Pf*12 domain 1 (chain A) K30 ε-amino group nitrogen and *Pf*41 domain 2 (chain B) K63 ε-amino group nitrogen. You can create a similar command for each cross-link you wish to visualize and execute all commands at once.

Acknowledgements

This work was supported by Genome Canada and Genome BC through the funding of the Science and Technology Innovation Centre at the University of Victoria—Genome BC Proteomics Centre.

References

1. Back JW, De Jong L, Muijsers AO, De Koster CG (2003) Chemical cross-linking and mass spectrometry for protein structural modeling. J Mol Biol 331:303–313
2. Schulz DM, Sinz A (2004) Chemical cross-linking and mass spectrometry for structural and functional characterization of proteins. BIOspektrum 10:45–48
3. Novak P, Giannakopulos AE (2007) Chemical cross-linking and mass spectrometry as structure determination tools. Eur J Mass Spectrom 13:105–113
4. Sinz A (2003) Chemical cross-linking and mass spectrometry for mapping three-dimensional structures of proteins and protein complexes. J Mass Spectrom 38:1225–1237
5. Petrotchenko EV, Borchers CH (2010) Crosslinking combined with mass spectrometry for structural proteomics. Mass Spectrom Rev 29:862–876
6. Leitner A, Walzthoeni T, Kahraman A, Herzog F et al (2010) Probing native protein structures by chemical cross-linking, mass spectrometry, and bioinformatics. Mol Cell Proteomics 9:1634–1649
7. Leitner A, Reischl R, Walzthoeni T, Herzog F et al (2012) Expanding the chemical cross-linking toolbox by the use of multiple proteases

and enrichment by size exclusion chromatography. Mol Cell Proteomics 11(M111):014126

8. Chen ZA, Jawhari A, Fischer L, Buchen C et al (2010) Architecture of the RNA polymerase II-TFIIF complex revealed by cross-linking and mass spectrometry. EMBO J 29:717–726
9. Chu F, Mahrus S, Craik CS, Burlingame AL (2006) Isotope-coded and affinity-tagged cross-linking (ICATXL): an efficient strategy to probe protein interaction surfaces. J Am Chem Soc 128:10362–10363
10. Hurst GB, Lankford TK, Kennel SJ (2004) Mass spectrometric detection of affinity purified crosslinked peptides. J Am Soc Mass Spectrom 15:832–839
11. Fujii N, Jacobsen RB, Wood NL, Schoeniger JS, Guy RK (2004) A novel protein crosslinking reagent for the determination of moderate resolution protein structures by mass spectrometry (MS3-D). Bioorg Med Chem Lett 14:427–429
12. Petrotchenko EV, Thomas JM, Borchers CH (2008) A collection of novel isotopically-coded crosslinkers for structural proteomics. Paper presented at the 56th ASMS conference on mass spectrometry and allied topics, Denver, CO June 1–5 2008
13. Kang S, Mou L, Lanman J, Velu S et al (2009) Synthesis of biotin-tagged chemical cross-linkers and their applications for mass spectrometry. Rapid Comm Mass Spectrom 23: 1719–1726
14. Nessen MA, Kramer G, Back J, Baskin JM et al (2009) Selective enrichment of azide-containing peptides from complex mixtures. J Proteome Res 8:3702–3711
15. Sebela M, Stosová T, Havlis J, Wielsch N et al (2006) Thermostable trypsin conjugates for high-throughput proteomics: synthesis and performance evaluation. Proteomics 6: 2959–2963
16. Sarbolouki MN, Mahnam K, Rafiee-Pour H-A (2004) Determination of pore/protein size via electrophoresis and slit sieve model. Electrophoresis 25:2907–2911
17. Müller DR, Schindler P, Towbin H, Wirth U et al (2001) Isotope-tagged cross-linking reagents. A new tool in mass spectrometric protein interaction analysis. Anal Chem 73:1927–1934
18. Petrotchenko EV, Xiao K, Cable J, Chen Y et al (2009) BiPS, a photocleavable, isotopically coded, fluorescent cross-linker for structural proteomics. Mol Cell Proteomics 8:273–286
19. Soderblom EJ, Bobay BG, Cavanagh J, Goshe MB (2007) Tandem mass spectrometry acquisition approaches to enhance identification of protein-protein interactions using low-energy collision-induced dissociative chemical cross-linking reagents. Rapid Comm Mass Spectrom 21:3395–3408
20. Soderblom EJ, Goshe MB (2006) Collision-induced dissociative chemical cross-linking reagents and methodology: applications to protein structural characterization using tandem mass spectrometry analysis. Anal Chem 78:8059–8068
21. Lu Y, Tanasova M, Borhan B, Reid GE (2008) Ionic reagent for controlling the gas-phase fragmentation reactions of cross-linked peptides. Anal Chem 80:9279–9287
22. Petrotchenko EV, Serpa JJ, Borchers CH (2011) An Isotopically-coded CID-cleavable biotinylated crosslinker for structural proteomics. Mol Cell Proteomics. doi:10.1074/mcp.M110.001420
23. Petrotchenko EV, Borchers CH (2010) ICC-CLASS: isotopically-coded cleavable crosslinking analysis suite. BMC Bioinformatics 11:64
24. Petrotchenko EV, Borchers CH (2014, manuscript in preparation), Isotopically-coded cleavable crosslinking analysis software suite (ICC-CLASS) for the automated analysis of the MALDI- and ESI-LC-MS-MS/MS crosslinking data
25. PyMOL, http://www.pymol.org/ 2011.
26. Petrotchenko EV, Serpa JJ, Berjanskii M, Suriyamongkol BP et al (2012) Use of proteinase K non-specific digestion for selective and comprehensive identification of interpeptide crosslinks: application to prion proteins. Mol Cell Proteomics 11:M111.013524
27. Shevchenko A, Tomas H, Havlis J, Olsen JV, Mann M (2006) In-gel digestion for mass spectrometric characterization of proteins and proteomes. Nat Protoc 1:2856–2860
28. Petrotchenko EV, Serpa JJ, Cabecinha AN, Lesperance M, Borchers CH (2014) An "out-gel" tryptic digestion procedure for chemical crosslinking studies with mass spectrometric detection. J Proteome Res 13(2):527–535
29. Tonkin ML, Arredondo SA, Loveless BC, Serpa JJ et al (2013) Structural and biochemical characterization of Plasmodium falciparum 12 (Pf12) reveals a unique interdomain organization and the potential for an antiparallel arrangement with Pf41. J Biol Chem 288: 12805–12817

Chapter 31

Top-Down Proteomics by Means of Orbitrap Mass Spectrometry

Kai Scheffler

Abstract

Top-down proteomics has become a popular approach for the analysis of intact proteins. The term "top down" has been coined for the analysis of proteins not involving any enzymatic or chemical cleavage but rather the ionization of the protein as a sound molecule and mass analysis of intact species and fragment ions thereof produced upon dissociation inside a mass spectrometer. One or several charge states of the protein are mass-isolated and subjected to dissociation (MS/MS) in the gas phase. The obtained fragment masses, predominantly from cleavages of the protein along its amino acid backbone, are directly related to the intact protein. Using bioinformatics tools the fragment masses are matched against a known protein sequence or can alternatively be used for partial or full de novo sequencing, depending on the size of the protein and the number of fragment ions obtained. Moreover, this approach provides global information about modification states of proteins including the number and types of isoforms and their stoichiometry and allows for the precise localization of modifications within the amino acid sequence. Top-down analysis of a single, purified protein can be performed by matrix-assisted laser desorption ionization or electrospray ionization upon direct infusion without online chromatographic separation, whereas top-down analysis of complex protein mixtures makes pre-fractionation combined with an efficient front-end chromatographic separation coupled online to the mass spectrometer inevitable.

Key words Top-down proteomics, Protein fragmentation, Orbitrap, CAD/CID, ECD, ETD, HCD, High resolution, Mass spectrometry

1 Introduction

Top-down (TD) is still a fairly new technique in the field of mass spectrometry, even though the first experiments date back to 1990 when Loo and colleagues proposed the possibility to get sequence information from bovine ribonuclease A (14 kDa) [1] and from serum albumin from different species in 1991 [2] by collision of the multiply charged ions with an inert gas. The real breakthrough of top-down proteomics happened in the year 2000 with the introduction of electron capture dissociation (ECD) by Roman Zubarev [3]. Ever since top-down proteomics has gained significant space

Daniel Martins-de-Souza (ed.), *Shotgun Proteomics: Methods and Protocols*, Methods in Molecular Biology, vol. 1156, DOI 10.1007/978-1-4939-0685-7_31, © Springer Science+Business Media New York 2014

in proteomics it has evolved with the work of relatively few research groups from a promising approach to a solid, well-established technique.

Top-down proteomics is a method to elucidate or to confirm the sequence of a protein and moreover allows for the detection of different protein forms, including isoforms and those forms arising from posttranslational modifications (PTMs). One or several proteins are introduced into a mass spectrometer for the determination of their M_r values based on an isotopically resolved charge state or the whole envelope of different charge states. One or several charge states of a particular intact protein are subsequently mass-isolated in gas phase and dissociated resulting in product masses related directly to its M_r value. Proteins are most often ionized by electrospray ionization (ESI) and trapped in a Penning trap- or Paul trap-type mass spectrometer. Fragmentation of the protein can be accomplished by different fragmentation types: collision-induced dissociation (CID), higher energy collision-induced dissociation (HCD), and electron transfer or electron capture dissociation (ETD, ECD). Alternatively matrix-assisted laser desorption ionization (MALDI) can be performed and the protein is fragmented in the source region of the mass spectrometer supported by the usage of special matrices such as 1,5-diaminonaphthalene (1,5-DAN) [4–6].

The opposite approach to top-down is called bottom-up: Proteins are digested with proteases, most commonly trypsin, and the resulting peptides are typically separated chromatographically and fragmented in the mass spectrometer. The obtained peptide fragments are submitted together with the precursor mass of the intact peptides to a database search and result in the identification of peptides associated to one or several proteins. The bottom-up and top-down techniques represent two different trends in MS-based proteomics and have both their advantages and limitations: bottom-up proteomics is easy to perform with samples of low and high complexity, only little sample preparation is required since trypsin works well even in the presence of detergents and salts and is little pH sensitive, works reliably and reproducibly, and most proteins contain sufficient cleavage sites to produce peptides in a size working well with most mass spectrometers to detect and fragment. With the mass spectrometers of the latest generation very complex samples such as a whole-cell lysate can yield the identification of up to 8,000 proteins in a matter of a few hours of analysis time [7]. However, the majority of proteins is identified based on a very limited number of peptides and thus do neither provide information on the complete sequence of the intact protein nor on the full set of PTMs. In top-down analysis on the contrary, the first information obtained is the molecular weight of the intact protein, calculated on the basis of one multiply charged ion if isotopically resolved or else based on the entire charge envelope from the full MS spectrum. The comparison between the expected and obtained mass of the protein will alert about the presence of a mass difference due to an amino acid substitution, a sequence truncation, or PTMs. Assignment of the obtained fragment masses will then allow for allocating the precise location of sequence variations or modifications.

1.1 Mass Spectrometry Instrumentation

Top-down analyses to date have predominantly been performed with high-resolving-power Fourier transform-based instruments [8–12] since they provide the mass accuracy and resolution necessary, while maintaining high sensitivity, to resolve complex and overlapping isotope patterns commonly obtained in top-down spectra from proteins. Other types of instruments such as ion traps and time-of-flight mass spectrometers can also be applied to analyze intact proteins but provide far less resolving power than FT-ICR- and FT-Orbitrap™-based instruments. The latest-generation Orbitrap instruments are utilizing high-field Orbitrap analyzers with significantly improved resolving power of several hundreds of thousands and improved scan speed providing an excellent tool for top-down analysis and a competitive alternative to FT-ICR instruments [13]. A significant advantage of Orbitrap-based instruments compared to FT-ICR mass spectrometers is the fact that resolving power decreases as the square root of m/z not as its first power, making resolving power on Orbitraps less sensitive to m/z.

The majority of top-down studies has been performed using ESI; however, several groups utilized a MALDI-TOF approach for top-down analysis by employing the strategy of in-source decay (ISD) [14–18]. MALDI-ISD of intact proteins can as well be performed on a hybrid linear ion trap Orbitrap instrument, the Thermo Scientific LTQ Orbitrap XL, equipped with a vacuum MALDI source [19] with the advantage of higher resolution and accurate mass detection of the fragment ions compared to MALDI-TOF instruments. The limited detection of, inherent to the MALDI process, singly charged intact protein ions due to the restricted mass range on a MALDI-Orbitrap is overcome by the fact that in ISD experiments the intact protein is fragmented in the source region due to the use of special highly energetic matrices such as 1,5-diaminonaphthalene producing fragments with m/z values well detectable in the available mass range [20].

1.2 Fragmentation Techniques

CID, also referred to as collisionally-activated dissociation (CAD), is the most widely applied fragmentation technique available in a variety of different types of mass spectrometers [21]. Under CID conditions, the peptide or the protein precursor undergoes one or more collisions with neutral gas molecules such as helium or nitrogen, contributing to vibrational energy which will redistribute over the peptide/protein resulting in dissociation occurring at the amide bonds along the peptide or the protein backbone. Fragments of b- and y-type are obtained as well as losses of water and ammonia but also loss of labile PTMs such as phosphorylation and glycosylation are observed [22].

An alternative fragmentation type is the beam-type CID, referred to as higher energy collision-induced dissociation (HCD), and was first described in 2007 [23]. The fragmentation technique has been further refined and instrument generations introduced in 2008 (LTQ Orbitrap XL™) and later were equipped with a dedicated collision cell for HCD and provide a much improved sensitivity and fragmentation efficiency. HCD fragmentation is achieved by injections of ions in a collision cell filled with nitrogen gas generating fragments of b and y type [24, 25]. Upon the usage of higher collision energies fragmentation of second and third stages can occur resulting in further fragmentation of b ions to result in a ions and smaller species. HCD spectra do not suffer from a low mass cutoff inherent in ion trap CID spectra and have been applied successfully for peptide and terminal protein sequencing and in PTM studies due to characteristic diagnostic ions specific for HCD [26].

Top-down analysis for detailed characterization of proteins strongly benefits from ECD [3, 27–29], so far unique to FT-ICR, and ETD [30, 31] implemented on ion trap-based and time-of-flight (TOF) instruments. In both ECD and ETD low-energy electrons are transferred to a multiply protonated peptide/protein leading to the cleavage of the N-Cα backbone bonds to generate c- and z-type fragment ions. The advantage of both ECD and ETD is that often higher sequence information is obtained compared to CID spectra [32] and the labile PTMs such as phosphorylation and glycosylation are retained during protein fragmentation enabling the localization of the modification sites [33]. In ETD experiments fragmentation can be greatly influenced by extending or shortening the activation time, which is the time the electron donor such as the fluoranthene radical anion is allowed to react with and transfer the electrons onto the peptide/protein cations. The required reaction time is directly dependent on the size of the peptide/protein and its charge state. Whereas smaller doubly charged peptides require about 100 ms, higher charged peptides and proteins starting at 10 kDa and higher require only 1–50 ms activation to fragment efficiently. Shorter activation times result typically in fragmentation across the entire sequence, whereas prolonged reaction times result in fragment ions predominantly representing the protein termini. ETD also works very efficiently on large proteins such as intact, non-reduced antibodies [34] as well as on low-resolution-type instruments such as ion traps. Since the ETD reaction of proteins results also in large and higher charge fragments that cannot be isotopically resolved by an ion trap due to the lack of resolving power, the spectrum can be simplified by reacting the ETD fragment ions with a second anion that deprotonates the product ions and thus reduces their charge state [35]. The charge state of the resulting product ions can be well controlled as the ion–ion reaction rate increases proportionally with the square of the charge [36].

Whereas CID and HCD can be performed on ions generated either via electrospray or MALDI, ECD and ETD are only applicable to ESI since they require multiply charged precursors which are typically not observed under MALDI conditions. A fragmentation technique unique to MALDI is in-source fragmentation (MALDI-ISD). MALDI-ISD fragment ion production was first described in the mid-1990s upon studies employing delayed extraction in axial TOF systems by Brown et al. [37–39]. 1,5-DAN as matrix is able to rapidly reduce disulfide bonds and N-terminal and C-terminal sequencing of peptides and small proteins and to specifically produce a significant amount of "in-source decay" fragments on MALDI axial time-of-flight (MALDI TOF) systems [40]. Takayama proposed a model in 2001 explaining the underlying mechanism in MALDI-ISD highlighting parallels to ECD. In contrast to axial TOF systems most commonly used in combination with a MALDI source with relatively low pressure in front of the sample plate, the MALDI ion source used on the MALDI LTQ Orbitrap XL™ liberates the MALDI-produced fragment ions into a region of significantly higher pressure (75 mTorr), i.e., a collisional cooling regime [19]. ISD fragment ions are detected either in a linear ion trap or an Orbitrap detector. The major difference of fragments produced with 1,5-DAN over more commonly used MALDI matrices is that the entire sequence tags of all possible fragment ion types can be observed. a-, b-, c-, d-, w-, y-, and z-type fragment ions are obtained simultaneously in a single spectrum. In a middle-down or a top-down approach, complete sequences of large peptides (<4 kDa) and N-terminal or C-terminal sequence tags of intact proteins over a broad molecular weight range can be obtained and assigned. Moreover, the rarely observed d- and w-ions originating from side chain losses are the only ones allowing for distinguishing between leucine and isoleucine residues as part of a peptide or a protein.

1.3 Software for Data Analysis

The group of Neil Kelleher has pioneered the development of software for the top-down technology. Their algorithm for data processing is part of the software Prosight PTM [41–44] and is so far the most successful software package for analysis of top-down data of proteins with or without modifications (http://groups.molbiosci.northwestern.edu/kelleher/ software.html). The software has been commercialized by Thermo Fisher and is available under the name ProSightPC 3.0. It is the first stand-alone software for analyzing top-down proteomics data and has been enhanced to add support for middle-down and bottom-up experiments, making it an all-around tool for identification and characterization of both intact proteins and peptides. The software enables high-throughput processing of all accurate-mass MS/MS data, whether from top-down, middle-down, or bottom-up experiments including the characterization of proteins with known PTMs. ProSightPC 3.0 software uses multiple search modes to determine the exact protein sequence including modifications and alternative splicing. It is the only proteomics software that allows the user to search their tandem MS data against proteome warehouses containing the known biological complexity present in UniProt.

Another commercial software platform is BioTools™ from Bruker Daltonics which is using a mixture of automated calculations and interactive sequence assignments for MALDI-ISD spectra obtained from MALDI-TOF instruments. The approach of MALDI top-down sequencing is abbreviated with "MALDI-TDS."

The lab of Catherine Costello at the Boston University makes available a Web-based protein identifier software package, BUPID top down [45, 46] for analysis of top-down data. The software is capable of identifying b-, y-, c-, and z-ions as well as internal ions arising from the relevant activation techniques. BUPID top down uses a sequence tag approach to identify potential protein candidates from a protein sequence database. Tags are generated by comparing the mass difference between pairs of experimental masses obtained from MS/MS spectra. The tags are then used to identify protein candidates in a sequence database.

Another software tool is provided by the group of John Yates called YADA [47] which can deisotope and decharge high-resolution mass spectra from large peptide molecules, link the precursor monoisotopic peak information to the corresponding tandem mass spectrum, and account for different co-fragmenting ion species (multiplexed spectra). YADA also enables a pipeline consisting of ProLuCID and DTASelect for analyzing large-scale middle-down proteomics data.

Two more software tools that go hand in hand are MS-Align + [48] and MS-Deconv [49] provided by the Department of Computer Science and Engineering University of California in San Diego. First MS-Deconv is applied for top-down spectral deconvolution. An algorithm is used to generate a set of candidate isotopomer envelopes for a spectrum and then selects the highest scoring envelope. It produces the input file for MSAlign+, a software tool for top-down protein identification based on spectral alignment that enables searches for unexpected PTMs. MS-Align + is fast in identifying unexpected PTMs.

1.4 Sample Preparation

The top-down workflow for samples of different complexities is outlined in Fig. 1.

Full characterization of intact proteins by top-down mass spectrometry is most commonly performed on proteins that have been enriched or purified. A single-protein sample can be analyzed with either MALDI-ISD or ESI with direct infusion but requires a minimum of salt in the sample. Whereas MALDI is less sensitive to

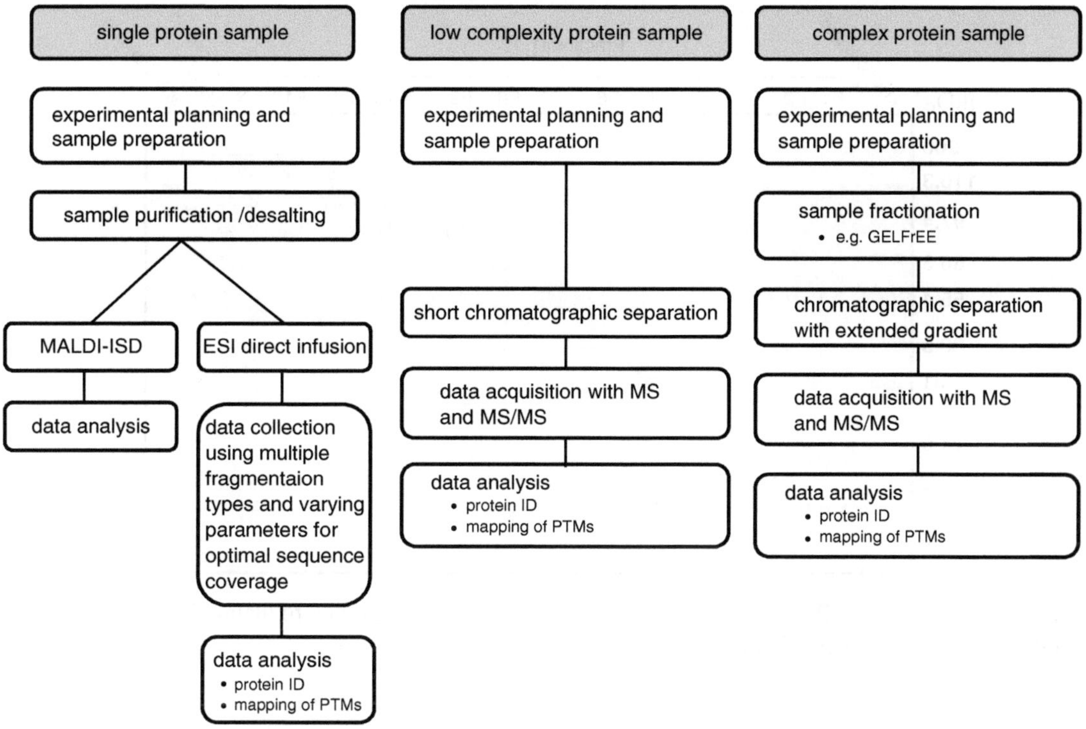

Fig. 1 Schematic of the workflow for top-down analysis of samples with different complexity: single-protein sample, low-complexity protein mixture, and complex protein mixture

salts the analysis via ESI is suffering significantly already with small amounts of salt in the sample [50]. If an off-line sample cleanup cannot be performed or the sample is a heterogeneous mixture of several proteins online LC-MS analysis is recommended. Direct infusion provides more time for signal averaging and facilitates the use of multiple fragmentation techniques.

For the top-down analysis of highly complex mixtures such as whole-cell lysate pre-fractionation is required to reduce precursor spectral complexity and minimize ion suppression.

The available chromatography methods for protein separation span a wide variety of techniques and stationary phases [51] such as 2D reversed-phase chromatography, size exclusion (SEC) and ion exchange (IEC), hydrophilic (HILIC) and hydrophobic interaction, immobilized metal ion affinity, and chromatofocusing [52]. In addition, also electrophoretic techniques, e.g., isoelectrofocusing (IEF), gel electrophoresis, capillary electrophoresis (CE), and capillary electrochromatography (CEC), are currently used [53]. Even though some of these methods cannot be connected online with ESI-MS they may be used as a first step in an off-line/online 2D separation method. A technology that has recently become available includes the approach of Gel-Eluted Liquid Fraction Entrapment Electrophoresis (GELFrEE®) to separate proteins

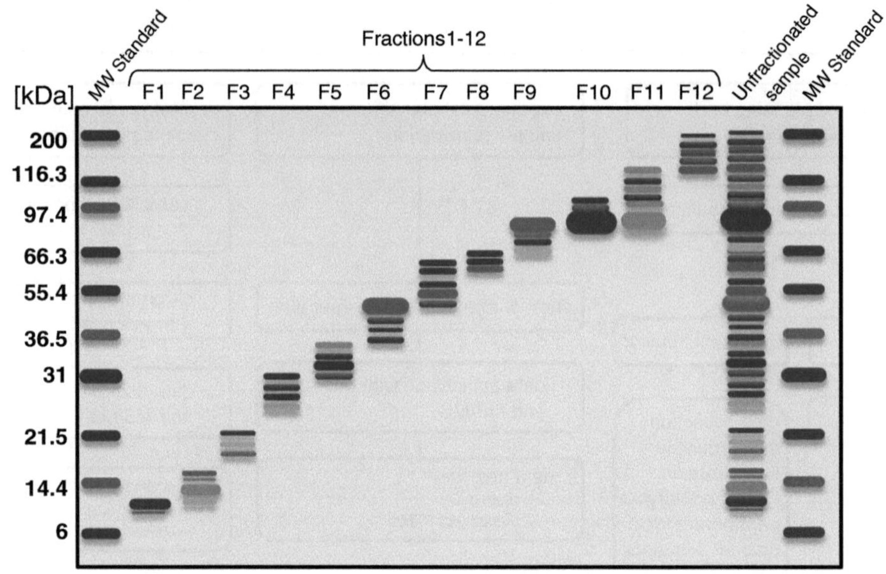

Fig. 2 Schematic of an SDS-PAGE analysis of a complex protein sample before (unfractionated sample) and after (fractions 1–12) GELFrEE® fractionation

over a broad MW range and then utilize online-chromatographic separation combined with high-resolution mass spectrometry on the fractions that contain 7–200 kDa proteins [54, 55]. The fractionation can be assessed and visualized by applying an aliquot of each fraction to SDS-PAGE followed by silver staining. Typically a pattern as shown in the schematic in Fig. 2 is obtained.

For top-down analysis there is unfortunately not one single protocol that applies to every instrument and every sample. Thus we describe in this chapter the basic steps to perform top-down experiments with the Thermo Scientific MALDI LTQ Orbitrap, the Thermo Scientific Q Exactive benchtop Orbitrap, and the Thermo Scientific Orbitrap Elite hybrid mass spectrometers for single-protein analysis, a low-complexity protein mixture, and a complex cell lysate including pre-fractionation via the GELFrEE® system. All experiments are described for the analysis of proteins under reducing (nonnative) conditions.

2 Materials

2.1 Sample Preparation for MALDI-ISD

1. MALDI matrix: Saturated solution of 1,5-DAN in 50 %/50 % v/v water/acetonitrile (both LC-MS grade) with 0.1 % trifluoroacetic acid (TFA), prepared fresh and centrifuged before usage.
2. Protein sample: For example insulin B chain or cytochrome C, dissolved in water, 5 pmol/μL (LC-MS grade).

3. MALDI calibration and tuning kit MSCAL4 (Sigma-Aldrich).
4. Vortexer.

2.2 Sample Preparation for ESI Direct Infusion

1. Protein sample: For example carbonic anhydrase, dissolved in water (LC-MS grade)/0.1 % formic acid (ultrapure grade) at a concentration of 10 pmol/μL.
2. 50 %/50 % v/v water/acetonitrile (both LC-MS grade) with 0.1 % formic acid (ultrapure grade).
3. Zeba™ micro spin desalting columns, 7 K MWCO, 0.5 mL (Thermo Scientific Pierce), or detergent removal spin column (Thermo Scientific Pierce).
4. 0.01 % formic acid (pH 4).
5. Vortexer, centrifuge, 1.5 and 2 mL microcentrifuge collection tubes.

2.3 Sample Preparation for Fractionation of *E. coli* Cell Lysate

1. 2 mg *E. coli* whole-cell lysate.
2. 20 mM and 1 M dithiothreitol (DTT) solutions in 18 MΩ water.
3. Removal of cell debris and insoluble proteins: 8 μm, 0.8 μm, and then 0.2 μm pore filters.

Sample Desalting:

4. Zeba™ Micro Spin Desalting Columns, 2 mL resin bed for 200–700 μL, 7K MWCO (Thermo Scientific Pierce).
5. Coomassie Plus Bradford Protein reagent assay.
6. Requires spectrophotometer with 595 nm absorbance.
7. 18 MΩ water.

Sample Fractionation:

8. GELFrEE® 8100 Fractionation System (Expedeon, Knoxville).
9. 10 % Tris–Acetate GELFrEE® cartridge (Expedeon, Knoxville).
10. 30 μL GELFrEE® sample buffer.
11. GELFrEE® HEPES running buffer.

SDS-PAGE:

12. SDS polyacrylamide mini gels, e.g., precast Novex 10–20 % Tris-Glycine gels (Life Technologies).
13. Lithium dodecyl sulfate (LDS) sample buffer.
14. 500 mM DTT solution.
15. 500 mL SDS-PAGE Tris-Glycine running buffer: 25 mM Tris, 192 mM glycine, 0.1 % SDS, pH 8.3.
16. Premixed molecular weight standard mix, 2.5–200 kDa. For example Mark12™ Unstained Standard (Life Technologies).
17. Mini gel tank SDS-PAGE electrophoresis device and power supply.

Silver Staining for SDS Gels:

18. Clean glass or polypropylene container.
19. 200 mL 30 % ethanol (non-denatured)/10 % acetic acid.
20. 400 mL 20 % ethanol (non-denatured).
21. 18 MΩ water.
22. Sensitizing solution: 200 mL 0.8 M thiosulfate in 18 MΩ water.
23. 200 mL 12 mM silver nitrate solution in water.
24. Staining developer solution: 200 mL 2 % potassium carbonate (K_2CO_3) with 0.025 % formalin (250 μL per liter) and 0.0125 % sodium thiosulfate (125 μL per liter).
25. 200 mL each of staining stop and storage solutions: 5 % acetic acid and 1 % acetic acid.

SDS Removal:

26. Detergent Removal Spin Column (Thermo Scientific Pierce) [56].
27. 0.01 % formic acid (ultrapure grade), pH 4.
28. 0.1 % formic acid (ultrapure grade).
29. 1.5 and 2 mL collection tubes.
30. Centrifuge (1,500 × *g*).

2.4 Chromatography

1. Mobile-phase solution A: 0.1 % formic acid (ultrapure grade) in water (LC-MS grade).
2. Mobile-phase solution B: 0.1 % formic acid (ultrapure grade) in acetonitrile (LC-MS grade).
3. ESI column setup:
   Column: Thermo Scientific™ Hypersil™ BioBasic™ C8 reversed-phase 1 mm inner diameter (i.d.) × 100 mm column packed with 5 μm particles or a ProSwift RP-10R 1 mm i.d. × 50 mm monolithic column.
4. Nano-ESI column setup:
   150 μm i.d. × 2 cm PLRP-S (5 μm particle size) trapping column (New Objective, Inc., Woburn, MA) with a 75 μm i.d. × 10 cm PLRP-S analytical column (New Objective).
5. Chromatography system: High pressure Thermo Scientific UltiMate 3000 RSLC HPLC for normal flow and UltiMate 3000 RSLCnano HPLC or EASY-nLC 1000 (Thermo Scientific) for nanoliter flow rates.

2.5 Mass Spectrometers

1. MALDI LTQ Orbitrap XL hybrid mass spectrometer (Thermo Scientific, Bremen, Germany).
2. Q Exactive™ Quadrupole Benchtop Orbitrap Mass Spectrometer (Thermo Scientific, Bremen, Germany).

3. Orbitrap Elite™ linear ion trap Orbitrap hybrid mass spectrometer (Thermo Scientific, Bremen, Germany) equipped with an ETD source.
4. Ionization sources: Ion Max source equipped with the ESI or HESI (heated electrospray ionization) probe for flow rates higher than 3 μL/min and Nanospray Flex ion source for flow rates lower than 1 μL/min.

2.6 Software for Data Acquisition

1. MALDI LTQ Orbitrap XL: Thermo Scientific Xcalibur software version 2.1 SP1 with Thermo Foundation 1.0.2. SP2.
2. Q Exactive: Xcalibur™ version 2.2 SP1 with Thermo Foundation 2.0 SP1 and Exactive Series Tune 2.2 SP1.
3. Orbitrap Elite equipped with an ETD source: Xcalibur version 2.2 SP1 with Thermo Foundation 2.0 SP1 and LTQ Tune 2.7 SP1.

2.7 Software for Data Acquisition

1. Analysis software: ProSight PTM and ProSightPC 3.0 (Thermo Fisher Scientific).

3 Methods

3.1 Top-Down Analysis of a Single Protein by MALDI Top Down

1. MALDI-ISD sample preparation:
   The 1,5-DAN solution is prepared shortly before the MS experiments because of the instability of 1,5-DAN in acetonitrile. The protein solution, e.g., insulin B chain or cytochrome C (5 pmol/μL), is mixed in a 1:1 ratio with the MALDI matrix solution of 1,5-DAN. 0.5 μL of the protein/matrix mixture is spotted on a stainless steel target and air-dried at ambient temperature.
2. Mass calibration of Orbitrap mass analyzer 1,5-DAN matrix and the peptides provided in the instrument calibration and tuning kit MSCAL4 (Sigma-Aldrich) are mixed according to the protocol provided by Sigma and spotted on a stainless steel plate as a regular sample preparation. Signals produced upon desorption/ionization of 1,5-DAN (m/z 158.0839, m/z 159.0950) and upon MALDI of the peptides therein (m/z 524.2650, m/z 1060.5687, m/z 1936.9850, m/z 2845.7614) are used to mass calibrate the Orbitrap mass analyzer via the FT manual tab in the Thermo Scientific Tune software.
3. *MALD-ISD data acquisition* experiments are performed on a MALDI LTQ Orbitrap XL mass spectrometer (Bremen, Germany), operating at intermediate pressure (75 mTorr) in front of the sample plate. Eight spectra are acquired from single scans or as averages from up to 3 μscans. The instrument is operated in positive mode, and full MS scan data are acquired over an m/z range of 1,000–4,000 Th. Averages of up to

10–25 scans are displayed; a single µscan can easily meet a requested "number of charges" of 1e6 in the Orbitrap detector with several tens of laser shots on a given position. Mass resolution is set to 100,000 at an m/z 400 to obtain the necessary mass resolution across the whole mass range.

3.2 Top-Down Analysis of a Single Protein by Direct Infusion and ESI-MS

Sample introduction by direct infusion provides the benefit of the data acquisition not being time restricted and only limited by the amount of sample and the flow rate. However, nonvolatile salts and detergents have a strong negative effect on the analysis of proteins by direct infusion by suppressing ionization resulting in noisy and low-intensity spectra. Thus removal of salts and detergents is strongly suggested using either desalting or detergent removal spin columns. Since the described protocol focuses on the analysis under reducing (nonnative) conditions, an organic content of 30–70 % such as acetonitrile in the final sample is suggested to support the desolvation and ionization process. Also, direct infusion works best on a clean, single protein or at most a rather simple mixture of very few proteins.

3.2.1 Sample Preparation: Desalting Using Zeba™ Spin Columns (7k MWCO, 0.5 mL)

1. Remove column's bottom closure, and loosen cap.
2. Place column in a 1.5 or a 2.0 mL collection tube. Centrifuge at 1,500 × g for 1 min to remove storage solution.
3. Blot the bottom of the column to remove excess liquid.
4. Place a mark on the side of the column where the compacted resin is slanted upward. Place column in the microcentrifuge with the mark facing outward in all subsequent centrifugation steps. Improper orientation will result in reduced desalting efficiency.
5. Place column in a new tube, remove cap, and slowly apply 30–130 µL of sample to the center of the compacted resin bed.
6. For sample volumes <70 µL apply a 15 µL stacker of ultrapure water to the top of the resin bed after the sample has fully absorbed to ensure maximal protein recovery.
7. Centrifuge at 1,500 × g for 2 min to collect desalted sample.

3.2.2 Detergent Removal Cleanup Using Spin Columns

1. Remove cartridges from refrigerator, twist off bottom, and place cartridge into 2 mL tube.
2. Centrifuge for 1 min at 1,500 × g to remove the storage buffer.
3. Add 0.4 mL of 0.01 % formic acid (pH 4), centrifuge at 1,500 × g for 1 min, discard the flow-through, and repeat two additional times.
4. Place cartridge in a clean 1.5 mL tube.
5. Add 25–100 µL of detergent-containing protein sample to the cartridge, and incubate for 2 min at room temperature.

   Note: Recovery is best if sample concentration is at least 100 µg/mL.

6. Centrifuge at 1,500 × *g* for 2 min.
7. Collect detergent-free sample form the 1.5 mL microcentrifuge tube.

3.2.3 Sample Dilution for Direct Infusion

1. Dilute purified protein sample (commercially available proteins such as carbonic anhydrase typically do not require desalting and can be directly dissolved in water) in 50 %/50 % v/v water/acetonitrile with 0.1 % formic acid to a protein concentration of 5–10 pmol/μL (*see* **Note 1**). The solution is supplied via a syringe pump with a flow rate of 5 μL/min to the ion source of the mass spectrometer.

3.2.4 Mass Spectrometric Analysis by Direct Infusion

The Q Exactive benchtop Orbitrap or the Orbitrap Elite ion trap Orbitrap hybrid mass spectrometer is equipped with an electrospray (ESI) or heated electrospray ionization (HESI) probe installed in the Ion Max source.

1. The ESI probe is not equipped with a temperature control, for the HESI probe the probe temperature is set to 30 °C, and auxiliary and sheath gas supply is set to 0.
2. The capillary temperature is set to 250 °C, and the spray voltage is set to 3.5–4.0 kV.
3. All data are acquired with external calibration, and the analysis is performed in positive ionization mode.
4. Automatic Gain Control™ (AGC) settings are 3e6 (Q Exactive) and 1e6 (Orbitrap Elite) for full MS and 1e6 for MS/MS. All spectra in all experiments are detected in the Orbitrap mass analyzer.
5. Maximum injection times are not critical in direct infusion experiments and can thus be set to 100–500 ms both for full MS and MS/MS.
6. Full MS spectra are acquired with 15 % source CID (SID) (*see* **Note 2**) for 1 min with the mass range set to m/z 400–2,000. The resolution is set to 140.000 on the Q Exactive and 240.000 on the Orbitrap Elite mass spectrometer (*see* **Note 3**).
7. HCD top-down MS/MS spectra are acquired in a separate acquisition by selecting the most intense charge state with an isolation window of 10 Da and the microscan count set to 5. 15–20 % normalized collision energy (NCE) is applied and can be stepwise optimized aiming at the precursor being reduced to ~5 % total intensity, and fragment ions of high and low charge states across the mass range are present as pictured in Fig. 3. Resolution for top-down MS/MS spectra is set to 140,000 on the Q Exactive and 120,000 or 240,000 on the Orbitrap Elite (*see* **Note 4**). Data are acquired for ~1 min or longer if signal-to-noise benefits from longer spectrum averaging (*see* **Note 5**).

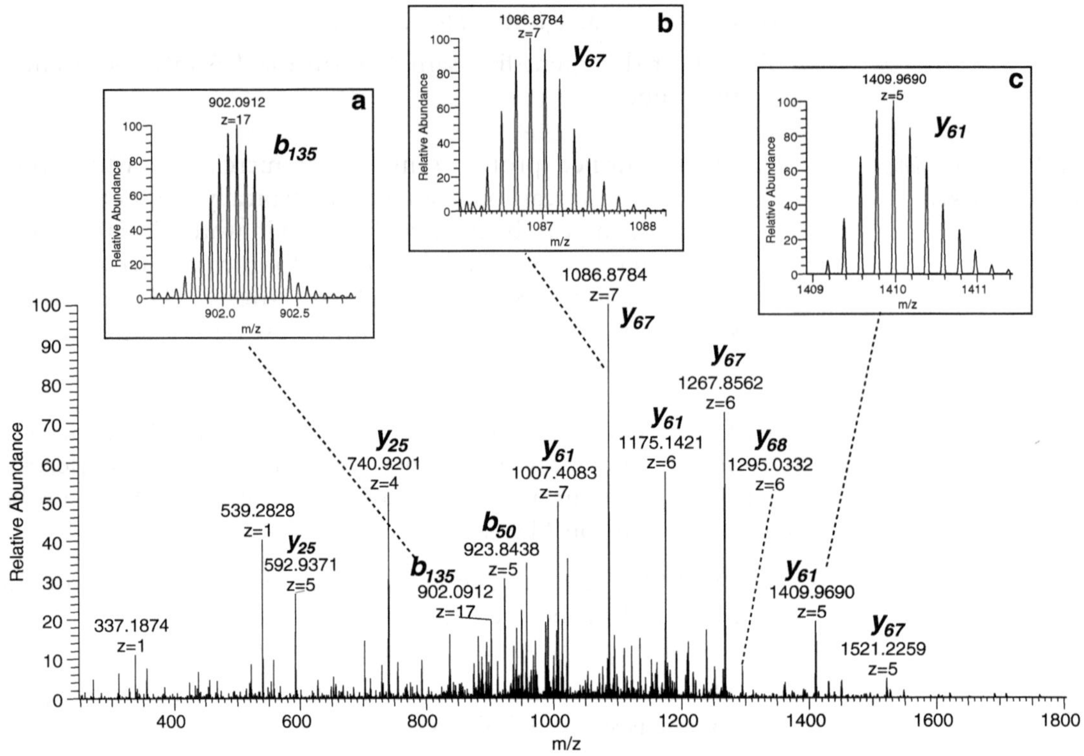

Fig. 3 Top-down HCD spectrum of carbonic anhydrase, acquired on a Q Exactive mass spectrometer with a resolution setting of 140,000. The *inserts* (**a–c**) highlight the isotope clusters of obtained fragment ions

8. CID top-down MS/MS spectra are acquired on the Orbitrap Elite by selecting the most intense charge state in an isolation window of 10 Da and applying 35 % NCE with 5 µscans. Data are acquired for ~1 min or longer if signal-to-noise benefits from longer spectrum averaging.
9. ETD top-down MS/MS spectra are acquired by selecting the highest charge state at lowest *m/z* with an intensity not lower than 20 % of the most abundant charge state. An isolation window of 10 Da is applied with an activation time of 1–15 ms (*see* **Note 6**) and the microscan count set to 5. Resolution for top-down MS/MS spectra is set to 140,000 on the Q Exactive and 120,000 or 240,000 on the Orbitrap Elite (*see* **Note 7**). Data are acquired for 1 min or longer if signal-to-noise benefits from longer spectrum averaging.

3.3 Top-Down Analysis of Low-Complexity Protein Mixtures

Protein mixtures of low complexity (<30 proteins) can be analyzed without desalting prior to analysis and with rather short chromatographic separation times of 30 min or less. For samples of high salt concentration the use of the divert valve is recommended directing the solvent at the beginning of the gradient into waste instead of the source of the mass spectrometer.

3.3.1 Chromatography

Chromatographic separations can be performed on, e.g., Thermo Scientific™ Hypersil™ BioBasic™ C8 reversed-phase 1 mm × 100 mm column packed with 5 μm particles or a ProSwift RP-10R 1 mm i.d. × 50 mm monolithic column with a high pressure UltiMate 3000 RSLC capillary HPLC. The monolithic column is recommended to perform at elevated temperature of 55–80 °C. 0.1–5 μg total protein is injected via the autosampler onto the column for separation.

A short gradient of 20 min total is employed with a flow rate of 150 μL for the C8 column or 60 μL for the monolithic column: the mobile-phase separation starts at 90 % solvent A and decreases to 85 % over a period of 0.6 min and then reduces from 85 to 70 % solvent A in 1.4 min and then further to 50 % solvent A in 12.8 min and to 2 % in 1 min. It is then maintained at isocratic flow for another 1.1 min; the gradient then increases from 2 to 90 % solvent A in 1.1 min and is kept for another 2 min prior to the next separation.

3.3.2 Mass Spectrometry

1. The Q Exactive benchtop Orbitrap or the Orbitrap Elite ion trap Orbitrap hybrid mass spectrometer is equipped with an electrospray (ESI) or heated electrospray ionization (HESI) probe installed in the Ion Max source.
2. The ESI probe is not equipped with a temperature control, for the HESI probe the heater temperature is set to 30 °C, and sheath gas supply is adapted to a flow rate of 150 μL/min and set to 12 U.
3. The capillary temperature is set to 250 °C, and the spray voltage is set to 3.5 kV.
4. The ion optics S-lens level is set to 60.
5. All data are acquired with external calibration, and the analysis is performed in positive ionization and profile mode.
6. AGC settings are 3e6 (Q Exactive) and 1e6 (Orbitrap Elite) for full MS and 2e5 for MS/MS. All spectra in all experiments are detected with the FTMS Orbitrap detector.
7. Full MS spectra are acquired without source CID for small proteins up to 30 kDa and with 15–20 % SID for proteins up to 60 kDa. Data are acquired with a method acquiring a full MS scan at 140,000 resolution (Q Exactive) or 120,000 resolution (Orbitrap Elite) for the mass range m/z 1,000–3,000 (<30 kDa) or m/z 400–2,000 (20–50 kDa). Target values are 3e6 for the full scan with a maximum injection time of 500 ms. The top five most intense ions are selected and isolated one after the other with a 4 Da isolation window and fragmented with HCD (20 % NCE) on either the Q Exactive or the Orbitrap Elite or with ETD and 15 ms activation time on the Orbitrap Elite. For smaller proteins <20 kDa a resolution setting of 35,000 or 30,000 with 2 μscans is sufficient, whereas for proteins >20 kDa a higher resolution of 140,000 on the Q Exactive and 120,000 with 5 μscans on the Orbitrap Elite is chosen (*see* **Note 8**).

3.4 Top Down of High-Complexity Protein Mixtures

Complex protein mixtures such as whole-cell lysates require sophisticated pre-fractionation. Here the preparation of an *E. coli* whole-cell lysate is described using the GELFrEE® approach and LC-MS analysis in the Orbitrap Elite hybrid mass spectrometer.

3.4.1 Sample Preparation

1. 2 mg dried *E. coli* lysate protein pellet is resuspended in 1 mL 20 mM DTT, vortexed rigorously for 5 min, and then filtered through 8 μm, 0.8 μm, and then 0.2 μm pore filters.
2. 200 μL of the sample is desalted using a Zeba™ Micro Spin Desalting Column (Thermo Scientific Pierce) with a 7K molecular weight cutoff (*see* Subheading 3.2).
3. The protein concentration is quantified using the Coomassie Plus Bradford Protein reagent assay (optional).
4. 200 μg of the desalted lysate is diluted with 18 MΩ water to a final volume of 112 μL, and 8 μL of 1 M DTT is added followed by 30 μL GELFrEE® (Expedeon) sample buffer such that the final sample volume for loading is 150 μL.

3.4.2 Sample Fractionation Using a GELFrEE® 8100 (Expedeon) Instrument with 10 % Tris–Acetate Cartridge (Expedeon)

1. Prepare the cartridge by removing the cartridge from the foil package, and remove the plate sealer.
2. Empty the storage reservoirs for cathode and anode of those channels intended for use.
3. Add 6 mL HEPES running buffer in each cathode reservoir and 8 mL running buffer in each anode reservoir.
4. Add 150 μL running buffer in the collection chamber. Program run method for 10 % cartridge in 13 steps over a time period of 160 min: Voltage is set to 50 V for **steps 1–5** and increased to 100 V for **steps 5–13**. Fraction interval times are 16, 36, 2, 3, 4, 3, 4, 5, 7, 10, 15, 20, and 35 min.
5. Place the cartridge in the instrument, and lower the electrodes. Remove any buffer from sample loading chamber, and immediately add 150 μL of sample. Close the instrument lid, and start the method.
6. After **step 1**, the instrument pauses for rinsing the sample loading chamber twice and to add another 2 mL of running buffer in the cathode buffer reservoir. Resume the method.
7. After **step 2**, the system will pause for collection of the first fraction. Transfer the sample from the sample collection chamber into a clean tube or if several samples are run use a multichannel pipette and transfer to a microtiter plate. Wash the sample collection chamber twice with fresh running buffer. Fill the fraction collection chamber with fresh running buffer, and resume the method.
8. Repeat process for all the remaining fractions. Replace the running buffer in the anode and cathode reservoirs after **steps 4**, **10**, and **12**.

3.4.3 Evaluation of Fractionation by SDS Gel Electrophoresis

1. Prepare a 10–20 % Tris-glycine gel by removing the precast gel from the packaging and inserting it into mini gel tank for SDS-PAGE electrophoresis.
2. Dilute a 10 μL aliquot of each GELFrEE® fraction with 8 μL water, 5 μL LDS sample buffer, and 2 μL 500 mM DTT.
3. Dilute a 1 μg aliquot of the unfractionated sample in running buffer to 13 μL, and add 5 μL LDS sample buffer.
4. Heat all samples at 50 °C for 10 min.
5. Load 5 μL of a 1:20 dilution of the premixed molecular weight standard mix Mark 12 (Life Technologies) in lanes 1 and 15.
6. Load 10 μL of each GELFrEE fraction in lanes 2–13 on the gel.
7. Load 1 μg of the unfractionated lysate in lane 14.
8. Run the gel until the dye front reaches the bottom of the gel.
9. Remove the gel from the cassette and silver stain: Place gels in a thoroughly cleaned glass or polypropylene container containing 30 % ethanol/10 % acetic acid and shake for at least 30 min. Rinse the gels twice in 20 % ethanol for 10 min each time and then rinse twice in water for 10 min each. Sensitize the gel by soaking it in 0.8 M sodium thiosulfate for 1 min. Rinse the gel twice for 1 min each in water. Add sufficient 12 mM silver nitrate solution to cover the gel well, and shake for 20 min. Rinse the gel for 10 s in water, drain, and then add the developer solution (2 % potassium carbonate with 0.025 % formaldehyde and 0.0125 % sodium thiosulfate). Shake gel in the developer solution until full pattern of gel bands has appeared. Add 5 % acetic acid to stop development. Shake in stop solution for 10 min, and then transfer gel in 1 % acetic acid for storage.

3.4.4 Detergent Cleanup

All fractions collected are cleaned up from detergent introduced for GELFrEE® fractionation using the detergent removal resin (Thermo Scientific Pierce):

1. Remove cartridges from refrigerator (one per fraction), twist off the bottom, and place cartridge into 2 mL tube.
2. Centrifuge for 1 min at 1,500 × *g* to remove the storage buffer.
3. Add 0.4 mL of 0.01 % formic acid (pH 4), centrifuge at 1,500 × *g* for 1 min, discard the flow-through, and repeat two additional times.
4. Place cartridge in a clean 1.5 mL tube.
5. Add a GELFrEE® fraction to the cartridge, and incubate for 2 min at room temperature.
6. Centrifuge at 1,500 × *g* for 2 min.
7. Dry down the detergent-free sample in a speed vac until ~5 μL remains. Be sure not to dry sample to dryness.

8. Bring up sample to 15 μL in 0.1 % formic acid and centrifuge for 3 min at maximum centrifuge speed. This removes any precipitate and prevents the column from clogging.
9. The sample is now ready for analysis by LC-MS (each fraction is enough for three 5 μL injections).

3.4.5 Chromatography

5 μL of each detergent-free fraction is injected per LC-MS run and analyzed online using a split-free nano-LC system (e.g., EASY-nLC 1000). The gradient consisted of 5–30 % solvent B over 10 min, 30–65 % solvent B over 45 min, and 65–90 % solvent B over 5 min. The proteins are separated using a 2 cm × 150 mm ID reversed-phase (PLRP-S) precolumn (5 μm particle size) and 10 cm × 75 mm ID PLRP-S analytical column (both New Objective).

3.4.6 Mass Spectrometry

1. The Orbitrap Elite hybrid mass spectrometer equipped with an ETD source is connected online to the nano-LC chromatography system using a Nanospray Flex ion source (Thermo Fisher Scientific), operated at 2 kV spray voltage with a stainless steel emitter.
2. The acquisition method is set to acquire a full scan over m/z 600–2,000 with resolution of 120,000 with 8 μscans and 1e6 AGC target.
3. The maximum injection time is 2 s for both full scan and MS^2 scan.
4. For MS/MS the resolution is set to 60,000 for all fragmentation types with 10 μscans and 2e6 AGC target.
5. ETD, HCD, or CID fragmentation is performed in separate acquisitions on the most intense ion with the m/z as masses option selected to only perform MS/MS on the most intense charge state per protein, applying 15 % SID and a precursor isolation in a 10 Da window.
6. For ETD 20 ms activation time is used and HCD spectra are acquired applying 30 % NCE, and for CID spectra 40 % NCE is used.

3.5 Data Analysis

Isotopically resolved single full MS and MS/MS spectra can be deconvoluted using the Xtract algorithm implemented in Thermo Scientific Xcalibur Qual Browser software. Deconvoluted MS/MS spectra (M values) can then be directly loaded in ProSightPC 3.0 software to perform a search against a single-protein sequence.

Data obtained from LC-MS analyses via Xcalibur™ software are processed using ProSightPC 3.0 software for identification of proteins based on the MS/MS top-down spectra:

1. Download the full UniProt flatfile database for *E. coli* and import into ProSightPC 3.0.

2. Index the database with no modifications or SNPs (signal peptides will be taken into account).
3. Search the MS/MS data using a 60 kDa precursor tolerance and 10 ppm fragment tolerance. These will identify proteins based on N-terminal or C-terminal sequence coverage and thus will indicate their presence.
4. Search strategies:

   Absolute Mass: Intact Protein IDs

   Absolute mass search: 1.02 Da precursor mass tolerance (to account for "off-by-one" errors in Xtract) and 10 ppm fragment tolerance.

   Identifies intact forms that match the annotation forms from the indexed flatfile database.

   Takes into account annotated modifications, signal peptides, and sequence variants.

   Truncated Proteins:

   Biomarker search: 10 ppm precursor mass tolerance and 10 ppm fragment tolerance.

   Identifies subsequences of intact forms in the indexed flatfile database.

   Search also incorporates known modifications, signal peptides, and sequence variants.

   Unexpected Forms:

   Absolute mass search: 25 kDa precursor mass tolerance, 10 ppm fragment tolerance, and Delta m mode "on."

   Identifies unannotated forms of intact and truncated proteins, but does not fully characterize the protein form.

   Search also incorporates known modifications, signal peptides, and sequence variants.

4 Notes

1. The optimum protein concentration for direct infusion experiments is typically in the range of 1–10 pmol/μL. Increasing the protein concentration significantly can cause the solution becoming viscous resulting in clogging of the electrospray probe needle. Also, after the sample analysis is completed it is recommended to flush the infusion line and source probe with excess of aqueous solution to remove residual protein in order to maintain source performance.
2. SID is a critical parameter for the analysis of intact proteins and helps desolvation. The level of required SID depends

strongly on the size of the protein. Smaller proteins (<30 kDa) most often do not require any SID. On the contrary, applying SID on small proteins results in a high level of fragmentation and should be used with care. Larger proteins such as intact antibodies require often a high SID level of 60–80 % in order to provide high-quality spectra.

3. The mass range and resolution of the full MS spectra need to be set according to the molecular weight of the protein. For proteins up to about 80 kDa an upper m/z of 2,000 will be sufficient. For large molecules such as antibodies the mass range has to be shifted to m/z 1,800–5,000 on the Q Exactive and to m/z 1,800–4,000 (high mass range) on the Orbitrap Elite. The resolution has to be adapted accordingly. The size limit of proteins that can be isotopically resolved in full MS spectra acquired at maximum resolution on both instruments is <30–50 kDa. Thus for proteins larger than that it is recommended to switch to the lowest resolution setting of 17,500 on the Q Exactive and 15,000 on the Orbitrap Elite. Isotopically unresolved spectra are obtained providing average masses after deconvolution.
4. Even for proteins that cannot be isotopically resolved as an intact molecule the maximum resolution is recommended for acquiring off-line top-down MS/MS spectra since smaller pieces of the protein are detected and often require the highest resolution to determine their charge state and/or resolve overlapping isotope patterns.
5. Spectrum averaging can be achieved in two ways: (a) acquisition of many single scans and averaging post-acquisition in Xcalibur Qual Browser software and (b) spectrum averaging in the Tune acquisition software and saving of the averaged spectrum as a single scan (access via Tune window-right mouse button-write to raw file). Advantage of the latter way is a better signal-to-noise ratio since the resulting spectrum is obtained by averaging the transients whereas the first method averages spectra.
6. In order to obtain the highest sequence coverage it may be advised to acquire data separately with different settings between 1 and 30 ms and combine identification results.
7. Implementation of AGC [57] ensures that a desired number of ions and charges are trapped and do not compromise instrument performance by the induction of space charging effects. Fast pre-scans are used to measure the total ion current generated, employed to calculate the optimal injection time to reach the target number of ions. Thus, the number of ions that reach the mass analyzer is kept constant while adapting the required time to collect ion changes for every scan, leading to measurements based on reproducible, optimal size ion populations.

8. The resolution setting and number of microscans have to be adapted according to the molecular weight of the proteins and the speed of chromatography. Ideally 5–10 μscans are recorded per MS/MS spectrum.

Acknowledgements

The author would like to thank Rosa Viner, Terry Zhang, Eugen Damoc, Kerstin Strupat and David Horn for valuable discussions and input for the manuscript.

References

1. Loo JA, Edmonds CG, Smith RD (1990) Primary sequence information from intact proteins by electrospray ionization tandem mass spectrometry. Science 248:201–204
2. Loo JA, Edmonds CG, Smith RD (1991) Tandem mass spectrometry of very large molecules: serum albumin sequence information from multiply charged ions formed by electrospray ionization. Anal Chem 63:2488–2499
3. Zubarev RA, Horn DM, Fridriksson EK et al (2000) Electron capture dissociation for structural characterization of multiply charged protein cations. Anal Chem 72:563–573
4. Demeure K, Quinton L, Gabelica V, De Pauw E (2007) Rational selection of the optimum MALDI matrix for top-down proteomics by in-source decay. Anal Chem 79(22):8678–8685
5. Reiber DC, Grover TA, Brown RS (1998) Identifying proteins using matrix-assisted laser desorption/ionization in-source fragmentation data combined with database searching. Anal Chem 70:673–683
6. Takayama M, Tsugita A (2000) Sequence information of peptides and proteins with in-source decay in matrix assisted laser desorption/ionization-time of flight-mass spectrometry. Electrophoresis 21:1670–1677
7. Mann M, Kulak NA, Nagaraj N, Cox J (2013) The coming age of complete, accurate, and ubiquitous proteomes. Mol Cell 49:583–590
8. Du Y, Parks BA, Sohn S et al (2006) Top-down approaches for measuring expression ratios of intact yeast proteins using fourier transform mass spectrometry to ions above 5 kDa, and eventually to global quantitative. Anal Chem 78:686–694
9. Macek B, Waanders LF, Olsen JV, Mann M (2006) Top-down protein sequencing and MS3 on a hybrid linear quadrupole ion trap-orbitrap mass spectrometer. Mol Cell Proteomics 5:949–958
10. Tran JC, Zamdborg L, Ahlf DR et al (2011) Mapping intact protein isoforms in discovery mode using top-down proteomics. Nature doi:10.1038/nature10575
11. Kellie JF, Catherman AD, Durbin KR et al (2012) Robust analysis of the yeast proteome under 50 kDa by molecular-mass-based fractionation and top-down mass spectrometry. Anal Chem 84:209–215
12. Ahlf DR, Compton PD, Tran JC et al (2012) Evaluation of the compact high-field orbitrap for top-down proteomics of human cells. J Proteome Res 11:4308–4314
13. Michalski A, Damoc E, Lange O et al (2012) Ultra high resolution linear ion trap Orbitrap mass spectrometer (Orbitrap Elite) facilitates top down LC MS/MS and versatile peptide fragmentation modes. Mol Cell Proteomics 11(3) doi: 10.1074/mcp.O111.013698
14. Resemann A, Wunderlich D, Rothbauer U et al (2010) Top-down de novo protein sequencing of a 13.6 kDa camelid single heavy chain antibody by matrix-assisted laser desorption ionization-time-of-flight/time-of-flight mass spectrometry. Anal Chem 82:3283–3292
15. Suckau D, Resemann A (2003) T3-Sequencing: targeted characterization of the N- and C-termini of undigested proteins by mass spectrometry. Anal Chem 75:5817–5824
16. Suckau D, Resemann A (2009) MALDI top-down sequencing: calling N- and C-terminal protein sequences with high confidence and speed. J Biomol Tech 20:258–262
17. Kovacech D, Kovac A, Suckau D, Asperger A (2011) Characterization of unexpected by-products in a recombinant TAU protein by capillary LC coupled to MALDI-top-down-sequencing (LCMALDI-TDS). J Biomol Tech 22(Suppl):S66, ABRF Poster Abstract
18. Evers W, Resemann A, Irrgang J et al (2010) SDS-PAGE-MALDI-TDS: mass spectrometric

top-down sequencing of proteins extracted from polyacrylamide gels. J Biomol Tech 21(3 Suppl):S48
19. Strupat K, Kovtoun V, Bui H et al (2009) MALDI produced ions inspected with a linear ion trap-Orbitrap hybrid mass analyzer. J Am Soc Mass Spectrom 20:1451–1463
20. Scheffler K, Strupat K (2010) MALDI in-source decay fragment ions of peptides and proteins generated in a collisional cooling interface and detected with an ion trap–Orbitrap hybrid mass analyzer. Thermo Scientific Application Note No. 30218
21. Laskin J, Futrell JH (2003) Collisional activation of peptide ions in FT-ICR mass spectrometry. Mass Spectrom Rev 22:158–181
22. Huang TY, McLuckey SA (2010) Gas-phase chemistry of multiply charged bioions in analytical mass spectrometry. Annu Rev Anal Chem 3:365–385
23. Olsen JV, Macek B, Lange O et al (2007) Higher-energy C-trap dissociation for peptide modification analysis. Nat Methods 4:709–712
24. Michalski A, Neuhauser N, Cox J, Mann M (2012) A systematic investigation into the nature of tryptic HCD spectra. J Prot Res 11:5479–5491
25. Frese CK, Altelaar AF, Hennrich ML et al (2011) Improved peptide identification by targeted fragmentation using CID, HCD and ETD on an LTQ-Orbitrap Velos. J Proteome Res 10:2377–2388
26. Hahne H, Kuster B (2012) Discovery of O-GlcNAc-modified proteins in published large-scale proteome data. Mol Cell Proteomics 11:1063–1069
27. Zubarev RA, Kelleher NL, McLafferty FW (1998) Electron capture dissociation of multiply charged protein cations. A nonergodic process. Am Chem Soc 120:3265–3266
28. Kruger NA, Zubarev RA, Horn DM, McLafferty FW (1999) Electron capture dissociation of multiply charged peptide cations. Int J Mass Spectrom 185(186/187):787–793
29. McLafferty FW, Horn DM, Breuker K et al (2001) Electron capture dissociation of gaseous multiply charged ions by Fourier-transform ion cyclotron resonance. J Am Soc Mass Spectrom 12:245–249
30. Syka JEP, Coon JJ, Schroeder MJ et al (2004) Peptide and protein sequence analysis by electron transfer dissociation mass spectrometry. Proc Natl Acad Sci U S A 101:9528–9533
31. Coon JJ, Ueberheide B, Syka JEP et al (2005) Protein identification using sequential ion/ion reactions and tandem mass spectrometry. Proc Natl Acad Sci U S A 102:9463–9468
32. Tian Z, Tolić N, Zhao R et al (2012) Enhanced top-down characterization of histone post-translational modifications. Genome Biol 13:R86
33. Elviri L (2012) ETD and ECD mass spectrometry fragmentation for the characterization of protein post translational modifications. Tandem Mass Spectrometry-Applications and Principles pp. 161–178
34. Fornelli L, Damoc E, Thomas PM et al (2012) Analysis of intact monoclonal antibody IgG1 by electron transfer dissociation orbitrap FTMS. Mol Cell Proteomics 11:1758–1767
35. Udeshi ND, Shabanowitz J, Hunt DF, Rose KL (2007) Analysis of proteins and peptides on a chromatographic timescale by electron-transfer dissociation MS. FEBS J 274:6269–6276
36. Stephenson JL, McLuckey SA (1996) Ion/ion reactions in the gas phase. Proton transfer reactions involving multiply-charged proteins. J Am Chem Soc 118:7390–7397
37. Brown RS, Lennon JJ (1995) Mass resolution improvement by incorporation of pulsed ion extraction in a linear time-of-flight mass spectrometer. Anal Chem 67:1998–2003
38. Brown RS, Lennon JJ (1995) Sequence specific fragmentation of matrix assisted laser desorbed protein/peptide ions. Anal Chem 67:3990–3999
39. Brown RS, Carr BL, Lennon JJ (1996) Factors influencing the observed fast fragmentation of peptides in matrix assisted laser desorption. J Am Soc Mass Spectrom 7:225–232
40. Fukuyama IS, Tanaka K (2006) Rapid sequencing and disulfide mapping of peptides containing disulfide bonds by using 1,5-DAN as a reductive matrix. J Mass Spectrom 41:191–201
41. Taylor GK, Kim Y-B, Forbes AJ et al (2003) Web and database software for identification of intact proteins using "top down" mass spectrometry. Anal Chem 75:4081–4086
42. LeDuc RD, Taylor GK, Kim Y-B et al (2004) ProSight PTM: an integrated environment for protein identification and characterization by top-down mass spectrometry. Nucleic Acids Res 32:W340–W345
43. LeDuc RD, Kelleher NL (2007) Using ProSight PTM and related tools for targeted protein identification and characterization with high mass accuracy tandem MS data. Curr Protoc Bioinform 13.6.1–13.6.28. doi: 10.1002/0471250953.bi1306s19
44. LeDuc RD, Zamdborg L, Glowacz KJ et al (2007) ProSight PTM 2.0: improved protein identification and characterization for top down mass spectrometry. Nucleic Acids Res 35, doi:10.1093/nar/gkm371

45. Tong RTW, Infusini G, Cui W et al (2009) BUPID top-down: database search and assignment of top-down MS/MS data. Proceedings of 57th annual conference of the American Society Conf for Mass Spectrometry, Philadelphia, PA
46. Theberge R, Infusini G, Tong W et al (2011) Top-down analysis of small plasma proteins using an LTQ-Orbitrap. Potential for mass spectrometry-based clinical assays for transthyretin and hemoglobin. Int J Mass Spectrom 300:130–142
47. Carvalho PC, Xu T, Han X et al (2009) YADA: a tool for taking the most out of high resolution mass spectra. Bioinformatics 25:2734–2736
48. http://bix.ucsd.edu/projects/msalign/
49. http://bix.ucsd.edu/projects/msdeconv/
50. Huber CG, Premstaller A, Kleindienst G (1999) Evaluation of volatile eluents and electrolytes for high-performance liquid chromatography-electrospray ionization mass spectrometry and capillary electrophoresis-electrospray ionization mass spectrometry of proteins. II. Capillary electrophoresis. J Chromatogr A 849:175–189
51. Capriotti AL, Cavaliere C, Foglia P et al (2011) Theoretical and practical aspects in the separation of large biomolecules, intact protein separation by chromatographic and/or electrophoretic techniques for top-down proteomics. J Chromatogr A 1218:8760–8776
52. Kastner M (2000) Protein liquid chromatography. Elsevier, Amsterdam
53. Fang X, Balgley BM, Lee CS (2009) Recent advances in capillary electrophoresis-based proteomic techniques for biomarker discovery. Electrophoresis 30:3998–4007
54. Tran JC, Doucette A (2008) Gel-eluted liquid fraction entrapment electrophoresis: an electrophoretic method for broad molecular weight range proteome separation. Anal Chem 80:1568–1573
55. Tran JC, Doucette A (2009) Multiplexed size separation of intact proteins in solution phase for mass spectrometry. Anal Chem 81: 6201–6209
56. Antharavally BS, Mallia KA, Rosenblatt M et al (2011) Efficient removal of detergents from proteins and peptides in a spin column format. Anal Biochem 416:39–44
57. Stafford GC, Taylor DM, Bradshaw SC, Syka JEP (1987) Enhanced sensitivity and dynamic range on an ion trap mass spectrometer with automatic gain control. Proceedings of 35th annual conference of the American Society for Mass Spectrometry, Denver, CO, pp 775–776

Index

Daniel Martins-de-Souza (ed.), *Shotgun Proteomics: Methods and Protocols*, Methods in Molecular Biology, vol. 1156, DOI 10.1007/978-1-4939-0685-7, © Springer Science+Business Media New York 2014

L

M

MIX
Papier aus verantwortungsvollen Quellen
Paper from responsible sources
FSC® C105338

If you have any concerns about our products,
you can contact us on
ProductSafety@springernature.com

In case Publisher is established outside the EU,
the EU authorized representative is:
Springer Nature Customer Service Center GmbH
Europaplatz 3, 69115 Heidelberg, Germany

Printed by Libri Plureos GmbH
in Hamburg, Germany